The Marvel of Maps

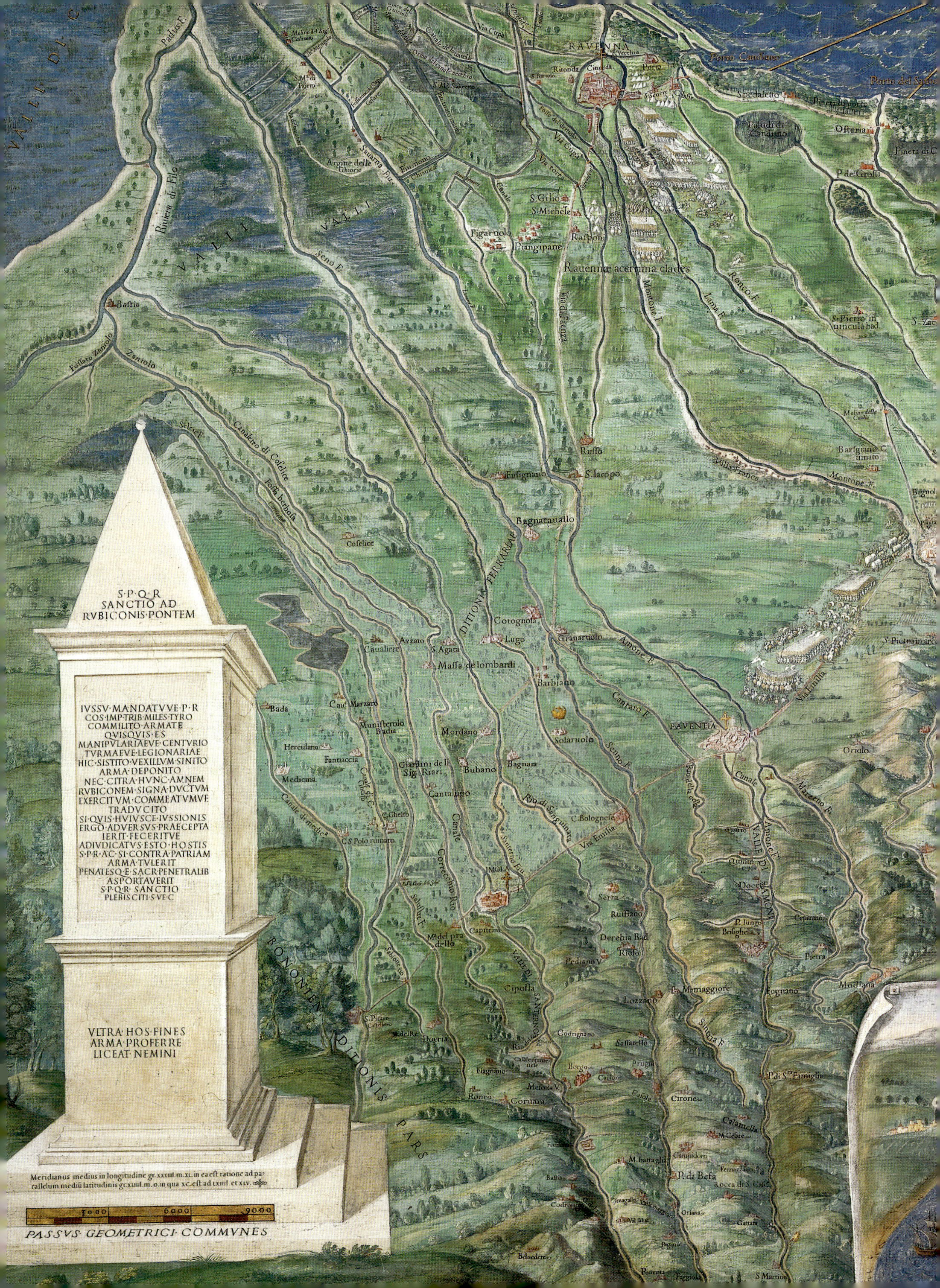
S·P·Q·R
SANCTIO AD
RVBICONIS·PONTEM
IVSSV·MANDATVVE·P·R
COS·IMP·TRIB·MILES·TYRO
COMMILITO·ARMATE
QVISQVIS·ES
MANIPVLARIAEVE·CENTVRIO
TVRMAEVE·LEGIONARIAE
HIC·SISTITO·VEXILLVM·SINITO
ARMA·DEPONITO
NEC·CITRA·HVNC·AMNEM
RVBICONEM·SIGNA·DVCTVM
EXERCITVM·COMMEATVMVE
TRADVCITO
SI·QVIS·HVIVSCE·IVSSIONIS
ERGO·ADVERSVS·PRAECEPTA
IERIT·FECERITVE
ADIVDICATVS·ESTO·HOSTIS
S·P·R·AC·SI·CONTRA·PATRIAM
ARMA·TVLERIT
PENATESQ·E·SACR·PENETRALIB
ASPORTAVERIT
S·P·Q·R·SANCTIO
PLEBISCITI·S·VE·C
VLTRA·HOS·FINES
ARMA·PROFERRE
LICEAT·NEMINI
Meridianus medius in longitudine gr.xxxiii.m.xi.in ea est ratione ad parallelum mediũ latitudinis gr.xliii.m.o.in qua xc.est ad lxiii.et xlv.
3000 6000 9000
PASSVS·GEOMETRICI·COMMVNES
RAVENNA
Rauennae acerrima clades
FAVENTIA
BONONIEN DITIONIS PARS
VALLI
Bagnacauallo
Cotognola
Lugo
Massa de lombardi
Barbiano
Solarolo
Mordano
Bubano
Bagnara
Cotignola
S. Agata
Conselice
Russi
S. Iacopo
Fusignano
Bastia
Oriolo

THE MARVEL OF MAPS

Art, Cartography and Politics in Renaissance Italy

Francesca Fiorani

Yale University Press
New Haven and London

Designed by Laura Bolick

Printed in China

Published with the support of the Vatican Museums
Con il patrocinio dei Musei Vaticani

לאלון

Library of Congress Cataloging-in-Publication Data

Fiorani, Francesca.
The Marvel of Maps : art, cartography and politics in Renaissance Italy / Francesca Fiorani.
p. cm.
Includes bibliographical references and index.
ISBN 0-300-10727-7 (cl : alk. paper)
1. Renaissance--Italy. 2. Cartography--Italy--History--16th century.
3. Art, Renaissance--Italy--Florence. 4. Art, Renaissance--Italy--Rome.
5. Florence (Italy)--Court and courtiers.
6. Rome (Italy)--Court and courtiers.
7. Cosimo I, Grand-Duke of Tuscany, 1519–1574.
8. Gregory XIII, Pope, 1502–1585.
I. Title: Painted map cycles in Renaissance and Counter Reformation courts. II. Title.
DG445.F56 2005
526'.0945'09031--dc22
2004023964

A catalogue record for this book is available from The British Library

frontispiece Egnazio Danti, *Flaminia* (detail of fig. 6.16), ca. 1580, fresco. Vatican, Gallery of Maps

Contents

1 Interpreting Renaissance Map Cycles 1

Part One
Maps as Worldly Art

2 The Guardaroba Nuova 17

3 'The Cosmos is Cosimo's Ornament' 33

4 The Ordering of Maps 61

5 The Mapping of the Modern World 93

Part Two
Maps as Sacred Art

6 The Sala Bologna of Gregory XIII 141

7 The Gallery of Maps 171

8 The Primacy of Italy 209

9 The Mapping of the Universal Church 233

Conclusion 253

Appendix A 256
Appendix B 266
Appendix C 273

Notes 279
Bibliography 318
Index 338
Photograph Credits 347

Acknowledgements

This book acquired a temporal and geographical extension that I could never have anticipated when I started it in Rome years ago. Along the way, I accumulated immense debts of gratitude towards the many friends and scholars who shared their insights and criticism. First and foremost, I thank the late Corrado Maltese who, in myriad Roman conversations, initiated me to the intricacies of art and science and left a lasting mark on the way I look at Renaissance art. Later, Marisa Dalai Emiliani read parts of the project and her comments stayed with me for a decade. In 1993 I was fortunate enough to spend a few intense months at the Warburg Institute in London, where the project further developed; there, I also met the historians of cartography with whom I maintained a lively dialogue throughout the years. I particularly benefited from the exchanges with Catherine Delano Smith, who sustained my work from the very early stages, with Peter Barber and Tony Campbell, who kept asking probing questions, and with Denis Cosgrove, whose perceptive insights inform large parts of this book. The late David Woodward offered his expertise and encouragement over the years, and I wished he could have seen the final result.

This book acquired its final form in Virginia, where the contingencies of life brought me ten years ago. At the University of Virginia I found a congenial group of colleagues and a community of Renaissance scholars who, thankfully, did not spare me their criticism. Paul Barolsky offered unfailing support and generously shared his time, his mind, and the writer's craft. David Summers's invaluable comments came at critical moments. Anne Schutte asked a particularly perceptive question, which I could answer only by reshaping half of the book. Mary McKinley read the manuscript in an earlier incarnation. Erik Middlefort clarified fundamental points in brief, pointed conversations. Lawrence Goedde, Chair of the Art Department, facilitated the completion of this book in more ways that he can imagine. Two successive Deans of the College of Arts and Science, Melvyn Leffler and Edward Ayers, and the Vice President for Research and Graduate Studies supported this project with numerous research grants over the years.

For its final polishing the book returned to Rome and London. Particularly inspiring have been the conversations with Nadia Cannata, Rossella Caruso, Claire Farago, Monica Martinat, and Tatiana String on the underlying assumptions of this book. Nicola Courtright, Judith Field, and Thomas Frangenberger shared their insight on the life and work of Egnazio Danti. With Giorgio Mangani I had fruitful conversation on moral cartography, while Lucia Nuti and Cesare De Seta helped to sharpen my thoughts on the language of cartographic representation. Evelyn Edson answered questions on medieval maps and Barbara Wisch was a guide on Gregory XIII. Konrad Eisenblicher generously translated an intractable poem. Claudia Cieri Via and Ingrid Rowland gave me the chance to bring my work back to Rome. My thanks go also to the four anonymous readers of the manuscript for their sharp comments.

It is a great pleasure to express my gratitude to the institutions and foundations that generously contributed to the making of this book. Fellowships from the Ministero dell'Universitá e della Ricerca Scientifica and the Consiglio Nazionale delle Ricerche in Italy enabled me to conduct the bulk of the research in the early 1990s. In 1997 my research was further supported by the American Council for the Learned Societies and the National Endowment for the Humanities. In 1993-1994 fellowships at the Warburg Institute (London), The Herzog August Bibliothek (Wolfenbüttel), the John Carter Brown Library (Brown University), and the Folger Shakespeare Library (Washington D.C.) provided ideal conditions for research and writing. A Postdoctoral Fellowship in the History of Art and the Humanities from the Getty Research Center in 2000-2001 gave me the peace of mind to complete the book.

My special thanks go to Francesco Buranelli, Director of the Vatican Museums, for greatly facilitating the production of this book. His support and the invaluable assistance of his staff at the Archivio Fotografico, especially Rosanna Di Pinto, made this book possible.

In addition, I wish to thank the the Archivio di Stato in Rome, Florence, Mantua, and Perugia for facilitating my research, and the numerous museums and libraries from around the world that granted me access to their collections and permission to reproduce their artifacts: in Basel, the Öffentliche Bibliothek, Universität; in Bologna, the Biblioteca Comunale dell'Archiginnasio, the Biblioteca Universitaria and the Pinacoteca Nazionale; in Boston, the Houghton Library at Harvard University; in Charlottesville, The University of Virginia Library; in Chicago, The Newberry Library; in Florence, the Biblioteca Nazionale Centrale, the Istituto e Museo di Storia della Scienza, the Comune di Firenze, and the Soprintendenza Speciale al Polo Fiorentino; in Leiden, the Universiteitsbibliotheek; in New Haven, the Yale University Library and the Yale Center for British Art; in New York, the Cooper-Hewitt National Design Museum; in London, the British Library and the British Museum; in Paris, the Bibliothèque Nationale; in Princeton, Princeton University Library; in Providence, the John Carter Brown Library; in Rome, the Biblioteca Apostolica Vaticana, the Musei Vaticani, the Biblioteca Hertziana, and the Museo Nazionale Preistorico Etnografico L. Pigorini; in Washington D.C., the Folger Shakespeare Library.

Parts of this book have appeared in shorter and altered form elsewhere: "Post-tridentine 'Geographia sacra': The Galleria delle Carte Geografiche in the Vatican Palace", in Imago Mundi, 48, 1996, 124-148; "Maps, Politics and the Grand Duke of Florence. The Sala della Guardaroba Nuova of Cosimo I de' Medici", in Renaissance Representations of the Prince. Basilike Eikon, eds. Roy Eriksen and Magne Malmanger, Rome, 2001, 73-102; "La Sala Bologna di Gregorio XIII", in Tra oriente e occidente. Iconografia delle citta' europee, ed. Cesare De Seta, Naples, 2004.

At Yale University Press, I was fortunate enough to enjoy the support of Gillian Malpass, who believed in this book from the very start, and to work with Laura Bolick, who sharply edited the manuscript, designed the book, and saw it into press.

Finally, my deepest gratitude goes to my family. My parents, Paola and Paolo Fiorani, were a constant source of personal support. With Vera, Laura, Brenno, and Lorenzo I shared the liberating good humor that cements siblings.

My sons, Paolo and Davidi, were born in the midst of this project. Their cheerfulness and demands kept things in perspective and, for this, I cannot thank them enough.

This book is dedicated to Alon Confino for reasons that are too deep and too many to enumerate.

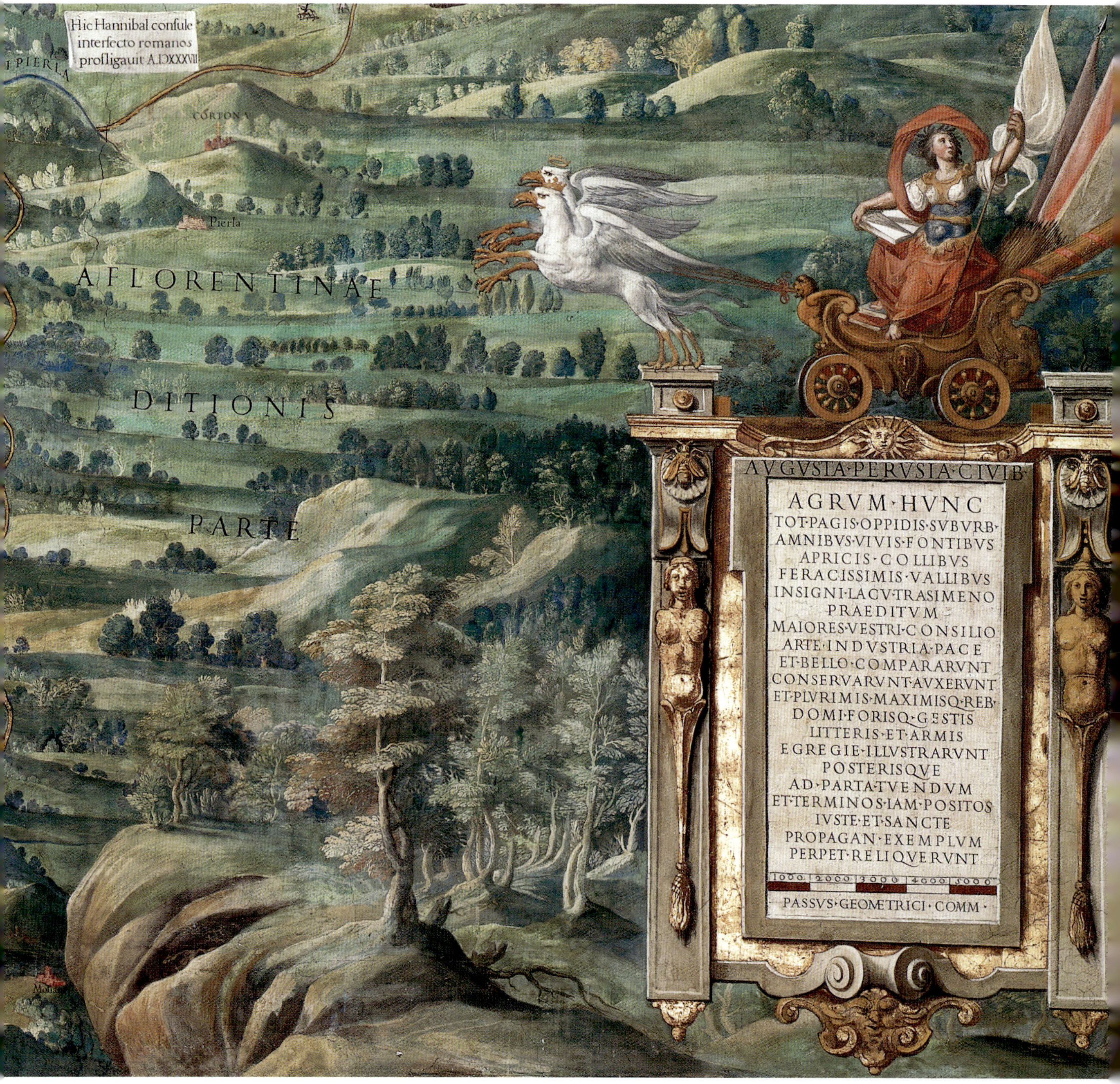
Hic Hannibal consule
interfecto romanos
profligauit A.DXXXVIII
PIERIA
CORTONA
Pierla
A FLORENTINAE
DITIONIS
PARTE
AVGVSTA·PERVSIA·CIVIB·
AGRVM·HVNC
TOT·PAGIS·OPPIDIS·SVBVRB·
AMNIBVS·VIVIS·FONTIBVS
APRICIS·COLLIBVS
FERACISSIMIS·VALLIBVS
INSIGNI·LACV·TRASIMENO
PRAEDITVM
MAIORES·VESTRI·CONSILIO
ARTE·INDVSTRIA·PACE
ET·BELLO·COMPARARVNT
CONSERVARVNT·AVXERVNT
ET·PLVRIMIS·MAXIMISQ·REB·
DOMI·FORISQ·GESTIS
LITTERIS·ET·ARMIS
EGREGIE·ILLVSTRARVNT
POSTERISQVE
AD·PARTA·TVENDVM
ET·TERMINOS·IAM·POSITOS
IVSTE·ET·SANCTE
PROPAGAN·EXEMPLVM
PERPET·RELIQVERVNT
1000 2000 3000 4000 5000
PASSVS·GEOMETRICI·COMM·

I

Interpreting Renaissance Map Cycles

This book is about the interactions of mapping with other forms of knowledge and representation in the Renaissance and Counter-Reformation. It examines the place of cartography in two famous courts, the court of Duke Cosimo I de' Medici (1537–1574) and the papal court of Gregory XIII Boncompagni (1572–85). These two rulers, like others of the period, used maps as a tool for the administration and defense of the state and for the collection and distribution of resources. As learned patrons, they also used the maps as visual aids in reading the Bible and the classics, and as a means of understanding the daily reports on the European wars. But these two rulers distinguished themselves among their contemporaries by the magnificent painted maps that they commissioned for their palaces. These map cycles were conceived to be displayed in appositely built rooms that still fascinate the modern visitor just as they must have the Renaissance viewer. Monumental in size and a magnificent displays of color, these map cycles, although useless as guides for the traveler, were among the most marvelous cartographic artifacts of the Renaissance. Because the Medici and papal courts are so well known for their art, life, and culture, these map rooms recapture the pervasiveness of cartographic images among Renaissance political and religious imagery, and, more generally, the relevance of cartography to Renaissance culture.

Cosimo I's Guardaroba Nuova presents an unusual iconography among the rooms of the Medici ducal palace. This large room is not decorated with the deeds of Medici forebears, ancient heroes, or pagan gods, but with colorful maps of the world, which are oddly painted on the doors of huge cupboards (Figs. 1.1 and 1.2). The visitor entering this room is surrounded by maps of the world. At the very center of this painted world, the viewer may contemplate the regions of the globe displayed around him. It is soon realized that, owing to the peculiar perspective, the maps represent the earth as seen from above though viewed from its center, thus generating the simultaneous experience of being inhabitants of the earth and outside onlookers. As the viewer adapts to this artifice of representation, it becomes apparent that the regions of the world are displayed as if they were depicted on the inner side of the globe, rather than on the outer side of the globe. The monumental globe in the middle of the room helps to make the quick mental move from the internal view of the globe

(*facing page*) Detail of fig.1.4

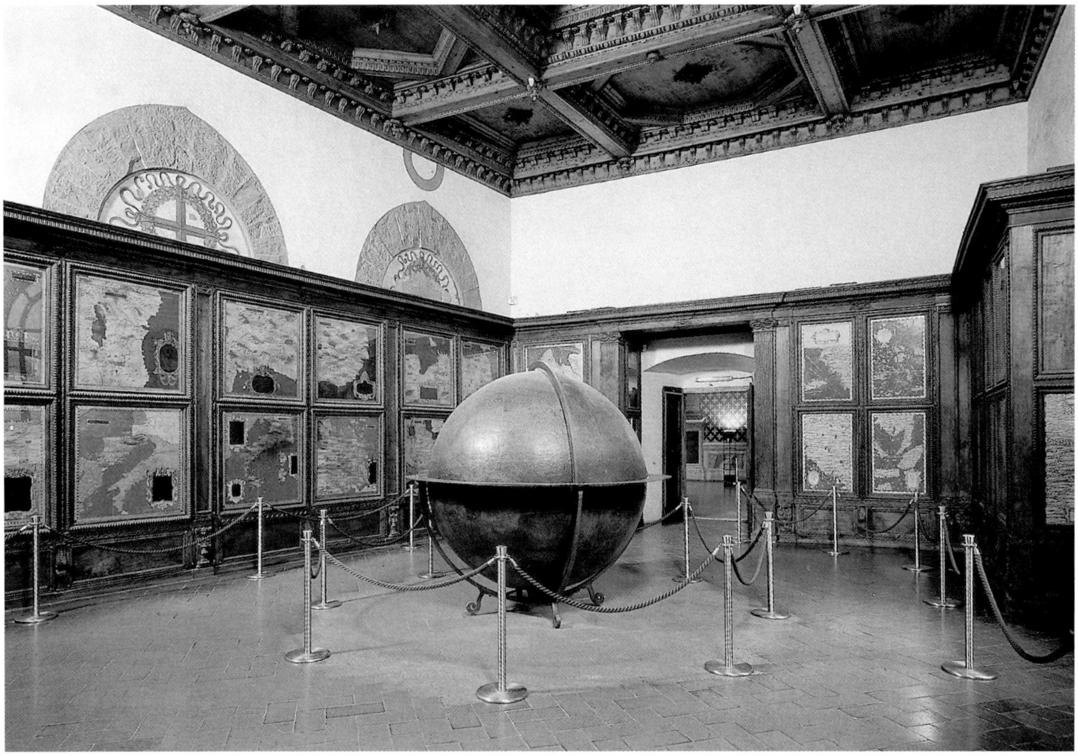

1.1 Guardaroba Nuova. Florence, Palazzo Vecchio

to the aerial view of the sphere. As one moves around the room each map is understood in connection with its neighboring regions. Renaissance visitors would have enjoyed the additional wonder of opening the doors of the cupboards, on which the maps are painted, to admire the beautiful objects from the Medici collections that were stored there; the origin of each object was described by the map. Further admiration would have been generated by the knowledge that no other ruler possessed such a unique display of modern maps, or a collection of artifacts from all over the world, organized geographically.

Similarly, the Gallery of Maps, a long corridor in the Vatican Palace, defies expectations of papal decorations. Instead of religious stories, saints, and miracles, which are relegated to the ceiling, the viewer is confronted with an unprecedented atlas of monumental maps of modern Italy depicted in bright colors along the walls (Figs. 1.3 and 1.4). The walk along the corridor reveals that the maps have been arranged as they would appear in an imaginary walk along the Apennine spine of the Italian peninsula. The maps progress from north to south, depicting the regions of Italy from above. The regional maps on the walls are subtley adjusted in their orientation to the imaginary trip from north to south. The maps on the left, which represent the regions along the Adriatic Sea, depict north on top, as was customary in Renaissance maps, but the maps on the right, which represent the regions along the Tyrrhenian Sea, show north at the bottom. The orientation of the maps is defined not by cartographic conventions but also by the projected movement along the Apennine spine.

Post-Tridentine Catholics might have pondered the role of Italy within the universal church knowing that they were admiring the first atlas of Italy ever produced. They probably would have also understood that the images on the vault of the corridor, which modern viewers tend to ignore, were depictions of the events of church history that happened in the lands mapped below.

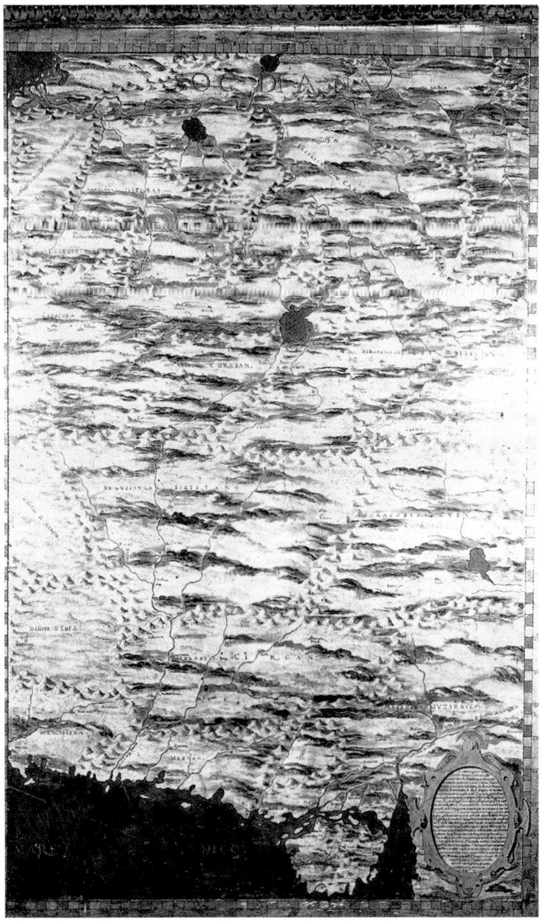

1.2 Egnazio Danti, *Sogdiana* (Central Asia), not dated, panel. Florence, Palazzo Vecchio, Guardaroba Nuova

These map cycles are challenging to modern interpreters of Renaissance court art. Countless art-historical studies have demonstrated that Renaissance rulers were acutely aware of the power of images to assert political and religious claims. The works of historians of cartography have shown that the themes of dominion and conquest were inherent in Renaissance mapping. But since Cosimo I did not rule over the world he depicted in the Guardaroba Nuova, and Gregory XIII had no political control of the regions of Italy he had painted in the Gallery of Maps, these two cycles elude a direct political interpretation of dominion and conquest. Furthermore, these two map cycles were cartographic enterprises in their own right: Cosimo I's Guardaroba Nuova displayed an atlas of the modern world before the very notion of the atlas was born, while Gregory XIII's Gallery of Maps represented the concept of modern Italy centuries before the peninsula was unified politically. Moreover, these cartographic achievements do not conform to the geographical theory of Ptolemy, the second-century astronomer whose *Geography* became the backbone of Renaissance mapping. Ptolemy wished for a mathematical description of the world projected onto a cartographic grid, while these Renaissance maps were surrounded and filled with myriad other images, such as views of cities and buildings, vignettes, animals, plants, trees, and inscriptions, features that Ptolemy would have regarded as extraneous to geography (Figs. 1.2 and 1.4). And finally, the rooms themselves and the individual maps that they contain pose an epistemological challenge, for they make use of the full resources of visual representation to evoke the three-dimensionality of the earth as effectively as possible. Scale, size, frames, colors, texture, projections, words, icons, symbols, grid, and architectural space interact creatively to produce a novel, informative, and pleasing representation of the earth and its parts. Indeed, the three-dimensional display of these enriched representations of the earth is fundamental to the understanding of the cycles

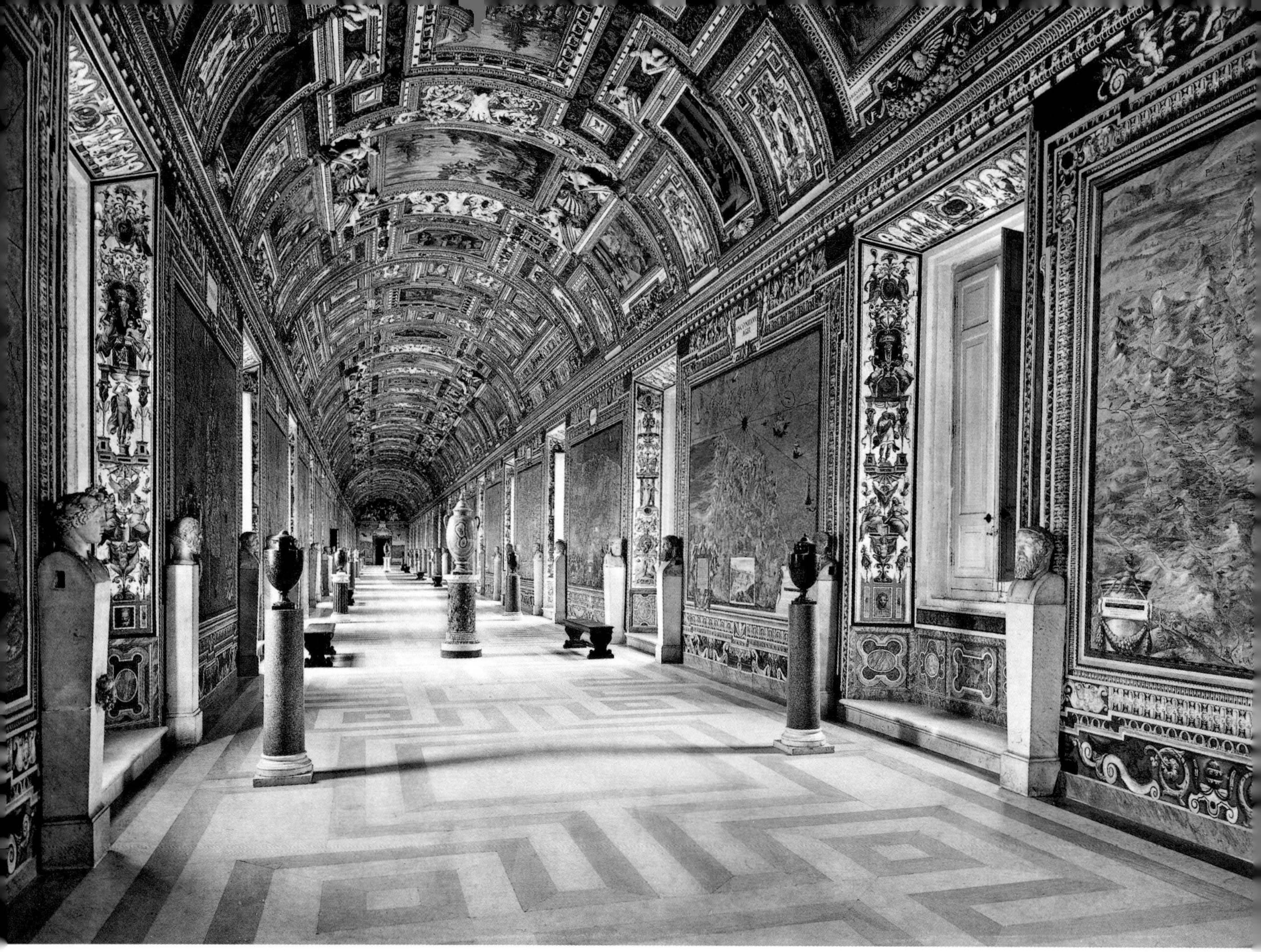

1.3 Gallery of Maps. Vatican

themselves: it strengthens the relationship between individual maps and the totality of the geographical unit (be it Italy or the earth itself), reinforces the perception of the maps as a cycle, and enhances the relations between the maps and other parts of the iconography.

In order to understand these map cycles an approach is needed that takes into account simultaneously their cartographic originality, their interaction with other forms of knowledge, their encompassing description of the world and its parts, their visual richness, and their three-dimensional display. Such an approach demands the fusion of different bodies of contemporary scholarship. First, the humanistic Renaissance was a movement that encompassed mathematics and philology, science and humanism, measurements and antiquarianism.[1] Anthony Grafton has explained how science and humanism, now considered and practiced as divergent forms of scholarship, shared a common history, principles of order, and methods of inquiry in the Renaissance, while Ingrid Rowland showed how computation, from the measurement of the universe to the length of the Roman foot, was part and parcel of Renaissance humanism. Their approach is essential for the interpretation of

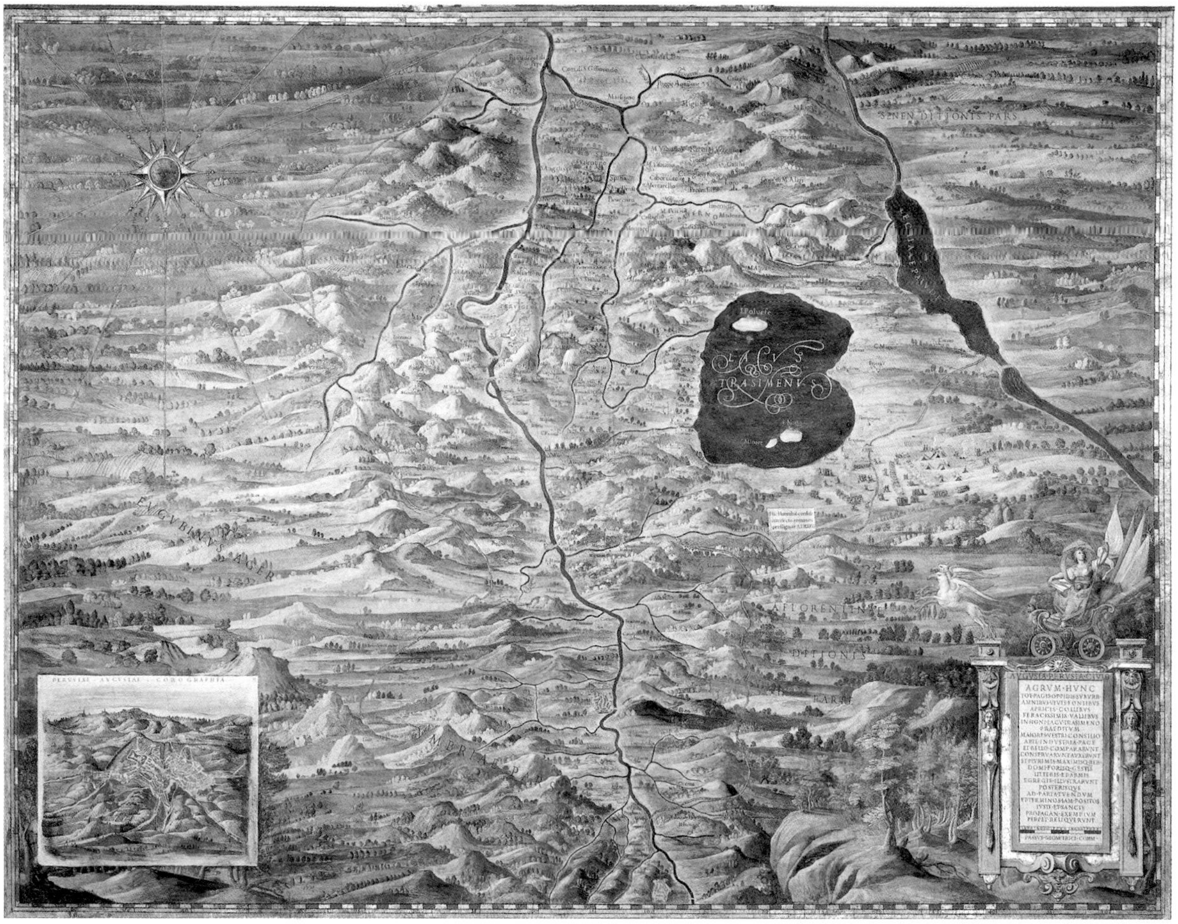

1.4 Egnazio Danti, *Perusinus ac Tifernas* (Map of the Territory of Perugia), ca. 1580, fresco. Vatican, Gallery of Maps

Renaissance mapping as art proposed in this book: mapping was an enterprise integrating textual analysis, computation, and visualization to which scholars, artists, merchants, and patrons contributed working side by side.

Second, I have used the work of Edward Casey on the notions of space, place, and site.[2] While modern philosophers from Immanuel Kant onward have studied the notion of space as an abstract and incorporeal entity and contrasted it to time, Casey focused on the relations between space and place. Like space, place is one of the conditions of our being in the world but, unlike space, it maintains some element of corporeality and refers to an individual locale represented with its unique characteristics. In this respect, place also differs from site, which is a "place reduced to location and position in space."[3] A site is thus a place that is mapped through the agency of the grid and deprived of its individual characteristics. The notions of place and site, which Casey himself applied to landscape painting and maps in general, are essential to explain the language of Renaissance maps: in these images a locale is not only positioned mathematically on the cartographic grid but also described with a plethora of qualitative details.

Third, I have followed the lead of art-historical approaches that bring the historical processes of visualization to the center of cultural history and the history of ideas. The work of Barbara Stafford, Horst Bredekamp, and Martin Kemp is not specifically concerned with maps but more generally concerned with the cognitive power of visualization in early modern Europe.[4] It is through the studies by Stafford on anatomy and scientific entertainment, the essay of Bredekamp on the theory of the Kunstkammer, and the work of Kemp on anatomical and botanical images that the visual logic of these three-dimensional map cycles comes to light, and that the visual synthesis between different branches of knowledge achieved in Cosimo I's and Gregory XIII's map cycles becomes evident.

Renaissance Mapping

Mapping has always been a selective cultural process that, as Denis Cosgrove has recently discussed, involves choices, reductions, omissions, and distortions in order to obtain the representation of the three-dimensional globe of the earth on a two-dimensional surface.[5] During the Renaissance, mapping became a particularly laborious endeavor: from a process based entirely on the visualization of texts, as it had been in the Middle Ages, it was transformed into a craft that required the mastery of both mathematics and philology, of computation and textual exegesis. In fact, a great deal of Renaissance mapping resulted from the scrutiny of an ancient text that was rediscovered in the Renaissance, Ptolemy's geographical manual. This text, which humanists, merchants, and artists passionately read, translated, and illustrated, made it possible to imagine a novel way to visualize the world: the earth's globe was covered with an imaginary spatial grid of coordinates of latitude and longitude, new projections allowed the proportions between the sphericity of the earth and the flat surface of the map to be maintained, and astronomical observations defined the position of terrestrial locations. Having reacquired the mathematical dimension it had lost in the Middle Ages, cartography became a mixed science (*scientia media*), that is, a discipline that applied the abstract principles of geometry and arithmetic to the solution of practical problems. As a mixed science, cartography participated in the larger history that gradually found in mathematics the explanation of the nature of things. By examining the increased importance of mapping in the culture and education at the Medici and papal courts I shall try to establish the place of cartography both in the investigation of the natural world and in the contemplation of the divine.

Parallel to the mathematization of mapping, which was fundamental, since it made possible the new visualization of the earth's globe, was the expansion of the notion of geographical description, which came to include not only the mathematical location of a place, as Ptolemy had suggested, but also whatever that place contained, namely people, animals, plants, rocks, and the vestiges of the ancient past. Indeed, imagined in Renaissance mapping was the antiquarian reconstruction of ancient culture and civilization, either through the definition of the location of events, cities, and buildings reported by ancient writers, or through the reconciliation between ancient geography and modern discoveries. As contradictory as it may seem to us, Renaissance cartographers regarded the reconstruction of ancient geography and the mapping of the modern world as part and parcel of the same process, a process that I shall analyze in detail. Although sailing unknown seas and discoveries of new lands were extending the Ptolemaic world, Renaissance map-makers perceived their work not as the subversion of Ptolemy's *Geography* but rather as the completion of it. As passionately as they debated the length of the Roman foot, so they engaged in equally lively discussion on the location of Ptolemy's mythical Golden Chersonese in modern maps. They were ultimately defenders of classical geography.[6]

This reading of Renaissance mapping as the fusion of mathematics and philology does not apply to every map produced in the Renaissance, but only to the printed maps made for a wide public. As it is becoming increasingly clear thanks to the studies of Brian Harley, Catherine Delano Smith, and David Woodward, among others, the reading public expected maps to include a variety of information pertinent to the mapped place: its history, ethnography, botany, and zoology.[7] Gradually but unmistakably, the cartographic language of printed maps came to differ from the language of other cartographic images, such as cadastral maps and navigational charts, which were made for specific utilitarian purposes, a fact that should be a warning not to confuse the epistemology of printed maps with that of other kinds of cartographic images. I shall analyze the cartographic language of Renaissance maps and their encyclopedic description of places through the maps that Cosimo I and Gregory XIII commissioned for their palaces. Although these maps were made for the restricted circle of Medici and papal courtiers and their visitors, they did share the cartographic language of printed maps, on which in fact they were based.

Like printed maps, the Florentine and Vatican maps were conceived as marvelous images on the boundaries between art and science that conveyed the knowledge of the world spatially and graphically. The maps represented the world according to the cartographic grid as well as according to the descriptions of ancient and modern geographers. They were the product of both the humanistic analysis of the geographical texts by Ptolemy, Strabo, and Mela, and the visual comparison of cartographic images made by modern map-makers. Consequently, they did not simply describe places mathematically, but rather reconciled, at times superbly, the contradictions between the tradition of classical geography and the discoveries of the modern voyages. In Renaissance maps different systems of representation (the plan view, the perspective view, and the bird's-eye view) and different modes of description (verbal and visual, cartographical and historical, mathematical and literal) coexist side by side. As a construct of modern observation, classical knowledge, and visual conventions, Renaissance maps aspired to an accurate record of the world informed by both the standards of cartography and other forms of knowledge. In this book I shall consider maps as scientific illustration, for maps are a form of record of the knowledge of the world.[8] Like the images of plants, animals, rocks, and of the human body that illustrate sixteenth-century books on botany, zoology, and anatomy, maps were visual fields, on which scholars, humanists, and artists joined the description of the modern world with the humanistic exegesis of classical texts.

As hybrid systems, Renaissance maps were polysemic structures, combining at once indexical, iconic, and symbolic signs.[9] The grid makes the map indexical, for to each point on the surface of the earth corresponds one place on the surface of the map. Other signs, however, are iconic, since they retain a resemblance to the way things look in the world, such as the trees identifying forests, the buildings marking cities, and the cones indicating mountains. Others still are symbolic, conventional representations, such as the cross marking a bishopric, or the double cross indicating an archbishopric. The myriad words that fill the map as toponyms, legends, and inscriptions were also symbols. These hybrid semiotic systems fit uneasily within the dichotomy art and cartography traditionally invoked to explain them.[10] The sole focus on the cartographic accuracy of Renaissance maps, which is generally understood as the accurate correspondence of locations between the earth and the map, regards these artifacts as modern scientific images but eliminates the need for analysis of their antiquarian origins and metaphorical meanings. By the same token, the exclusive focus on the contextual meaning of these images, which tends to interpret the maps as representations of power and dominion, leaves out their important role in the investigation of the natural world. Finally, the famous dichotomies between resemblance and difference, similitude and identity, pictorial description and geometric abstraction that Michel Foucault used to distinguish the pre-modern and modern investigation of the natural world are inadequate to grasp the all-encompassing geographical description to which Renaissance maps aspired.[11]

Mapping Sites as Places

The understanding of Renaissance maps as cognitive visual fields made up of different modes and techniques of representation is clarified by Casey's phenomenological investigation on the notions of place and site, which, in turn, are particularly pertinent to the Renaissance notion of space. Renaissance natural philosophers did not conceive space as the fully abstract entity of modern philosophy, but rather as an entity with some elements of corporeality. Their hybrid conception of space as both abstract and corporeal, as quantitative and qualitative, mathematical and descriptive, could accommodate the two opposing notions of site and place investigated by Casey. Renaissance maps should not be interpreted as visual representations of philosophical space, yet the Renaissance philosophical notion of space provided a theoretical framework that made it possible to map the land in terms of both site and place.

In fact, through the notions of place and site, the polysemic structure of Renaissance maps appears as an all-encompassing representation of a locale. The cartographic grid, mathematical and quantitative, establishes the univocal correspondence between sites on the earth and sites on the map. The grid also measures abstract, incorporeal space, since it implies a disembodied viewer with no fixed place on the map and no fixed viewpoint. But the cartographic grid coexists with the perspectival views of mountains, cities, and people, which instead are qualitative, descriptive, and corporeal: although the viewer is still disembodied, he has acquired a fixed place, a point of view from which to measure distances, judge directions, and evaluate sizes. These other features of the map extend the representation of the site as location to the site as place. The maps respond to the double, divergent, but not irreconcilable view of a mathematized space and a highly qualified place, of a homogeneous and heterogeneous space. What might appear to the modern eye as an unresolved combination of different systems of representation and modes of description, was in fact a more encompassing representation of space: bodiless and corporeal, absolute and limited, quantitative and qualitative. In short, a totality of space made up of locations of sites and fragments of places.

The mundane activity of walking is fundamental to Renaissance maps. Edmund Husserl considered walking to be the mediator between body and place, and Casey analyzed specifically as the primary means by which we experience place.[12] According to Husserl and Casey, by walking a coherent world is built up out of the fragmentary appearances that, taken in isolated groupings, would be merely kaleidoscopic. These appearances are both familiar and unfamiliar, near and unknown, but, when brought together by the basic action of walking, become the experience of a coherent organism. The unity of the body parts is supplied precisely by the kinesthetic feelings systematically associated with the actual movement of the body. The activity of walking through a place is implied in the spatial configuration of every cartographic image, Renaissance or otherwise. The represented territory, whether large or small, is mapped by imagining the viewer walking through its roads, mountains, and places. In two-dimensional maps, the bodily movement of walking through the mapped territory is either reduced to the movement of a finger on the surface of the map, or further limited to the movement of the eyes over the image. Exploring the map means, in most cases, transforming the actual movement of fingers and eyes into the mental movement of the entire body. But in the three-dimensional map cycles that constitute these case studies, the mental walking through a map is made vivid and concrete by the actual walking through the architectural spaces, in which the maps are displayed.

In the painted map cycles of Cosimo I and Gregory XIII the illusion of the three-dimensionality of the earth was enhanced by the display of the maps themselves in a specific architectural space. Not only were the maps inextricably connected to their original location, but they were conceived, ordered, and displayed in relation to that physical place. Viewers were meant to enter the room and literally walk into the assemblage of maps. The three-dimensionality of the architectural space represented the three-dimensionality of the world itself, so that the representation of the earth acquired a spatial dimension not afforded in two-dimensional maps. In contrast to the comprehension of two-dimensional maps, in three-dimensional map cycles the cognitive function of the body is mediated not only through the metonymical movement of eyes and fingers on the surface of the map, but it is complemented by the actual movement of walking through the real architectural space occupied by the maps as objects. The movement of walking in and across a room or a corridor mimics, microscopically, the movement of the entire body through the world. Conversely, the different parts of the map cycles are organized not only in relation to each other, but also in relation to the viewer's body and its position.

Furthermore, the sheer size of the maps contained in these cycles entailed the expansion of the scale of representation to such an extent that it effected profoundly the representation and perception of the mapped land, and even well-known places appeared as novel and unfamiliar.[13] At such a monumental scale, the mapped area appeared less distorted, more sites were located on the grid, and more topographical details were added in the forms of symbols, icons, and names. The combination of the monumental scale with the cartographic grid enhanced the possibilities of cartographic representation so that the individual geographical units were represented as both sites, that is, as abstract locations within the cartographic grid, and as places with individual characteristics. The physical surface of the maps as artifacts, their distinct texture as murals, panels, or canvases further contributed to the efficacy of the all-encompassing representation of the place.

Indeed, the meaning of these map cycles, created by the interaction of the painted maps with images of history, mythology, zoology, botany, and religion, could be fully grasped only through the firsthand experience of the rooms that contained them. In this book I shall demonstrate how these three-dimensional map cycles forced the viewer to make connections, create constructive interactions, and build analogies, primarily visual analogies. These analogies were not spelled out in inscriptions within the rooms or in accompanying booklets, but were made possible, indeed encouraged, by the very layout of the different parts of the iconography.

The construction of meanings that these map cycles made possible within their respective architectural spaces was determined by the cultural context of the Medici and papal courts, the ideals of their patrons, and the skills of their makers. The context of their respective courts shaped the representation of political structures, national borders, trading routes, settlements, religious organization, economic resources, history, and devotion. As the following pages shall show, these map cycles constructed meaning through the interaction of mapping with many facets of Renaissance culture. Well-established topics of Renaissance studies, such as the representation of power, the theory and practice of collecting, the fabric of courtly patronage, the rise of mathematics and mixed sciences in the investigation of the natural world, the relationship between images and observation in the explanation of the natural world, the interactions between science and religion, the spatial organization of knowledge, and the self-fashioning of a polymath, all form the backbone of the present investigation.[14]

Egnazio Danti and His Map Cycles

Cosimo I's Guardaroba Nuova and Gregory XIII's Gallery of Maps are the happy result of the cooperation between sophisticated patrons, who were keen to make the investigation of the world part of their own personal iconography, and an accomplished polymath, who also had great familiarity with the visual arts. Cosimo I and Gregory XIII must have decided which maps to represent in their respective map rooms, but it was their cosmographer, the Dominican Egnazio Danti (1536–1586), who conceived these map cycles as pioneering cartographic enterprises, having no parallel either in print or in paint. It was Danti who invented the exceptional arrangement of the selected maps within their architectural space. And it was Danti who, in collaboration with other courtiers, made these map cycles participate in the main themes of their patrons' politics, the legitimacy of the Medici duke, and the universal mission of the Boncompagni pope.

Egnazio Danti, a little-known Dominican, was able to please such different patrons because of his prismatic personality and his skills as a courtier. Known to specialists of artistic perspective for his edition of Giacomo Vignola's *Le due regole della prospettiva pratica* (Rome, 1583) and to historians of science for his signature under the Gregorian reform of the calendar, Egnazio Danti was an exceptional polymath. Extensive training in both the visual arts and mathematics made it possible for Danti to enjoy lifelong familiarity with artists and their practice, as well as with university professors and their theories, and with mathematicians and their procedures. In Florence, Bologna, and Rome, he moved with equal ease in artistic, humanistic, and scientific circles. His mathematical knowledge ranged from astronomy to geography, from optics to perspective, from map-making to the manufacture of instruments. As the cosmographer at the Medici court and later at the papal court, he actively participated in the shaping of the politics and ecclesiastical mission of his patrons.

By the end of his life, he had climbed the ecclesiastical ladder and was nominated Bishop of Alatri, a small town a hundred kilometers south of Rome, where, much to his dismay, he was forced to move in compliance with the Tridentine decree requiring bishops to reside in their dioceses. He was also a persuasive preacher, and in 1585 was invited to deliver the public homily, attended by numerous cardinals and prelates, to celebrate St Thomas Aquinas's day in Santa Maria sopra Minerva, the main Dominican church of Rome.

A passionate bibliophile and collector, Danti had a personal library rich with books, manuscripts, maps, and drawings. It is known that he owned precious manuscripts, including one of Giovanni Boccaccio's banned stories, a copy of Giacomo Vignola's treatise on perspective, and an abridged version of Leonardo da Vinci's *Treatise on Painting*. His collection of drawings was assembled in a book, and included drawings of machines by the Sienese architect Francesco di Giorgio Martini and images of the Virgin and Christ by the Bolognese painter Bartolomeo Passerotti.[15] Unfortunately, this collection is now lost. When Danti died in 1586, his books were offered for sale to the Duke of Guastalla, who declined the offer, and they have disappeared from public sight since then.[16] Danti's private papers have also been lost. His manuscripts, instead, were acquired by the Apostolic Chamber in compliance with another Tridentine decree mandating the acquisition of bishops' earthly goods by the papacy. What remains of his prolific life are fragments of his correspondence, sporadic entries in the state archives of Rome, Florence, Bologna, and Perugia, his treatises on applied mathematics, and, above all, the magnificent map cycles he designed and supervised in Florence and Rome.

Renaissance Map Cycles

Danti's map cycles for the Medici and the papacy were part of a distinct preference of Renaissance patrons for cartographic imagery to decorate their palaces and villas.[17] At the peak of their diffusion, in the second half of the sixteenth century, painted maps proliferated on the walls of royal, papal, and republican palaces, in monasteries, and in cardinals' villas. The most spectacular examples are documented in Italy, where they seem to have originated in the second half of the fifteenth century, and some of them are still well preserved in Rome, Florence, Parma, Naples, and at Caprarola. Most, however, are known only through documents describing them, such as the maps painted for the Gonzaga palaces in Mantua, or the marvelous series that embellished the Ducal Palace in Venice. Later, the fashion for monumental map cycles spread also to other parts of Europe, such as London, Madrid, and Salzburg. Today it is hard to appreciate the spread of painted map cycles in Renaissance Italy, but Jacob Burckhardt was so impressed by the wide use of cartographic images to decorate Renaissance residences and halls of power that he listed them among the artistic genres of the Italian Renaissance.[18] The cartographic content of these map cycles varied greatly. Civic authorities usually painted monumental maps illustrating the territory under their jurisdiction. Similarly, rulers and nobles selected cartographic representations (either maps or views) of their possessions in order to visualize the extent of their political and administrative power. In other cases, the map cycle included maps or views that did not correspond to any political entity, nor were they under the jurisdiction of a single ruler. Maps of the old and new world, or views of important European cities, represented a symbolic possession for the patron, stressing the borders of his religious, political, and economical influence. Extremely popular were cycles of city views, a subject that both men and women seemed to have liked for the decoration of private quarters, such as study rooms, balconies, and loggias.

These map rooms fit well in the decorative schemes of princely palaces. They added variety to the overall decoration; flattered their patron's interests in geography, astronomy, and cartography; were invented in accordance with the principle of decorum, that pillar of Renaissance art that dictated the suitability of the iconography to the function of the room; and were invaluable in the political and religious aggrandizement of their patron. In these exceptional rooms, the painted maps were so coherently selected in relation to each other and in conjunction with other non-cartographic images, such as allegories, religious scenes, and historical events, that they can be regarded as a cycle, that is as a coherent series of images organized around a central theme and around a single geographical unit, be it a region, a continent, or the earth itself.

The creation of painted map cycles in the Renaissance was part of an older tradition that, from the Middle Ages onward, had favored the display of the encyclopedic world map in monastic libraries, chambers of rulers, or papal dining halls. But Renaissance patrons and map-makers ignored the continuity with the medieval tradition, preferring instead to relate their predilection for the display of maps to ancient models. They learned from the Roman historian Livy that Tiberius Sempronius Gracchus celebrated his conquest of the island of Sardinia by exhibiting publicly a map in the shape of the island, on which the battles of the Roman conquest were marked. They read in the writings of Aelian, an authority on military art, that Socrates used maps to impart moral lessons on the relativity of earthly things, and gleaned from the revered Pliny that maps and city views adorned private houses and public sites in ancient Rome.[19]

More specifically, Pliny praised Ludius, a painter contemporary with the reign of Augustus, for bringing in "the fashion of painting seaside towns on the walls of open galleries, producing a delightful effect at a very small cost."[20] In addition, Pliny reported on Agrippa's map of the Roman world that Augustus had set up in a colonnade in the Campus Martius, a map that is now lost. Pliny gave a particularly detailed description of the large-scale plan of Rome, which was displayed on an outer wall of the library attached to the Temple of Peace in Rome and which was presumbaly an earlier version of the *Forma Urbis Romae*, the marble map carved between 203 and 208 and displayed in the same location.[21] Fragments of the *Forma urbis Romae* were surfacing in Renaissance Rome, demonstrating to antiquarians and map-makers that the monumentality of the map was germane to the richness of its description. Precisely because it was so monumental, the map could include a plethora of details, such as the division of the city into quarters, the routes of streets and their names, the location, name, function, and height of monuments and buildings. Though it mapped only the city of Rome, the *Forma urbis Romae* was the most authoritative surviving evidence for the display of monumental maps made of durable materials in public spaces.

Inspired by this Roman tradition and wishing to surpass the ancients, Renaissance patrons and authors transformed the ancient display of individual maps into a true Renaissance fashion. Paraphrasing Pliny, Leon Battista Alberti recommended the use of maps as mural decorations to instruct and delight. Paolo Cortesi regarded map murals as particularly suitable embellishment for the houses of cardinals. In the late sixteenth century Giovanni Battista Armenini and Gabriele Paleotti repeated the arguments of earlier art theorists, showing that the fascination with cartographic murals was still in full swing in post-Tridentine Europe.[22] John Dee made the fashion for cartographic decorations known in England, reporting that "some, to beautify their Halls, Parlors, Chambers, Galeries. Studies or Libraries . . . liketh, loveth, getteth and useth Mappes, Chartes, and Geographical Globes."[23] The widespread public display of maps in the Renaissance bespeaks not only the achievement of Renaissance mapping but also the wish of emulating Ludius's art by recreating ancient interiors in modern palaces.

Among the surviving map murals, Danti's map cycles for the Medici and the papacy stand out for their magnificence, their documented historical context, and their emblematic presentation of the full scope of Renaissance mapping. Although they were very different in content, meaning, and function, they clarify substantially the understanding of the two main branches of cartography defined by Ptolemy: geography, the representation of large regions of the worlds, and chorography, the representation of small areas. These two cycles also allow a comparison of the political and religious meanings of Renaissance maps, making it possible to see their functions within a collection of artifacts and within the recounting of church history. These extraordinary three-dimensional displays effectively connect cartography to politics, religion, encyclopedism, collecting, church history, spirituality, and much more.

The aim of this book is to recapture the context of these map cycles through a narrative reconstruction of the cultural nexus from which they emerged. This is an interpretative narrative that puts at the center the world of Renaissance maps and the people who commissioned, made, and enjoyed them. The narrative itself is an interpretative synthesis that reconstructs the interdependence of art, patronage, science, cartography, politics, and religion. Centering on two important courts, this book tells the story of how mapping became a major form of courtly art.

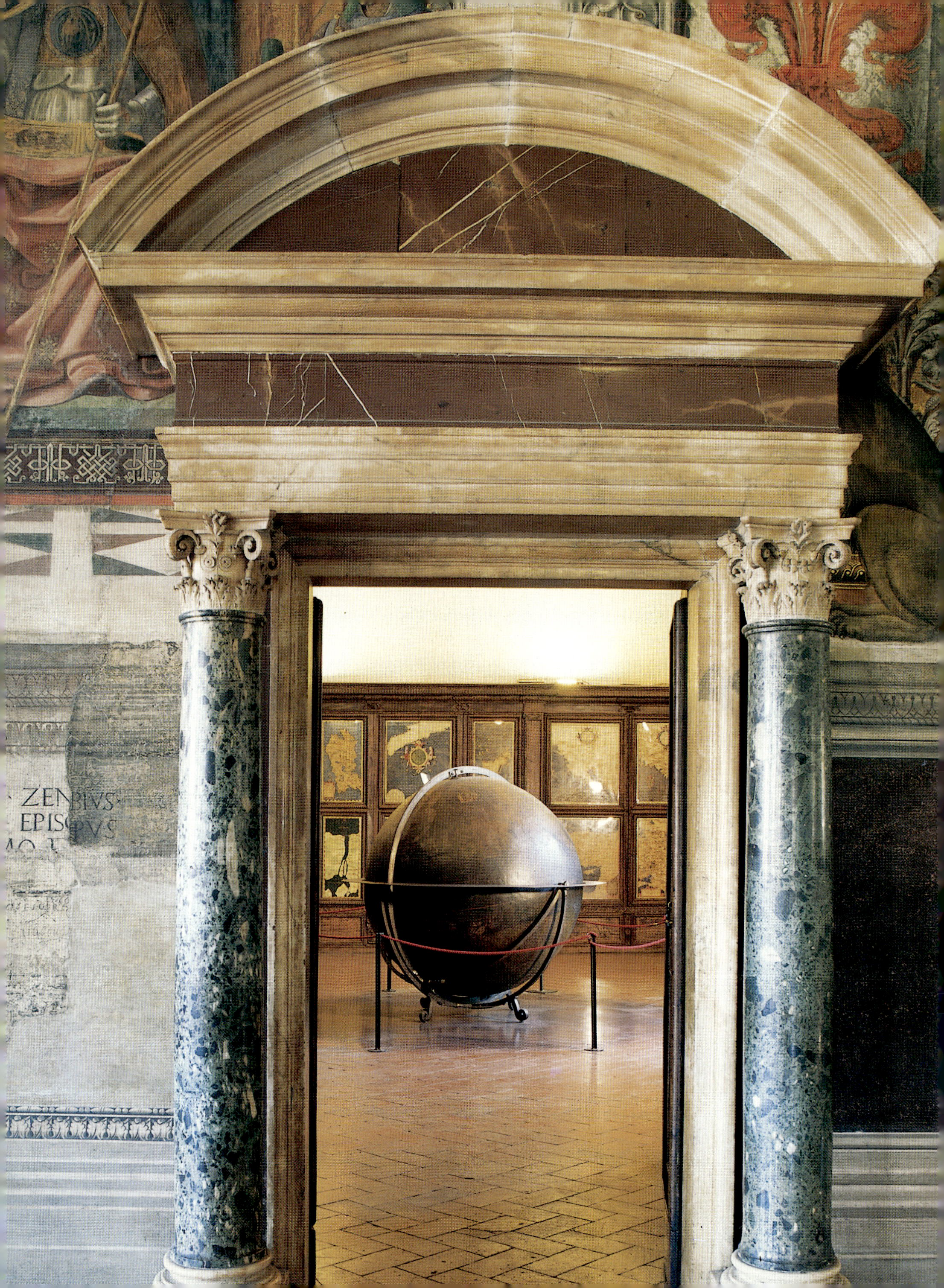
ZENBIVS
EPISCPVS

PART ONE

Maps as Worldly Art

(*facing page*) Detail of fig. 3.16

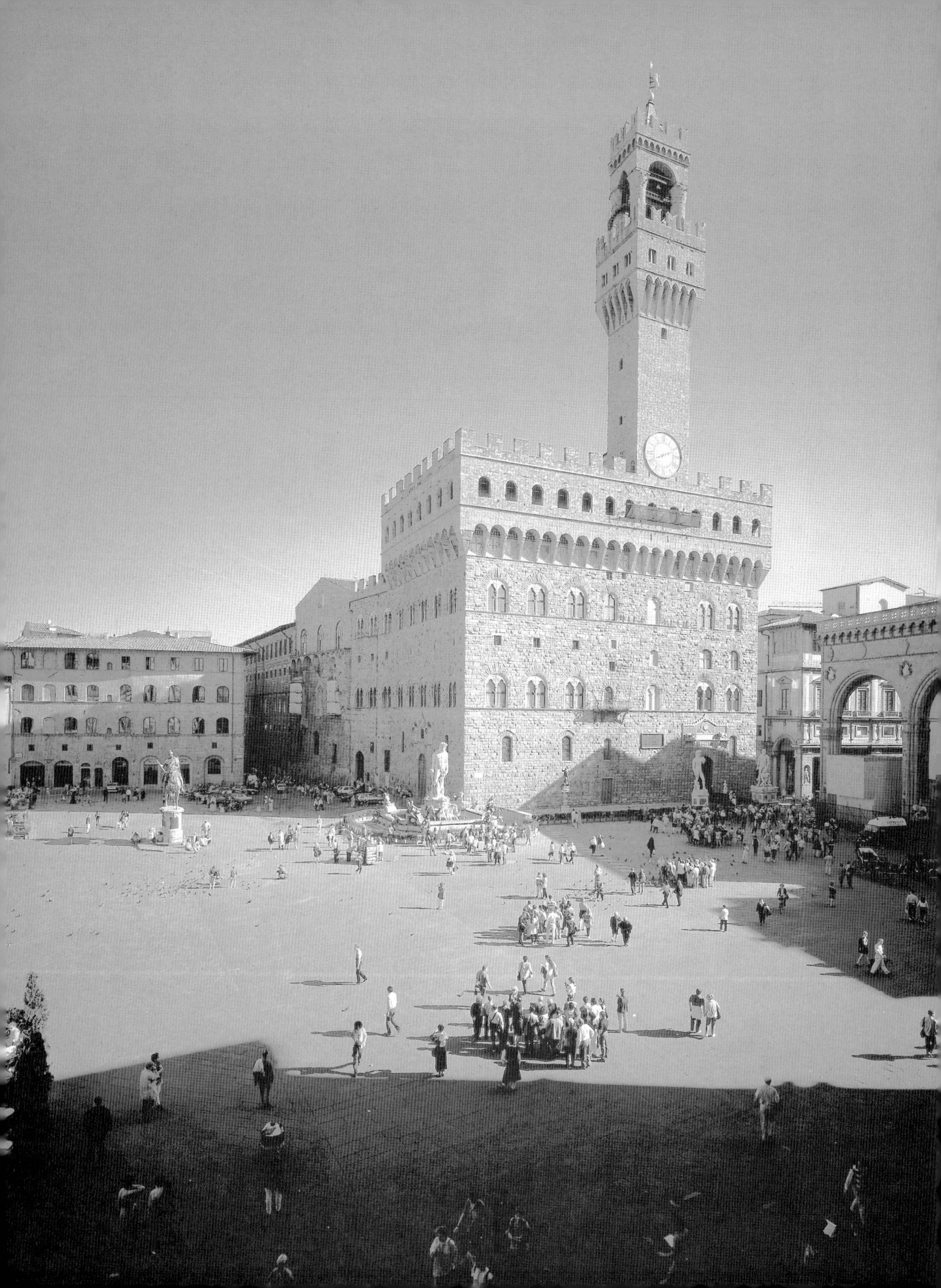

2

The Guardaroba Nuova

Cosimo I and His Ducal Palace

On 15 May 1540, three years after he had been elected Head of the Florentine Republic and had masterfully succeeded to keep his rule over Florence, Cosimo I de' Medici, his wife Eleanor of Toledo, and their newly born daughter Maria packed their belongings, left the Medici Palace, crossed the streets of Florence, and moved into the medieval seat of the Republic in the main square, an imposing building then simply known as the *palazzo di piazza*. From that day, the republican building became the Medici ducal palace and an extensive campaign of restorations, additions, remodeling, painting, and refurbishing transformed the austere medieval building into the sumptuous palace still admired today (Fig. 2.1).

Cosimo I's reasons for abandoning the Medici Palace, the massive family building with a glorious past, in favor of the medieval fortress-like seat of the Republic were both practical and political. The *palazzo di piazza* was easier to defend than the Medici Palace, a considerable advantage in such a volatile political situation. It had spacious halls to serve perfectly the ceremonial needs of the ducal court and an infinity of small rooms and apartments to accommodate the growing family. Moreover, the change of residence was meant to reinforce Cosimo I's legitimacy as the ruler of Florence. As Giovambattista Adriani, Cosimo I's secretary, commented later, the duke moved to the *palazzo di piazza* in order to show "that he was an absolute prince and the arbiter of Government; and also to set at rest the mind of those who thought that the government of the city was separated from the government of the Medici family."[1] To make sure no one would miss the political meaning of this move, Cosimo I selected as his living quarters the rooms that Pietro Soderini, the last prior of the Republic, had inhabited during his tenure 1502–12. The Medici ruler was, thus, the legitimate successor of the republican prior.

But Cosimo I also had personal, equally important, reasons for leaving the Medici Palace. Built by Cosimo the Elder in the mid-fifteenth century to replace the older and more modest building on the Via Larga, the new palace had set the Florentine tone in matters of architecture and decoration.[2] From their new palace the Medici had conducted their political and

2.1 Palazzo Vecchio, Florence

economic affairs for over a century, while at the same time embellishing it with one of the most astonishing collections of ancient and modern works of art. Although the palace had been ransacked in 1494, it was still in good condition and served as the ducal palace of Alessandro de' Medici, Cosimo I's predecessor. But for Cosimo I, the family palace was a constant, annoying reminder of his disputed legitimacy as the Medici heir and of his immense political debt to Emperor Charles V.

Cosimo had never lived in the new Medici Palace before his election as Head of the Republic in January 1537. The Medici Palace belonged to the main branch of the family descending from Cosimo the Elder, a branch to which Cosimo I belonged only through his mother Maria Salviati, the granddaughter of Lorenzo the Magnificent. His father, the famous, honored but indigent commander Giovanni dalle Bande Nere, belonged to a side branch of the family – Giovanni dalle Bande Nere's father was the brother of Cosimo the Elder – and owned one floor of the Medici Palace on the Via Larga, the old family building now lost. When the young Cosimo was not hunting in the Mugello, he lived in the old Medici Palace, which he had to share with his cousin Lorenzaccio, and only visited the Medici ducal palace as a guest of the reigning duke. Once he became duke, he also inherited the ducal palace. Cosimo had been chosen as the Medici heir not because of his genealogy, intellectual prestige, or political experience, but because he was the only legitimate adult male in the Medici family available at the time. His descent from both branches of the Medici family was transformed from a problem into a genealogical virtue and his name, after Cosimo the Elder, was regarded as an omen good enough to assure his brilliant political future. As the new heir of the Medici family, Cosimo I took possession of the Medici ducal palace and marked the occasion with a sumptuous lunch for his supporters and retinues. But, since the ducal palace was still occupied by Margarita of Austria, widow of the assassinated duke and daughter of Emperor Charles V, Cosimo I felt compelled to leave his ducal palace after lunch and to sleep in his old palace on the Via Larga.[3]

Moreover, the political situation was so uncertain, Cosimo I's legitimacy as the Medici heir so disputed, and his authority as the ruler of Florence so weak that Margarita of Austria could successfully proclaim herself the legitimate heir of the Medici possessions.[4] The imperial commander in the city, Alessandro Vitelli, seized the Fortezza da Basso in the name of the emperor, ordered the Spanish soldiers to ransack the Medici Palace, and convinced Margarita to move to the fortress together with the Medici collections. Cosimo negotiated skillfully with the emperor and his representative in Florence, but his political power depended so heavily on imperial support that he was forced to accept a rather humiliating situation: Charles V gave the Medici properties in usufruct to his daughter who, in turn, agreed to rent the Medici real estate back to Cosimo I for the amount of 7,500 scudi a year. Not only did Cosimo have to rent his ducal palace from the emperor's daughter but he also had to buy back some essential furnishings from Margarita to survive the chilly winter in the empty palace. As soon as his power seemed stable and his place among the Italian rulers established – his marriage to the daughter of Pedro of Toledo, the powerful viceroy of Naples, was crucial in enhancing his political recognition – Cosimo I turned his thoughts towards the search for a suitable princely palace. Like other Renaissance rulers, he expected a palace to reflect the perfect order of his dominion, illustrate the history of his family, and demonstrate his own intellectual refinement. The *palazzo di piazza* reconciled harmoniously political, practical, and symbolic concerns.

When Cosimo I and Eleanor moved into the *palazzo di piazza*, the original medieval palace had developed into a huge complex of heterogeneous buildings. The palace included majestic halls, but small and irregular rooms were still used as offices for the administration of the Republic and as living quarters for some of its representatives.[5] The palace obviously needed major renovations, some of which were undertaken immediately. However, its complete transformation into the superb Medici residence was accomplished over a period of more than three decades, first under the direction of the architect Battista del Tasso and later under the astute supervision of Giorgio Vasari, the skilled artist and devoted courtier who, from 1554 to 1574, interpreted Cosimo I's political aspirations in paintings and buildings. Battista del Tasso renovated the apartments for Cosimo, Eleanor, their children, and the small group of courtiers who accompanied them in the new residence. Artists such as Agnolo Bronzino, Francesco Salviati, and Ridolfo Ghirlandaio painted these apartments that were adorned with beautiful furniture, precious collectibles, paintings, and statues. Francesco Salviati painted the Sala dell'Udienza with the stories of Furius Camillus, the Roman alter ego of the new Medici duke. Baccio Bandinelli redesigned the Udienza del Salone, the north end of Cosimo's audience hall where the ducal throne was placed, and embellished it with statues of great Medici. Bronzino, Pontormo, and Salviati made cartoons for twenty tapestries to adorn the Salone dei Duecento. Vasari planned and decorated two magnificent apartments, the Apartment of the Elements on the second floor, and the Apartment of Leo X on the first. In the six rooms that made up the latter apartment, the deeds of Cosimo the Elder, Lorenzo the Magnificent, Pope Leo X, Pope Clement VII, Giovanni dalle Bande Nere, and Duke Cosimo I celebrated the Medici century-long *de facto* rule of Florence rather than the recent acquisition of the hereditary rights under Duke Alessandro de' Medici, whose quite inglorious actions were in fact not represented.

The culmination of Cosimo I's grandiose restorations was the complete renovation and decoration of the Sala Grande, the monumental meeting hall of the Republic, which had become the ducal audience hall (Fig. 2.2). Under Vasari's supervision, the Sala Grande was enlarged, adorned with a new magnificent ceiling and with thirty-nine canvases illustrating the history of Florence. In its very size, the renovation of the grandiose Sala Grande surpassed anything Cosimo I had previously commissioned for his ducal palace and, for that matter, "all the halls made by the Venetian Senate and all the kings and emperors and popes that ever were because – or so Vasari claimed – even if they had had the money, none of them had in their palace either a masonry fabric so grand and so magnificent or the indomitable spirit to know how to begin an enterprise so awesome and of such importance."[6]

2.2 Sala Grande. Florence, Palazzo Vecchio

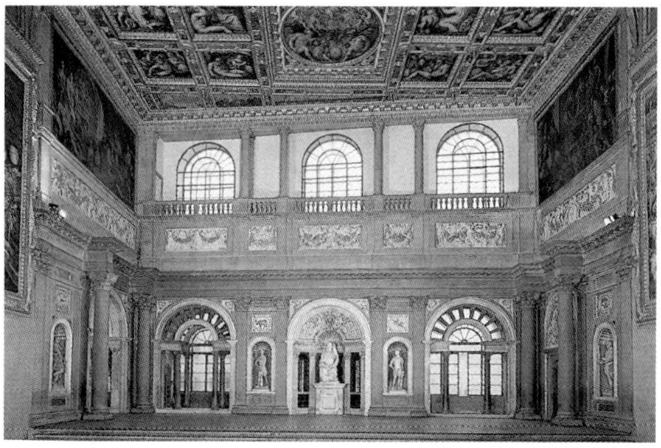

The project of the Sala Grande is rightly considered an intrinsic component of Cosimo's political success from the mid-1550s onward. Through careful and astute political moves, he had been able to acquire autonomy from the emperor, establish a reliable group of administrators, transform the Florentine merchant families into devoted courtiers, enlarge his duchy, and conquer the city of Siena that he had triumphantly entered in 1560. After the conquest of Siena, his international recognition as an absolute prince grew considerably. As the Duke of Florence and Siena, he contributed to the election of Pope Pius IV de' Medici, a member of the Milanese branch of the family. In 1569, after years of prolonged diplomatic skirmishes, he was finally granted the title of Grand Duke of Tuscany and Siena, a quasi-royal title that put him above any other Italian ruler. He also arranged the wedding of his son Francesco, heir to the ducal title, to Johanna of Austria, the emperor's daughter. The triumphal entry into Florence of the imperial bride culminated in the Sala Grande, which was also the stage of the wedding ceremonies.

The decoration of the Sala Grande presented a considerable challenge since the traditional schemes used in Renaissance princely halls of power, generally based on the history of the ruling family, did not adequately convey Cosimo I's political message in the mid-1550s: his ducal power was the fulfillment rather than the negation of the Republic. Like the official halls of many Renaissance rulers, the iconography of the Sala Grande had to transform the hall of power into a microcosm of the entire dominion and represent the prince as the force behind the well-being of the state. In the earlier iconography for the halls and apartments of his palace Cosimo I had largely followed Renaissance traditions. Like the Farnese in Rome and Caprarola or the Gonzaga in Mantua, he filled his palace with images celebrating himself through the dynastic history of his family, the mythological stories of the pagan gods or the deeds of the Roman heroes. But the Sala Grande required a more innovative iconography that would stress the continuity between the republican past and the Medici rule by representing the uninterrupted history of Florence, from the Roman foundation to the medieval enlargement until the war of the Florentine Republic against Pisa in the late fifteenth century and the recent conquest of Siena under Cosimo I.[7]

The project of the Sala Grande is documented in minute detail from its early conception to the final execution, as is the role of each of its protagonists: Duke Cosimo I controlled the content and meaning of each historical scene; Vasari invented the compositions, painted most of the canvases, and saw to their completion; the humanist Vincenzo Borghini was in charge of the archeological and historical accuracy of the paintings and their inscriptions; the official court historian Giovambattista Adriani conducted the historical research. From the intense collaboration of Cosimo, Vasari, Borghini, and Adriani the order of the ceiling emerged. The battles against Pisa and Siena that had brought the two cities under the dominion of Florence were depicted on the long sides of the ceiling and were flanked by allegorical representations of other Tuscan cities and their counties. At the center of the ceiling was a central tondo, which Vasari described as the "key and culmination of the historical scenes."[8]

As is well known, the iconography of the ceiling was modified numerous times. According to the initial plan for the ceiling, still documented in a preparatory drawing, Florence had to be represented as a personification in the center of the ceiling. It had to be surrounded by episodes of civic history and by the republican institutions: the corporations, guilds, quarters, and gonfalons. This was a particularly fitting choice considering the city's role as the administrative and political center of the dominion, and a particularly fitting location within the ceiling considering the representation of other subject cities at both ends of the ceiling. But as the iconography of the entire ceiling was being refined in response to Cosimo I's wish, Florence shared the central tondo with the duke. At the center of the ceiling, Cosimo is seated on a curule chair and surrounded by the symbols of his power, the cross of the Order of

Santo Stefano that he himself had founded in 1562 to fight against the Turks, the chain of the Order of the Golden Fleece that Charles v had given him in 1545, and the ducal coronet. The personification of Florence crowns him with the laurel wreath, the ultimate reward the Roman Republic granted to its grandest citizens in antiquity, while shields with the insignia of the Florentine guilds frame the entire composition.

The choice to represent Cosimo I and Florence together in the central tondo has been related to the duke's intention to link his own rule with the city's republican past. Indeed, the inscription around the central tondo did not mention Cosimo's hereditary title as duke of Florence but rather presented his political life as stemming from the Republic: he was the Head of the Florentine Republic, the organizer of the city of Florence, the conqueror of Pisa and Siena, and the patron of peace in Tuscany.[9]

Perceptively, it has also been suggested that the ceiling of the Sala Grande has "a centrifugal movement outward across the territorial state of Tuscany to its boundaries and beyond."[10] From the central tondo, the force emanates simultaneously from the administrative center of the state and the persona of the duke, from the center of the dominion to its periphery. Cosimo had suggested the depiction, in one single painting, of "our states all together" in order to illustrate "their extension and acquisition." Although this scheme was not realized on one canvas alone, it was achieved with great subtlety at the two ends of the ceiling, which reflected the geographical, territorial, and administrative organization of the ducal dominion rooted in the administration of the Republic but centralized and unified under Cosimo I.[11] Thus, the cities of the Tuscan state that were not included in the initial plan were eventually represented at the sides of the ceiling and identified by an inscription that included their date of annexation to the Medici state. With sophistication the individual scenes combined allegorical personifications of cities, regions, and virtues with accurate topographical and historical representations of buildings, places, and people. More importantly, their position on the ceiling was carefully determined by the Florentine quarter to which they were administratively subject and by their compass direction from Florence. In this respect, the ceiling was a diagram of the administration of the state and a schematic map of its territory.

In short, the Sala Grande has been interpreted within the context of Cosimo I's political aspirations, stressing alternatively his personal glorification, the continuity between the city's republican past and the Medici rule, or the schematic representation of the entire dominion and its institutions. Yet what has escaped attention is the fact that the political and symbolic meaning of the Sala Grande comes into better focus when Cosimo's hall of power is considered in connection with the Guardaroba Nuova, a neglected room located near the grandiose audience hall, which, in 1563, was redesigned as one of the duke's cabinets of art and curiosities. For this purpose, he had the room furnished with new cupboards, which are still in place today, and planned a complex iconography made up of maps of the world, images of plants and animals, portraits of famous people, scientific instruments, and ancient busts (Fig. 2.3).[12]

The Sala Grande and the Guardaroba Nuova were planned and decorated simultaneously and, between 1563 and 1565, Vasari, Borghini, and Cosimo I exchanged letters dealing with the two projects. The Sala Grande and the Guardaroba Nuova are entirely unrelated in terms of iconography, function, and use. Whereas the Sala Grande is decorated with magniloquent historical paintings, the Guardaroba Nuova is filled with images of the earth's globe. The Sala Grande is the most official space of the ducal palace, but the Guardaroba Nuova is a

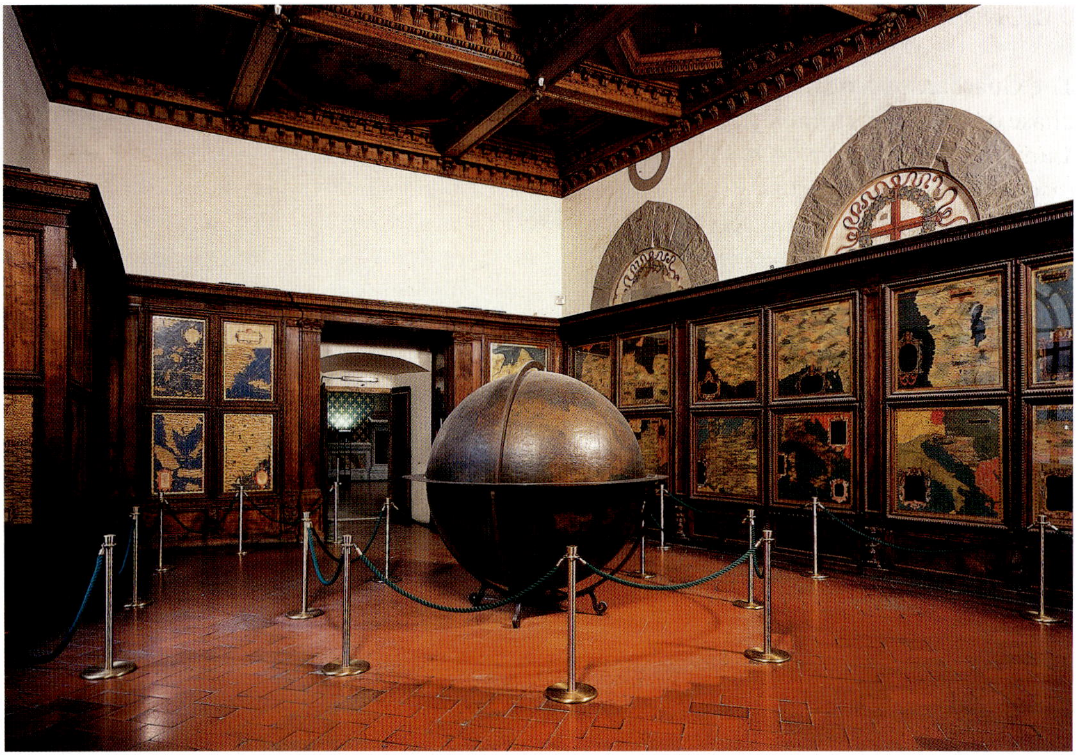

2.3 Guardaroba Nuova. Florence, Palazzo Vecchio

secluded cabinet of art and curiosity. The Sala Grande is the stage of ducal audiences; the Guardaroba Nuova is the destination of selected visitors. And yet, in spite of all these differences, the spatial and temporal proximity of these two rooms forces us to interpret them as connected and their iconography as complementary. Although the Guardaroba Nuova has been briefly considered in relation to Renaissance *studioli* or in connection with other map cycles, its complex cultural context still remains elusive and its significance in relation to Cosimo I's iconography uncharted. The fact that the decoration of the Guardaroba Nuova was never fully executed did not facilitate the understanding of its relation to the rest of the palace and to Cosimo I himself. But, once the Guardaroba Nuova is examined in connection with Cosimo's political imagery, within the larger context of the European cabinets of art and curiosities, and in relation to the scientific and geographic knowledge it embodied, is it possible to understand how eloquently it epitomized his interests and how deeply it participated in the glorification of the duke depicted on the walls of the Sala Grande. First, however, the planned, though unrealized, decoration of the Guardaroba Nuova has to be reconstructed, a task that can be accomplished through the careful assemblage of archival sources, correspondence, and objects from the Medici collections currently kept in Florentine museums.

* * *

The Room

The Guardaroba Nuova, located on the second floor of the ducal palace, stems from the last phase of Vasari's restorations, particularly from the ambitious and costly project of building a unified system of stairways for the entire palace. Originally the landing of an earlier staircase, the room still documents the mishaps of its building through its irregular, trapezoid plan. Vasari, who had been the chief artistic adviser in the remodeling of the ducal palace after 1554, was responsible for the addition of this obsolete landing loggia to the ducal guardaroba, the series of rooms used to store Medici possessions.[13]

The original function of this new room as part of the ducal guardaroba is concealed by its modern name, Room of Maps (Sala delle Carte Geografiche) or Room of the Globe (Sala del Mappamondo), but it is made clear by its original name, Guardaroba Nuova, recorded by printed and archival sources.[14] The main function of the *guardaroba* (*guarda roba*, literally keep goods) was to keep track and count of the portable objects that the duke possessed, such as tapestries, carpets, pillows, bed linens, pottery, sculptures, paintings, books, gems, scientific instruments, arms, medals, silverware, tables, chairs, stools, chests, and cups, to name only a few items. A *guardarobiere* and a few helpers rearranged these objects according to the constantly changing needs of the court: the passing of seasons, the setting of banquets, audiences, and spectacles, the travels of the ducal family, and the stays of numerous dignitaries of varying rank and nobility. The guardaroba was also the place to repair damaged items and was equipped with laboratories to restore, and soon also to produce, tapestry, silverware, pottery, and woodworks. The Medici guardaroba grew as steadily as Cosimo I's holdings and, by the time the Guardaroba Nuova was added, it occupied many rooms, and its contents needed to be inventoried regularly to check that nothing was missing.

Like other rooms of the ducal guardaroba, the Guardaroba Nuova was furnished all around with large wood cupboards to store a selection of the Medici possessions. Unlike any other room of the ducal guardaroba, however, the new room had to be lavishly decorated with a complex system of interrelated images representing the cosmos. In the 1568 edition of his important book, *The Lives of the Most Famous Architects, Painters, and Sculptors*, Vasari described the planned decoration in detail:

> His Excellency, under the direction of Vasari, has walled a new hall of some size expressly as an addition to the guardaroba, on the second floor of the apartments in the Ducal Palace; and this he has furnished all around with cupboards seven braccia high, with rich carvings of walnut-wood, in order to deposit in them the most important, precious, and beautiful things that he possesses. Over the doors of those cupboards, within their ornaments, Fra Egnazio has distributed fifty-seven pictures about two braccia high and wide in proportion, in which are painted in oils on the wood with the greatest diligence, after the manner of miniatures, the Tables of Ptolemy, all measured with perfect accuracy and corrected after the most recent authorities, with exact charts of navigation and their scales for measuring and degrees, done with supreme diligence; and with these are all the names, both ancient and modern.[15]

Though Vasari's *Lives* must be treated with a degree of scepticism, this particular passage on the Guardaroba Nuova does read as the accurate description of a work in progress.[16] According to Vasari's description, the doors of the cupboards had to be painted with fifty-

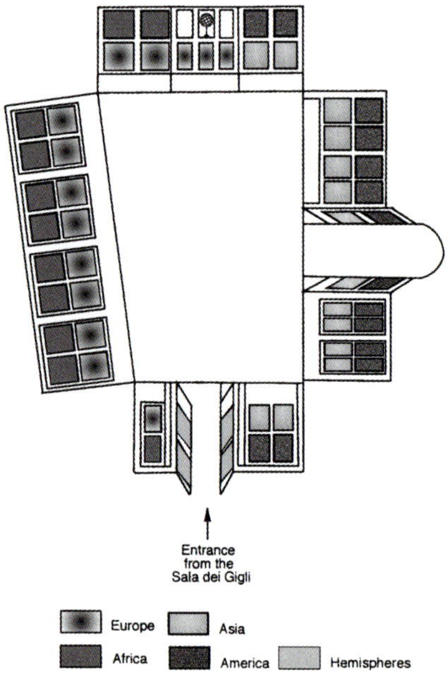

2.4 Reconstruction of the original sequence of the map panels in the Guardaroba Nuova

seven maps of the world, which he called the "Tables of Ptolemy" in reference to the maps that illustrated the Renaissance edition of Ptolemy's *Geography* (Fig. 2.4). Four hemispheres representing the earth and the sky were at the side of the main entrance. On the lower level, starting from the left, there were fourteen maps of Europe followed by fourteen maps of Asia. On the upper level, starting again from the left, there were eleven maps of Africa, the armillary sphere by Lorenzo della Volpaia, and fourteen maps of the West Indies. Spatially and conceptually related to the maps, Vasari reported, were images of plants and animals native to the mapped countries, and portraits of ancient and modern rulers who possessed those lands. Above the cupboards were to be located marble busts of emperors and princes, while between the cupboards and the ceiling, on the remaining portion of the walls, were to be displayed portraits of famous people. The Ptolemaic constellations were to be represented on the ceiling. Two monumental globes, one representing the earth, the other the sky, were to be located in the room, concealed behind ceiling panels that could be lowered to pedestals on the floor.

Vasari's description vividly suggests to the modern reader the first impression that the elaborate system of images of the Guardaroba Nuova would have made on a contemporary viewer: the celestial map, the panel maps, the images of plants and animals, the busts of ancient rulers, the portraits of famous people, the armillary sphere, the two monumental globes, all in one room, would have been like a walk into the cosmos, with the regions of the world displayed on all sides, and with the sky depicted on the ceiling. Indeed, the Guardaroba Nuova provided an explo-

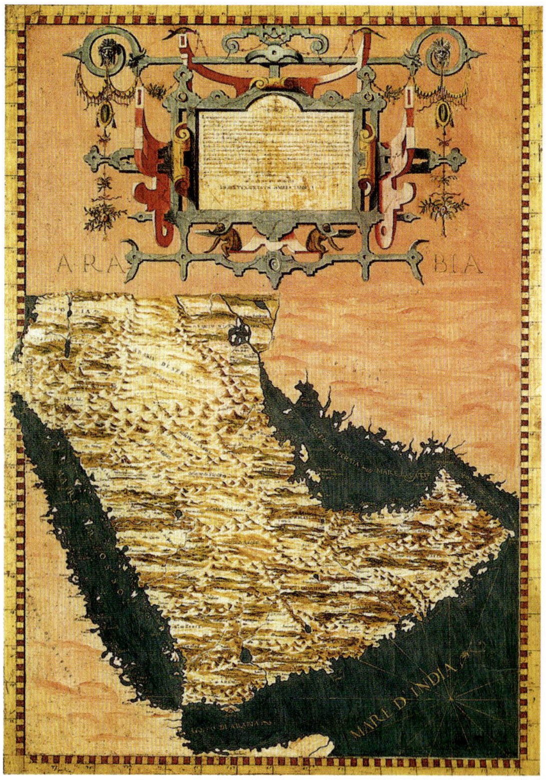

2.5 Egnazio Danti, *Arabia* (Map of Arabia), 1575, panel. Florence, Palazzo Vecchio, Guardaroba Nuova ration of the cosmos aptly reduced to room-size dimensions. It would have been enriched by the observation and study of the images of plants, animals, and people that populate the world, and by the added wonder of being able to handle fragments of the world's treasures and curiosities kept in the cupboards.

Only fragments of this fanciful decoration were ever realized, and today the room corresponds only roughly to Vasari's description. Fifty-four of the fifty-seven planned maps were executed by two different map-makers over a period of twenty years: the Dominican Egnazio Danti made thirty-one maps from 1563 to 1575, while the Olivetan Stefano Bonsignori made twenty-three maps from 1577 to 1586 (Figs. 2.5 and 2.6). Some changes were introduced while the decoration was in progress and therefore the sequence of the maps reported by Vasari was altered. For instance, instead of the four hemispheres representing the earth and the sky, there are four maps of the Polar areas painted neither by Danti nor Bonsignori, but perhaps by the mathematician Antonio Santucci around 1590. Vasari's description also does not match with the physical space of the room and must be amended by substituting the word "right" with "left" at the beginning of his description. Furthermore, the maps, which are the best preserved part of the room, are displayed today according to an order that does not reflect Vasari's description, not even its amended version. The original sequence of the maps can be reconstructed according to the geographical order of the Renaissance editions of Ptolemy's *Geography*,[17] but the reconstruction of other parts of the decoration, which were either removed or never executed, is more demanding.

The monumental terrestrial globe (204 cm in diameter) was built, between 1563 and 1568, by Egnazio Danti who proudly reported on its technical features. Its interior was, "entirely armed in iron, because such a large globe could not sustain itself," and it could be easily "moved in any direction by the touch of a single finger" thanks to a grand mechanism that Danti himself had devised.[18] It sat on an elaborate iron pedestal, which alone cost 400 scudi. Danti even planned an elaborate mechanism to

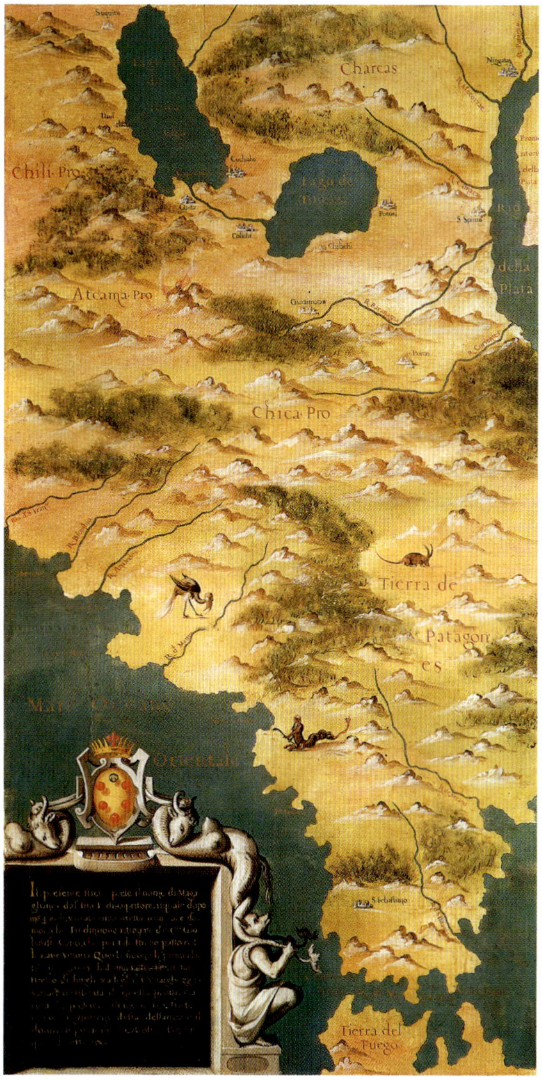

2.6 Stefano Bonsignori, The Strait of Magellan, not dated, panel. Florence, Palazzo Vecchio, Guardaroba Nuova

lower the globe from its hidden compartment behind the ceiling to the floor, but this project, possibly the most outstanding feature of the room, was postponed by Prince Francesco, Cosimo I's son and successor, who explicitly said that "it does not matter."[19] The terrestrial globe was more conventionally displayed at the center of the room (where it has been recently relocated), but in 1589 it was transferred to the Loggia della Cosmografia in the Uffizi Palace, where it was irremediably damaged by exposure to the elements. The globe still amazes viewers with its monumentality but, dark and illegible, it is now a shadow of its original splendor. It has lost its brilliant colors, most of the geographical details, and its original pedestal, and no longer performs the smooth movements, of which Danti was so proud.

The celestial globe, pendant to the terrestrial globe, was also planned. Although it was thought until recently that Danti never began work on it, new archival evidence has surfaced showing that he undertook the project, had the leather bought and cut into parts, and perhaps even started to paint it, but certainly did not assemble it. The unfinished globe was kept in the Medici guardaroba for many decades awaiting proper use, which never came. Although Danti did not complete the celestial globe and although the Guardaroba Nuova was already partially disassembled in 1577, the original idea of a monumental celestial globe as a companion of Danti's terrestrial globe was not completely forgotten. In 1588 Grand Duke Ferdinando I, another of Cosimo's sons, commissioned Antonio Santucci to make a monumental armillary sphere. Its close relationship with Danti's terrestrial globe is indirectly acknowledged by the fact that, as soon as Santucci completed his armillary sphere in 1593, he convinced the grand duke of the need to restore Danti's terrestrial globe, a task he undertook two years later. From the late sixteenth century onward, Danti's terrestrial globe and Santucci's armillary sphere were displayed together in the Loggia della Cosmografia in the Uffizi Palace.[20]

Complementary to the two monumental globes was the famous armillary sphere built by Lorenzo della Volpaia in the 1490s, which was moved into the Guardaroba Nuova perhaps as early as 1570. This armillary sphere, now lost, suited the room not only because of its function as an astronomical instrument to investigate the sky but also as a prestigious collectible with an illustrious history. Built for Lorenzo the Magnificent, the armillary sphere was later acquired by the Parte Guelfa in the attempt to save it from leaving Florence, and it was finally donated to the Florentine Republic, which displayed it as a symbol of its grandeur in the Sala dei Gigli, the meeting room adjacent to the Guardaroba Nuova. The inventories of the palace and Vasari's description agree: this precious armillary sphere was displayed prominently on the upper level of the cupboards, facing the main entrance from the Sala dei Gigli, and it stood on a wood pedestal made by the skilled woodworker Dionigi di Matteo Nigetti. In 1589, it was relocated to the Uffizi Palace.[21]

The 237 portraits of famous people have been displayed in the room since at least 1570, when the inventory of the ducal palace for that year recorded them hanging according to Vasari's description, on the walls above the cupboards. The inventory further specifies that they were aligned in two rows, framed by similar wood frames, and protected by four green curtains. Vasari, who listed the portraits in a separate appendix to the 1568 edition of his *Lives* and identified them as the portraits copied at Cosimo I's request by Cristofano dell'Altissimo from the renowned collection of the historian Paolo Giovio, provides additional information concerning their organization by professions: popes, kings and emperors, dukes, heroes, cardinals, scholars, poets, condottieri, and illustrious Medici (Fig. 2.7).

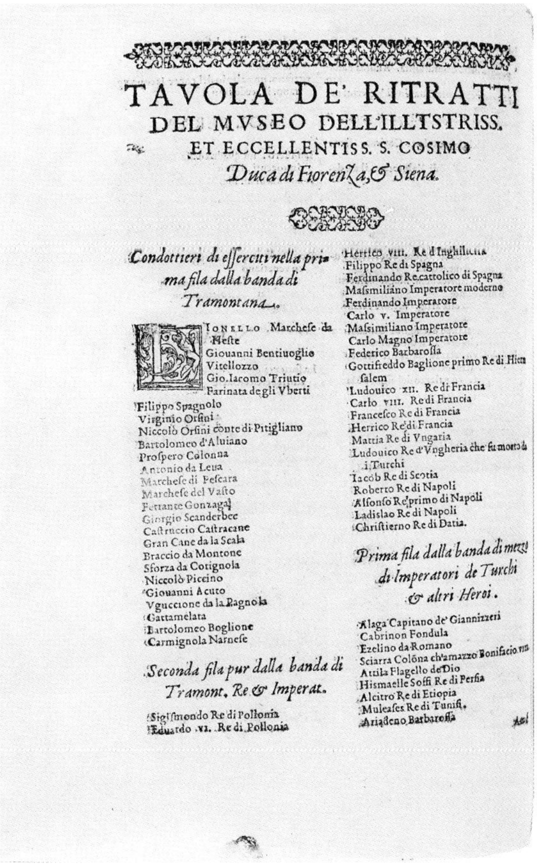
TAVOLA DE' RITRATTI
DEL MVSEO DELL'ILLTSTRISS.
ET ECCELLENTISS. S. COSIMO
Duca di Fiorenza, & Siena.

Condottieri di esserciti nella prima fila dalla banda di Tramontana.

DIONELLO Marchese da Heste
Giouanni Bentiuoglio
Vitellozzo
Gio. Iacomo Triutio
Farinata degli Vberti
Filippo Spagnolo
Virginio Orsini
Niccolò Orsini conte di Pitigliano
Bartolomeo d'Aluiano
Prospero Colonna
Antonio da Leua
Marchese di Pescara
Marchese del Vasto
Ferrante Gonzaga
Giorgio Scanderbec
Castruccio Castracane
Gran Cane da la Scala
Braccio da Montone
Sforza da Cotignola
Niccolò Piccino
Giouanni Acuto
Vguccione da la Ragnola
Gattamelata
Bartolomeo Boglione
Carmignola Narnese

Seconda fila pur dalla banda di Tramont. Re & Imperat.

Sigismondo Re di Pollonia
Eduardo VI. Re di Pollonia
Herrico VIII. Re d'Inghilterra
Filippo Re di Spagna
Ferdinando Re cattolico di Spagna
Massimiliano Imperatore moderno
Ferdinando Imperatore
Carlo V. Imperatore
Massimiliano Imperatore
Carlo Magno Imperatore
Federico Barbarossa
Gottifreddo Baglione primo Re di Hierusalem
Ludouico XII. Re di Francia
Carlo VIII. Re di Francia
Francesco Re di Francia
Herrico Re di Francia
Mattia Re di Vngaria
Ludouico Re d'Vngheria che fu morto da i Turchi
Iacob Re di Scotia
Roberto Re di Napoli
Alfonso Re primo di Napoli
Ladislao Re di Napoli
Christierno Re di Datia.

Prima fila dalla banda di mezzo di Imperatori de Turchi & altri Heroi.

Alaga Capitano de' Giannizzeri
Cabrinon Fondula
Ezelino da Romano
Sciarra Colõna ch'amazzo Bonifacio VIII.
Attila Flagello de Dio
Hismaelle Soffi Re di Persia
Alcitro Re di Etiopia
Muleasses Re di Tunisi.
Ariadeno Barbarossa

The marble busts of ancient and modern rulers that were to be located above the cupboards were never, it would seem, displayed in the room.[22]

The images of flora and fauna, which were to be painted below the maps, were never realized. It is unclear whether drawings were ever made or collected to fulfill this part of the project. It is, however, well known that Cosimo I had a genuine interest in the flora and fauna of the world: he commissioned herbariums, actively assembled samples of plants and animals for his botanical gardens and menageries, and collected drawings of such plants and animals. Among his holdings were the book of animals (*libro d'animali*), the now-lost sketchbook made by the painter Piero di Cosimo that Cosimo had received as a gift from his courtier Cosimo Bartoli, and more than 100 paintings of birds made by Giovanni Neri, an artist who specialized in botanical and zoological illustrations and worked extensively for the Bolognese scholar Ulisse Aldrovandi (Fig. 2.8). Piero di Cosimo's sketchbook, the drawings by Aldrovandi's painter, the herbariums, and the samples of plants and animals would have served as an important visual source for the frieze of the Guardaroba Nuova. The

2.7 *Tavola de Ritratti del Museo dell'Illustrissimo et Eccellentissimo Cosimo Duca di Fiorenza et Siena* (List of Portraits in the Museum of the Most Illustrious and Excellent Signor Cosimo, Duke of Florence and Siena), from Giorgio Vasari, *Le Vite de' più eccellenti pittori scultori et architettori*, Florence 1568. Yale Center for British Art, Paul Mellon Collection

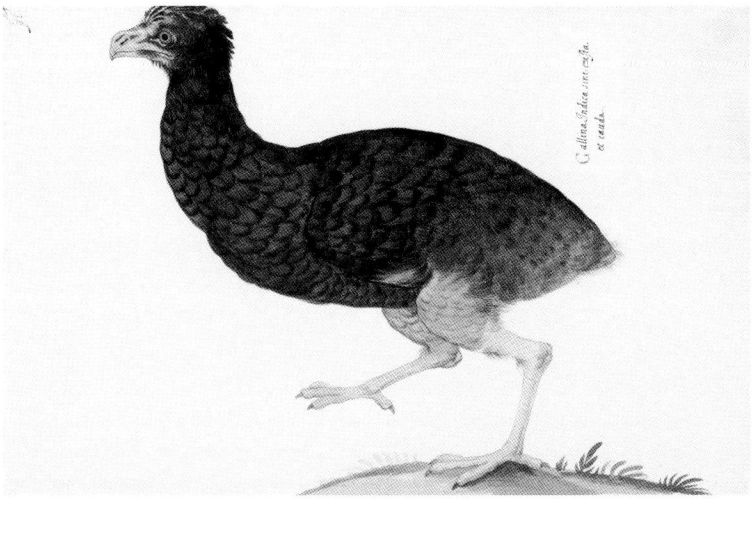

2.8 Giovanni Neri, *Gallina Indica* (Indian Chicken). Bologna, Biblioteca Universitaria, MS Aldrovandi, *Tavole di animali*, vol. 2, fol. 91

systematic collecting of drawings after nature was also favored by Francesco I, who convinced Jacopo Ligozzi to join the Medici court, and commissioned from him the numerous drawings of plants and animals that are still kept in the Uffizi. There is no evidence that Ligozzi, who arrived in Florence in 1577, made his botanical and zoological drawings in connection with the Guardaroba Nuova, but it is well known that his drawings were later used to paint a similar frieze for the Tribuna in the Uffizi Palace.[23]

The representation of the world populated by animals, plants, rulers, and famous people had to be placed under the sky and its constellations that were to be painted on the ceiling. The original wood ceiling, divided into the twelve large panels described by Vasari, is still in the Guardaroba Nuova, but it was never painted with the Ptolemaic constellations. Although Vasari specified that in each panel there would have been "four celestial signs, making in all forty-eight, and little less than life-size," there is no way to imagine how the 48 constellations would have been organized coherently into this awkward partition of the ceiling, whose main purpose was to conceal with its geometric design the irregular plan of the room, a solution Vasari adopted extensively in the ducal palace. On the other hand, it is tempting to imagine that the celestial map would have included some reference to the astrological destiny of the duke, if not his astrological chart, which, as for any Renaissance ruler, was the object of intense study and commentary at the Medici court.[24]

Remembering that at one time the room was meant to be filled with decorations that were either removed or never put in place, it must be acknowledged that the maps, which are the only part of the iconography extant in the room, were one element of a very elaborate system of images that was meant to represent, in its entirety, the whole cosmos. Although representations of the cosmos had been used extensively since antiquity to decorate the walls of public and private buildings, the overwhelming presence of the maps in the Guardaroba Nuova transformed traditional cosmic imagery into quintessential Renaissance invention.

The Invention of the Iconography

According to Vasari, the invention of the cosmic imagery of the Guardaroba Nuova "came from Duke Cosimo" himself, and it was executed by the cosmographer Egnazio Danti, who was, as Vasari said, "most excellent in matters of cosmography and of rare genius."[25] Vasari's statement is certainly complimentary to the duke, but also plausible in light of the knowledge concerning Cosimo's constant involvement in the decoration and reshaping of his ducal palace.[26] The correspondence between Cosimo and Vasari shows that they exchanged numerous ideas, comments, and views on each of the ducal projects, as they no doubt did on the iconography of this new room, although no record has survived on the iconography of the Guardaroba Nuova. The correspondence between the duke and Vasari, however, recorded the role played by Egnazio Danti in the map cycle, which is different from that given in Vasari's *Lives*. In fact, in these letters Vasari discussed the early stages of the Guardaroba Nuova not with Egnazio Danti, but rather with his lifelong friend Miniato Pitti, the abbot of the monastery of San Miniato al Monte and an expert in mathematics and geography.[27] From these letters it is clear that Vasari had initially enlisted Miniato Pitti as his cartographic adviser for the map cycle and Egnazio Danti, who would eventually become an accomplished polymath but was then only a young and unknown Dominican, as an assistant to Pitti.

The son of a goldsmith and instrument-maker from Perugia, Egnazio Danti was introduced to the Medici court owing to the favors that the powerful ducal chamberlain, the Perugian Sforza Almeni, reserved to his fellow citizens, but he must have also acquired a reputation in matters of geography and map-making.[28] Following his family tradition, he had received training in both mathematics and the visual arts, a fact that, according to Serafino Razzi, a Dominican contemporary of Egnazio and close to Vasari who wrote a detailed history of his order, the *Cronica*, contributed decisively to his success as a cosmographer and to his participation in the decoration of the Guardaroba Nuova.[29]

But the correspondence between Vasari, Cosimo I, and Miniato Pitti is important for another reason that, so far, has escaped attention: the connection made by the correspondents between the Florentine map cycle and the map murals that Pope Pius IV commissioned for the Terza Loggia in the Vatican Palace (Fig. 2.9).[30] This great example of the Renaissance vogue for cartographic images, which had been underway since the early 1560s, included painted maps of the modern world arranged according to Ptolemy's geographical order. The maps, made by the French cartographer Etienne Du Perac and painted by Giovanni Antonio da Varese, better known as Vanosino, were connected with the representations of the good and bad life painted on the ceiling under the supervision of the Bolognese painter Lorenzo Sabatini.

Although the Terza Loggia and the Guardaroba Nuova differed fundamentally in meaning – the Vatican map cycle subsumed the representation of the earth to religion, while the Florentine map cycle included the world in the astronomical cosmos – they shared an exceptional iconographic feature: the connection between the maps of the world and the images of plants and animals native to the mapped countries. In fact, this connection between maps and images of the world's flora and fauna was so highly regarded that the famous naturalist Ulisse Aldrovandi was asked to provide samples of his drawings of plants and animals for the Terza Loggia. The letter documenting this request is worth quoting in full, since the relevance of the Vatican map cycle to the invention of the Guardaroba Nuova has not been previously observed. Antonio Compagnoni of Macerata, a cleric at the papal court, wrote to the scholar Ulisse Aldrovandi, requesting a copy of his drawings of birds and plants and reporting the following:

> Today, in the papal palace, there was a conversation about those who delight in minerals, plants, and every kind of animals, and almost everybody said that Matthiolo [Pietro Andrea Mattioli] is rare. Being together with those learned men, I talked about Your Lordship to someone who is in charge of painting a loggia with the cosmography of the entire world. Since he wishes to adorn some areas above the maps of Cosmography, I told him that Your Lordship could provide plants and birds painted after nature, and I described in detail your study. He asked insistently that I would request from Your Lordship a plant and a bird painted after nature, with their respective names, so that it would be possible to show them to Cardinal Amulio [Cardinal Marcantonio da Mula, then superintendent to the Vatican Palace]. Wishing to decorate this loggia as the most beautiful in the world, he might talk to His Holiness if he likes the drawings. This could be the occasion for His Holiness to learn about Your Lordship, and even if nothing came of it, it would be possible to ask for some favors for your study, to augment your post, and there is nothing to lose. Therefore, I ask Your Lordship to send me the portrait of a plant and a bird among the most beautiful you have so that when I bring them to Cardinal Amulio he will be

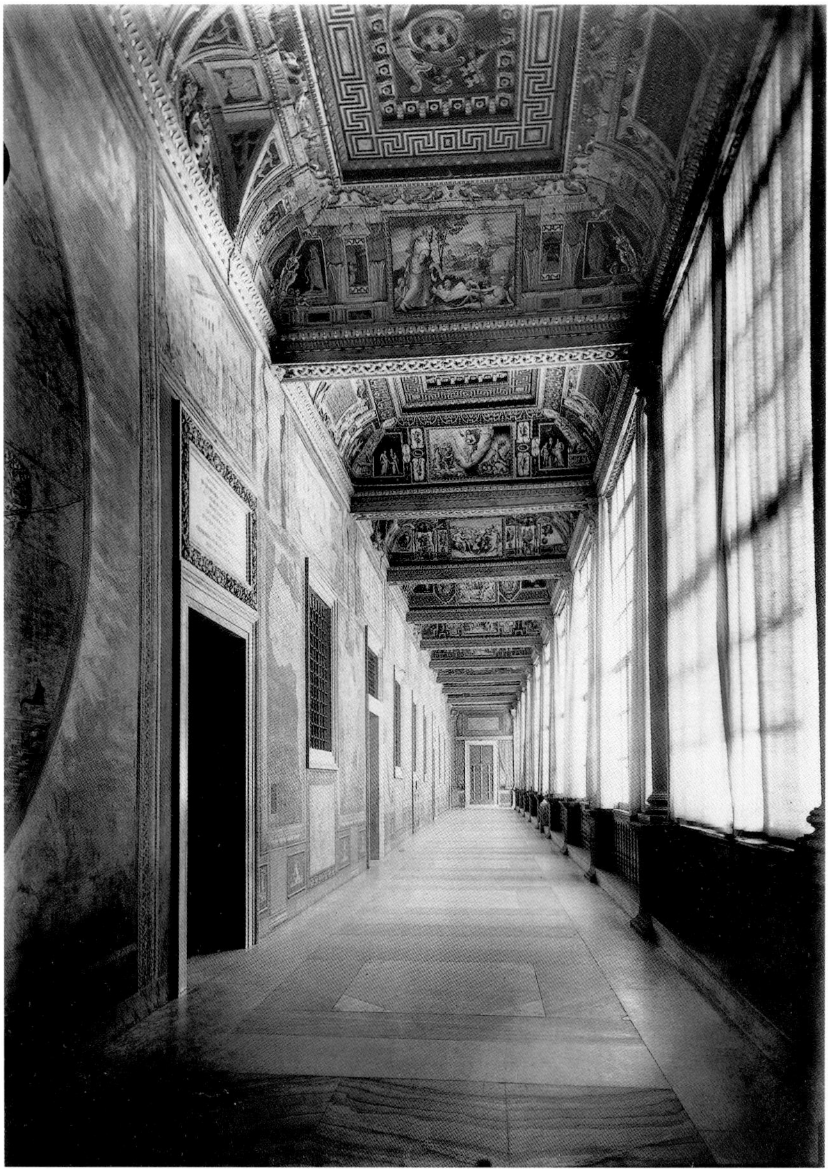

2.9 Terza Loggia. Vatican

delighted. Enclose them in a parcel and address them to me in San Pietro, in the presbytery. As soon as the occasion arises, I will do whatever it will be possible.[31]

This letter shows that the connection between maps and images of the world's flora and fauna had been originally envisaged for the Terza Loggia, although it is not known whether such images were actually depicted because this loggia was heavily repainted in the nineteenth century. The letter clearly suggests that the Terza Loggia served as a model for the

Guardaroba Nuova, although Cosimo and his advisers soon aspired to surpass the geographical accuracy of the papal model. In a letter to Cosimo, Miniato Pitti acknowledged the relevance of the Vatican map murals to the Florentine map cycle, but also wrote that the Vatican maps "were made and are still made sloppily by unskilled artists: but they are good for one single thing: being painted a secco and exposed to the air, soon they will be ruined and will be corrected."[32]

Whoever originally had the idea to represent the cosmos in the Guardaroba Nuova, whether it was Cosimo I, Vasari, or someone else (the issue cannot be resolved), the invention surely met the enthusiastic approval of both the duke and Vasari. Extensive archival evidence shows that, from 1563 to 1565, the decoration of the Guardaroba Nuova proceeded expeditiously. Vasari supervised every single detail of the execution of this room, as he normally did for the restoration of the ducal palace: he translated the conceptual program into visual images, consulted with friends and humanists, and supervised an army of artists and craftsmen. Dionigi di Matteo Nigetti, a skilled woodworker and Vasari's *sottoarchitettore*, built the cupboards and a wood model of the room to facilitate the accurate realization of the decorative scheme.[33] Miniato Pitti "divide[d] the maps of Ptolemy for the guardaroba,"[34] and "helped [Danti] and taught him many things."[35] By the end of November 1563, Miniato Pitti had compiled the sequence of the maps, submitted the entire project for Cosimo's approval, and obtained his permission to continue the cartographic work without need for further drawings.[36] Meanwhile, Danti "worked vigorously": he drew the cartoons, painted the maps in his cell, and received a salary from the duke together with a regular supply of colors and materials.

Initially, Danti's task concentrated on the preparation of the cartoons for the maps, a considerable undertaking in which even the Dominican authorities in Florence were forced to cooperate. The prior of the monastery of San Marco had to agree to let Danti move from his monastery to that of Sta Maria Novella in order to be closer to the ducal palace,[37] while the prior of Sta Maria Novella had to assign the young Dominican additional space, "certain small rooms," that Vasari promptly renovated with ducal funds in order that friar Egnazio could "work there comfortably and healthily."[38] While there is little doubt that Danti prepared the cartoons for the maps, it is debatable whether he also painted the maps. But the seemingly contradictory evidence can be reconciled by suggesting that Danti prepared the cartoons for the maps in his cell and at first also painted them. Later, especially after 1566, his role at the ducal court increased so considerably that it is likely he delegated to others the actual painting of the maps.[39]

Vasari praised Danti's performance highly and wrote to Cosimo with satisfaction: "The friar from San Marco, who is painting the Cosmography for you, is proceeding very well, and it was a good choice; I don't fail to help him and communicate with him about what he needs to serve Your Illustrious Excellency; I am sure you will be satisfied and well served."[40] Danti also had the opportunity to meet the duke personally. In February 1564 Vasari sent Danti to Pisa to show Cosimo the model of the Guardaroba Nuova and to inform him of the details concerning its decoration and the scientific basis of the maps. In January 1565, Vasari proudly reported to the duke that the maps for one of the walls had been completed together with their elaborate wood frames, and that the construction of the terrestrial globe was well underway.[41] In the following months, however, the work on the room's decoration came almost completely to a halt. Vasari concentrated the efforts of every painter, sculptor, and carpenter available in Florence on the construction and decoration of the ephemeral apparatus to celebrate the marriage of Cosimo's son Francesco to Johanna of Austria. Even Egnazio Danti worked on the task, although the specifics of his involvement can only be speculated about.[42]

After 1565 the work on the decoration resumed, although at a slower pace. By then Danti had gained the duke's personal favor and appreciation and enjoyed a distinguished position at the ducal court, which culminated in 1571 when he was officially appointed cosmographer of the grand duke. Because of his increased activities at the ducal court Danti never completed the maps of the Guardaroba Nuova. In 1575, when he was forced to leave Florence, the huge cupboards were in place but only 31 of the 57 maps were painted. Whole parts of the decoration had not even been started, and some never would be. In fact, soon after Cosimo I's death in 1574, the whole project lost momentum. Although Stefano Bonsignori was commissioned to complete the maps, the intriguing and complex cosmic imagery of the room had been abandoned, and the unity of the cosmic design disrupted. The maps still fascinate with their bright colors, but they are isolated relics of the cosmic imagery Vasari so eloquently described. Yet, the unrealized Guardaroba Nuova with its ambitious cosmographic scheme deserves closer scrutiny not only for its deep foundations in the cultural history of Europe, but also for its emblematic fusion of art, collecting, and cartography in the celebration of its patron.

The reconstruction of the decoration of the Guardaroba Nuova as originally planned, obtained through the integrated analysis of surviving artifacts, archival sources, and printed documents, demonstrates the care and attention that Cosimo I and his advisers put into this new addition to the Medici guardaroba. Such concentrated efforts were not unusual at the Medici court, especially when the glorification of Duke Cosimo was at stake. From the first day of his rule, Cosimo had showed a keen awareness of the meaning of his gestures, the significance of the places he selected for living, eating, and sleeping, the symbolic value of the art he collected, and the power of the images he commissioned. The Guardaroba Nuova, like buildings, statues, medals, seals, gems, emblems, portraits, pageantry, festivities, and historical and mythological paintings, must have been envisaged as contributing to the aggrandizement of Cosimo I's persona. But, how exactly the cosmography of the Guardaroba Nuova participated in the celebration of this ruler and how closely it related to the scientific and cultural enterprises of his rule remains to be explained.

3

"The Cosmos is Cosimo's Ornament"

The iconography of the Guardaroba Nuova was devised in the last decade of Cosimo I's rule, when the ducal imagery, which had been inspired by the model of Emperor Charles V, showed an unprecedented preference for those imperial attributes that explicitly carried a cosmic, universal meaning. Cosimo's rule depended on Charles V and, as it was customary at the time, he flattered his political patron by including features of the imperial iconography in his own emblems, medals, and portraits.[1] As his personal device, the Medici duke adopted the Capricorn, the astrological sign that since ancient times was considered essential to imperial power and that was indeed the personal device of both Augustus and Charles V.[2] While flattering the emperor was among the duke's political duties, the appropriation of the imperial cosmic theme, referring to both the emperor's worldly dominion and his divine recognition, was rather presumptuous. Cosimo I was only a minor monarch, who ruled over a small state that could hardly compare with the imperial dominion, which under Charles V extended from Spain to Holland, from North America to the Caribbean. And yet, in the last decade of his rule, the traditional elements of Cosimo's iconography centered on a new effective motto proclaiming the cosmos as Cosimo's ornament. Harnessing the assonance between the duke's name and the Greek word designating the universe, the theme proclaiming the cosmos as Cosimo's ornament found its visualization in triumphal entries, in statues, paintings, and emblems, in the writing of Vasari as well as in the adaptation of the traditional iconography of Hercules to describe Cosimo. But, above all, the duke's cosmic theme was fully expressed in the actual study of the cosmos, which indeed flourished in the last years of Cosimo I's rule. Cosmography at the Medici court aspired to nothing less than the reform of the calendar, the reform that had, in the past, been the domain of emperors, kings, and popes.

Cosimo I's Imperial Iconography

The first occurrence of Cosimo's cosmic theme is in Vasari's *Ragionamenti sopra le invenzioni delle storie dipinte ne le stanze nuove del Palazzo Ducale*, a celebratory treatise in the form of a dialogue between Vasari and Prince Francesco on the meaning of the paintings in

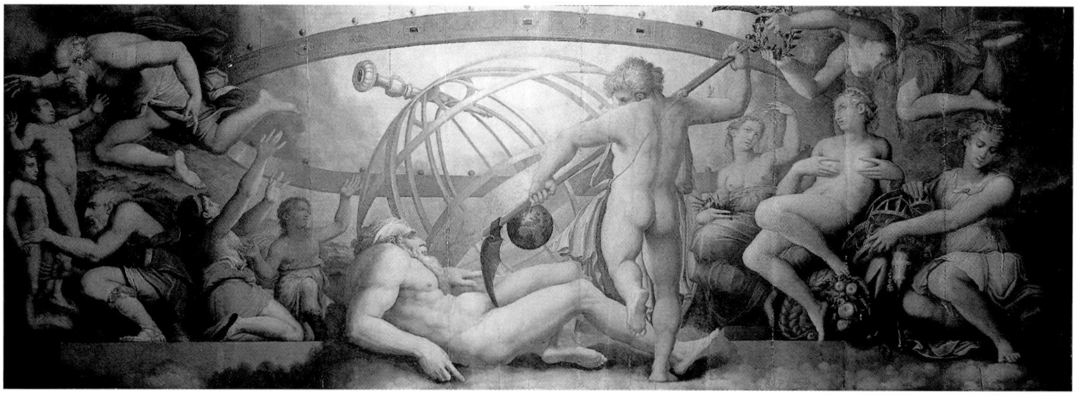

3.1 Giorgio Vasari, *Saturn Castrating Heaven*, ca. 1558, ceiling panel. Florence, Palazzo Vecchio, Sala degli Elementi

the ducal palace. Written in about 1558 but published posthumously, the *Ragionamenti* was an *ex post facto* attempt to present the iconography of the Apartment of Leo x on the first floor and of the Apartment of the Elements on the second floor as a coherent unity conveying a Medicean meaning. Correctly dismissed in modern scholarship as irrelevant to the study of the genesis of Medici iconography, Vasari's *Ragionamenti* is, however, an invaluable document for the understanding of the visual exegesis practiced at the Medici court: among his duties as a devoted courtier Vasari was expected to explain the painted images in the context of Cosimo I's political claims, even though these same images had been originally selected only as illustrations of historical events and ancient myths.[3] Like the published pamphlets of Medici pageantry, Vasari's *Ragionamenti* had the crucial function of disseminating the decoration of the Medici Palace beyond the restricted circle of people admitted to the ducal apartments and, more importantly, of suggesting the correct interpretation of historical and mythological stories in connection with the duke's rule, or as Vasari simply put it, "according to our meaning" (*secondo il senso nostro*).

Vasari's discussion of the iconography in the Apartment of the Elements exemplifies the way paintings were attributed a Medicean meaning that they had not had at the moment of their invention.[4] Vasari had designed the iconography of this apartment in consultation with Cosimo Bartoli, another of Cosimo's trusted advisers, and painted it with his workshop between 1556 and 1557.[5] Inspired by Giovanni Boccaccio's *Genealogy of the Gods*, the apartment included a large hall, the Sala degli Elementi, which was painted with mythological stories pertaining to the four elements (this room gave the name to the entire apartment), and five rooms, which were each decorated with the deeds of an ancient god: Saturn, Ops, Ceres, Jupiter, and Hercules. Because the Medicean interpretation of these mythological stories was added after the paintings were painted, many details of the paintings did not lend themselves to such a dynastic reading. Indeed Vasari's explanations are at times so contrived that even his listener Francesco, well disposed toward Medici aggrandizement, was perplexed. However, Vasari was successful in interpreting the apartment in relation to the cosmic theme of the ducal iconography, a fact that is particularly evident in the Sala degli Elementi.

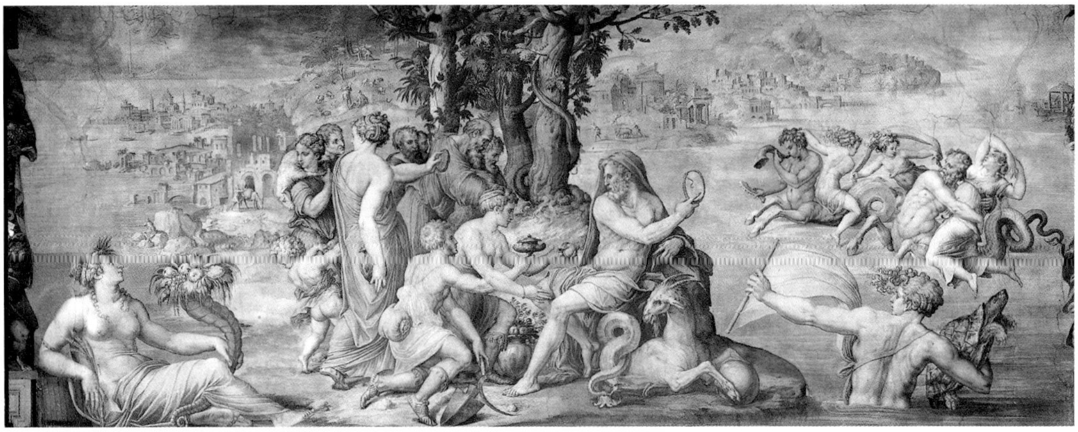

3.2 Giorgio Vasari, *Saturn Receiving the Tribute of the Earth*, ca. 1558, fresco. Florence, Palazzo Vecchio, Sala degli Elementi

Regardless of the genesis of the four elements, Vasari interpreted the iconography of the Sala degli Elementi as an allegory of Cosimo's dominion. In particular, explaining to Francesco *il senso nostro* of the ceiling panel, *Saturn Castrating Heaven* (Fig. 3.1), he commented on the globe represented at the center of the composition: "That is a cosmic body, which is called the world by astrologers, and which is precisely the name of our lord, the duke, who had been made master of this state; and Saturn, his planet, touches his Capricorn ascendant, and by their aspects they cast a benign light on the globe of the earth and, in particular, Tuscany, and, as capital of Tuscany, Florence, which is ruled today with such justice and such good government by His Excellency."[6] Continuing his dialogue with Francesco, Vasari revealed to him the meaning of the wall fresco *Saturn Receiving the Tribute of the Earth* (Fig. 3.2) in the same room, interpreting the globe between the hooves of the Capricorn as, "a red *palla* of the arms of your house, which symbolizes the body of the world, which is the globe."[7] To the eyes of the skilled courtier and iconographic adviser to the duke, the cosmic body was "precisely the name of our lord, the duke," and the traditional Medici *palle* symbolized "the body of the world." In the Sala degli Elementi the assonance between Cosimo I and the cosmos, and the symbolic connection between the Medici *palle* and the world were based on small details from these elaborate images which only Vasari's erudite, at times contrived, revelatory explanations revealed. However, as the years went by, the convoluted analogies between Cosimo I and the cosmos, and between the Medici *palle* and the globe, became more explicit, and gigantic globes started to appear everywhere.

Cosimo I and the Globe

For Cosimo the globe was not simply a representation of the earth and the universe or the ancient symbol of sovereignty, but it had instead an irresistible appeal as a synthesis of numerous themes of his political imagery.[8] Hercules, the mythological hero of the Florentine Republic, of the early Medici, and of Cosimo himself, is an exemplary case, for even he was equipped with a Medicean globe.

Hercules was considered to be the founder of Florence, at least until Vincenzo Borghini revised this traditional account and replaced Hercules with Emperor Augustus. The identification of Hercules with the Florentine Republic was so deep that his effigy appeared on the city's official seal. Moreover, because of this association, Hercules and his labors had been a favorite subject for the art commissioned by the early Medici, especially by Lorenzo the Magnificent. Needing to emphasize the ties between his own rule, the city's republican tradition, and his Medici forebears, Duke Cosimo I and his advisers resorted endlessly to the Herculean iconography to shape the ducal political imagery. Cosimo's personal seal, like the republican seal, represented the hero; one of the rooms of the Apartment of the Elements was entirely dedicated to his labors; busts and prints represented him as Hercules; ancient large statues of the Greek hero were displayed in the ducal palace; modern bronzettes by famous artists adorned the ducal studioli; a monumental fountain for a Medici villa included large statues representing his labors; and even a colossal equestrian statue made by Vincenzo Danti, Egnazio's brother, for the ephemeral decorations celebrating the marriage of Francesco I and Johanna of Austria in 1565 was interpreted by contemporary viewers as a symbol of Herculean virtue.[9]

Around 1560 Cosimo's political imagery started to include an episode of Hercules's life that had hardly been represented in Florence before: Hercules holding the globe. According to the most popular version of the myth, it was not Hercules who actually fetched the golden apples in the Garden of the Hesperides but rather the titan Atlas, whose daughters also sought the apples. While Hercules held up the celestial sphere on his shoulders, Atlas went to acquire the apples. Obviously, the iconography of Atlas holding the globe overlaps with that of Hercules. And, indeed, it is Atlas who was first represented in Cosimo I's palace.

In the Sala di Ercole the ceiling is decorated with nine panels illustrating Hercules's deeds painted by Vasari in 1556 to 1557, while a series of seven tapestries illustrating additional deeds was made for the walls of the same room a few years later.[10] In his *Ragionamenti* Vasari explained the deeds of Hercules as parallel to the events of the duke's life. In the present discussion it is the frieze under the ceiling that is particularly relevant. It represents landscapes and putti holding a globe that, according to Vasari, are a reference to Atlas (Fig. 3.3). Disregarding the discrepancy in size and strength between the bodies of the putti and the body of the titan, Vasari made the painted images fit his explanation: "Like Hercules, who put the globe of the world on Atlas's shoulders, which refers to the help of princes in their government, princes who are made similar to God for piety, clemency, justice, and other virtues, whose very strong limbs held the globe of the earth, so you will give similar help to our duke in the government of this state."[11] The gigantic effort Atlas makes to bear the globe is compared to that made by princes to govern their states. There is no doubt that contemporary readers would have understood Vasari's generic reference to the rule of princes as a more circumstantial reference to Cosimo I, a reference that would have been reinforced visually by the frieze with Capricorn, now lost, originally placed under the ceiling. More explicitly, Cosimo Bartoli, who had helped Vasari with

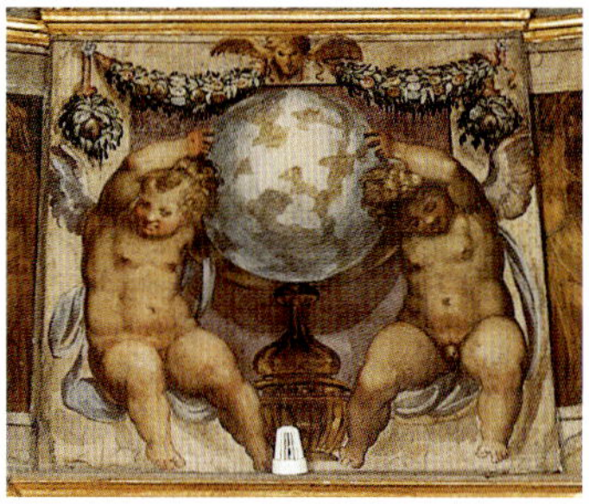

3.3 Giorgio Vasari and Workshop, *Putti Holding a Globe*, 1556–1557. Florence, Palazzo Vecchio, Sala di Ercole

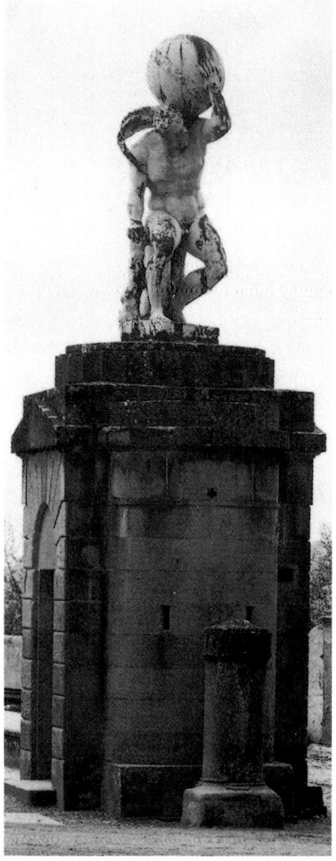

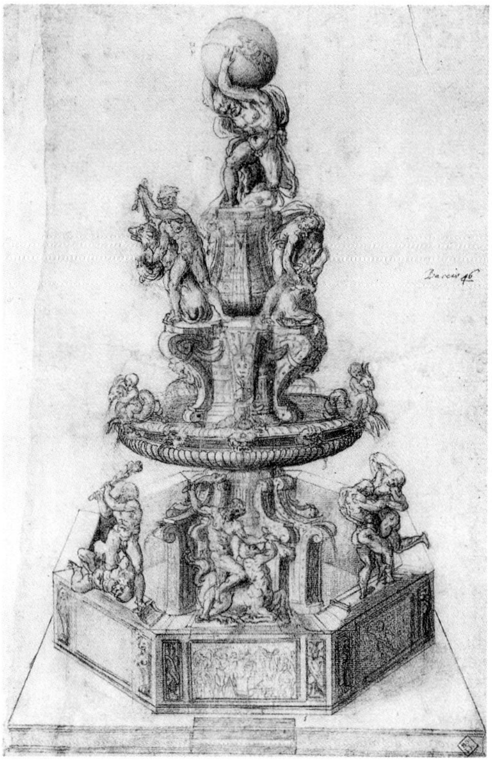

3.4 (left) Vincenzo di Raffaello de' Rossi, *Hercules Holding the Globe*, designed 1561, executed before 1584, marble. Florence, Villa Medici at Poggio Imperiale

3.5 (right) Vincenzo di Raffaello de' Rossi, *Design for a Fountain with the Labors of Hercules*, ca. 1560, black chalk on off-white laid paper. Cooper-Hewitt, National Design Museum Smithsonian Institution

the invention of the apartment's iconography, compared Cosimo to Atlas in the dedication letter of his *Discorsi historici universali* (Venice, 1569). Atlas was also among the masques parading the streets of Florence for the pageantry *La genealogia degli dei*, performed for the wedding of Francesco and Johanna.[12] Although Atlas could provide a moral allegory paralleling his physical strength to the huge task of government, and had even been considered the founder of Fiesole, his myth was alien to Florentine and Medici iconography. It did not provide the multiple literal and visual associations that Medici courtiers loved to conceive for their patron. But it did offer an established iconography that could readily be transferred to the more popular Hercules.

When Hercules holding the globe appeared in Cosimo I's art, the image took the form of a larger-than-life-size statue crowning the monumental fountain (about 15 m high) designed by Vincenzo de' Rossi in 1561 after consultation with Vincenzo Borghini, who indeed might have suggested the transfer of iconography from Atlas to Hercules.[13] A sketch documents the original design of the fountain dedicated to Hercules's labors (Fig. 3.5). The fountain had to be a Florentine candelabrum type on four levels. At the bottom an octagonal balustrade included bronze reliefs representing Hercules's deeds and many larger-than-life-size groups. Three monumental statues of the Capricorn adorned the basin above, while three additional monumental groups were arranged one level up. The gigantic lone statue, *Hercules Holding the Globe* crowned the fountain at the top (Fig. 3.4). Although it is still unclear where the fountain was to be located, its iconography and monumentality were an unmistakable reference to the duke. The three large sculptures of the Capricorn, which adorned the basin, would have suggested and encouraged the visual connection between the duke and the statue at the top. About a century later another over-life-size statue of *Hercules Holding the Globe* was made to celebrate another grand duke named Cosimo. In 1661 the wedding between Cosimo III de Medici (1670–1723) and Marguerite-Louise d'Orléans, the cousin of Louis XIV,

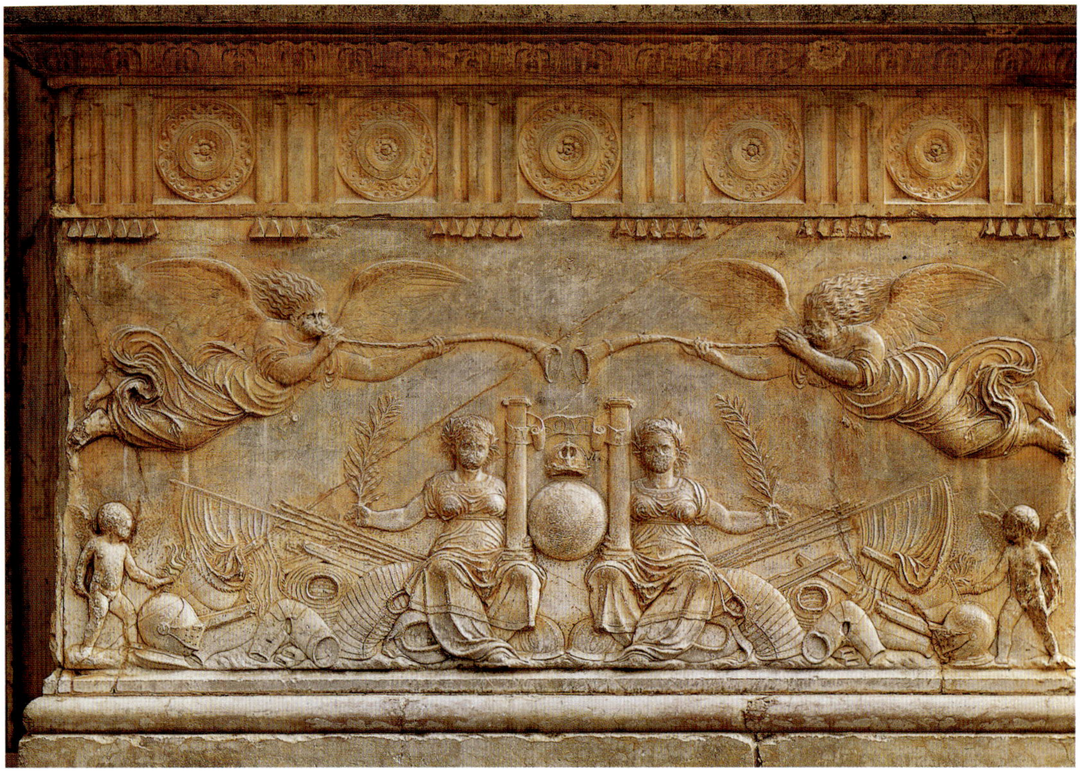

3.6 *Allegorical Relief.* Granada, New Palace of Charles v, west gate

King of France, was marked by numerous celebrations and festivities, culminating in the equestrian ballet *Il Mondo Festeggiante* which, as Stefano della Bella's engraving beautifully recorded, was opened by a gigantic stage machine representing Hercules holding the globe.[14]

Significantly, the motif of Hercules and the globe also included a rather precise reference to Emperor Charles v, who used the globe as the traditional symbol of imperial power but also added to it an additional, literary meaning. As an accurate, scientific representation of the world, the globe illustrated his patronage of map-making and cosmography. Charles v had many globes and a keen interest in cosmography, map-making, and topography. In 1530 he had secured the exclusive rights to reproduce the famous world map by Matteo da Verona. The next year he received the first terrestrial globe made by Gemma Frisius, who soon became the imperial cosmographer. Another globe by Gemma was added to the imperial collection a few years later (this second globe is mentioned in an imperial privilege granted to him in 1536 during his visit to Naples). In 1553 Charles received two miniature globes of the heavens and the earth, one of blown crystal, the other of wood, and two sets of astronomical instruments made by Gerard Mercator, the famous map-maker who had been trained by Gemma Frisius. Some of the most important books on cosmography were dedicated to him or his closest advisers. Peter Apian dedicated his *Astronomicum Caesareum* (Ingolstadt, 1540), one of the most beautifully illustrated books of the Renaissance, to Charles v and actually presented a copy to the emperor in person at Ingolstadt in 1541. Apian's *Quadrans Apiani Astronomicus* was dedicated to the emperor's secretary Matthias Zimmermann.[15]

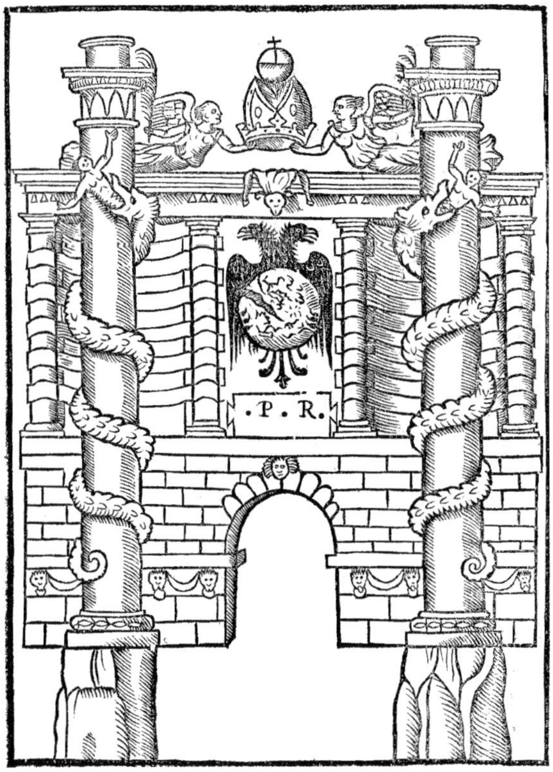

That the globe carried for Charles v the conflated meaning of worldly authority and geographical accuracy is further suggested by its recurrent presence in conjunction with his distinctive device, the columns of Hercules and the accompanying motto, "Plus Ultra" (Fig. 3.6). The meaning of Charles v's device and motto was both moral and geographical. It stressed the emperor's desire to expand his domains and power to the entire world, but it could also acquire a more specific meaning according to the symbol placed in the central field between the columns. Often the central symbol was a terrestrial globe copied from one of the most detailed globes of the time illustrating the specific region that Charles v was targeting for expansion. The Herculean columns and the globe were featured on fireplaces, pedestals, and portals in Charles v's palace in Granada, and on medals, portraits, and in ephemeral decorations for his triumphal entries into many European cities. Although the decoration of the palace in Granada may not have been familiar to the Florentine court, other images were surely well known. The temporary decorations for Charles v's entry into Genoa in 1530 on his way to Bologna to be crowned by the pope included two magnificent triumphal arches built under the direction of Perino del Vaga as well as a monumental globe surmounted by the imperial eagle expressing perfumed water which, according to Luigi Gonzaga, a witness to the parade and a reporter of Charles v's sojourn in Italy, was the most astonishing feature of the entire pageantry. The pageantry at Milan in 1541 included a triumphal arch surmounted by a huge globe (Fig. 3.7), a feature that appeared again in the triumphal entry of his son, the future King Philip ii, into Antwerp in 1549.[16]

3.7 Triumphal Arch at Porta Romana for Charles v's entry into Milan, woodcut, from Giovanni Alberto Albicante, *Trattato del'intrar in Milano di Carlo v*, Milan 1541. Bibliothèque nationale de France

Charles v's device symbolized both his desires to expand his domain and the accurate cartographic representation of the world he hoped to rule. It conflated the political meaning with the cartographic accuracy of the globe, the construction of which he patronized extensively. The fact that for the emperor the symbolic imperial orb was acquiring the additional function of illustrating the world accurately is further documented by the papal bull that Clement vii signed on 1 March 1530, only a week after the coronation of the emperor, to sanction the imperial orb as, "the image of the entire universe" (*totius mundi imago*).[17] This very *imago mundi* features prominently in the foreground of the emperor's allegorical portrait painted by Parmigianino in Bologna on the occasion of the coronation (Fig. 3.8). Indeed, Parmigianino effectively combined the traditional elements of the imperial iconography, such as the winged Fame, the laurel, and the palm, with the themes particular to Charles v, such as Hercules's deeds and the globe. With his characteristic ability to elucidate the symbolic meaning of art, Vasari reported that Parmigianino painted, "a boy in the form of a little Hercules who offered him [Charles v] the earth's globe, giving him, as it were, the dominion over it."[18]

The imperial connotations of Cosimo i's globes became explicit in the temporary decorations for the duke's triumphal entry into Siena on 28 October 1560. Trained to associate the Augustan iconography with the reigning ruler, Medici courtiers would not have failed to identify Augustus with Cosimo, and the successes of the Roman emperor with the duke's recent annexation of Siena. Augustus's first appearance on the arch at the city gate bore the inscription *Et Temperat Iras*, stressing the clemency of the new ruler toward the subjected city. His second appearance was much more striking. A lone statue with the Capricorn represented at the base, Augustus held a golden globe with the inscription *Aurea Condet Saecula*, referring to the duke's peace as a renewal of the Golden Age. But the cosmic theme of the duke achieved its climax at the second arch near the Piazza del Duomo that featured the statues of the Medici popes Leo x and Clement vii under the personifications of the virtues of Prudence and Abundance, and a statue of the young Cardinal Ferdinando de' Medici, Cosimo i's son, for whom a brilliant future was predicted as "one of the happy and holy successors of St. Peter."[19] The arch was crowned by a monumental globe behind which was the Greek inscription: "Cosmos Cosmoi Cosmos" (The Cosmos is Cosimo's Ornament). Antonio Francesco Cirni, a witness to the parade and a Medici courtier versed in Medici iconography, explained the motto as follows: "Duke Cosimo honors the world and the world him, or rather, the world is Cosimo's and he is the world's."[20] The arch was dedicated to past and future Medici popes, but in fact it expressed the universal power of the absolute prince. The crowning globe conflated the worldwide ecclesiastical and secular influence of the Medici. Playing on the double meaning of the Greek word *cosmos* as referring to both universe and ornament, the inscription presented the monumental globe, a very appropriate symbol of papal authority, as Cosimo i's ornament and symbolic possession.

As the ancient symbol of worldly dominion, the cosmos made multiple references to Augustus and

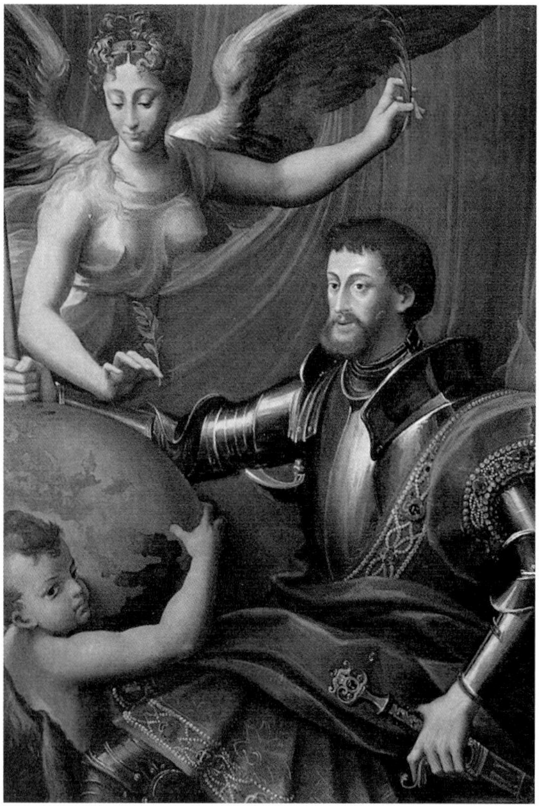

3.8 Francesco Mazzola, known as Parmigianino, *Allegorical Portrait of Charles v*, 1529–30, canvas. Courtesy of Stiebel, Ltd

Charles v. It also provided reference to Hercules, the hero of the Florentine Republic and of the early Medici. But, unlike any other element of the imperial or Herculean iconography, for Cosimo I the cosmos had the invaluable advantage of enriching the imperial imagery with a more personal element that no other ruler could exploit so poignantly. As a pun on the name of the duke, the cosmos proved an unsurpassable visualization of Cosimo's desire for political autonomy and worldwide influence. The wordplay Cosimo–Cosmos, like many other puns of the time, fulfilled the Renaissance assumptions that names are the consequence of things, having the prophetic and magic components of omens, and therefore fitting and witty additions to emblems, portraits, and encomiastic writings. This pun had been used extensively in connection with Cosimo the Elder, but only under Cosimo I did it acquire a full visual dimension. Not only did the *palla del mondo*, the globe of the world, recall the Medici *palle*, but it also visually blended Cosimo's iconography with that of the emperor, of his Medici ancestors, and of the Florentine Republic. Because of his very name, Cosimo was able to wed the globe of the cosmos to his persona with striking clarity and poignancy.

As the imagery of the Medici grand duke was being refined according to the imperial model, it became apparent that Cosimo's name was an omen to justify much more than the inclusion of imperial globes in the ducal imagery. His very name destined him to patronize the study of the universe, a patronage that would put him among the greatest rulers in history.

Cosmography and Renaissance Courts

The greatest kings and emperors of ancient times had regarded cosmography as a defining task of their rule, and often employed natural philosophers, astronomers, mathematicians, and later theologians to study its nature, describe its structure, and measure its size. Literally the description of the universe, which the Greeks called *cosmos* and the Romans *mundus*, cosmography had long been regarded as a means to universal knowledge and global control: to understand and explain the universe was a claim to its possession and conquest. Conceived as a material entity that had to correspond to the laws of physics, the nature of the universe was of primary concern for natural philosophers and theologians. They debated whether it was finite and eternal, as Aristotle taught, or infinite and temporal, as Christian theologians preferred; whether it was filled with matter or it contained the void; and how it related to the infinity of divine power, a topic particularly important for Christian theologians. Philosophers also imagined the universe as made up of a series of concentric orbs, which according to Aristotle moved uniformly in one fixed direction. Although the movement of each orb could be different, they all shared the same center, believed to align with the earth's center. But Aristotelian philosophers could not account for the different distances of the planets from the earth that astronomers had observed since antiquity, and they left to astronomers and mathematicians the geometric description of the universe. Astronomers made diagrams of the geocentric universe that contradicted Aristotle's uniform and circular motions of the planetary orbs but that had the advantage of making it possible to predict celestial phenomena. Geographers, instead, measured the earth's globe in relation to the universe, made maps of the earth and its regions, and described the earth's inhabitants, its animals, plants, and climates.

In antiquity, examples of imperial and royal patronage of the different branches of cosmography abounded, and many were well known in the Renaissance. Julius Caesar sponsored the astronomer Sosigenes. Augustus had the world map of his friend Agrippa displayed publicly in Rome; he envisaged too a new imperial planning of Rome that reflected the perfection of the cosmos, and sponsored Vitruvius's architectural text, revered in the Renaissance, to build this ideal city. In Alexandria the Egyptian kings assembled astronomers and mathematicians to make diagrams of the universe and invent projections to represent its spherical body on a flat surface. Emperor Alexander the Great took cosmographers and map-makers on his expedition to the east with the specific purpose of mapping the world, particularly those unknown areas.

Like their illustrious predecessors, Renaissance rulers created centers for cosmographical studies at their courts and hired professional cosmographers to address a variety of problems. Emperor Maximilian, political heir to the Roman emperors and a great patron of cosmography, hired Stabius as his imperial cosmographer. Later Charles v selected the famous cosmographer Gemma Frisius for the same post. Other rulers followed the imperial model. Elizabeth i of England appointed John Dee as her court cosmographer. The King of Denmark supported Tycho Brahe's research. Rudolf ii had Tycho and Kepler at his court in Prague. The Montefeltro of Urbino, Wilhelm of Hesse, and Albert of Saxony also sponsored cosmographical research at their courts. The tasks and talents of the cosmographers varied from court to court, but generally they were expected to build scientific instruments, make maps, survey the ruler's land, write original scientific treatises, translate classical scientific texts, improve water supply, and make astronomical observations. As courtiers, they were also expected to participate in court life by attending official ceremonies and by initiating scientific debates to entertain the court. Among their duties was also the celebration of their patron, either through the dedication of their books to him and his family or through their advanced cosmographical research, which would have implicitly suggested the comparison of their patron with the great kings and emperors of ancient times.

Indeed, Renaissance cosmography faced new challenges to old problems. The structure of the universe came under new scrutiny, since modern observations revealed varying movements of stars and planets. In order to maintain the Aristotelian uniformity of planetary motion, Copernicus entirely revised the structure of the universe, and the sun, rather than the earth, came to occupy its center. Even the perfect harmony of the four elements (water, earth, fire, and air) that Aristotle had attributed to the universe was about to be disrupted, since modern voyages showed an imbalance between water and land. As for the earth's globe, the discovery of new lands demanded the modern mapping of both the old and the new continents in order to find lucrative and secure trading routes and establish empires.

Following this long-standing tradition that regarded cosmography as a princely pursuit, in 1571 Cosimo i appointed Egnazio Danti "Cosmographer of the Grand Duke of Tuscany," assigning him "a chair in mathematics, which seems to us much needed in Florence."[22] Medici courtiers commented on this appointment as the culmination of the study of cosmography at the ducal court, which indeed was made possible by the converging interests of Cosimo and his cosmographer.

* * *

Cosimo I and Egnazio Danti

Cosimo must have enjoyed the company of Egnazio Danti. Occasionally, at night, he went with the Dominican friar to observe the sky. Equipped with a celestial globe or a star map and in the company of other learned men, the duke found it particularly amusing to climb to a very high place, usually the campanile of Sta Maria del Fiore, to observe the stars and the constellations, and try to match the lights in the sky with the stars on the map and the globe. These nocturnal observations of the sky became such an entertaining pastime at the Medici court that the duke decided to provide an observatory station more easily accessible to the court and, as Danti himself reported, built "the terrace . . . above the roof of the Casino at the Belvedere in the Pitti [gardens], which is at the same level of [the Florentine campanile]."[22] Cosmography at the Medici court was an entertainment appropriate for a duke, who wished to model his interests, patronage, and public persona after that of ancient and modern emperors.

Even though Cosimo's dominion was small and his power limited to the local affairs of the Italian peninsula, the breadth of cosmological studies pursued at his court put him on equal footing with the greatest emperors of all times. His own name destined him to the patronage of advanced astronomical research. In Egnazio Danti, Cosimo found a versatile polymath able to meet his aspirations. As a mathematician with great interests in the observation of the universe and in the applied mathematics that made such observation possible, Danti initiated a series of astronomical observations aimed at the measurement of the length of the solar year in order to propose nothing less than the reform of the calendar. As a consummate courtier, Danti connected the investigation of the cosmos to the public persona of the duke. Magisterially, he praised his duke as a new Julius Caesar and a modern Egyptian king, while at the same time succeeding in enhancing his own standing in the profession of cosmography. According to Danti, Cosimo I not only "supports me so that I can continue these studies of mine . . . but he also helps me with the best advice and suggestions, which specifically belong to this activity."[23]

Pleased with the services of his cosmographer, Cosimo conceded him numerous privileges. Although he did not donate an island to Egnazio Danti as the king of Denmark had done for Tycho Brahe, he none the less gave him a generous dowry of 300 florins for his sister, paid Egnazio's medical bills, defended him against numerous attacks, wrote letters for him to the pope and the Dominican authorities and even interceded, at Egnazio's request, to secure a position for his father in Perugia.[24] He also entrusted him with the delicate mission to smuggle the Etruscan statue known as the *Arringatore* from the Papal States to Florence.[25] But first and foremost he expected Danti to teach cosmography to his offspring and ducal administrators, who were all in need of a sound knowledge of cosmography, mathematics, and its applications.

The Teaching of Cosmography

Cosimo I attached great importance to the mathematical and cosmological training of his courtiers. The University of Pisa, which he had reopened in 1544 and forced the Tuscan youth to attend by ducal decree, offered a curriculum in mathematics, which was quite advanced in the panorama of sixteenth-century university teaching. A stronghold of conservative Aristotelianism in natural philosophy, the University of Pisa had been much more receptive to outside innovations in botany, anatomy, and mathematics, innovations which eventually also influenced the teaching of natural philosophy. The statutes at Pisa established that the mathematical curriculum had to be based on Giovanni Sacrobosco in the first year, Euclid in the second, and Ptolemy in the third. Filippo Fantoni and Giuseppe Nozzolino, the professors of mathematics at Pisa while Danti was teaching at court, indeed taught the foundational texts by these authors, but they also gave special attention to practical mathematics. Fantoni regularly taught Ptolemy's *Geography* and even prepared a book for publication in Latin and Italian that combined Ptolemaic astronomy and geography. Danti's numerous publications in the late 1560s and early 1570s, which were connected to his teaching job, confirm his adherence to the Pisa curriculum.[26]

Danti's class developed over a period of three years. The first year was dedicated to basic notions of cosmography, the study of the spheres, sky, planets, and constellations. Danti adopted as textbooks his own *Trattato della sfera ridotto in cinque tavole* (Florence, 1571), an outline for his lectures rather than a proper treatise, and Giovanni Sacrobosco and Proclus Lyceus's texts, which he translated into Italian. The second and third years were devoted to the applications of mathematics, the *scientiae mediae*, as Danti called them, on which he wrote extensively outlining their importance in the investigation of the natural world.[27] He published translations of Euclid's *Optics* and of miscellaneous Greek texts on optics, then attributed to Heliodorus. He annotated Giacomo Vignola's treatise on artistic perspective, a text that would have been the logical follow-up to his teaching on ancient optics.[28] He left a manuscript on military architecture, a topic of great interest to the aristocratic warriors who attended his classes.[29]

Geography, though, was the subject of the day. While Danti was painting the maps of the world for the Guardaroba Nuova, he may have taught his students how to make a map. He even planned a new edition of Ptolemy's *Geography*, which, however, was never published.[30] The comprehensiveness of Danti's teaching in applied mathematics is best inferred from his treatise *Le Scienze matematiche ridotte in tavole* (Bologna, 1577), which includes a series of diagrams outlining the *scientiae mediae* derived from mathematics: cosmography, astronomy, geography, geometry, military engineering, architecture, optics, painting, and artistic perspective.[31] Indeed, the predominance of the *scientiae mediae* seems a defining characteristic of Danti's teaching at Cosimo's court, a fact that reflects both the practical training that ducal administrators required as well as Danti's own ideas on the role of the *scientiae mediae* in the investigation of the natural world.

In his numerous treatises Danti showed awareness of the contemporary debate on the role of mathematics in the investigation of the natural world, a debate that came to be known as the debate on the certainty of mathematics and that was to have crucial consequences for the rise of modern science.[32] In the course of the sixteenth century, mathematics underwent a slow but radical epistemological shift: from a tool to study the natural world it transformed into the ultimate explanation of natural phenomena. According to the traditional Aristotelian hierarchy of knowledge, mathematics was considered to be one of the liberal arts, albeit one inferior to natural philosophy. While natural philosophy explained the substance and causes of natural phenomena, mathematics could only formulate hypotheses on the same phenomena but not explain their true causes. But the spread of neoplatonic philosophy in Renaissance courts and universities encouraged a revision of Aristotelian philosophy that made it possible to envisage mathematics as a science based on abstractions of the soul (points, lines, and planes), a discipline that could explain the phenomena not simply describe them. A favorable view of mathematics in the acquisition of knowledge profoundly affected every *scientia media*, but astronomy was the discipline that could demonstrate better than others the superiority of mathematics over natural philosophy. Significantly, the debate on the certainty of mathematics emerged in response to Copernicus's *De revolutionibus orbium coelestium* (1543).[33] Copernicus's early critics were not bothered by the heterodoxy of his heliocentric system, since numerous biblical passages and patristic commentaries, including Augustine's, could validate the movement of the earth. Instead, they opposed his cosmology because it was based on the abstract entities of the mixed science of mathematics rather than on the principles of the higher science of physics. Contradicting the method of enquiry of Aristotelian philosophy, Copernicus's astronomy could not explain the actual motions of the heavens. Many others, however, agreed with his calculations on the prediction of planetary motions and his view of mathematics as an explanatory tool of the structure of the universe, even though they did not accept his premise that the sun stood at the center of the universe, which, incidentally, Copernicus had introduced because he could not give up the Aristotelian homogeneity of planetary motions.

Danti himself debated the role of the *scientiae mediae* in his treatises written at the Medici court. In the preface of his edition of Euclidean *Optics*, he articulated his views on the importance of optics (*perspectiva naturalis*) as a proper science for the investigation of the universe. In spite of the fact that optics was not taught in the schools of the philosophers but left in the hands of craftsmen and practitioners, Danti argued that it was crucial to the natural philosopher "in order to know well the motion, rest, location, size, and quality of natural phenomena, of which his entire speculation consists."[34] Danti reminded his readers that optics allowed the cosmographer to know the position of the stars and planets, "to reduce two-dimensionally, either in an oval or circular shape, and in other different ways proportionally, the site of the entire earth, and of its particular provinces . . . to have perfect understanding of the distances between places, making known the proportion and location of sites that one region has not only in relation to another [region], but also [in relation to] the sky."[35] In Danti's view, the *scientia media* of optics became a tool to represent the world and interpret its nature. The full implications of Danti's view, shared also by other cosmographers, became evident in the early seventeenth century, when Galileo used the telescope and optics to interpret the dark areas on the moon's surface as shadows of its mountains and valleys.[36]

Contemporary writers reported that Danti's classes were well attended by "the most noble youth, and even [by] some illustrious ladies."[37] Among his students it is safe to assume were Cosimo I's sons: Cardinal Ferdinando, who was the dedicatee of one of Danti's treatises, and remained his affectionate patron for years;[38] and Francesco, the future duke, who, on the other hand, developed a deep and personal dislike for both cosmography and Danti. The illustrious ladies included Cosimo's daughter Isabella Medici Orsini, Duchess of Bracciano, who was "so fond of this most noble and delightful science" that Danti dedicated two of his books on cosmography to her.[39] Also other professionals attended Danti's classes, as the dedication letters of his treatises suggest. Cesare della Penna, the dedicatee of Danti's *La Sfera di messer Giovanni Sacrobosco*, a cadet of the noble Perugian family and one of Cosimo's commanders in the fight against the Turks, was a representative of the growing group of professional warriors of aristocratic origins who needed a sound knowledge of practical mathematics in order to calculate a cannon's trajectory and build artillery-resistant bastions.[40] The Knights of the Order of Santo Stefano, explicitly mentioned as potential readers of Danti's *Trattato dell'uso et della fabbrica dell'astrolabio* (Florence, 1569), were exemplary of naval commanders, who would find the astrolabe and its use "extremely important things."[41] Gherardo Spini, who attended Danti's summer class of 1569 and found the topic so interesting that he felt compelled to publish his classroom notes, was a typical Medici courtier from a distinguished Florentine family who became a trusted administrator of the Medici state, as well as an author on military architecture.[42] Moreover, there is a slight possibility, undocumented yet plausible, that Danti also taught the young artists in training at the Accademia del Disegno in Florence, which stressed the theoretical education of artists and included the teaching of perspective as an integral part of the curriculum.[43]

To this varied group of students Danti delivered his intensive program in applied mathematics in Italian. For their benefit he also published an array of textbooks in the vernacular rather than in Latin, the international language of high education and scholarship. While the use of the Italian language may indicate the wide audience to whom Danti's classes and books were directed, it is also a unmistakable indication of Danti's participation to one of Cosimo I's dearest cultural campaigns, the promotion of the vernacular as the official language of literature, history, and natural philosophy.

Cosmography in the Vernacular

Cosimo I had made the use of the Italian language in scholarship and literature one of the pillars of his cultural policy. His desire was to influence the way people wrote and spoke beyond the borders of his own duchy, in the entire Italian peninsula.[44] He had founded the Florentine Academy with the specific purpose of editing the works of the great authors who wrote in Italian, and of sponsoring the writing of new works in any possible field of knowledge. He even convinced the famous Florentine humanist and historian Benedetto Varchi to put aside his anti-Medici feelings, return to Florence, and direct the Florentine Academy. Medici courtiers took to heart the fact that to write for the duke meant to write in Italian, and produced important works in Italian in many branches of knowledge. For instance, the historian Adriani wrote a history of Florence in the vernacular, Cosimo Bartoli translated into Italian the books that Leon Battista Alberti had written in Latin, and Vasari put together the first series of lives of artists, sculptors, and architects. In line with Cosimo's desires, Danti made the textbooks on cosmography and applied mathematics available in Italian. As the first translator of important scientific texts, Danti coined new words and idioms in the vernacular that were later accepted into the vocabulary of the Accademia della Crusca, the official vocabulary of the Italian language promoted by a later Medici grand duke.

Danti's edition of *La Sfera di messer Giovanni Sacrobosco* (Florence, 1571) was the first annotated translation into the vernacular of the compendium of cosmography written in Paris around 1230 and since then the most read, copied, and translated cosmography textbook. In 1570, only one year before Danti's translation, Christoph Clavius, the Jesuit mathematician at the Collegio Romano, published an influential Latin edition of Sacrobosco's text complete with his own lengthy comments that transformed the short original text into a veritable encyclopedia of cosmography, used in Jesuit colleges and other institutions throughout Europe for many decades.[45] But Danti's contribution was to publish the revered cosmographical text in Italian thus addressing it to a wider readership. Danti probably based his translation on that made by his grandfather Piervincenzo in 1498 and provided it with an extensive commentary. To further emphasize the Perugian origin of the project, he dedicated it to Cesare della Penna, but did not forget to pay due homage to Cosimo, who is remembered for his support for the wars against the infidels and the institution of lectures in mathematics and cosmography. Indeed, this book exemplifies the diplomatic skills Danti had developed as a Medici courtier. In a single stroke he was able to celebrate his native town of Perugia by dedicating the book to a member of the Perugian aristocracy; to honor his patron Grand Duke Cosimo I by praising his patronage of cosmography; and to promote himself as a courtier able to support the ducal policy in favor of vernacular participation in the celebration of the duke, and in the cultivation of relations with neighboring cities.

Danti's translations of ancient and medieval texts into the vernacular did not only adhere to Cosimo's cultural policy but also coincided with Danti's own aspirations as a cosmographer. Indeed, his translation from Greek into Italian of Proclus Lyceus's book on cosmography, Euclid's *Optics*, and Heliodorus's *Optics* marked a significant shift in the cosmographer's career, which facilitated the expansion of his intellectual network beyond the Medici court. Euclid's and Proclus Lyceus's Greek texts were available in the Medici library, but a copy of Heliodorus's *Optics* had to come from the Vatican library, sent directly to Danti by the Vatican librarian, the influential and powerful Cardinal Guglielmo Sirleto. The cardinal of Lorraine, who had visited Danti at work in Sta Maria Novella, and had urged him to do the translation, was the dedicatee of his work on Heliodorus.[46] In these translations, instead of summarizing and translating medieval scientific texts, Danti went back to the Greek originals. Conceptually and graphically, these books follow the editorial conventions of such Renaissance editions of ancient scientific writings as those of Euclid, Ptolemy, and Archimedes, and should be regarded as part of the efforts of humanist scholars to edit and publish the scientific writings of the ancients. Although the philological rigor of Renaissance humanists can be so seriously questioned that today we have to undo what they had done to the ancient texts, Danti and other humanists were genuinely committed to the restitution of uncorrupted scientific texts of the ancients. These humanistic publications of scientific texts in Greek and Latin, as well as their translations into the vernacular, eliminated the mistakes of medieval commentaries and translations from Arabic rather than the original Greek, and soon became a pillar of Renaissance scientific culture. Cosmographers from Regiomontanus onward were at the forefront of translating ancient texts and, in this context, Sebastian Münster's Latin edition of Ptolemy's *Geography*, or Federico Commandino's Latin edition of Ptolemy's *Planisphaerium* and *Analemma* might be recalled.[47]

The knowledge of Latin and Greek and the publication of scientific treatises written by ancient authors had remained the only major distinction between the mathematical practitioners and the humanist mathematicians who, more often than not, also came from different social backgrounds.[48] The mathematical practitioners had been acquiring increasing respect and responsibility in many crucial facets of sixteenth-century life. From teaching in the urban abacus schools the mathematicians' expertise extended to military and civic engineering, building restoration, and the publication of popular scientific treatises, thus

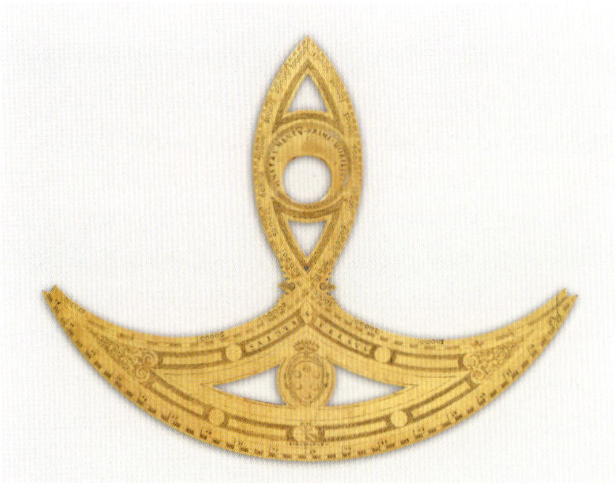

3.9 Egnazio Danti, *Instrumentum Primi Mobili* (Instrument of the First Mover), 1568, brass. Florence, Museo di Storia della Scienza

threatening and eroding the social privileges of university mathematicians and of aristocratic warriors. Socially, Danti belonged to the group of the mathematical practitioners. He came from a family of craftsmen, acquired his basic mathematical knowledge partly at home and partly in an abacus school. By training in the Dominican school, he later expanded his mathematical knowledge, became a teacher of students of high rank, and gained the prestigious title of Cosmographer of the Grand Duke. Yet his role at court was still largely defined by applied mathematics, the *scientiae mediae* of instrument and map-making, building water channels, and surveying. In order to move from the world of the practitioners to the more socially prestigious circles of the humanist mathematicians, Danti wed his social aspirations to Cosimo's cultural policy through his translations and editions of Greek scientific texts.

If social status was important, other aspects of Danti's activities as ducal cosmographer were decisively connected to the manual crafts. Such was the case of the making of scientific instruments, in which Danti excelled thanks to his early training in his father's goldsmith workshop.

The Instruments of Cosmography

Like his forebears, Cosimo cultivated a passion for scientific instruments, which he regarded as precious and rare collectibles, worthy of being displayed alongside his sculptures, paintings, exotica, and naturalia in the very best rooms of his palaces; it may be recalled that the armillary sphere made by Lorenzo della Volpaia was proudly displayed in the Guardaroba Nuova. His sons, particularly Cardinal Ferdinando, shared this passion, and the Medici collection of scientific instruments became so extensive that later in the century a special room, significantly named *Stanza delle Matematiche*, was set aside in the Uffizi Palace for their display. Like most Renaissance scientific instruments that have survived, the Medici instruments were made of precious materials, but were also too small in size to be of any use in obtaining accurate astronomical measurements. Indeed, they were used instead as teaching aids or as demonstration pieces to illustrate the structure of the universe and how to measure it. They were "intellectual ornaments," to use Martin Kemp's effective definition, made in order to encourage and celebrate the interests in astronomy of "diverse Lords and gentlemen."[49] Among the most famous pieces associated with Cosimo's patronage are the hemicylindrical perspective machines made by Baldassarre Lanci (1557), the many armillary spheres made by Girolamo della Volpaia, and the spectacular pieces built by his court cosmographer (Figs 3.9 and 3.10).

For Cosimo and other members of the Medici family, occasionally also for their close associates, Danti built globes, astrolabes, quadrants, and anemoscopes, of which only few have survived.[50] Indeed the duke's personal appreciation of Danti's work as a cosmographer was reflected in his collection of scientific instruments and maps, which became highly prized artifacts. People connected with the Medici court, eager to emulate the duke in their interests and possessions, lobbied to obtain a globe, a map, or an instrument by Danti. The cosmographer was very selective in accepting commissions outside the ducal family, conscious of the privilege he would lavish on the aspiring patrons. Exemplary is the case of Francesco Bolognetti, son of Camilla Paleotti from the distinguished Bolognese family, himself a prolific author, whose work was not printed but circulated widely in literary circles, including at the Medici court. Bolognetti had to resort to his extensive network at the Medici court and in the religious establishment to convince Danti to make him a globe, a smaller version of Cosimo's monumental globe for the Guardaroba Nuova. Only after endless mediated negotiations concerning payments, colors, decorative features, and cartographic content did Danti agree to make him a globe, but insisted on setting the price.[51]

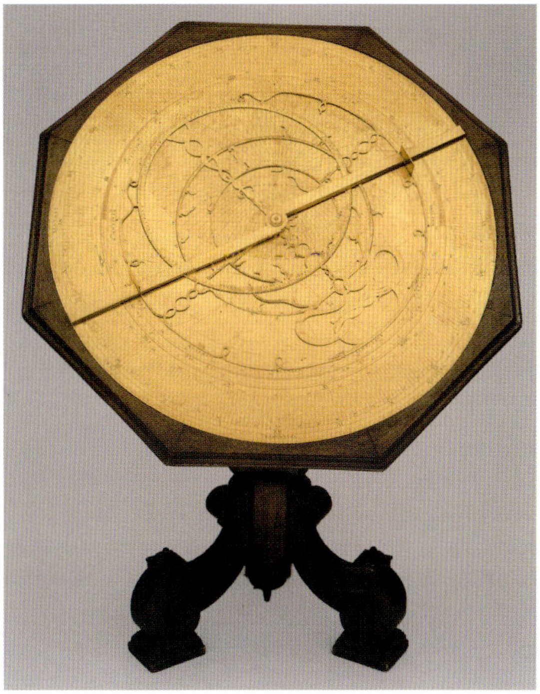

3.10 Egnazio Danti, Astrolabe, ca. 1568, brass and wood. Florence, Museo di Storia della Scienza

Another of Danti's instruments that learned patrons prized for its ancient pedigree and delightful use was the anemoscope, an instrument to record the direction of the winds. From a Greek source, according to Danti, or more likely from the very accessible Vitruvius (*De architectura*, I, 6, 4), he learned that Andronicus of Cirrha had built an octagon tower decorated with personifications of the winds, having at its top "a metal Triton holding a stick in its right hand, with which, being moved around by the wind, and stopping on one of those heads, he showed what wind was blowing."[52] The description inspired Danti to rebuild the ancient instrument which he named, "vertical anemoscope" (*anemoscopio verticale*), paying homage to the innovation that he himself had introduced to the ancient instrument. Danti's vertical anemoscope made it possible to register the movements of the winds not only on a plane parallel to the ground, but also on a plane perpendicular to the ground. Apparently Danti's anemoscope was so successful that it came to be regarded as a learned and enjoyable ornament for country villas and urban palaces. In Florence, and later in Bologna, Perugia, and Rome, aristocrats, cardinals, and even a pope built an additional storey to their properties, pierced holes in walls and ceilings, and adjusted the mural decoration to the theme of the winds in order to have the instrument properly installed. Most of Danti's anemoscopes have disappeared but their mechanism and functioning, of which he was very proud, are documented in his writings and drawings.[53] It is unknown whether he ever designed an anemoscope for the Medici, but a similar instrument was part of the display of the Tribuna in the Uffizi Palace, which was planned under Ferdinando's rule.

Danti's scientific artifacts were regarded as intellectual ornaments that increased the social prestige of their maker and of their owners. The beautifully crafted and accurately executed scientific artifacts Danti made for the Medici were tangible signs of their patronage of the study of the natural world. The copies of the Medici scientific artifacts that Danti made for other courtiers and distant patrons not only reproduced the accuracy of the originals, but also became symbols of the political, cultural or familial proximity of its owner to the Medici duke. The court cosmographer was the one who actually dispensed such symbols of prestige.

The most spectacular pieces, however, those closely related to observational astronomy, were made exclusively for the duke. Danti explained the need "to make these instruments as large as possible, because as they are larger so they can be divided more minutely, and one can use them better and more easily. It is possible to see this in a metal astrolabe, with the diameter of four piedi, which I am having made at the moment for His Excellency the Duke, my Lord."[54] The astrolabe mentioned by Danti may be identified with the large astrolabe mounted on a wood table and equipped with a mechanism that Danti himself had devised so that a single person could operate it properly (Fig. 3.10). With this astrolabe, the second largest surviving astrolabe of the period, it was possible to obtain accurate measurements owing to its size and graduation system, which divided each degree in multiple exact parts. Danti also introduced some variations to the traditional instrument, which was still the most common astronomical instrument in the late sixteenth century, and dedicated to it a hefty treatise, *Trattato dell'uso et della fabbrica dell'astrolabio* (Florence, 1569). This famous treatise by Danti is a cursory exposition on cosmography, in which the author refers the reader to specialized books such as, "the Geometry by Oronce Finé, whose book has been successfully translated into our language by the Magnificent Messer Cosimo Bartoli,"[55] a book that Bartoli translated during his tenure as the Florentine ambassador in Venice, and was published posthumously in 1587, but evidently known to Danti before publication, possibly directly from Bartoli. Danti did not dwell on the basic but essential discussion of the Ptolemaic projections but rather referred his readers to, "Ptolemy, and Giordano, whose Planispheres are illustrated not unsatisfactory in Commandino's commentary and succinctly summarized in the sixth part of Barbaro's perspective."[56] On the other hand, he reported in detail his innovations on the traditional astrolabe made possible by Cosimo, who made available to him rare materials from the Medici collections: a thirteenth-century Arab astrolabe "made maybe 500 years ago," and its accompanying treatise, "a book of instruments which Alfonso, King of Spain, had translated from Arabic and Chaldean into Castilian, and then long ago was translated from Castilian into the Florentine vernacular, a book that I keep with me."[57] Danti also integrated the medieval Arab sources with modern published books, especially with the treatise by Juan Rojas de Sarmiento, who had revised the traditional astrolabe with an instrument featuring a new projection known as the Rojas projection.[58] Danti's astrolabes combined the traditional projections with Rojas's projections and included two other instruments – the *specchio geographico* invented by Peter Apian and the surveying instrument by Oronce Finé – therefore expanding considerably the functions and uses of the astronomical instrument.

Giorgio Spini reported that Danti showed the Arab astrolabe in his classes in conjunction with a modern, large astrolabe, possibly the monumental astrolabe still in Florence. Spini's eyewitness account is significant in not only suggesting Danti's hands-on approach to cosmography, but also to show that these two astrolabes were discussed in relation to the motion of trepidation.[59] The motion of trepidation, the imaginary movement of the stars over time that had been introduced by medieval astronomers, changed the number of spheres that made up the cosmos. Renaissance cosmographers added a tenth sphere to the nine the medieval astronomer Sacrobosco had originally imagined and made the motion of trepidation, not Copernicus's heliocentric theory of the cosmos, their main preoccupation. Not accepting the opinion of other modern cosmographers, such as Oronce Finé who refuted trepidation and did not extend the cosmos behind the eighth sphere, Danti subscribed to the motion of trepidation and consequently considered the cosmos made up of ten spheres. His reasoning was grounded in a typical sixteenth-century blend of observational astronomy and classical authority. From his own observations made with a large astronomical quadrant over a period of two years, which he discussed in his lectures (as Spini reported), he confirmed the motion of trepidation, but from Aristotle's authority he determined the unique movement of each sphere. Danti exhibited the two astrolabes in his class in order to demonstrate the motion of trepidation through the different positions of the stars in each instrument.

Additionally, Cosimo I financed the construction of four monumental instruments: a quadrant, an armillary sphere, and two meridian lines, which Danti placed on the façade of Sta. Maria Novella, the church of his monastery, between 1572 and 1574. These instruments, all widely used by ancient astronomers, stemmed from Danti's teaching activities as well as from his careful reading of ancient astronomical texts, but unlike small-size instruments, they were also actual tools in observational astronomy. Danti explained the use and function of each instrument in short treatises accompanying them that were written concomitantly with the building of the instruments, but published only in 1578 in the second edition of his *Trattato dell'uso et della fabbrica dell'astrolabio* (Florence, 1578). He also made clear that these instruments were built with the specific aim of calculating the length of the tropical year, namely the passage of time from one vernal equinox to the next, which since antiquity was regarded as the measurement germane to the accurate reckoning of time and, thus, to the reform of the calendar.

Cosimo I and the Reform of the Calendar

Since antiquity, the measurement of time and the promulgation of systems of time reckoning were imperial and royal prerogatives. Julius Caesar had instituted the Julian calendar, which calculated the tropical year too generously, a fact that caused the delay of the solar year over the Julian calendar. The Egyptian kings had sponsored astronomers, who were routinely consulted from all over the world on the prediction of planetary motions. From early Christianity onward, the astronomical research aimed at the accurate reckoning of time became a primary concern also for the popes, since only astronomy could predict the motion of the sun and the moon and consequently establish the correct date to celebrate Easter.[60]

Although a century-old problem, the measurement of the tropical year, on which to base a reform of the calendar, acquired new urgency in the Renaissance. Not only had the Julian calendar accumulated a delay of ten days over the solar year, which made it difficult to guarantee that the day of Easter would be celebrated at the same time everywhere in the Christian world, but also the Christian world had been fragmented and divided. Renaissance popes did not fail to recognize the symbolic value of a reform of the calendar as a way of reuniting the Christian world in the celebration of Easter on the same date and sponsored the most advanced astronomical research of the time. Cardinals who were expert in astronomy spoke at church councils: Pierre d'Ailly addressed the Council of Constance in 1415. Nicholas of Cusa spoke at the Council of Basle in 1434. The famous astronomer Regiomontanus came to Rome to address Sixtus IV on the matter in 1474, but died shortly after his arrival. Paul of Middleburg, bishop of Fossombrone, presided over the committee for the reform of the calendar instituted by Leo X, and conducted discussions on the topic in the Fifth Lateran Council of 1512. Copernicus started his computation of planetary motions in response to the problems raised in the Fifth Lateran Council, and eventually dedicated his *De revolutionibus orbium coelestium* (1543) to Pope Paul III. The new calendar, to which Danti also contributed, was finally promulgated in 1582 and named after Pope Gregory XIII, who had made the reform of the calendar one of the goals of his mission for the universal church. In the decade before the promulgation of the Gregorian calendar there was a general interest in the matter across Europe in anticipation of the discussion on the proposal prepared by the papal committee that Gregory XIII instituted in 1575.[61]

With his characteristic enthusiasm for cosmographical projects, and his unfailing wish to make an impact on the world culturally, Cosimo felt that it was appropriate for someone with his political standing to become involved with such a great enterprise and offered his patronage to achieve the reform of the calendar. Once again, in Danti he found a most suitable cosmographer to carry out the astronomical observations and a devoted courtier to celebrate properly the magnitude of the task: "Because Grand Duke Cosimo always had a heroic disposition inclined toward grand undertakings, having always emulated the notable deeds of ancient heroes, and knowing how much glory (beside universal good) had brought to Julius Caesar the reform of the calendar, which he introduced even though it was not perfect, he decided to do whatever was in his power toward this most noble deed, which (having proposed it to His Holiness) he would have undoubtedly brought to perfection, had he not been prevented by inopportune, long illness and later by death."[62] Even though Danti conceded that the reform of the calendar had to be discussed with the pope, he compared his duke to Julius Caesar, mentioning explicitly that Cosimo aspired to the glory that the first Caesar had achieved by the promulgation of the Julian calendar. The duke never achieved the promulgation of a new "Medici" calendar, an honor that after all had to be left to a pope, but he promoted and financed the construction of four monumental instruments built by Danti specifically to measure the tropical year.

The astronomical quadrant, which was placed on the west portal of the church in 1572, had been inspired by Ptolemy's description of the instrument in his *Almagest* (Book I, chapter XI) (Fig. 3.11).[63] This ancient instrument for measuring altitudes, consisting of a graduated arc with a movable radius for measuring angles, was still widely used in the Renaissance, even though Renaissance astronomers tried to achieve more accuracy by building it in gigantic sizes and by devising new techniques to draw its transversals, the divisions of the ten-degree units into smaller parts. The Danish astronomer Tycho Brahe had built a monumental quadrant in his observatory and printed it on the frontispiece of one of his books, and Danti himself had used a large one in his lectures.[64] Whether Danti knew of Tycho Brahe's enormous quadrant is unknown, but a preparatory drawing of his instrument, now at the Uffizi, marks the transversals according to a procedure that precedes that of Tycho Brahe. Danti remarked that his quadrant was essential for "the

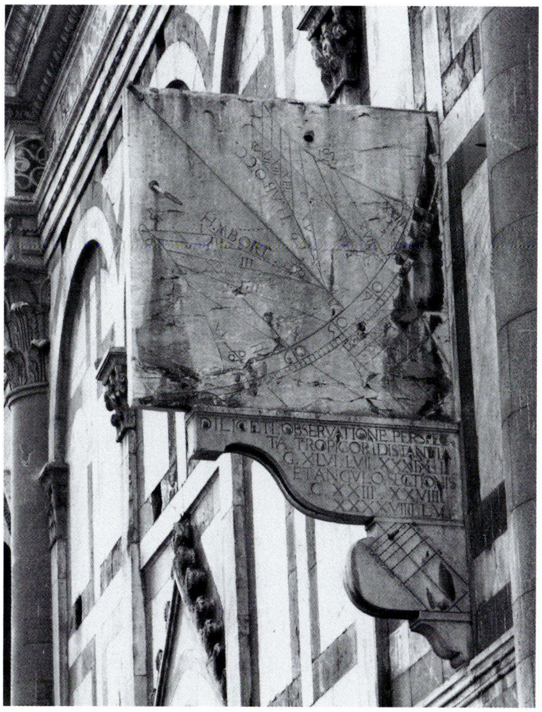

3.11 (*right*) Egnazio Danti, Astronomical Quadrant, 1572, marble. Florence, Sta. Maria Novella, façade

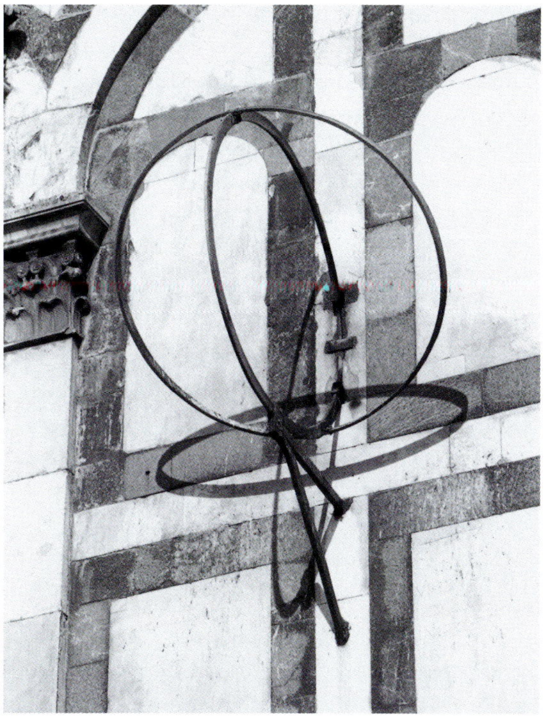

3.12 (*below*) Egnazio Danti, Armillary Sphere, 1574, metal. Florence, Sta. Maria Novella, façade

observation of the equinox, in order to calculate from it the correct length of the year."[65] But placed seven meters above the present street level, Danti's quadrant was too small in size and too high in location to be used comfortably and repeatedly. It soon became obsolete and worked simply as a multiple solar clock in conjunction with the equinoctial armillary placed on the façade in 1574, "so that, in Danti's words, when one uses the armillary [sphere] for observations it is possible to check immediately the exact time on the quadrant."[66]

On 11 March 1574, a few minutes before the sun was on the equinoctial line, in a public ceremony attended by his students and other notables, Danti placed the armillary sphere on the façade of the church (Fig. 3.12). By then Cosimo was very ill and unable to move, but still he financed the project and discussed with Danti the detail of the instrument's construction and placement. As Danti wrote, the equinoctial armillary was used by "Callippus, and later Hipparcos, Sosigenes and Ptolemy . . . for their most important observations pertaining to the length and measurement of the year."[67] Cosimo I and his cosmographer regarded the armillary sphere as the most appropriate instrument for checking and eventually revising the measurements of ancient astronomers. They also agreed on the need to have the instrument fixed in the same place to assure the accuracy of the measurements, which had to be taken exactly at the same time, year after year. Danti duly mentioned Cosimo's patronage in the instrument's inscription and pompously compared it with the royal patronage of the ancient Egyptian kings who had promoted the measurements of the year: "Cosimo de' Medici Grand Duke of Etruria being the first enthusiastic of astronomy after the ancient kings of Egypt put in place [the armillary sphere]." The instrument was never completed and Danti continued to care about it in the following years, although he soon focused on the construction of meridian lines, the best instrument in measuring the tropical year.

Meridian lines were built before in many places. In Florence, there existed the meridian line made by Paolo Toscanelli in the dome of Sta Maria del Fiore, of which Danti was surely aware. But he must have soon realized that Toscanelli's meridian line was unsuitable for year-long measurements and planned to replace it with a new meridian line in Sta Maria Novella. Documentary evidence shows that Danti planned and partially built the meridian line and even asked for permission from the Ricci family to pierce the church's façade that had been the object of their patronage since the fifteenth century. Much smaller than Toscanelli's instrument, Danti's meridian line presented the advantage of year-long usage. Although Danti was unable to pursue his observational work in Florence, he later built meridian lines wherever he happened to be, one in the convent of San Domenico in Bologna, another in the church of San Petronio in the same city, yet another in the Tower of Winds in the Vatican Palace, and even one in the church of St. Peter's in Rome. Additionally, in 1576 or in 1577 Danti wrote a one-page essay on the use of meridian lines to establish the year's length.[68]

The measurement of the tropical year remained one of Danti's main occupations even after his departure from the Medici court, first as a professor of mathematics at the University of Bologna and later as Gregory XIII's papal cosmographer in Rome. Eventually he participated in the papal committee for the reform of the calendar and undersigned the final document. But as long as Danti remained Cosimo I's cosmographer he joined with other Medici courtiers in comparing Cosimo I's cosmographical undertaking to the deeds of kings and emperors, both ancient and modern. Danti himself never failed to recognize that from Cosimo I, "proceeds all that I achieved in this gentle and pleasant profession, which is Cosmography."[68] Gherardo Spini compared Cosimo's patronage of the reform of the calendar to that of Julius Caesar, the Egyptian kings, Alfonso King of Spain, and Emperor Charles V. The archdeacon of Pisa, Bernardino Martini, another of the duke's close associates, wrote a telling sonnet, published in the introductory pages of Danti's *Trattato dell'uso et della fabbrica dell'astrolabio.*[70] Praising Danti's measurement of the cosmos, Martini considered Florence to share honours with Alexandria, and the Arno with the Nile. Even though the Medici state was rather small and Cosimo's role in Renaissance politics rather modest, his patronage of cosmography was imperial in scope and consequence.

* * *

The Guardaroba Nuova as Cosimo's Ornament

Bernardino Martini's sonnet referred explicitly to the glories derived from the measurement of the cosmos and suggested the association between the duke and the universe by capitalizing the words GRAN COSMO. But it was an impresa by Danti, published together with Martini's sonnet, that established the connection between the investigation of the cosmos and the persona of the grand duke (Fig. 3.13).[71]

Danti's impresa employs the usual elements of Cosimo I's cosmic imagery: the globe, the Medici *palle*, and the Greek inscription that had already appeared in the temporary decorations for the duke's triumphal entry into Siena in 1560. But these features are rearranged in a new and poignant image. The six traditional Medici *palle* are not depicted against a neutral background but are superimposed on a terrestrial globe, an image certainly fitting with the topic of the treatise, the construction of the astrolabe. The globe is not simply a reference to cosmography or merely an attribute of Cosimo's rule, but is explicitly identified with the duke. It contains the emblems of his rule: it bears the Medici *palle*, is surrounded by the chain of the Order of the Golden Fleece, and is crowned with the newly acquired ducal coronet. And, to eliminate ambiguous interpretations of the impresa, Danti included a cartouche bearing the Greek inscription "Cosmos Cosmoi Cosmos."

Danti's impresa employs the usual elements of Cosimo I's cosmic imagery: the globe, the Medici *palle*, and the Greek inscription that had already appeared in the temporary decorations for the duke's triumphal entry into Siena in 1560. But these features are rearranged in

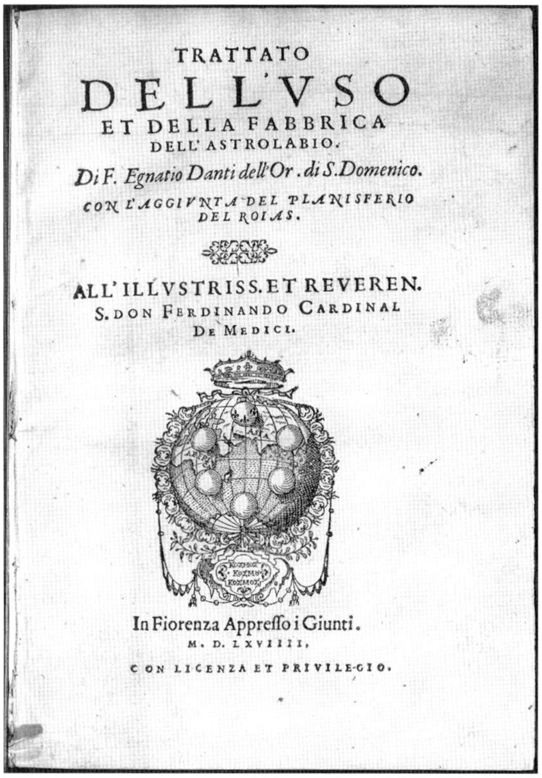
TRATTATO
DELL'VSO
ET DELLA FABBRICA
DELL'ASTROLABIO.
Di F. Egnatio Danti dell'Or. di S. Domenico.
CON L'AGGIVNTA DEL PLANISFERIO DEL ROIAS.
ALL'ILLVSTRISS. ET REVEREN.
S. DON FERDINANDO CARDINAL DE MEDICI.
In Fiorenza Appresso i Giunti.
M. D. LXVIIII.
CON LICENZA ET PRIVILEGIO.

3.13 (*left*) Egnazio Danti, *Trattato dell'uso et della fabrica dell'astrolabio*, Florence 1569, frontispiece. Houghton Library, Harvard University (IC5. D2361 596 tb)

3.14 (*below*) Detail of fig. 3.15: Medici Coat of Arms

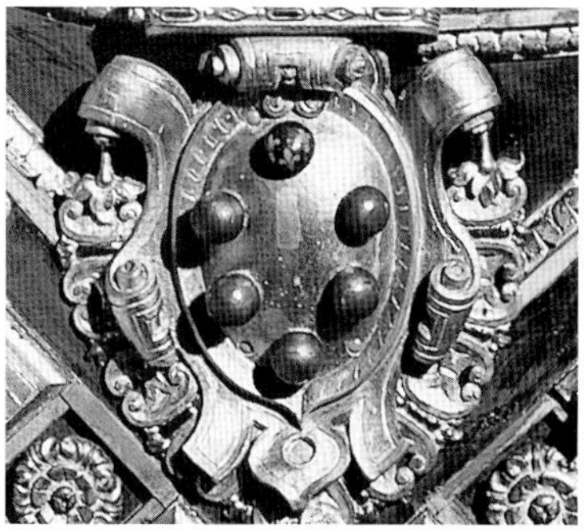

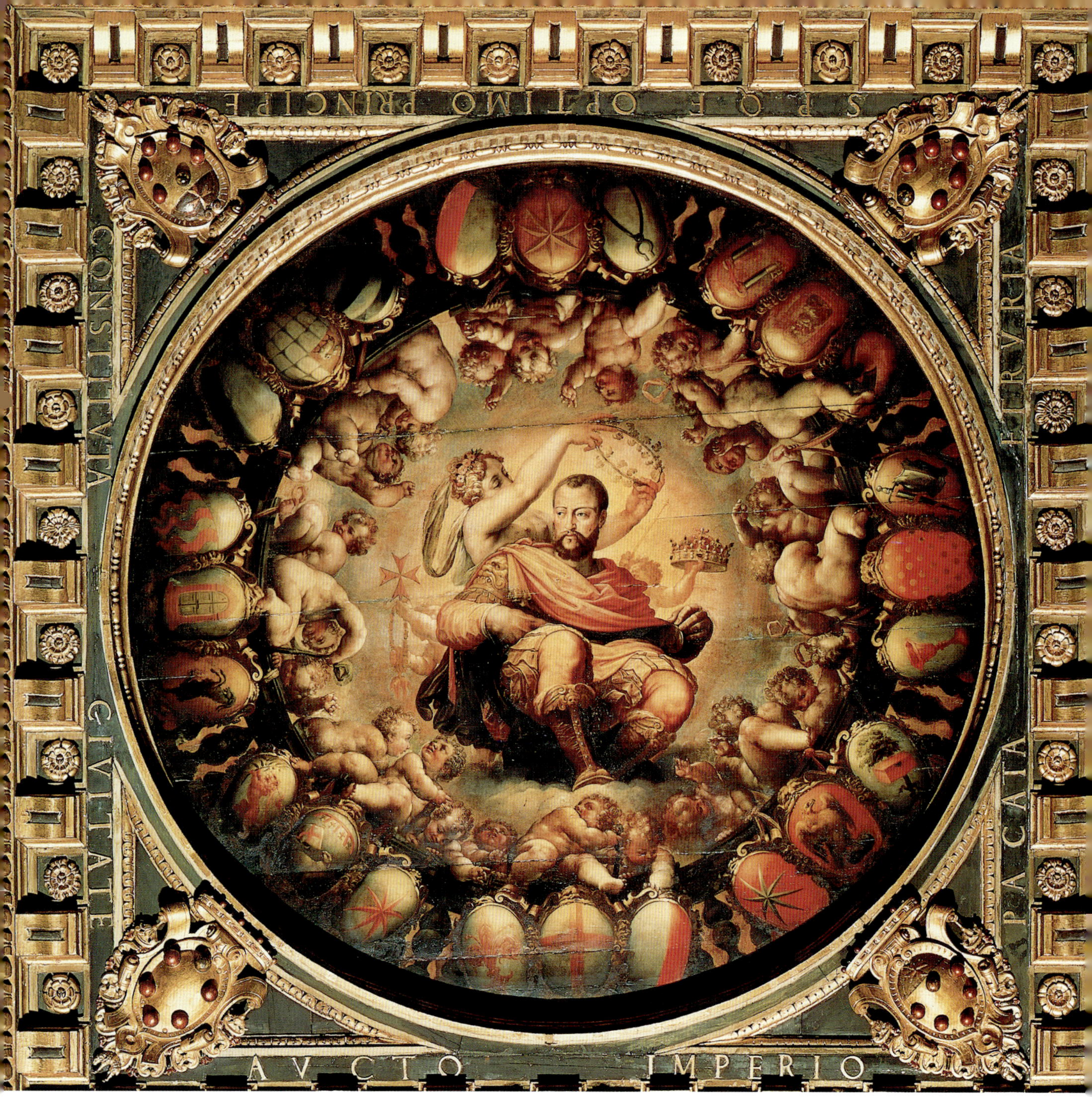

3.15 Giorgio Vasari, *The Incoronation of Cosimo I*, 1565, ceiling panel. Florence, Palazzo Vecchio, Sala Grande

a new and poignant image. The six traditional Medici *palle* are not depicted against a neutral background but are superimposed on a terrestrial globe, an image certainly fitting with the topic of the treatise, the construction of the astrolabe. The globe is not simply a reference to cosmography or merely an attribute of Cosimo's rule, but is explicitly identified with the

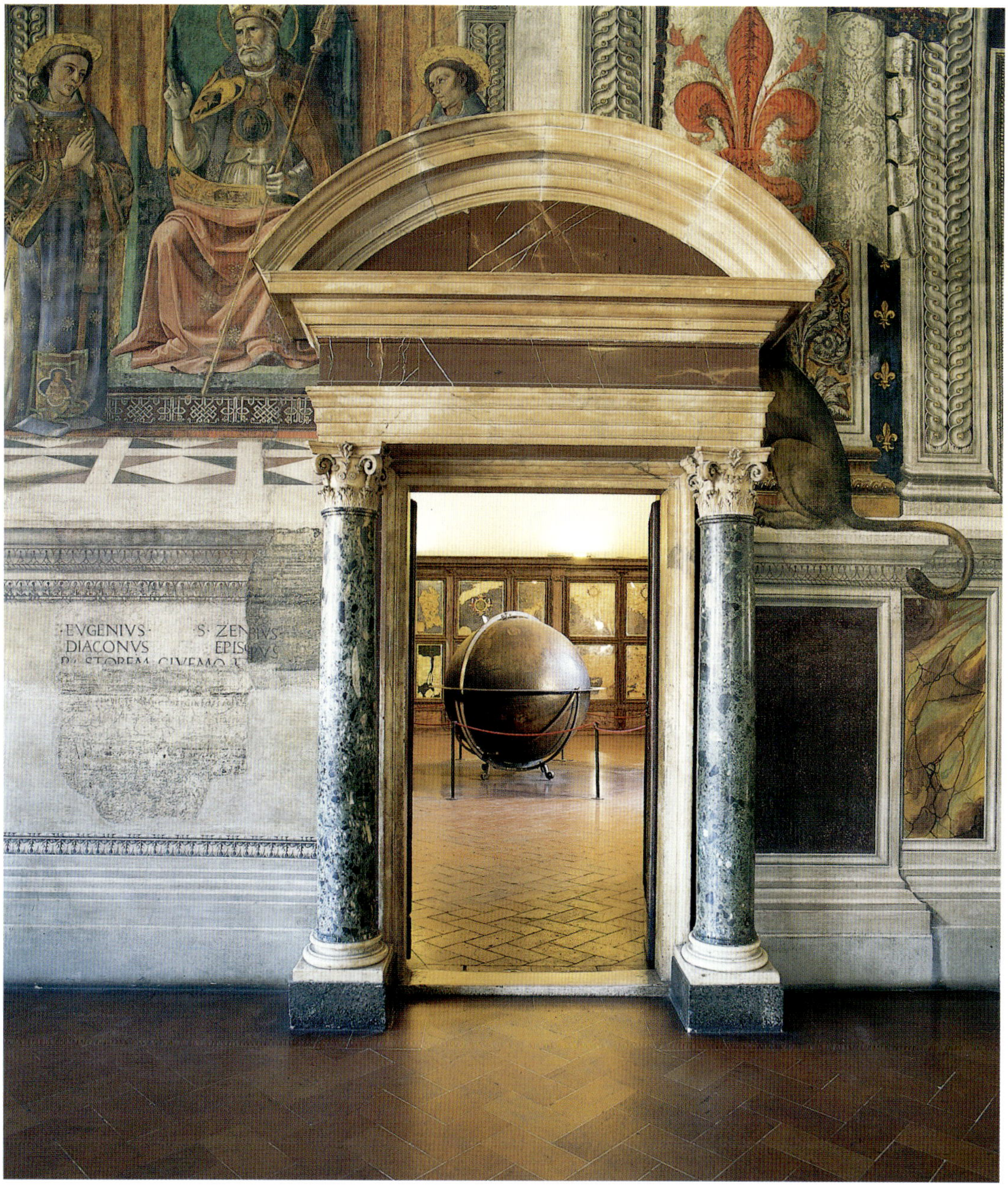

3.16 Guardaroba Nuova, view from the Sala dei Gigli. Florence, Palazzo Vecchio

duke. It contains the emblems of his rule: it bears the Medici *palle*, is surrounded by the chain of the Order of the Golden Fleece, and is crowned with the newly acquired ducal coronet. And, to eliminate ambiguous interpretations of the impresa, Danti included a cartouche bearing the Greek inscription "Cosmos Cosmoi Cosmos."

The identification between the duke and the globe is reiterated in a poem written by Giovambattista Strozzi the Elder, which also accompanied Danti's impresa and Martini's sonnet. Giovambattista Strozzi, one the most eminent madrigalists in Florence, was a good friend of the duke even though he was the nephew of Filippo Strozzi, the Medici archenemy. He was the author, among other things, of the *intermezzi* for the play *Commodo* by Antonio Landi, performed in the rear courtyard of the Medici Palace on the occasion of Cosimo's wedding, as well as the author of the *intermezzi* for the *Trionfo degli dei* performed during the Carnival of 1566. His poem strongly suggests that Medici courtiers would not have failed to connect the cosmos with their ruler. In Strozzi's poem Cosimo I was equated with the sun as the light source of Florence (*Flora*), Tuscany (*terren Tosco*), and the COSMO, written again in capital letters to hint to its double reference to the universe and the duke:

Non pur la bella Flora
Sua dolce figlia di bei cerchi adorna,
Ma come il Sol che in ogni lido aggiorna,
Mentre di gemme il terren Tosco infiora,
Questo e quello altro Polo,
E tutto il Mondo illustra un COSMO solo.

Not only the beautiful Flora,
His sweet daughter with beautiful circles adorned,
But as the sun that brings the daylight to every shore,
While it blossoms the Tuscan land with gems,
This and the other Pole
And the entire universe one single COSMO brightens.[73]

The theme of Strozzi's poem that suggestively connected the universe, Tuscany, Florence, and Cosimo I recalls the similar topos that Vasari adopted in the description of the Sala degli Elementi: "That is a cosmic body, which is called the world by astrologers, and which is precisely the name of our lord, the duke, who had been made master of this state; and Saturn, his planet, touches his Capricorn ascendant, and by their aspects they cast a benign light on the globe of the earth and, in particular, Tuscany, and, as capital of Tuscany, Florence, which is ruled today with such justice and such good government by His Excellency."[74] In other words, the cosmic body, which Medici courtiers identified with the name of their duke, implied the reference to Tuscany, Florence, and Cosimo I. The sequence cosmos, Tuscany, and Florence was so pervasive in images and poems of the last decade of Cosimo's rule that one is tempted to ask whether the iconography of the entire Medici Palace could not be subjected to this cosmic interpretation. In his *Ragionamenti*, Vasari attempted to present the palace's decoration as a coherent program carrying a moral and a Medicean meaning. But that book was completed before the cosmic imagery of the duke was fully developed and before the study of the universe had become a major occupation at the Medici court. Playing the courtier to Cosimo, as Vasari did, and using the same kind of visual and literal associations that Vasari used to explain the iconography of the duke's apartments, it is possible to detect the relationship between the cosmos, Tuscany, and Florence in Cosimo's palace.

The Medici dominion and its capital were illustrated in the Sala Grande. A microcosm of the dominion, the Sala Grande was also a diagram of the administration of the state and a schematic map of its territory. From the initial planning in 1563 to the final realization in December 1565, the emphasis of its iconography switched subtly, but unmistakably, from the theme of Florence as the unique republic to the theme of the territorial extension of the Medici state. The accent changed from civic to regional and the perspective from Florentine to Tuscan. Indeed, the culminating image of the central tondo contained in itself the city, its quarters, the regions of the state, and the absolute ruler. From the central tondo, which was also the center of the state, Cosimo I was the motor that administered Florence and, through its civic institutions, the dominion of Tuscany (Fig. 3.16). In the official space of the Sala Grande Cosimo I rules over his real dominion.

The cosmos that Strozzi and Vasari related to Tuscany and Florence was illustrated in the Guardaroba Nuova. As a description of the cosmos, the Guardaroba Nuova was intimately related to the persona of the grand duke. The room was a three-dimensional and powerful display of the motto, "Cosmos is Cosimo's ornament," literally and symbolically. Literally, the room provided for the safe keeping of the world's ornaments that Cosimo possessed. Symbolically, the room referred to the persona of the grand duke through his very name. Medici courtiers, illustrious visitors, and contemporary viewers well versed in the visual exegesis practiced at the ducal court, would not have failed to recognize the cosmography of the Guardaroba Nuova as a gigantic emblem of Cosimo I. Trained to interpret Hercules's globe as a reference to their duke, to regard cosmography as part of the destiny of their ruler, to read Martini's sonnet and Strozzi's poem as a simultaneous praise of the universe and their ruler, and to decode Danti's emblem as an illustration of the motto, "Cosmo is Cosimo's ornament," Medici courtiers would have understood that in the Guardaroba Nuova the cosmographic description of the world was conflated or, better still, coincided, with Cosimo I. In the secluded space of the Guardaroba Nuova, the duke's influence expanded beyond the walls of Florence and beyond the borders of Tuscany. In the tondo of the Sala Grande, he was represented through his likeness and surrounded by the insignia of his power. In the Guardaroba Nuova, he was alluded to in the highly simplified form of a globe which, at this stage in the development of Cosimo's imagery, was unmistakably identified with his body (Fig. 3.15).

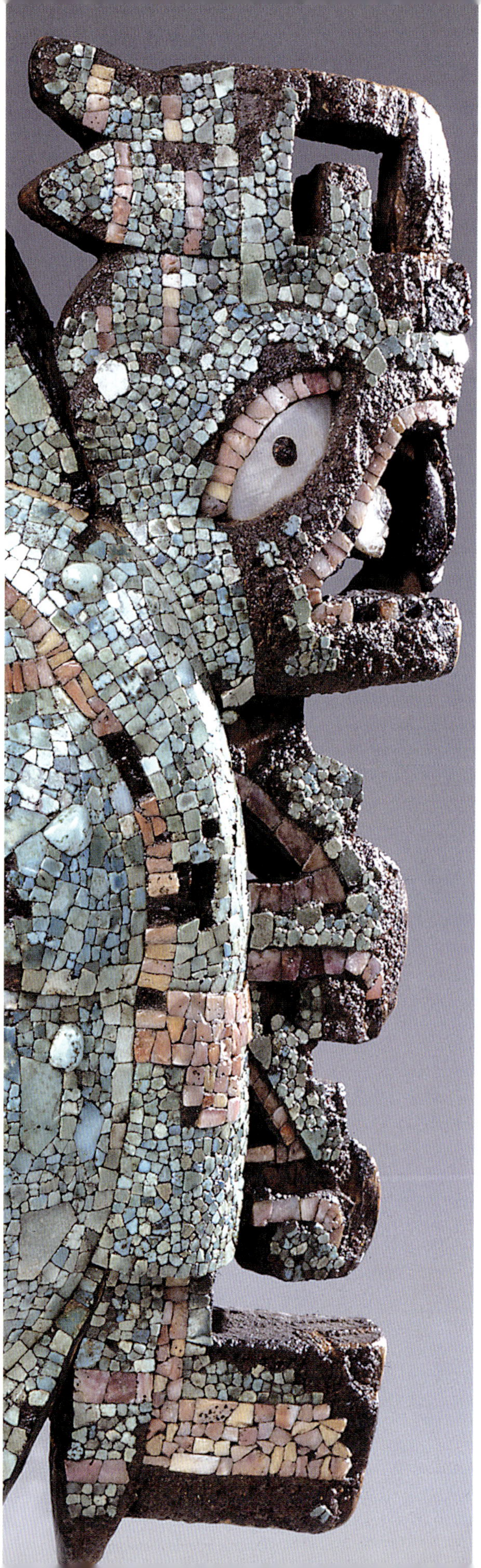

4

The Ordering of Maps

The cosmic imagery of the Guardaroba Nuova epitomized Cosimo I's personal iconography, but it was also appropriate to the main function of the room as a cabinet of art and curiosities for the most beautiful objects that he possessed. If, in many respects, the Guardaroba Nuova was similar to the princely cabinets that had become common in European courts, it was also exceptional for its unprecedented series of maps of the world, maps that plausibly not only related to the artifacts kept in the room but also provided a novel organizing principle for their display. Indeed, the Guardaroba Nuova holds a unique place in the history of the Medici collection and, more generally, in the history of collecting in early modern Europe. A cartographic theater of the world, it adopted the geographical order that Renaissance cosmography books used to organize universal knowledge, adjusting it to the display of an encyclopedic collection of artifacts. With his impeccable nose for novel imagery, Cosimo selected a geographical display that would have shown off his exotica, the products of unusual crafts from distant lands that were eagerly sought and exchanged among the rulers of Europe.

Cosimo I's Collection

Like other rulers, Cosimo regarded collecting as an expression of princely magnificence, intellectual sophistication, and, above all, as a manifestation of diplomatic ties and dynastic continuity.[1] As the Medici heir, he knew well that dynastic families passed their collection down through the generations, but he also knew that the famous collection of earlier Medici had been ransacked and dispersed.[2] At the time of his election, Cosimo I did not possess even a chest of medals, the most indispensable collectible of a Renaissance ruler, but he immediately invested a large portion of his exiguous funds in the acquisition of a marvelous set of medals, placing it in his bedroom. Ingrained in him was the notion that the history of his family collection, its acquisition, dispersal, and reconstitution were part and parcel of the Medici political fortune. He relentlessly reacquired the holdings of earlier Medici,

(*facing page*) Detail of fig. 4.12

either directly from Margarita of Austria or indirectly from whomever happened to have them. For Cosimo the presence of the early Medici's collection in his own ducal collection was a tangible demonstration of the continuity of Medici power from the early fifteenth century to his own rule, a theme that dominated his politics and inspired the depiction of Medici deeds in the Apartment of Leo X. In addition, Cosimo devoted immense funds to the acquisition of a stunning collection that could rival those of emperors and kings. By the time the Guardaroba Nuova was planned, he had assembled a body of Medici heirlooms for his successors. He had secured outstanding pieces on the art market. Other objects that he desired at all costs were smuggled into Florence, including the Etruscan statue that Egnazio Danti brought illegally from the Papal States. Many objects he had received as wedding presents and diplomatic gifts. The most spectacular gift came from Pope Pius IV in the form of a group of ancient life-size statues in appreciation for the ducal support in the papal election.

Like other rulers, Cosimo used his collection in different ways. Exceptional pieces, such as works by famous Florentine artists, distinguished diplomatic gifts, portraits of family members and political allies, and furniture in rare woods and in semi-precious stones embellished Cosimo's private apartment, those of his wife and children, the ceremonial spaces of his palace, and the guest rooms.[3] Used as furnishing, the collectibles in these rooms conveyed the sophisticated taste and the political alliances of the Medici dynasty. Many other objects of precious material and artistry, which had to be moved constantly to satisfy the needs of the court, were kept in the ducal guardaroba, where they were stored according to type in order to facilitate their retrieval for daily use.[4] Periodically, rare Chinese pottery, vases of oriental crystals in the shape of birds and flowers, damascene bowls and vessels, bowls in precious and semi-precious stones, and centerpieces in gold and coral were removed from the guardaroba to adorn the ostentatious Medici dinner table. Other objects, such as Mameluk carpets, Persian drapery, portraits, and religious paintings were constantly moved around to set up the guest rooms, each furnishing tailored to the nationality, interests, and rank of individual guests. Some objects, however, were singled out as especially precious and rare. Even though these exceptional pieces were used for practical needs on special occasions, they were kept in a separate room of the ducal guardaroba identified as the secret guardaroba (*guardaroba segreta*), where they were stored by type in fourteen large cupboards (the twelfth cupboard held pieces closely connected with Medici power).[5]

The inventories of the ducal palace recorded Cosimo's trove of *naturalia*, *artificialia*, and *exotica* of all times and places, and still today these lists convey a sense of amazement at the sheer quantity of the Medici collections, their preciousness, and rarity. But these same inventories, particularly one taken in 1553, suggest that if the size, content, and preciousness of Cosimo's collection could rival the finest European collections, its display was anything but impressive. Beautiful and rare artifacts were spread all over the palace, but no special room had been designed for the display of the most outstanding pieces, or for providing the quiet and seclusion indispensable for their study and contemplation. Even the secret guardaroba, a fairly large room with imposing cupboards and constant traffic of people and goods, was essentially a storage place, in which artifacts were either amassed in closed cupboards or assembled on top of them or hung randomly on the walls. In 1554 the duke who enjoyed the private contemplation of his collection, solved the problem of private space by having a small study built near the guardaroba with the specific purpose of "retiring in a private room."[6]

In this little, private space, alone or in the company of selected friends and artists, Cosimo indulged in the study of his collectibles, which he would retrieve a few at a time from the secret guardaroba. There he personally cleaned with a chisel the Etruscan statuettes recently discovered in Arezzo with the help of the sculptor Benvenuto Cellini.

But obviously the Medici collections demanded a more compelling display, a display that would present them as symbols of the duke's intellectual and political prestige and would integrate them within the ubiquitous theme of the palace, that of Medici glorification and dynastic continuity.

Cosimo I's Studies

Acutely aware of the power of representation, Cosimo must have realized that the symbolic meaning of the collection he so passionately assembled could fully be expressed only in rooms appositely designed for its display. In the late 1550s, in conjunction with the remodeling of his palace, he asked Vasari to transform some small rooms of the palace into collector's studies. Such rooms had become a common feature in the residences of rulers and aristocrats as well as in the houses of scholars, merchants, and professionals, and had been a favorite of the earlier Medici. Lorenzo the Magnificent had a collector's study filled with artifacts that he prized not only for their intrinsic monetary value, but also as gifts of important people and as examples of unusual crafts. Cosimo made a point of reacquiring as many of Lorenzo the Magnificent's collectibles as he could.

Eclectic in character, encyclopedic in scope, and elegant in furnishing, Cosimo I's cabinets of art and curiosities were not ceremonial spaces but rather secluded parts of the princely palace. No record has survived of how his collector's studies were actually used, but it is plausible that, like similar, better documented studies, they were made primarily for his own enjoyment. Some selected guests, usually fellow rulers, ambassadors, and high-ranking collectors, might have been invited to visit them as an acknowledgment of their high distinction. The public at large knew of Cosimo's studies through literary descriptions, diplomatic reports, and hearsay, but especially through the reading of Vasari's *Lives*, where their content and decoration were minutely recorded.

4.1 Glass Window. Florence, Palazzo Vecchio, Studiolo di Calliope

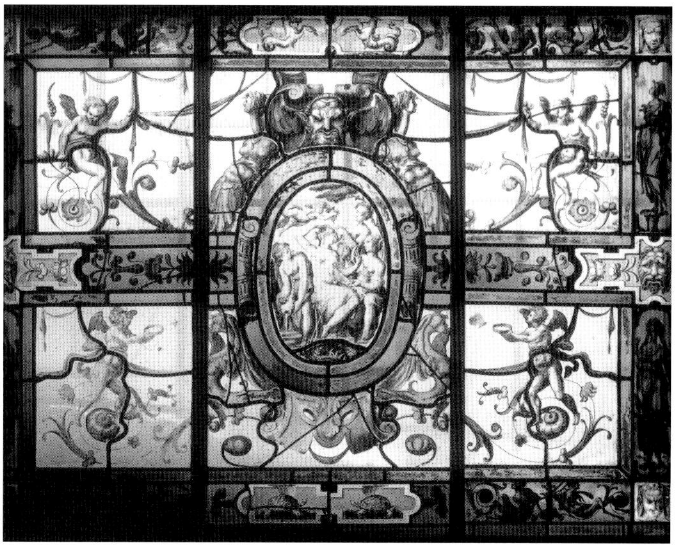

Cosimo's intention to integrate his collection with his cultural claims is particularly explicit in the Studiolo di Calliope, a small room in the Apartment of the Elements (Fig. 4.1).[7] For this study, Vasari painted a sophisticated ceiling panel, designed a stained-glass window, had

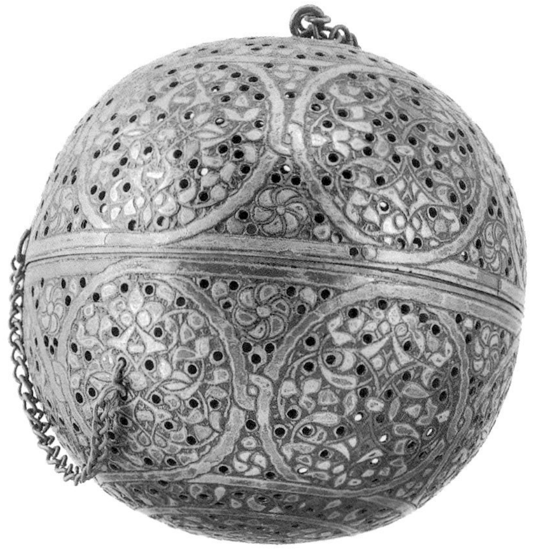

4.2 Damascene spherical perfume-burner, Syria, fifteenth century, brass, silver, and gold. Florence, Museo Nazionale del Bargello

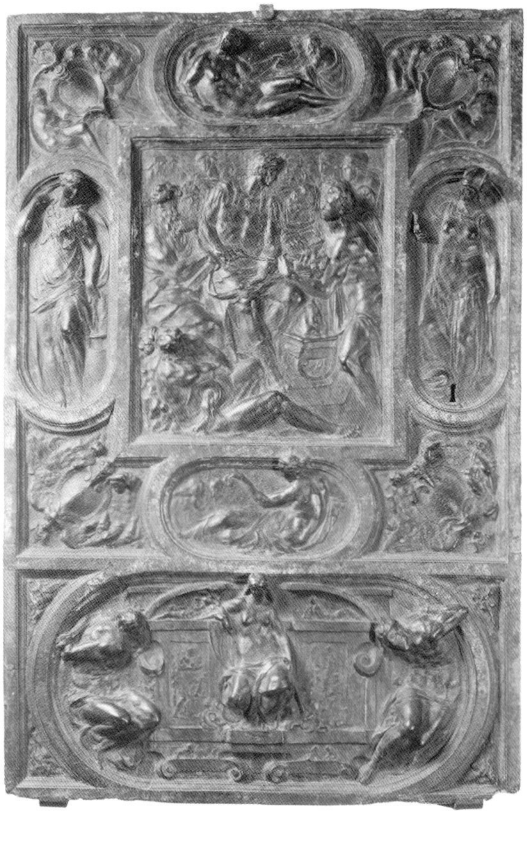

4.3 Vincenzo Danti, Sportello for one of Cosimo's studies, ca. 1560, bronze. Florence, Museo Nazionale del Bargello

lavish furniture built, and crimson velvet applied to the wall. The Studiolo di Calliope contained mainly small statuettes and reliefs in bronze and it was complementary to the Studiolo di Minerva, another small room in the same apartment, which instead displayed small objects in marble.[8] The objects displayed in the Studiolo di Calliope had been removed from the twelfth cupboard of the secret guardaroba and selected according to the primary criteria adopted in most European collections: their small size and the material from which they were made. The organization of a collection by material enjoyed a very distinguished precedent in Pliny's *Natural History* and it was used in the royal collections at Ambras, Prague, Munich, as well as in the collections of sixteenth-century scholars and apothecaries. Traditional though it was, the organization of a heterogeneous collection by material presented the advantage of blurring the boundaries between natural and artificial, carved and uncarved, near and exotic, modern and ancient. It also provided a visual experience that, in due course, would facilitate the understanding of the history of the natural world.[9]

But a more attentive analysis of the beautiful little objects selected for the Studiolo di Calliope reveals that they were displayed in such a way as to evoke Cosimo I's dynastic and cultural claims. The Etruscan statuettes were displayed alongside the medieval heads by Nicola Pisano removed from his pulpit in Pisa Cathedral, a bronze relief of the Virgin by Leonardo da Vinci, and numerous works by Donatello, including the death mask of a Medici lady.[10] This display of Etruscan, medieval, and Renaissance bronzettes was a visualization of the cultural and artistic continuity between the old and the new kingdoms of Etruria, a theme that court humanists, such as Giambattista Gelli and Pier Francesco Giambullari, developed in their historical writing, and Giorgio Vasari applied to his interpretation of Tuscan art.[11] Similarly, the twenty-four miniature portraits of the Medici family made by Bronzino's workshop, which hung on the door of the studiolo that led to a secret passage, did not simply represent family members but, by dividing them according to the two different branches of the family, restated once again the legitimacy of Cosimo I's rule.[12] Also the precious artifacts from exotic lands that were included in the Studiolo di Calliope may have exemplified Cosimo I's political claims, such as the direct connection with Lorenzo the Magnificent who had once owned the Islamic incense-burners still today kept in Florence (Fig. 4.2), or the relationship he forged with Spanish political patrons who almost certainly gave him the eight miniature Aztec animal heads that Benvenuto Cellini had mounted in gold.[13] The relevance of the Studiolo di Calliope to Cosimo's persona would have been even more evident if Vasari's proposal to display there the Etruscan statue *Chimaera* had been feasible (the statue was too large to be included in the studiolo). Interpreted by Vasari as an allusion to the Capricorn of the duke's impresa, this famous Etruscan bronze would have fused the interrelated themes of the Studiolo di Calliope: the artistic and cultural continuity between the new and old kingdom of Etruria, and Cosimo I's legitimacy as the heir of the Medici dynasty.[14]

Less public, but equally important to the relationship between Cosimo's persona and his own collection, were the two other studies that the duke had refurbished in his private apartment around 1560. The very small room known as the Tesoretto was furnished and decorated as the safe for the papers of a just and learned ruler (Fig. 4.4). Its ceiling fresco, a collaborative effort of Cosimo Bartoli and Vincenzo Borghini, was modeled on Raphael's Stanza della Segnatura and, like the Vatican room, it suggested the harmony of divine and terrestrial knowledge as the supreme attribute of an inspired ruler.[15] Cosimo I's second study, located below the Tesoretto, was even more secret. Its main feature was a safe where, according to Vasari, "important papers of the Duke" (*scritture d'importanza del Duca*) were kept.[16] The centrality of the safe was marked by the bronze door made by Vincenzo Danti (Fig. 4.3). Although the subject of Danti's relief has been variously interpreted as *Augustus Burning the False Books of the Sibyls* or as *The Burning of Numa's Books*, it is generally acknowledged that its iconography referred to the classified contents of the papers Cosimo kept in the safe and, more importantly, to illustrious Roman models of rule.

The duke's efforts to organize his collections coherently did not stop at the display of his small studies in the ducal palace, but also involved another major project pertaining to large-scale ancient statues, the most prestigious collectibles of any Renaissance ruler. Around 1560 he planned and realized an *antiquarium* in the great hall on the first floor of the Pitti Palace, the property acquired with Eleanor's dowry in order to accommodate the growing family and court. The *antiquarium* in the Pitti Palace conformed to the sixteenth-century fashion of displaying ancient statues in the courtyard of aristocratic residences. Pope Innocent VIII had envisaged the first of such courtyards in the Belvedere Villa in the Vatican Palace, and noble families had followed the fashion throughout Italy. The statues displayed in this grandiose hall during Cosimo's rule were recorded in the 1568 edition of Vasari's *Lives*, specifically in the appendix entitled, "Antiquities that are in the hall of the Pitti Palace" (*Anticaglie, che sono nella sala del Palazzo de' Pitti*).[17] There were twenty-five life-size or over-life-size classical statues Cosimo had either inherited, acquired, or received as diplomatic gifts, and

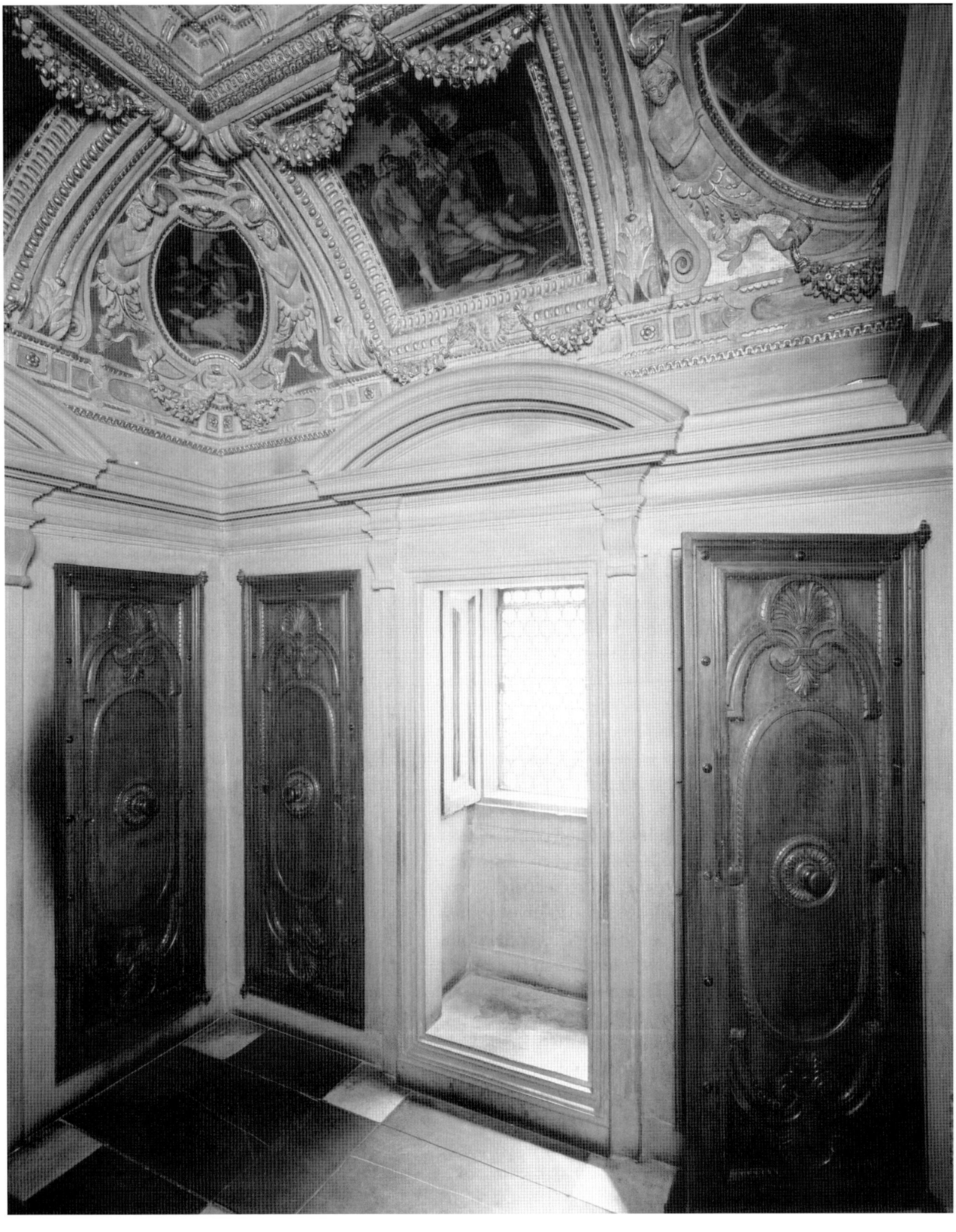

4.4 Tesoretto. Florence, Palazzo Vecchio

this collection included some renowned pieces such as the Medici Venus. The great hall in the Pitti Palace was as much a magniloquent display of the Medici collections of ancient statues as it was a demonstration of the political power and diplomatic ties that had made those acquisitions possible.

Around 1560 Cosimo's collection was effectively distributed in specialized rooms that enhanced the beauty and preciousness of the artifacts while, at the same time, reinforcing his public persona. The small artifacts tastefully displayed in collector's studies, the personal papers secured in private studies, and the life-size statues located in the Pitti hall: all converged to present him as a sophisticated collector, a virtuous and inspired ruler, a talented diplomat, and a legitimate heir of the Medici dynasty. It is from this reorganization of the Medici collection in specialized rooms that emerged the novel idea of having a cabinet of art and curiosities arranged geographically.

Collecting According to Maps

Unquestionably, the Guardaroba Nuova was meant for the display of a coherent group of distinguished pieces selected from Cosimo I's impressive holdings. Its function as a cabinet of art and curiosities is documented in its original name and confirmed by Vasari, who, in the 1568 edition of his *Lives*, published the list of Cristofano dell'Altissimo's portraits in the Guardaroba Nuova immediately following the list of the statues in the Pitti *antiquarium*, thus implying the similar function of the two halls as showpieces for Cosimo's collections. But how did the Guardaroba Nuova fit in the organization and display of the Medici collections around 1560? Which artifacts were kept in the room? Did the Guardaroba Nuova depart from or conform to the organizing principles of collections in Florence and elsewhere?

When one looks for an explicit answer to these questions the documentary evidence is regrettably absent. Vasari vividly described the planned iconography of the Guardaroba Nuova, but he was extremely vague about the objects to be kept in the room, "the most important, precious, and beautiful things," and inexorably silent on the organizing principles of their display. Even the inventories of the Medici guardaroba under Cosimo, which are so instrumental in the historical reconstruction of other rooms of the ducal palace, are of little help in understanding the Guardaroba Nuova. The two most relevant inventories, one made in 1570 when the courtier in charge of the guardaroba changed and another made in 1574 at the duke's death, listed the objects kept in the room, but these inventories did not reflect any particular order.[18] According to these inventories, the richly decorated Guardaroba Nuova that had demanded so much effort in terms of funds, planning, and cartography, was finally used as an ordinary room of the general guardaroba: to store objects that, although valuable for their material and artistry, were used mainly as furnishing for the ducal court. Pillows, tapestries, hangings, fabrics, and occasionally semi-precious stones fill the cupboards with no apparent order. Did these inventories reflect the intended organization of the room? Or were they compiled after the intended cosmographic meaning of the room had already been abandoned? The evidence drawn from these inventories would seem to confirm this. But such an interpretation is untypical of the very practical and parsimonious Cosimo and in sharp contrast to his attitude towards his own collection around 1560.

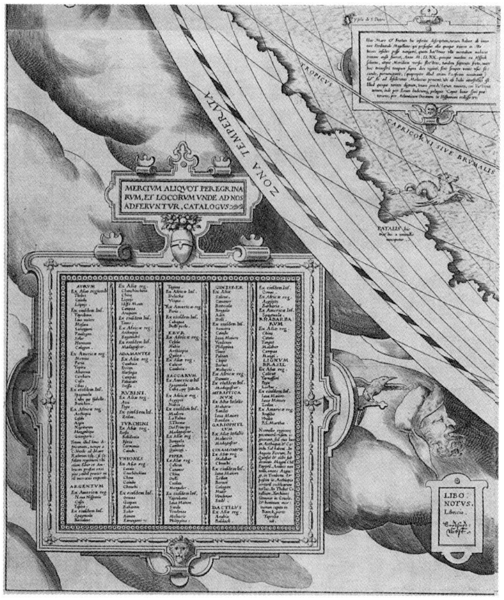

4.5 Abraham Ortelius, detail of fig. 5.22: *Mercium aliquot peregrinarum, et locorum unde ad nos adferuntur, catalogus* (Catalogue of Some Foreign Goods, and of the Places from which They Are Brought to Us)

4.6 (*facing page*) Egnazio Danti, *Indostan fuori il Gange* (Map of Hindustan Outside the Ganges), not dated, panel. Florence, Palazzo Vecchio, Guardaroba Nuova

Instead, if attention is focused not on the documentary evidence provided by the Medici inventories but on the visual evidence provided by the room's iconography, a very different picture emerges. The iconography of the Guardaroba Nuova strongly suggests that the objects had to be stored geographically, namely their place in the cupboards related to the geographical places represented by the maps. The maps on the cupboards' doors were to be a visual catalogue of the collection. They were labels to locate easily the objects both in the room's cupboards and in the world's geography. As the maps dictated the overall order of the iconography, the frieze of plants and animals, and the busts of ancient rulers, they also dictated the display of the samples of goods and curiosities of the world. In this respect, the room was an exceptional attempt to organize a major collection of artifacts according to maps.

In the Guardaroba Nuova the maps themselves hold the interpretative key to the organizing system of the artifacts of the world. The legends of the maps succinctly narrate the history of a place and the characteristics of its people, but they also list the objects for which the mapped place was renowned. The products, artifacts, and resources of a place were customarily reported in brief form in the cartouches of Renaissance printed maps. The Dutch cartographer Abraham Ortelius even included in his world map of 1564 a catalogue of precious goods and the places where they could be found (Fig. 4.5).[19] In the legends of Danti's maps, however, the products of each place were so meticulously listed that the viewer is encouraged to interpret them not only as an illustration of the mapped place but also as a reference to the artifacts to be placed behind the map panels. Such congruence is even more evident once Danti's legends are read in conjunction with those by Bonsignori. Written after 1577, Bonsignori's legends make no mention of the natural and artificial products of the mapped lands, perhaps an indirect indication that the original function of the maps as a visual catalogue of the artifacts kept in the room had already been abandoned. But Danti's legends tell a very different story.

His legend of the map *Hindustan Outside the Ganges* (*Indostan fuori il Gange*), for instance, exemplifies how precious products and marvelous animals were associated with the mapped region (Fig. 4.6):

> This part of India outside the Ganges extends up to the Cathay and it includes many provinces, which contain numerous notable things. In Ergimut in the said province one can find the most perfect musk. In the mountains of Ava one can find very beautiful rubies, and in the mountains of Satgaym, beside rubies, one can also find diamonds. In Caindu and the nearby cities people spend fifteen corals in exchange for money, and in

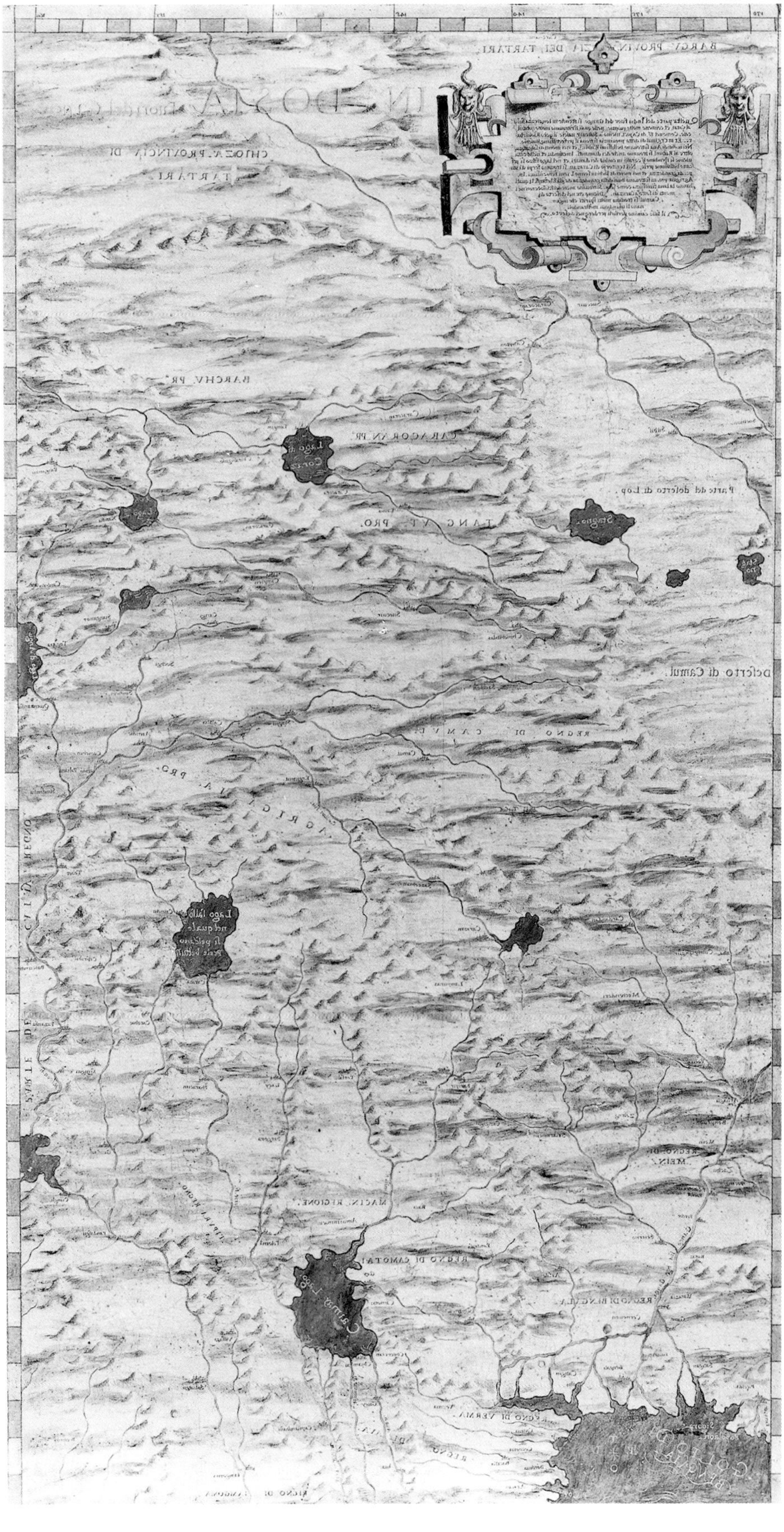

CHIOZA PROVINCIA DI TARTARI
BARCHU PR.
CARACORAN PR.
TANGVT PRO.
Parte del deserto di Lop
deserto di Camul
REGNO DI CAMUL
MACIN REGIONE
REGNO DI BENGALA

> the Salty Lake one can fetch very beautiful pearls. In the territory of Carazan one can find snakes of amazing length, and in the mountains of Lachi there are very ferocious lions. In the province of Agrigaia there are oxen as large as elephants, which have wool as fine as silk. In the mountains below Carazan one can find also rhinoceros. It is said that in the Desert of Camul there are spirits that deceive travelers by showing them the wrong way in order that they will get lost in the desert.[20]

If the spirit of the desert could hardly have been stored behind this map, the musk, pearls, corals, and precious stones, with which Cosimo I's collection was rich, could be easily found there. Also the exotic and marvelous animals of this region could have been present in the room, either metonymically, through a horn, a tooth, a piece of skin, or by derivation, through a product from the animal in question, such as the very soft wool of the "oxen as large as elephants" (*buoi grandi come elefanti*), or by depiction in the planned but never realized frieze below the map. In the legend of the map of China it can be read that the region was renowned for its porcelain (the legend even provides a recipe for the making of porcelain), for the printing press, which "they had before us" (*hanno avuta prima di noi*), for drapery in silk, cotton, and wool, and for powerful artillery, while the Chinese sea is rich in pearls.[21] The legend for the map of Indochina explains that it was an area rich in gold, silver, and precious stones which the King of Portugal brought to Europe in great quantity. India was celebrated for spices and precious stones, Ceylon for cinnamon, rubies, sapphire, topaz, and granites. The mountains of the province of Bedane in central Asia produced "beautiful balas" (*balasschi bellissimi*), orange rubies. The province of Arachosia in Asia was acclaimed for lavender and myrrh. Persia was very fertile and well known for the variety and abundance of flowers, samples of which could either be kept dried in the cupboard or depicted in the planned frieze below the map. Arabia had spices, mines of gold, silver, and many other metals. Anatolia and Syria were famous for their ancient scholars of arithmetic and astronomy, whose books and instruments could be kept in the corresponding cupboard. Livonia and Lithuania were celebrated for leather, sables, and ermines. The British Isles for gold, silver, and especially for the "very refined tin that looks like silver" (*stagno finissimo che pare argento*), Scotland for, "a certain stone that burns amazingly like coal and it is said to be a kind of agate"(*una sorta di pietra, che arde mirabilmente come il carbone, qual dica(n)o ch è una specie di Gagate*). The Island of Madagascar was rich in silver, amber, saffron, cloves, red sandalwood, beside many other spices, flowers, and animals that could have been depicted in the frieze below. The Cape of Good Hope was well known for its monkeys, peacocks, and ivory. New Spain was rich in minerals, gold, silver, copper, tin, and iron, as well as forests of pine trees, cypresses, oaks, and cedars.

The interpretation of the maps of the Guardaroba Nuova as a visual catalogue of the artifacts kept in the room and of the geographical order as a system for their retrieval is suggested by the very layout of the Guardaroba Nuova and by the maps themselves. Such an interpretation is further confirmed by the fundamental analogy between the Guardaroba Nuova and the Studiolo of Francesco I, a cabinet much better documented in all its planning stages, which was designed after the Guardaroba Nuova and which, it should be emphasized, was connected to it by a secret stairway (Fig. 4.7).[22] The comparative analysis of the two Medici cabinets of art and curiosities is germane to the very idea of displaying a collection of artifacts geographically, or simply, of collecting according to maps.

Vincenzo Borghini invented the program for Francesco I's Studiolo at Vasari's request.

4.7 Studiolo of Francesco I. Florence, Palazzo Vecchio

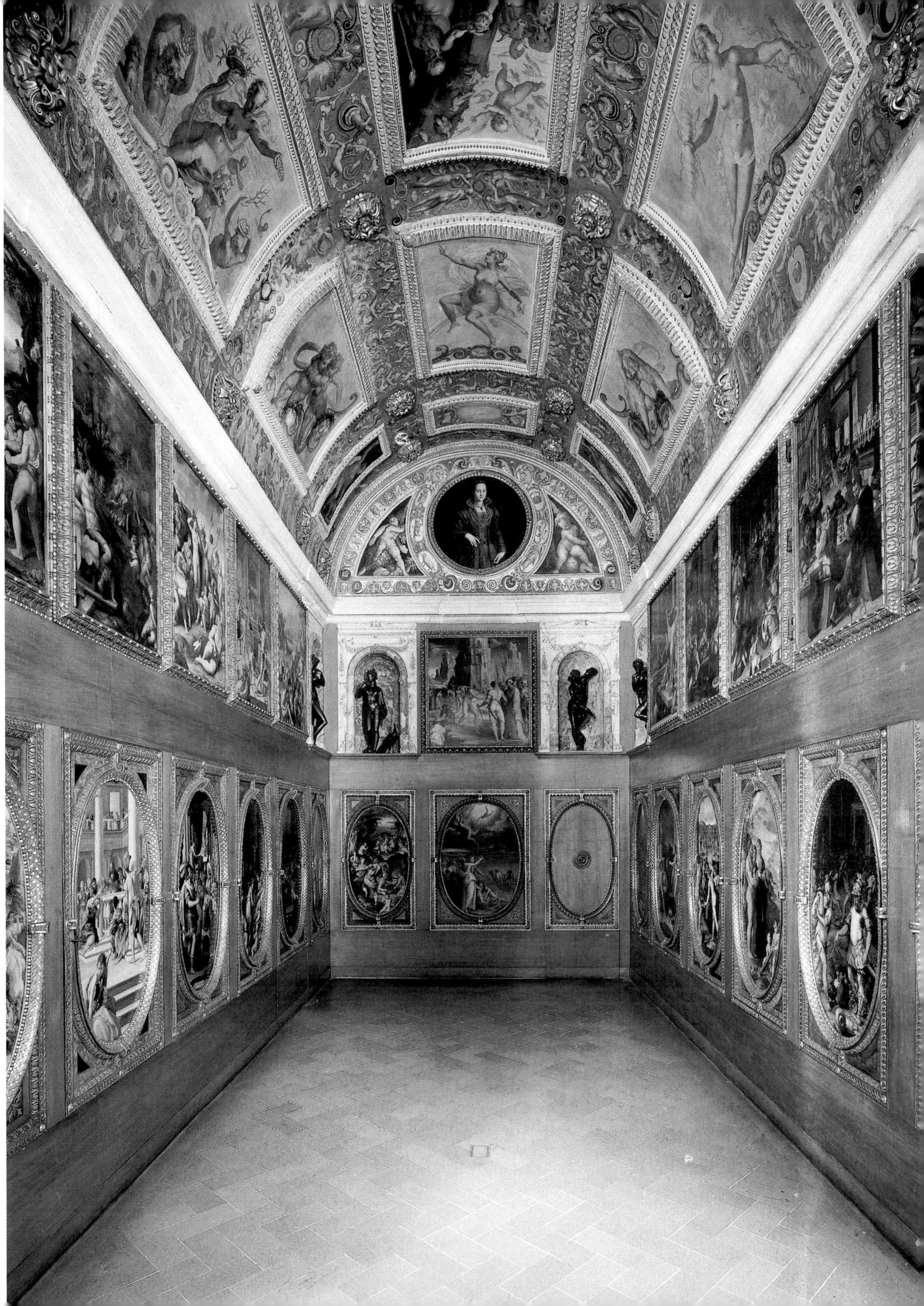

In two famous letters to Vasari, Borghini not only discussed the iconography of the Studiolo based on the four elements, a scheme often used in European collections, but also explained the criteria of his iconographic invention: "It seems to me that the invention ought to conform to the material and quality of the things that have to be stored there, so that it [the invention] will make the room pleasing, and it would not be entirely inappropriate, in fact that it would be a sign, a sort of inventory to retrieve things, referring the figures and the paintings, which are above, around, and on the cupboards, to what the cupboards contain."[23] The iconography of the Studiolo had to be closely related to the objects that were to be kept there. Moreover, the painted doors were a visual catalogue of the collection or, to use Borghini's own words, the invention was, "a sign, a sort of inventory to retrieve things." Obviously, Francesco I's Studiolo and the Guardaroba Nuova shared the idea of connecting the artifacts with the painted doors visually and conceptually.

Borghini's invention for Francesco's Studiolo was inspired by the interaction between nature and art, and was based on the traditional scheme of the four elements, earth, air, fire, and water. Different kinds of objects were connected with each of the elements. Crystals, porcelains, glasses, rings, and precious stones were to be associated with the element of air; horns and bones of animals, medicines and mixtures (*composti*), iron and other metals related to fire; pearls and corals to water; marbles such as porphyry, jasper, chalcedonies, and agates, as well as precious woods such as ebony were connected to earth. This scheme was not unusual for displaying collections of *naturalia*, *artificialia*, and *exotica* in the second half of the Cinquecento. But what made Borghini's invention highly original was the fact that the traditional scheme of the four elements was illustrated by panels representing the making, effects, and myths concerning each element. Moreover, Borghini made clear that the artifacts selected for the Studiolo had to be memorable and rare, "referring always to singular things that bring with them a certain marvel, because common things ought not to be placed in such rooms."[24] In a subsequent letter to Vasari, he made clear that the content of the collection dictated the iconography, not vice versa. Borghini wrote:

> Above all I need to know that the objects, which have to go in the cupboards, have been defined (take note of all this part that I am about to say) so that I can adjust the exterior paintings to the things inside; if the cupboards correspond to the things that are above, they would come even better, that is: things relating to fire, such as distillations and artifacts molded with fire in iron, steel, glass, and similar materials, would be under the statue of Vulcan and the element of fire; medals, artifacts in gold, rubies and minerals of similar things under [the statue] of Pluto and Ops and the element of earth; similarly, jewels under Juno, crystals under Zephyrus, pearls under Venus, and so forth; having established this partition, I need to know how many stories go in each cupboard and what it will contain. For instance, in the first cupboard, which has two exterior panels to be painted, will be located precious medicines, such as balsams, unicorn's corn, remedies against poisons, and similar things; in another [cupboard] there is only one panel and inside there will be medals, and so forth.[25]

In Borghini's invention, the four elements and the objects were further inserted in a cosmographical scheme of correspondence, which included the four cardinal points, pagan gods and goddesses, the dichotomy male–female, and much more. But in the letter dated

3 October 1570, Borghini also discussed another element, which is germane to the iconographic invention of the Guardaroba Nuova. In the effort to provide both a coherent program and a variety of subjects that would please Francesco I, Borghini suggested painting, "female personifications of the provinces: for example, in Germany there are excellent clocks and machines, and in France beautiful works made out of gold; on the cupboards where similar objects were to be kept there could be painted, in a little panel, a personification of Germany holding these clocks and machines in her hands, as if it were the reverse of a medal or other similar things."[26] Later Borghini discarded the idea of representing personifications of the provinces on the panels as unsuitable to the physical space of Francesco's Studiolo. But his association of precious objects with regions of the world, which incidentally also had ancient precedents in Strabo and Pliny, further supports the interpretation suggested here that at the Medici court artifacts could be grouped according to place and, consequently, displayed according to maps.

In the Guardaroba Nuova Cosimo I imagined a unique cabinet of art and curiosities in the panorama of Renaissance collections, one that not only was organized geographically but that would have also given unprecedented display to his exotica. In fact, Cosimo and his advisers must have had in mind a specific group of artifacts when they envisaged the geographical arrangement of the room. Borghini's letters on Francesco I's Studiolo clearly stated the mode of proceeding: first to select a body of objects to be stored in a cabinet of art and curiosities and then devise the iconography based on the nature and character of the selected objects, not vice versa. The geographical order of the Guardaroba Nuova presenting 39 maps of non-European regions and only 14 maps of Europe must have seemed particularly suitable to the proper display of Cosimo's artifacts from remote regions. Indeed, it would have greatly enhanced the visibility and relevance of his exotica within the Medici collection while, at the same time, integrating them in the order of the world.

Cosimo I's Exotica

Exotic goods and artifacts featured in every European collection of some distinction. Loads of exotica were brought to the ports of Europe by Portuguese ships sailing the new sea-route to Africa, Brazil, India, Indochina, and the islands of South East Indies: spices and slaves as well as beautifully crafted objects in metal, ivory, and precious stones, unusual plants and animals, and foreign arms. Also the Spanish ships brought a variety of artifacts from the West Indies: gold and silverware, which were systematically melted down and converted into money; jewels in precious and semi-precious stones, which were almost invariably dismantled and mounted anew to conform to European fashion; and feather artifacts. Feather works, oliphants, damascene metalworks, Meso-American masks, but also Chinese porcelain, and Persian carpets were among the ultimate exotica that filtered into European collections. Coming from distant lands and made out of unusual materials or unknown crafts, these exotica carried the components of the marvelous. They were collected and displayed along with the marvels of the ancient and pagan world, and praised for their craftsmanship, strangeness, and rarity. Metonyms of a distant world, such exotica evoked fragmentarily but tangibly lands far away.[27]

Exotica were also showcased as symbols of their owner's network of friendships, political alliances, and diplomatic ties. Some exotica, such as featherworks, acquired value, at least initially, because of their close connection to the emperor. The Spanish conquistador

Hernando Cortez had sent feather artifacts in his first shipment from the New World to the emperor who, in turn, donated them to select allies and friends. From the imperial collections, these unusual artifacts spread to the royal collections in Munich, Dresden, Prague, Ambras, Florence, and in the private cabinets of university professors and scholars.

Cosimo shared this wide interest in exotica. He imported samples of unusual plants for his botanical garden, and strange animals for his menageries; he went to a great deal of trouble to acquire manuscript maps and travel reports documenting the new voyages and discoveries, and used his best diplomatic and political skills to gather artifacts from remote lands. His collection of exotica became one of the most extensive of the time, a fact particularly significant since he had no direct control of the trading routes that brought these items to Europe. The Medici duke had inherited some objects; others he obtained as diplomatic gifts or as wedding presents; still others he acquired through his agents in Lisbon and Seville, and some he purchased in Egypt, such as Chinese porcelain and Islamic artifacts. His collection was so well stocked that he, like the emperor, could donate precious items to others, including a rare "idol in green stone" (*idolo in pietra verde*) he sent to the Duke of Bavaria. He shared the passion for exotica with his sons. Francesco shrewdly seized any opportunity to increase the family collection. In 1572, when a Spanish ship docked accidentally in Livorno on its way back from the West Indies, Francesco forced the captain to sell the entire load of exotica. He kept for himself some of the peacocks, parrots, monkeys, birds, llamas, feather artifacts, and objects in precious and semi-precious stones, but a significant portion he sent as a gift to Duke Albrecht of Bavaria. In the 1580s, Ferdinando acquired important Mexican manuscripts and used their imagery to decorate one of the rooms of the Uffizi Palace.[28]

Inventories of Cosimo's ducal palace recorded myriad marvelous non-European artifacts, variously and unsystematically labeled as oriental, Moorish, Indian, damascene, and Turkish (*orientali*, *moreschi*, *indiani*, *domaschini*, and *turcheschi*). Many of his exotica were frequently used to furnish the ducal residences and embellish the Medici dinner table, and therefore they were stored in the general guardaroba. But rare pieces of Chinese porcelain, Turkish weapons and armory, African oliphants, feather works, Mexican masks, and Aztec statuettes were treated as collector's items and kept in the secret guardaroba. These highly prized objects would have found a more appropriate and effective display in the Guardaroba Nuova. Although most of Cosimo's exotica have not survived, the few surviving pieces still preserved in Florentine museums suggest what kinds of marvelous artifacts might conceivably have been displayed in the Guardaroba Nuova.

Among the four hundred pieces of Chinese porcelain that Cosimo owned, a few items stood out as spectacular collector's items. The celadon plate and vase, for instance, now in the Museo degli Argenti in Florence, were exceptional both for their material and previous ownership (Fig. 4.8). Not only were they rare examples of the Chinese gray-green porcelain of which only a few pieces were available in Europe, but they belonged to Lorenzo the Magnificent who, in turn, had received them as a gift from the Sultan of Egypt in 1487. Among the items ransacked in 1494, these objects had been reacquired by the Medici, possibly by Cosimo I himself, who jealously kept them in his secret guardaroba. Samples of Chinese porcelain of high quality and distinction such as these pieces could conceivably have been displayed more effectively in the Guardaroba Nuova, specifically behind the map of China, the cartouche of which, it, may be recalled,, supplied the recipe to manufacture porcelain (Fig. 4.9).[29]

Cosimo's collection was also rich in non-Chinese oriental artifacts that were variously labeled in the inventories as Turkish or damascene. Turkish knives, weapons, and arms were kept in the ducal armory, but many oriental artifacts were kept in the secret guardaroba, some were even stored in the twelfth cupboard, such as the two incense burners that he had earlier selected for display in the Studiolo di Calliope. Among Cosimo's large selection of Turkish artifacts, particularly fascinating is the falcon hood mounted on an ivory stand with the Medici–Toledo coat of arms (Fig. 4.10). Although this was not a unique piece in the collection – the 1553 inventory recorded 40 of them – its precious ivory stand bearing the Medici–Toledo coat of arms is an unmistakable sign that the piece was regarded as worthy of display. Objects like this falcon hood would have found an appropriate location in the Guardaroba Nuova behind the maps of Anatolia, Armenia, and Persia, perhaps together with samples of oriental precious and semi-precious stones, or even with samples of spices kept in beautiful vases (Fig. 4.11).[30]

Exotica from the West Indies were also well represented in Cosimo's collection. No record exists of how the duke acquired his American exotica, but the fact that they were featured among his collection in 1539, the date of his marriage to Eleanor, suggests that they might have been prestigious wedding gifts from the Spanish. Cosimo had a few cloaks of feathers, now lost (those presently in Florentine collections were added later by his sons and successors), and masks, one example of which still survives. It is a wood Mexican mask covered with a mosaic of tesserae in turquoise, mother-of-pearl, shells, and jadeite (Fig. 4.12). In 1553, this mask was kept in the twelfth cupboard of the secret guardaroba together with Aztec animal heads in semi-precious stones, Turkish objects, and statuettes from many times and places. While the Aztec miniature animal heads were too small to be located in the Guardaroba Nuova and were in fact selected for display in the smaller Studiolo di Calliope, the other larger artifacts, masks, and feather garments and covers could have been effectively preserved behind one of the map panels illustrating the West Indies (Fig. 4.13).[31]

Among the duke's exotica were also numerous oliphants that the 1553 inventory simply recorded as horns "after the Moorish manner" (*alla moresca*), but which are actually the oldest Portuguese–African oliphants documented in European collections. These oliphants made from an elephant's tusk were crafted in the Congo or Central Africa and adapted to Portuguese fashion by inserting holes to carry them. They must have arrived in Florence as gifts from the Portuguese, possibly as diplomatic gifts to the Medici Pope Leo x, whose financial support the Portuguese sought in 1514. One oliphant bearing the Medici–Toledo coat of arms (Fig. 3.14) may have come as a gift from Eleonor's side of the family. Eleonor herself enjoyed the daily use of Portuguese–African artifacts and kept in her rooms ivory spoons, typical artifacts that the Portuguese commissioned from local craftsmen from the city of Benin in Nigeria, which they ruled. Later these oliphants were displayed in the armory of the Uffizi. Although the Medici inventories never suggested their African provenance, thus making it difficult to assert that they were regarded as African artifacts, still it is possible that their material and craft might have been sufficient reasons for an educated collector, if not the compiler of the inventory, to differentiate the provenance of the Moorish horns from the Moorish featherworks, and thus assure their display behind the maps of Africa with similar artifacts (Fig. 4.15).[32]

Indeed, the order of the maps in the Guardaroba Nuova appears suitable to accommodate the loose geographical provenance of artifacts recorded in Medici inventories. In the

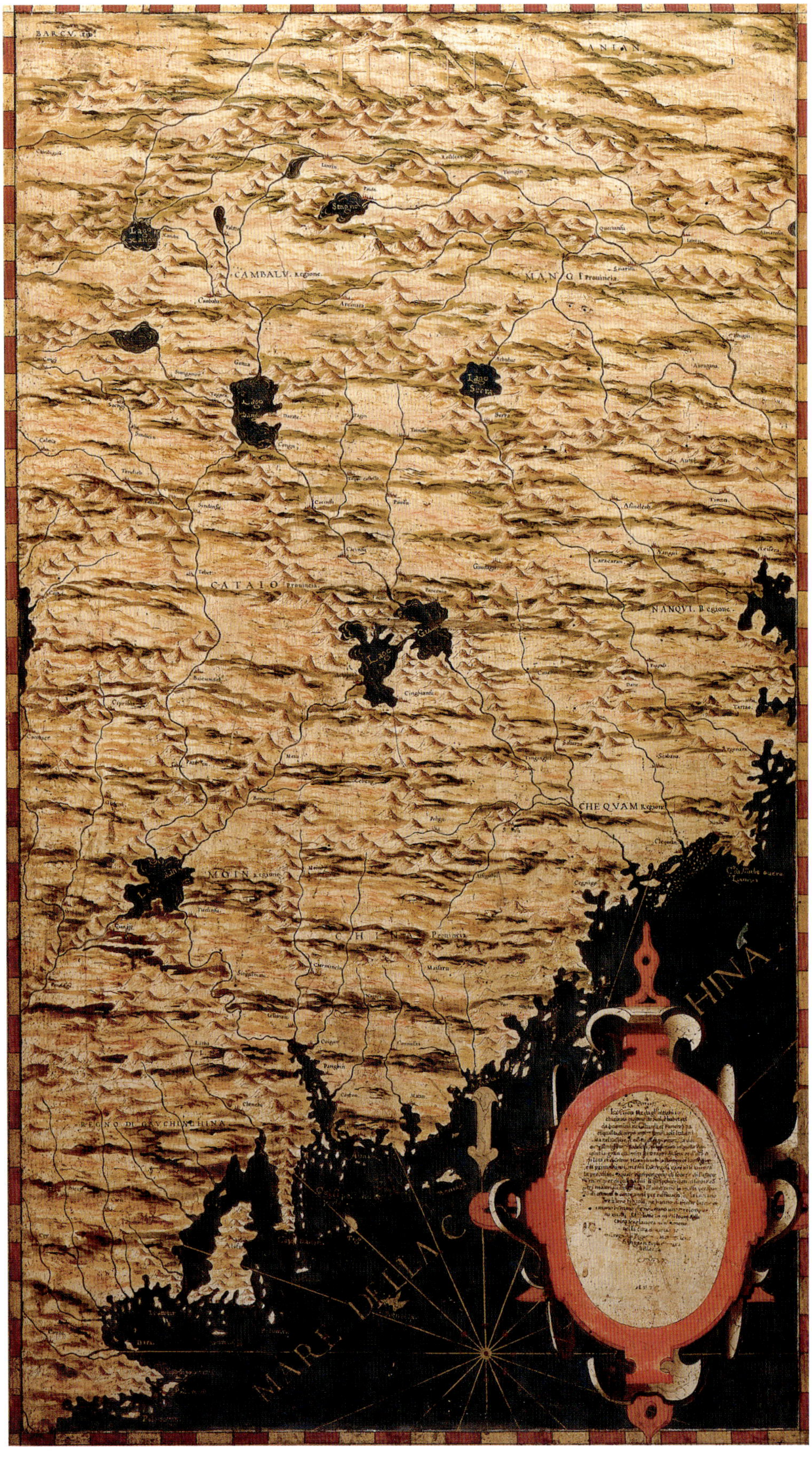
CAMBALV Regione
MANGI Provincia
CATAIO Provincia
NANQVI Regione
CHEQVAM Regione

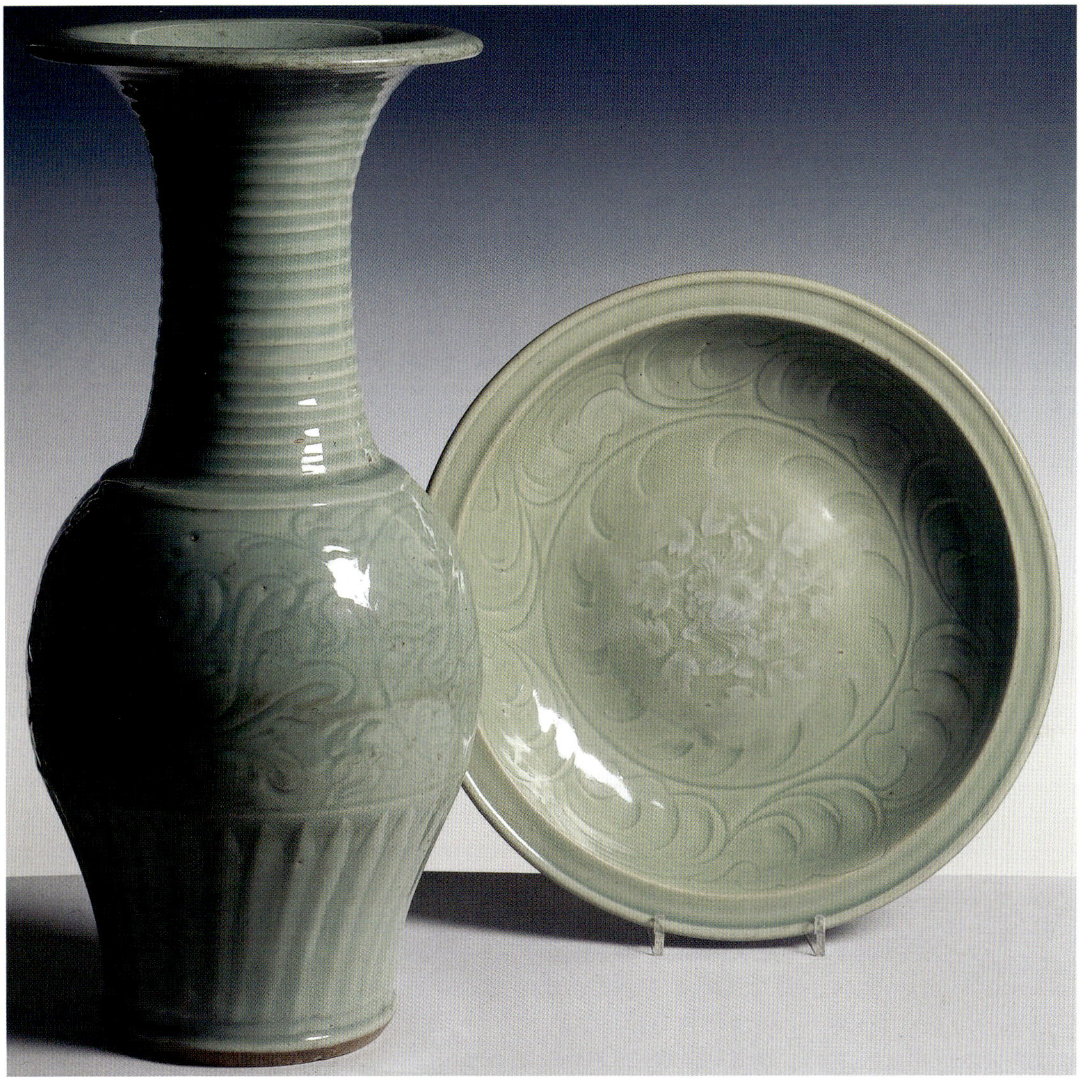

4.8 Celadon Plate and Vase, China fourteenth to fifteenth century, porcelain. Florence, Museo degli Argenti

Guardaroba Nuova, the maps of Asia and America appeared on the doors of the same cupboards so that any object coming "from the Indies" (*dalle Indie*) could be stored there with no further specification of whether it came from the East or the West Indies. Maps of Europe and Africa shared the same cupboards. Also in this case a rather loose arrangement must be imagined, although the European artifacts would have been distinguished immediately from African ones. On the other hand, the connection between some artifacts and a region of the Old World might have facilitated a more rigorous arrangement for some European products, such as the connection between glassware and France, or scientific instruments and Germany mentioned by Borghini. The arrangement of the four continents in the

4.9 (*facing page*) Egnazio Danti, *China* (Map of China), 1575, panel. Florence, Palazzo Vecchio, Guardaroba Nuova

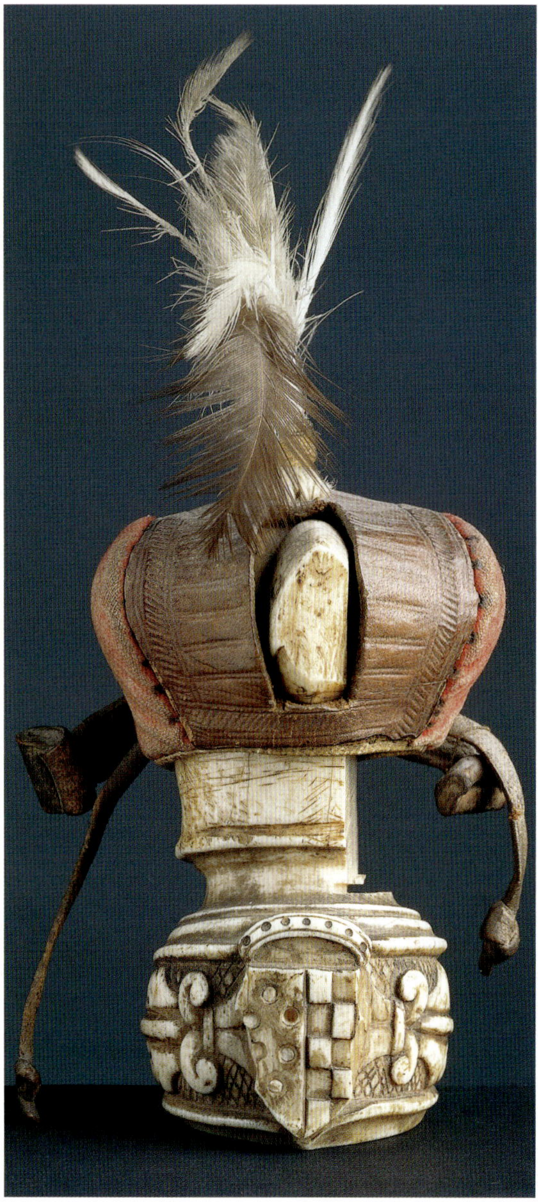

Guardaroba Nuova corresponded to the way the world was represented in Renaissance world maps: Europe and Africa shared the center while Asia and America occupied the periphery to the east and the west respectively.

The geographical display adopted in the Guardaroba Nuova and the use of maps as a visual catalogue to the artifacts kept in the room was a novelty in European cabinets of art and curiosities, but a novelty that relied on the Renaissance use of Ptolemy's geographical order to organize encyclopedic knowledge.

Ptolemy's Geographical Order

Ptolemy's geographical order was well known in the fifteenth and sixteenth centuries thanks to the diffusion of his *Geography*, the most authoritative geographical text in Renaissance Europe. Unknown to the Latin West during the Middle Ages, Greek manuscripts of Ptolemy's *Geography* arrived in Florence at the very end of the fourteenth century together with Manuel Chrysoloras, the scholar of the Greek language, who planned a Latin translation which was eventually completed by his pupil Jacopo d'Angelo in 1406. The eight books of Ptolemy's geographical manual aroused an immediate interest among scholars, humanists, and merchants, who were all obsessed with the measurement of the universe, the philology of ancient texts, and the reconstruction of the classical world. Ultimately, Ptolemy's geographical manual had a lasting influence on Renaissance mapping comparable only to the significance that Pliny's *Natural History* had on natural philosophy and Vitruvius's *Ten Books on Architecture* on Renaissance architecture.[33]

4.10 Falcon hood with ivory stand bearing the Medici-Toledo coat of arms, ca. 1550, leather, ivory and feathers. Florence, Museo Nazionale del Bargello

4.11 (*facing page*) Egnazio Danti, *Armenia* (Map of Armenia), not dated, panel. Florence, Palazzo Vecchio, Guardaroba Nuova

The major contribution of this second-century manual of geography to map-making consisted in connecting terrestrial locations to astronomical observations: places on earth were located on an imaginary spatial grid that covered the earth's globe according to theoretical coordinates of latitude and longitude. Book One started with a theoretical discussion of geography, distinguishing geography, the graphic representation of large parts of the world,

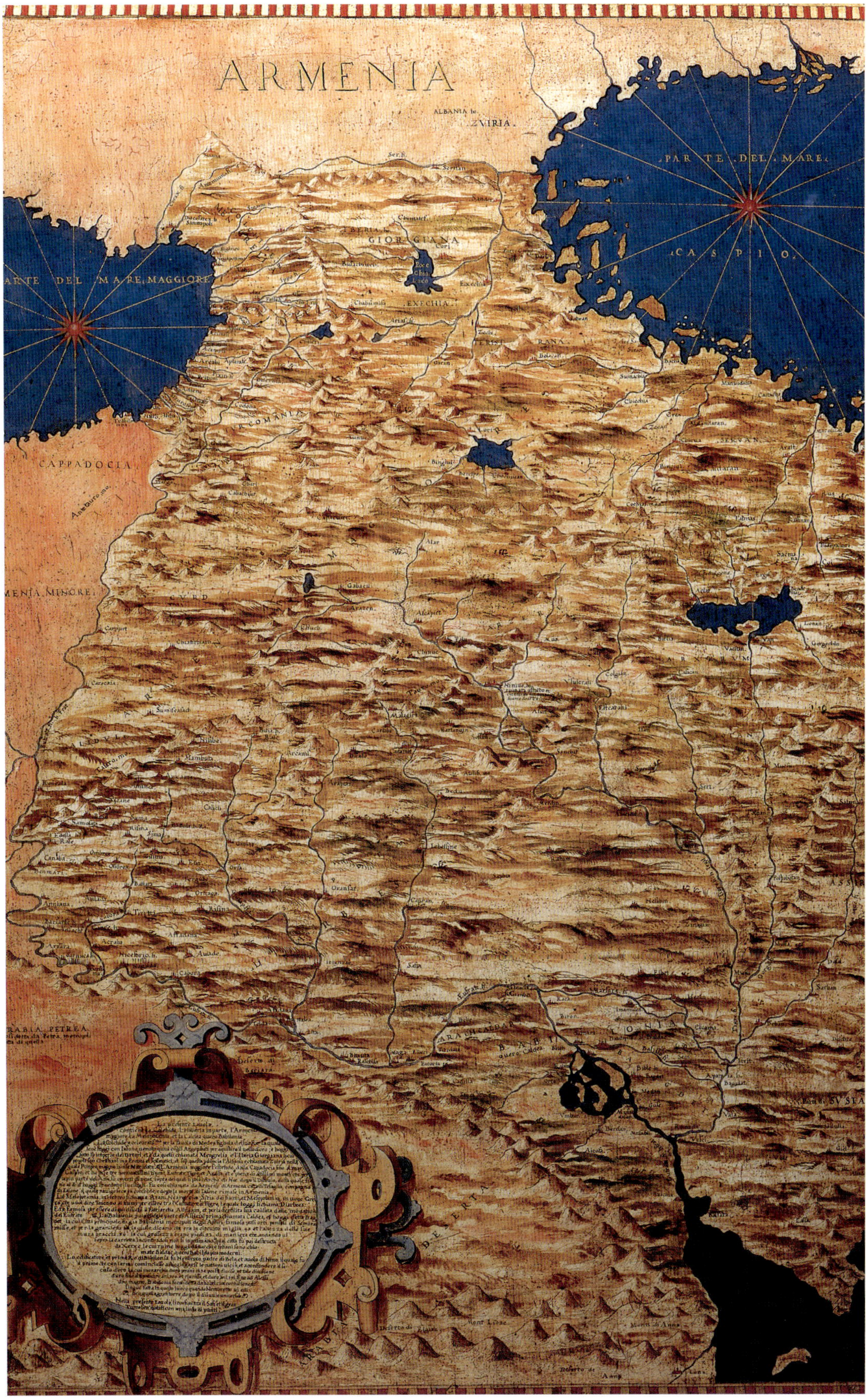
ARMENIA
ALBANIA
ZVIRIA
PARTE DEL MARE
CASPIO
PARTE DEL MARE MAGGIORE
GIORGIANA
CAPPADOCIA
ARMENIA MINORE
SERVAN

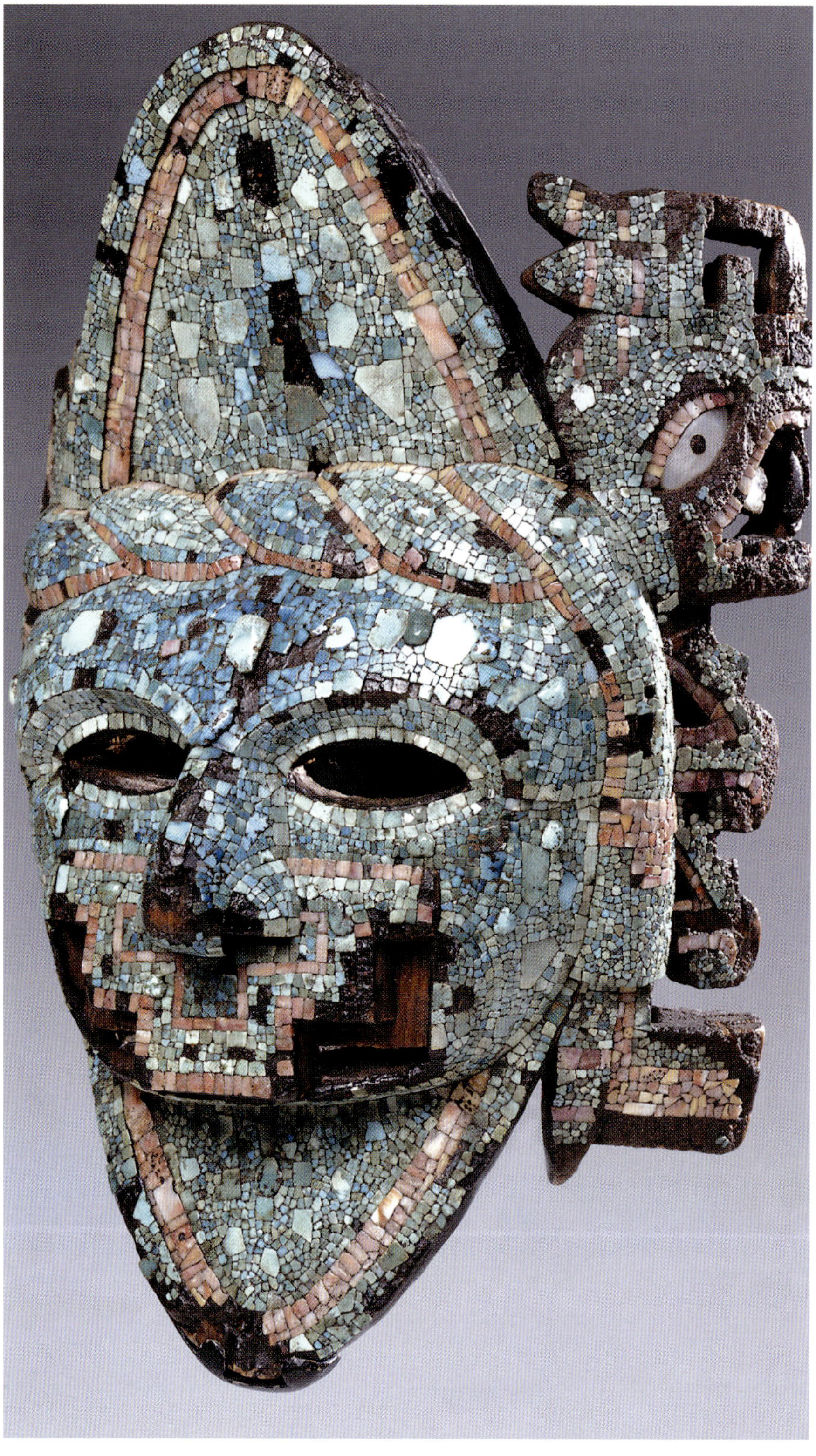

4.12 Mixteco-Pueblo Mask, wood and mosaic in turquoise and shell. Rome, Museo Nazionale Preistorico Etnografico L. Pigorini (MPE 4213)

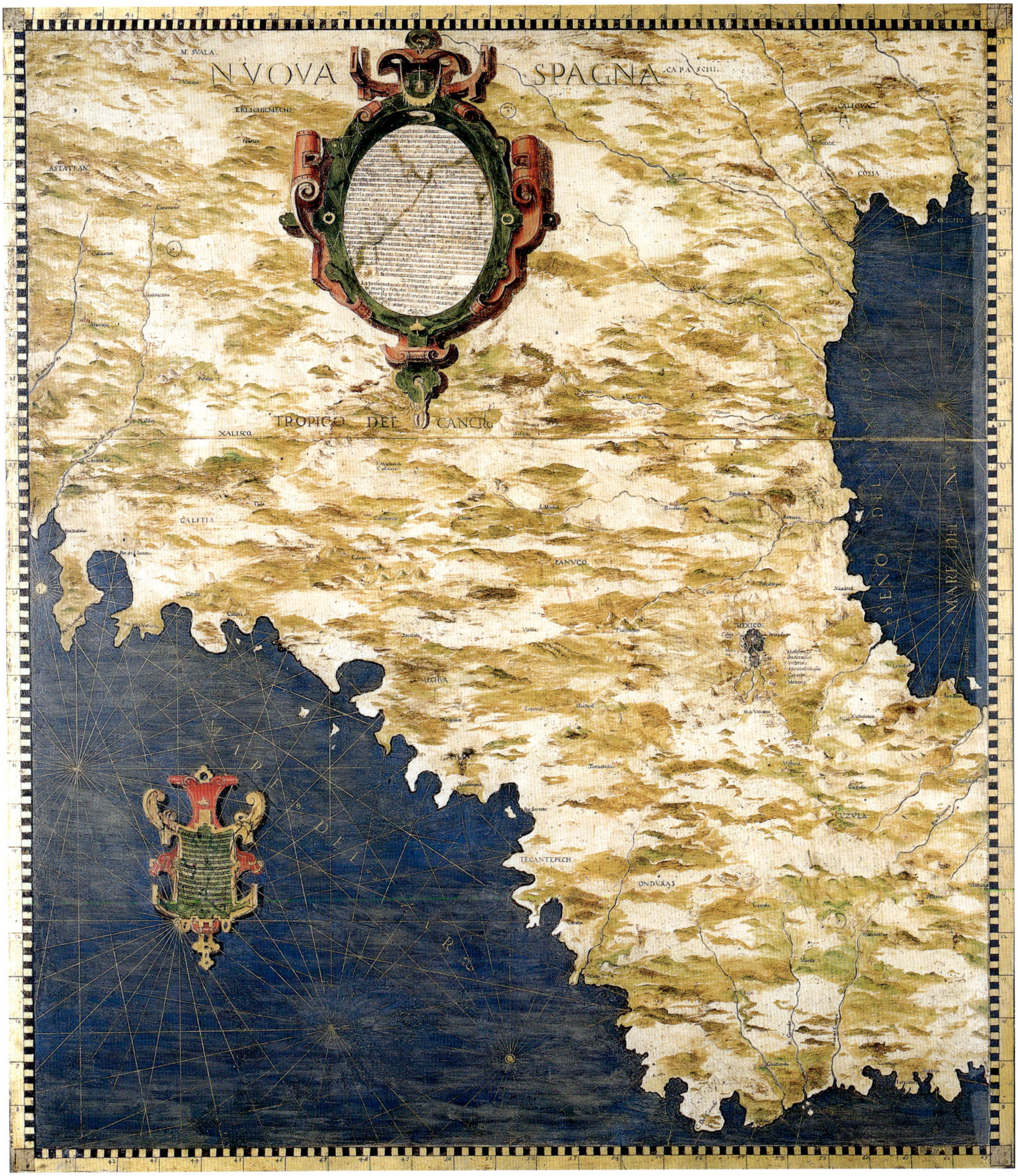

4.13 Egnazio Danti, *Nuova Spagna* (Map of Mexico), 1564, panel. Florence, Palazzo Vecchio, Guardaroba Nuova

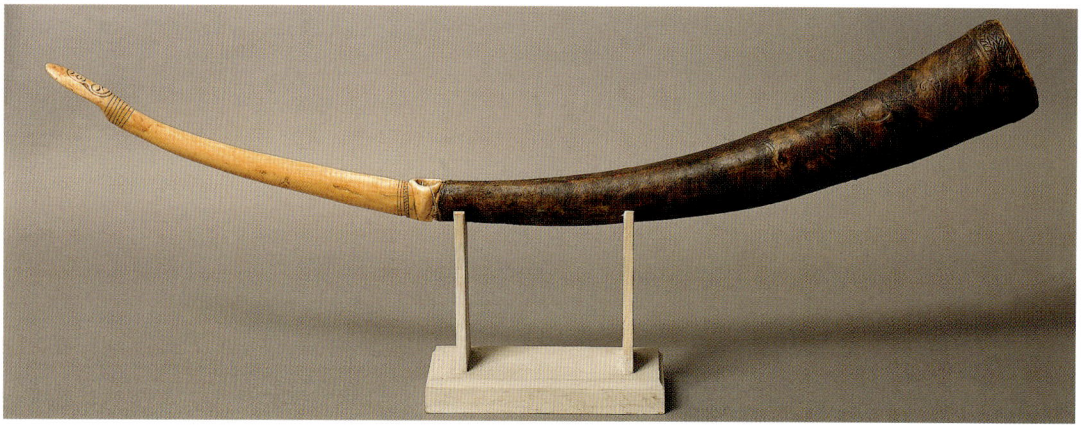

4.14 Oliphant bearing the Medici-Toledo coat of arms, Central Africa and Florence, ca. 1539. Florence, Museo di Etnografia e Antropologia

from chorography, the graphic representation of smaller parts of the world. It included an explanation of different projections that maintained the proportions of distances between the sphericity of the earth and the flat surface of the map. The remaining seven books included a brief verbal description of the world divided into three parts and a list of geographical coordinates for about 8000 places of the known world, which also had a tripartite organization.

Indeed, Ptolemy's geographical manual was organized according to the three continents that made up the known world: Europe, Africa, and Asia. Ptolemy discussed the known world as a whole in a general table, which he called the "table of our world" (*tabula nostri orbis*) at the end of Book Seven, but then divided it into three continents, which in turn were further divided into smaller parts. In Book Eight, the last book of the manual, each region of the world was discussed independently and introduced by a separate heading: "First Table of Europe," "Second Table of Europe," and so forth for each continent. Ptolemy did not number the "table of our world," but he inserted the tables that followed it in a tight sequence. So the first table of Europe always corresponded to the British Isles, the second table of Europe to Spain, the third to France, and so forth. Similarly, the first table of Asia always represented Anatolia, and the last table of Africa always represented central Africa.

Ptolemy's geographical order moved from the general to the particular, from the world map to the maps of continents and regions, and in this respect, it was structurally different from the circular order of a periplus, a text describing the circumnavigation of islands and continents, that started and ended in the same place. Ptolemy's order fostered the progressive move from the global to the local and the syncretistic view of the relationships between the whole and its parts. It also composed the different parts of the earth's globe in the coherent architecture of the universe.

Ptolemy's sequence of tables for the world's regions was fixed in the text and thus fully conveyed in the manuscripts that filtered into the West, including in those that did not include maps. Renaissance editors, however, interpreted Ptolemy's "tables" as maps, even though the word "table" could refer to both a text and an image. Thus they illustrated their

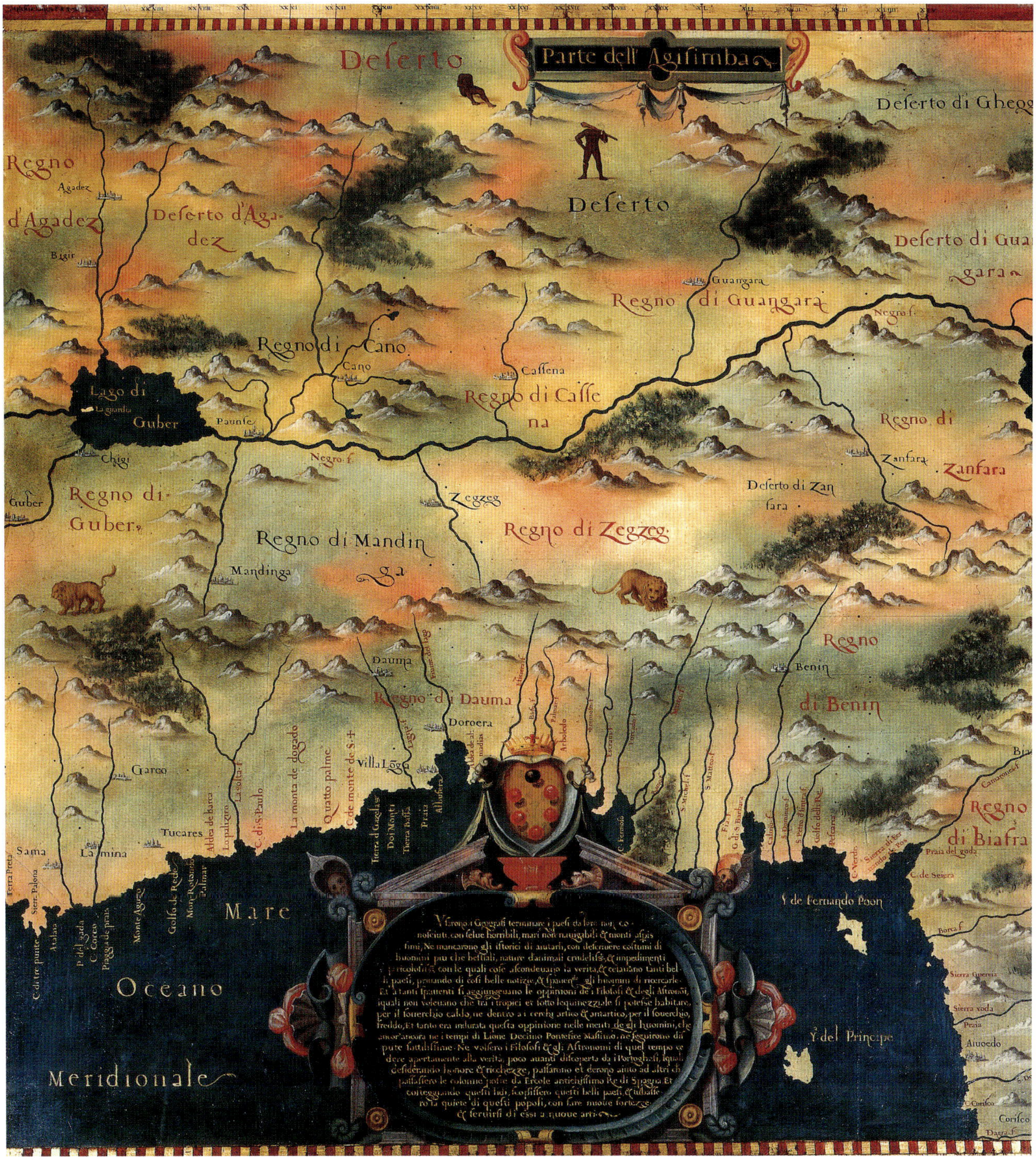

4.15 Stefano Bonsignori, *Parte dell'Agisamba* (Map of Côte d'Avoire), 1580, panel. Florence, Palazzo Vecchio, Guardaroba Nuova

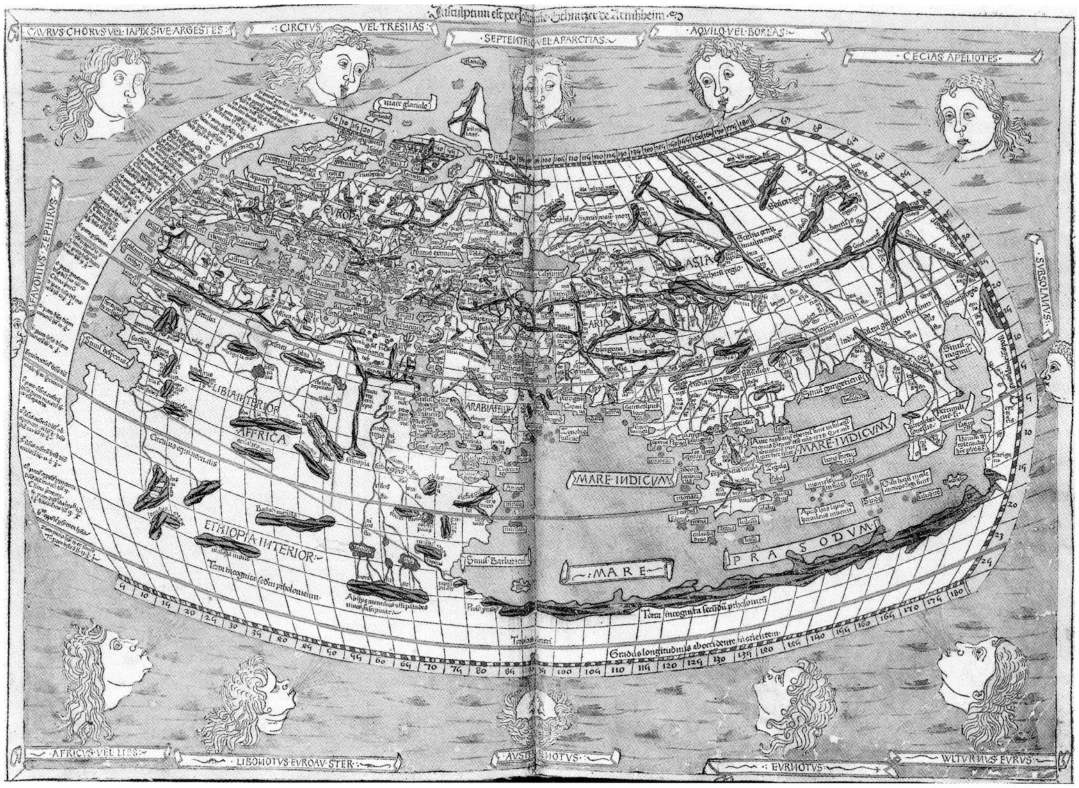

4.16 Map of the Known World, woodcut, from Ptolemy, *Geographia*, Ulm 1482. Princeton University Library

editions of Ptolemy's *Geography* with one world map (Fig. 4.16), which was not numbered, and with twenty-six regional maps: ten maps of Europe, four of Africa, and twelve of Asia. Moreover, the Renaissance editors reinforced the connection between text and maps that was implicit in Ptolemy's manual: they inscribed the pertinent text from Book Eight on the back of each map, and used the heading of each section in that book as the title for each corresponding regional map.[34]

Significantly, however, Ptolemy's sequence of maps was adopted not only for the Renaissance editions of his text, but also for the mapping of the world Ptolemy did not know about. As new lands were discovered in the fifteenth and sixteenth century, they were interpreted as extensions of the Ptolemaic world to the north and the east, and maps illustrating Northern Europe, Central and South Africa, Southeast Asia, and the West Indies were integrated within the Ptolemaic world. Later, when sailors, scholars, and map-makers realized that the West Indies were physically separated from Asia and so large as to constitute a continent in itself, they added a fourth part to Ptolemy's traditional three. Ptolemy's rigid order was expanded to include the continent of America and modern maps were added to the canonical 26 maps (in addition to the world map) that Ptolemy had made.[35]

It is this modified Ptolemaic order integrating ancient geography with modern discoveries that was widely used in the sixteenth century, both in the editions of Ptolemy's *Geography* as well as in cosmography books, atlases, and in the Guardaroba Nuova.

Indeed, the Renaissance editions of Ptolemy's *Geography*, which were highly praised as reprints of an ancient text, were also valued as the repository of the mapping of the modern world. Although the geographical information provided by the modern voyages and discoveries was disseminated in the numerous printed maps that appeared in great numbers during the fifteenth and sixteenth centuries, the most comprehensive assemblage of modern maps was to be found in the manuscript and printed editions of Ptolemy's text.[36] In 1428 the French Cardinal Guillaume Fillastre added a modern map of northern Europe, a region unknown to Ptolemy, to his manuscript copy. A few years later, the Florentine artist Pietro del Massaio started to make seven additional modern maps for the copies of Ptolemy's text that he had prepared, a practice that became routine in his workshop. In 1482 the Florentine poet Francesco Berlinghieri added for the first time four modern maps (*tabulae novellae*) to the canonical twenty-six Ptolemaic maps in the printed editions of Ptolemy's *Geography* (Florence, 1482).[37] In the following decades, the modern maps (*tabulae novellae* or *tabulae modernae*) increased steadily from edition to edition. Eventually they outnumbered the Ptolemaic maps, while at the same time Ptolemy's text was complemented with modern commentaries conveying new information on projections, instruments, travels, and lands. For example, to mention only the Italian translations closest to the planning of the Guardaroba Nuova, the pocket-size publication edited by Giacomo Gastaldi in Venice in 1548 included 34 modern maps and an extensive commentary.[38] The Venetian edition of 1561 published by Girolamo Ruscelli included 37 modern maps and a verbose commentary, remaining for a long time the most complete set of modern maps available in print.[39]

The way the *tabulae modernae* were added to the original ancient text is extremely telling in understanding how Ptolemy's geographical order was adapted to the expanding world of the Renaissance (Figs. 4.17 and 4.18). First, the modern maps mimicked Ptolemy's originals in their visual rendition as well as in their order. They adopted the Ptolemaic grid and one of the Ptolemaic projections, and followed his order from the general to the particular (the general map of a continent came before its regional maps). Within each continent the modern maps kept Ptolemy's numbers (the first modern map of Europe represented the British Isles, and so forth). In most cases, the modern maps also included descriptions of the mapped country printed on their back, although these description tended to be more verbose and eclectic than those of their Ptolemaic counterparts. Second, in some editions, instead of being published as a separate appendix at the end of the book, the *tabulae modernae* were placed in correspondence with the pertinent Ptolemaic map, so that, for example, the modern map of northern Europe followed Ptolemy's maps of the old continent, and the map of the newly discovered Moluccas Islands completed Ptolemy's maps of Asia. In this respect, Gastaldi's edition of 1548 is significant. By intercalating the modern maps with the Ptolemaic maps Gastaldi endowed modern cartography graphically and conceptually with Ptolemy's authority.

In the course of the sixteenth century, Ptolemy's expanded geographical order started to be used extensively to organize encyclopedic collections of texts. Sebastian Münster adopted it in his *Cosmographia universalis* (Basel, 1544), the bestseller of Renaissance cosmography books (Fig. 4.19). This was an example of the ultimate Renaissance encyclopedia that adapted the traditional scheme of medieval compendiums of universal knowledge to the new passion for geography. In the century after its publication over 40 editions were produced in German, Latin, English, French, Italian, and Bohemian. From edition to edition, it grew in pages, images, diagrams, and indices. Even the author's great emphasis on Germany (about half of

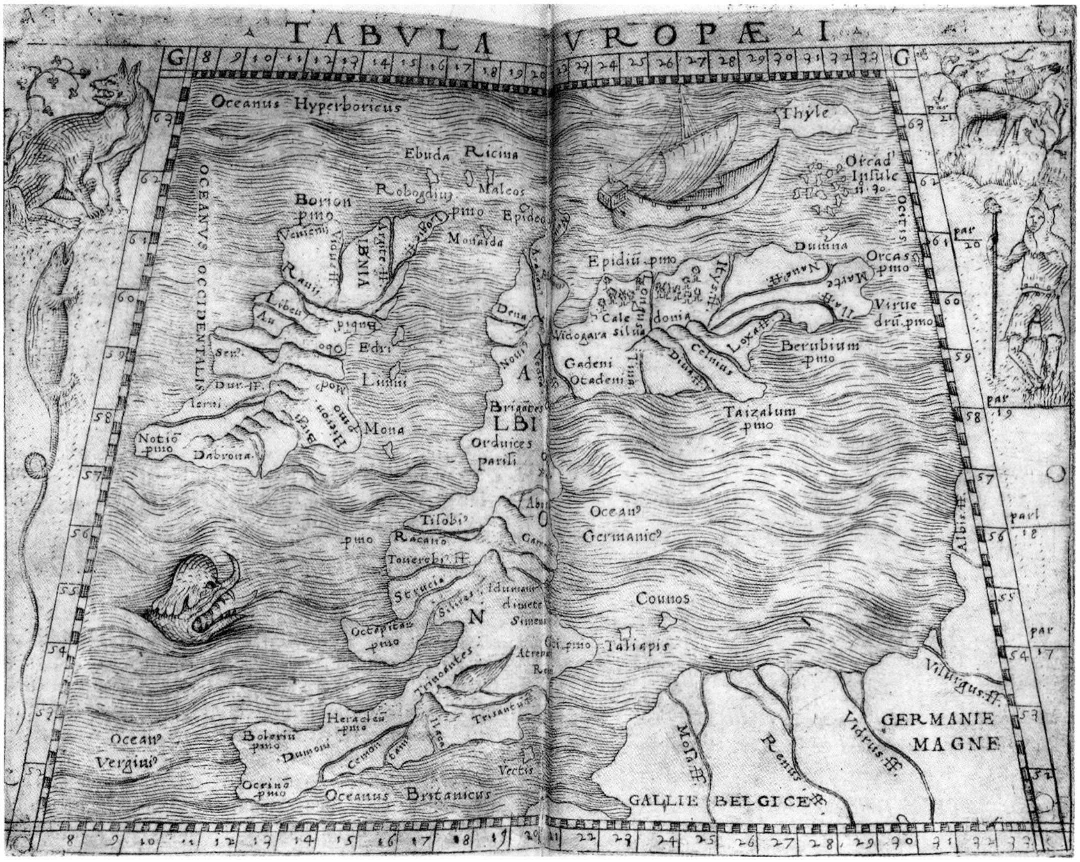

4.17 *Tabula Europae I* (First Map of Europe), engraving, from Ptolemy, *Geografia*, Venice 1548. Houghton Library, Harvard University (Typ. 525.48.715 vol. 2)

the book is dedicated to Germany), and his numerous anti-Roman and anti-Catholic remarks did not limit its circulation. In 1575, 25 years after its original publication, the papal administrators opted to censure it rather than ban it altogether. Münster's *Cosmographia universalis* had become the tool of those who wished to learn about the world without the inconvenience of traveling, as well as the textbook of descriptive geography for young men and women who prepared for their political and social lives, and the desired travel companion of those who, like Michel de Montaigne, traveled in foreign lands. Because of its wide success, Münster's formula for organizing encyclopedic knowledge reveals the full implications of Ptolemy's geographical order for Renaissance culture.[40]

Like other compilers of encyclopedias, Münster ransacked the writings of earlier authors on history, monsters, marvels, ethnography, zoology, botany, and much more. But unlike his predecessors, he imaginatively reorganized the selected fragments of universal knowledge geographically, adopting Ptolemy's order. From the cursory description of the creation of the cosmos and its division into climatic zones, Münster moved to the description of the regions of the world: Europe, Asia, Africa, and America. He complemented his geographi-

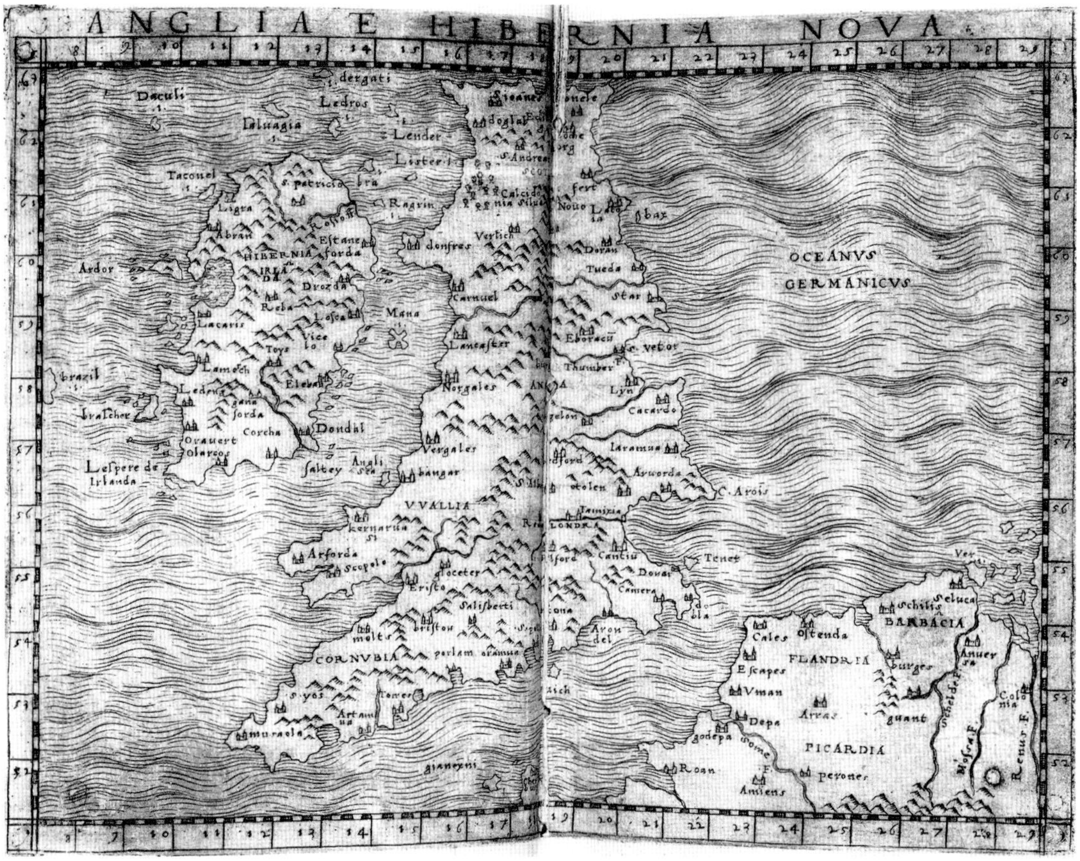

4.18 Giacomo Gastaldi, *Anglia et Hibernia Nova* (Modern Map of the British Isles) from Ptolemy, *Geografia*, Venice 1548. Houghton Library, Harvard University (Typ. 525.48.715 vol. 2)

cal narration with an ample visual apparatus of maps, city views, portraits, coats of arms, genealogical trees, ancient inscriptions, coins, and medals. In Münster's *Cosmographia universalis*, geography explained the actions of people, and included the descriptions of the world's inhabitants, their buildings, customs, animals, plants, and products. Geography was not ancillary to history, but rather, it was "a discourse of memory in relation to history" or, as might be said today, the very structure of history.[41] In this book, Ptolemy's geographical order worked as a system to organize universal knowledge, while the maps functioned as a visual aid to its memorization and retrieval.

Similarly, Abraham Ortelius adopted Ptolemy's geographical order in his book *Theatrum orbis terrarum* (Antwerp, 1570). His geographical order was slightly different from Ptolemy's sequence, since Asia was listed before Africa, but still he regarded it as truly Ptolemaic because it moved from the general to the particular, from the map of the world to the maps of continents and regions. Unlike Münster, Ortelius relied primarily on maps, eliminating other kinds of images and limiting the verbal descriptions of places to the length of a page or two, which were printed on the back of each map. For this innovative map book, gener-

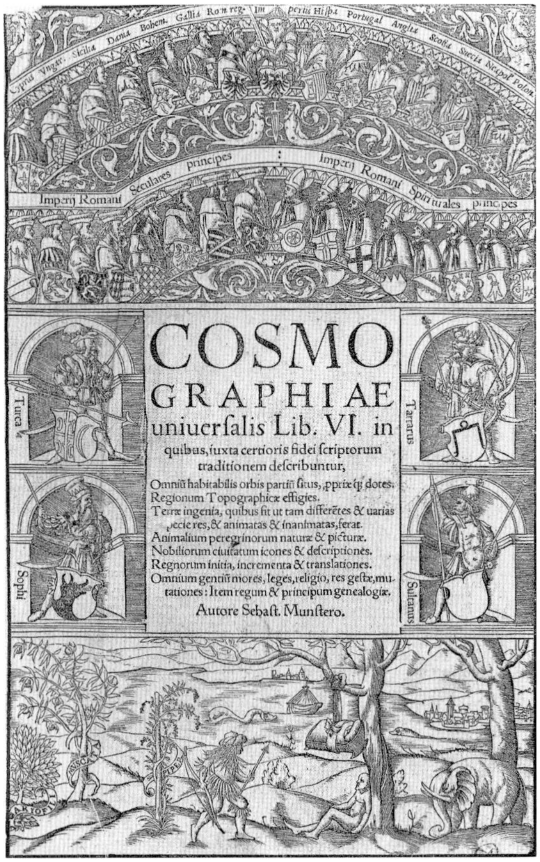

4.19 Sebastian Münster, *Cosmographia universalis*, Basel 1550, frontispiece. Houghton Library, Harvard University (Typ. 565.50.584 F)

ally regarded as the first modern atlas, he chose the traditional title for sixteenth-century encyclopedias, *Theatrum orbis terrarum* (Fig. 4.20). The title suggested that Ortelius's book was an encyclopedic, albeit brief, description of the entire world, which, like Münster, he augmented from edition to edition. But Ortelius's theater of the world, which was organized geographically and visualized by maps, also hinted at the function of the geographical order as a system of organization and retrieval of knowledge.[42]

In the introduction of his book Ortelius explained the mnemonic function of maps. In a passage often quoted he wrote: "whatever we read accompanied with maps placed in front of our eyes like mirrors, remains in our memory longer." Then he added that maps not only make reading more pleasant and fruitful, but also they "bring in front of the eyes the actions and the places where the deeds happened, as if they were present."[43] In both passages Ortelius referred to maps as visual aids to the recollection of events and, more importantly, to their capability to exercise the memory. Ortelius suggested that the reading of Exodus along with a map of the Holy Land would have fixed more firmly in one's memory the biblical wanderings of the Jews. Similarly, by reading Herodotus together with a map the scholar would have moved within the mapped land as the Greeks and Persians moved through the region during the Persian wars. Because maps offered a syncretistic view of the world, both spatial and temporal, they made history present.

Not only were maps often added to Bibles and history books, or even to literary texts such as Ludovico Ariosto's *Orlando furioso*, but Ptolemy's geographical order was also widely used by printers and scholars. The Venetian and Roman printers who assembled composite atlases at the request of individual patrons did so according to Ptolemy's order.[44] The number and selection of maps varied from copy to copy, but their order remained unchanged: they were "collected and arranged according to the order of Ptolemy," as the frontispiece that Antonio Lafrery engraved for one of these atlases proclaimed under the effigy of Atlas (Fig. 4.21). Federico Borromeo, the erudite cardinal of Milan, wrote with admiration about a professor who could converse competently about the world because he had memorized Münster's maps and his geographical order.[45] Ulisse Aldrovandi, who spent his life studying, growing, collecting, and illustrating specimens of the world's flora and fauna, used geography in some of his catalogues, a practice shared by many other scholars. Jean Bodin wrote about the role of geography as the organizing principle of the world and its history. In his *Methodus ad facilem historiarum cognitionem* (Paris, 1572), cosmography gave to history the order of exposition from the general image of the world map to the particulars of continents, regions, and cities. Cosmography also provided history with the interpretative key to the deeds and manners of people. In Bodin's historical method the order of geography was the organizing principle of the world and its history, not the eye of history.[46]

Finally, Ptolemy's geographical order was used to organize the Guardaroba Nuova, an encyclopedic collections of artifacts that, like similar rooms, was regarded as a microcosm of the world. Like the Renaissance editions of Ptolemy's *Geography*, the Florentine map cycle started with a world map which, according to Vasari, had to be represented in two *mezze palle*, or hemispheres, perhaps an indication that Girolamo Ruscelli's edition of 1561 was consulted since this is the only Renaissance edition to include a map of the world representing the earth's globe as divided into two hemispheres, (such a representation was, however, common in printed maps). The *mezze palle* of the earth's globe were followed by two celestial maps depicting the two halves of the sky, a feature not included in the Renaissance editions of Ptolemy's book but often present in individual world maps or in other geographical manuals. Then the maps of the continents followed: fourteen maps of Europe were followed by eleven maps of Africa, fourteen maps of Asia, and fourteen maps of the West Indies. Although no edition of Ptolemy's *Geography* matches exactly the number of maps planned for the Guardaroba Nuova (most likely their number was dictated by the perimeter of the room and the size of the cupboards), Gastaldi's edition of 1548 integrating graphically and conceptually Ptolemy's order with modern cartography must be regarded as fundamental to the conception of Cosimo I's cabinet of art and curiosities. As in Gastaldi's edition of 1548, the Florentine maps of the regions known to Ptolemy were intercalated with the modern maps that Danti and Bonsignori made to complete the mapping project of the ancient geographer. Like contemporary cosmography books, the maps of the Guardaroba Nuova ordered the images of plants and animals of the mapped territories as well as the artifacts peculiar to each region. Ultimately, the Guardaroba Nuova was a three-dimensional display of Gastaldi's edition of Ptolemy's *Geography* and Münster's *Cosmographia universalis*. As in these contemporary books, the modern world was structured according to the authority of the ancient geographer, just as the natural world was organized according to Pliny and architecture measured according to Vitruvius.[47]

The Geographical Order as a Heuristic System

The persistent use of Ptolemy's geographical order in such diverse encyclopedias as Münster's *Cosmographia universalis*, Ortelius's *Theatrum orbis terrarum*, Jean Bodin's *Methodus*, the Roman and Venetian composite atlases, and Cosimo's Guardaroba Nuova suggests that this order was regarded not only as an effective system to find one's position in the world, but also as a heuristic system to organize and retrieve the knowledge of the world.

Since antiquity the invention of heuristic schemes had been regarded as foundational to rhetoric, especially to the art of memory – the part of rhetoric that taught students the elementary techniques to store and retrieve the information they acquired from reading and listening to authors. Ancient rhetoricians and their Renaissance followers believed that visual coding was the most effective way to retrieve the information in the order in which it was stored, but also, more importantly, to move around freely within it so that one could retrieve

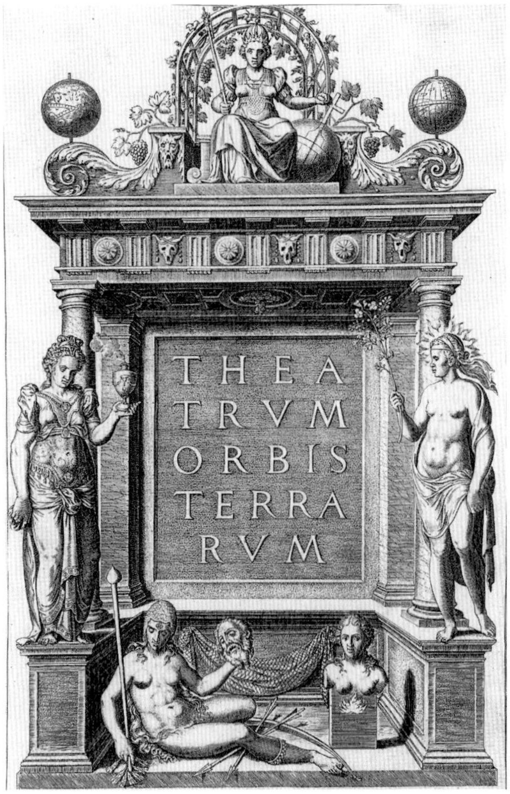

4.20 Abraham Ortelius, *Theatrum orbis terrarum*, Antwerp 1570, frontispiece. The University of Virginia Library

only what was pertinent to the speech one was delivering, or the composition one was writing. Consequently, students were trained to turn words, ideas, and concepts into visual form, and to invent effective heuristic systems made up of images. These systems could vary in iconography but their success was determined by the homogeneity of their images and the rigidity of their order. Successful heuristic systems could be based on architectural settings, alphabetical sets, human figures, numerical series, the Zodiac, the Bestiary, or maps of the world.[48]

Ptolemy's geographical order contained all the elements that ancient and Renaissance scholars regarded as essential for a perfect heuristic system: the move from the global to the particular; the organization of a series of images in the form of maps, into a tight sequence; and the possibility of expansion by adding new images, the new maps that were created, as universal knowledge expanded when new lands were explored.

The importance of the geographical order as a heuristic system of universal knowledge is evident in Renaissance treatises on memory. In his *Tipocosmia* (Venice, 1561), Alessandro Citolini summarized his rather tedious journey of universal knowledge with two emblematic geographical objects. One was an enormous book (*un libro di estrema grandezza*) that suggestively recalls a typical sixteenth-century cosmography book. Like Münster's *Cosmographia universalis*, Citolini's *libro di estrema grandezza* organized the encyclopedia of the world geographically, illustrating it with diagrams, images, and maps:

> In the earth they [the readers] saw its parts, the countries of the entire world both in general and in particular, and they even entered at the center of the earth . . . Then it seemed to them that they were entering the most complete garden they had ever seen, since every herb and tree was planted there . . . In another part they saw animals that live on earth, fly, and swim . . . and seeing them in this way, they arrived at understanding not only their mere names but their true essence and form, and not only of them but also of all the other things that have form.[49]

Citolini's second geographical object strikingly resembled Cosimo's Guardaroba Nuova: it was "an enormous sphere [*palla*], in which [the visitors] could enter; and when they entered, they saw the heavens around them; and in the middle, they saw the earth, and they saw things ordered there in a way more pleasing to the eye of the body than to the eye of the mind. And after having heard long discussions and reasoning from the Count [the host], they all concluded that these things were more childish than worth knowing."[50] The visitors judged the walk-through sphere too simple a divertissement, *cose da fanciulli.* On the other hand, Citolini's sphere conformed not only to the idea of the Guardaroba Nuova but, more generally, to other three-dimensional heuristic systems of the period.

The most famous three-dimensional system of information retrieval was Giulio Camillo's theater, a structure actually built in Venice in the 1540s (and perhaps also in Paris a few years later), a copy at a reduced scale of a Roman amphitheater, which survives only through verbal descriptions. Standing on the stage of Camillo's theater, the viewer would have looked at universal knowledge organized in the arena according to a system that combined the planets and the elements. Walking through the theater the visitor would have encountered the places of memory marked by innumerable little drawers (*cassettini*) containing small rolls of paper with inscribed quotations from different authors. Camillo reified ancient and medieval heuristic schemes into the actual copy of a Roman amphitheater, the images of memory into statues and paintings, the mental compartments of the storehouse of memory into *cassettini*, and the fragmented texts that filled the memory of ancient orators and medieval preachers into rolls of paper stored in drawers.[51]

Another typical three-dimensional system of classification was the Kunstkammer. Renaissance authors realized that cabinets of art and curiosities shared with memory the need to organize encyclopedic knowledge, albeit one made up of artifacts rather than texts, and applied the model of Camillo's theater to the organization and retrieval of collections. The Flemish physician Samuel Quicchelberg wrote a short theoretical treatise on collecting which was inspired by the Habsburg Kunstkammer in Munich, one of the greatest collections of the period where he worked. In Quicchelberg's treatise, the storehouse of memory turned into a real gallery containing a universal collection of art and curiosities.[52]

The Guardaroba Nuova, thus, applied the geographical order of Renaissance atlases and cosmography books to the three-dimensional space of a cabinet of art and curiosities. In the Guardaroba Nuova, the maps illustrated the world systematically but, more generally, they were the images of a memory theater to see, organize, and remember a collection of artifacts, tangible fragments of the encyclopedia of the world. The maps were not simply labels to store and locate artifacts. Rather they constituted the iconography of the geographical order, the images of the heuristic system, just as numbers, letters, and planets were the images of earlier mnemonic systems based on numerical and alphabetical sets, the Zodiac or the Bestiary. Just as Camillo had turned his heuristic system into a

4.21 Antonio Lafrery, *Geografia. Tavole moderne di geografia . . . raccolte et messe secondo l'ordine di Tolomeo* (Geography. Modern Geographical Maps . . . collected and arranged according to Ptolemy's order), ca. 1575, engraving. The Newberry Library, Edward E. Ayer Collection (Novacco 2 F O)

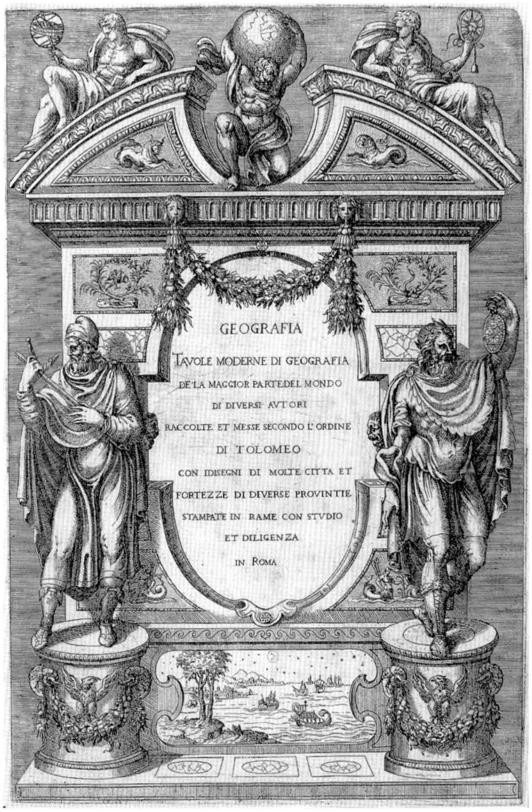

three-dimensional amphitheater, and Quicchelberg had transformed the warehouse of the mind into an actual museum, so did the Guardaroba Nuova render three-dimensionally a system of collecting according to maps, indeed the very ordering of maps of contemporary cosmography books.

The project of the Guardaroba Nuova was never fully implemented and, as far as it can be judged , Cosimo I's collections were never arranged geographically. Had it been realized, the geographical arrangement of the Guardaroba Nuova would have added variety to Cosimo's display of the Medici collections. Small artifacts were tastefully displayed in his studies, large ancient statues were located in the hall of the Pitti Palace, his personal papers were secured in secluded rooms, and exotica would have been displayed in the Guardaroba Nuova. As a theater of the world, the Guardaroba Nuova was inherently similar to other European Kunstkammern in conveying symbolically the patron's control over the world through its indoor microscopic reproduction. But the iconography based on maps of the world and its geographical display set it apart from other European examples. The visibility given to the exotica within the Medici collections was also unprecedented. While exotica were prized items in European collections, no ruler had yet set aside a room for them, let alone one that attempted their integration within the geography of the world.

The reconstruction of the organizing principle in the Guardaroba Nuova is at times suggestive and speculative, but the ordering of maps is the only criterion that can explain the physical layout of the room itself, the geographical order of the maps, their content, and their functional analogy with the paintings for Francesco I's Studiolo. Indeed, the device of the painted doors as a visual catalogue of the artifacts kept in the cabinet was immediately adopted for Francesco's Studiolo, although the maps were substituted there with the more familiar scheme of the four elements. Even more importantly, the ordering of maps adopted in Cosimo's cabinet of art and curiosities deeply resonates with the contemporary use of Ptolemy's geographical order as a heuristic system to organize and retrieve encyclopedic knowledge. But this use was possible because Ptolemy's notion of geography had gone through a process of revision in the Renaissance. From the mathematical representation of large areas of the world that Ptolemy had defined, geography transformed into an encyclopedic description of a place projected onto a mathematical framework. The revision of Ptolemy's *Geography*, which incidentally took place in the very commentaries that were supposed to explain his geographical theory, profoundly affected the cartographic language of Renaissance maps, a language that will be analyzed in the next chapter in relation to the maps of the Guardaroba Nuova.

5

The Mapping of the Modern World

The maps of the Guardaroba Nuova ordered a selection of Cosimo I's collections geographically, providing both the principle of organization of the artifacts and a visual catalogue for their storage and retrieval. But the Florentine maps also constituted a masterpiece of cartography, an unprecedented representation of the world in 57 maps that had no equals either on the walls of a princely palace or in the pages of a printed book. The regions of the world had never been represented on the large scale dictated by the size of the cupboards' panels nor with such richness of cartographic detail. Also the systematic connection of the modern maps of the entire world with other images illustrating different aspects of the mapped region was a novelty, which Pius IV had considered for his map cycle in the Vatican Palace but then abandoned. The boldness of this cartographic enterprise becomes evident once it is realized that the very idea of the atlas was still in formation, and that the only cycles of maps of the world available were the small-scale maps illustrating the Renaissance editions of Ptolemy's *Geography*. Lacking any existing model, the map cycle of the Guardaroba Nuova had to be made by comparing, collating, and unifying heterogeneous printed and manuscript maps made by different authors in different periods and according to different projections. Cosimo and his advisers were well aware that the Florentine map cycle they were crafting was a daunting cartographic challenge, "a work greater and more perfect than any that has ever been done at any time in the profession [of cosmography]," as Vasari boldly put it.[1] Paradoxically, however, Vasari also identified the modern maps of the Guardaroba Nuova as the "Tables of Ptolemy": the maps made by Ptolemy to illustrate his geographical manual. How could these Florentine maps be regarded simultaneously as the "Tables of Ptolemy" and as the most detailed modern maps? Or, to put the question in more general terms, what was the relation between ancient geography and modern discoveries in the mapping of the modern world?

* * *

Ptolemy's Geographical Maps

The maps of the Guardaroba Nuova could be regarded as Ptolemaic maps for a number of reasons. They were made according to the Ptolemaic grid of meridians and parallels, even though the mathematical grid coexisted with the windrose, the directional device of medieval nautical charts. They were also based on the trapezoidal projection, which Renaissance scholars and map-makers regarded as the most suitable to interpret Ptolemy's intention: the parallels are equidistant, the meridians converge to the pole, and the spatial relationships on earth are maintained. Indeed, this projection appeared around 1460, when the German scholar Nicolaus Germanus realized that the maps contained in the Greek manuscripts of Ptolemy's *Geography* were made according to the cylindrical rectangular projection of Marinus of Tyra, the very projection Ptolemy had explicitly criticized at the beginning of his text; since then Germanus's projection became standard for Ptolemy's maps as well as for Renaissance maps in general (Fig. 5.1).[2] Furthermore, the Florentine maps were ordered according to Ptolemy's geographical order, which, as previously described, integrated ancient geography with modern discoveries. Thus, in Cosimo's and Vasari's view, the Florentine maps were the "Tables of Ptolemy" on technical grounds since they adopted his grid, his pro-

5.1 *Tabula Europae x* (Tenth Map of Europe), woodcut, from Ptolemy, *Geographia*, Rome 1490. Princeton University Library

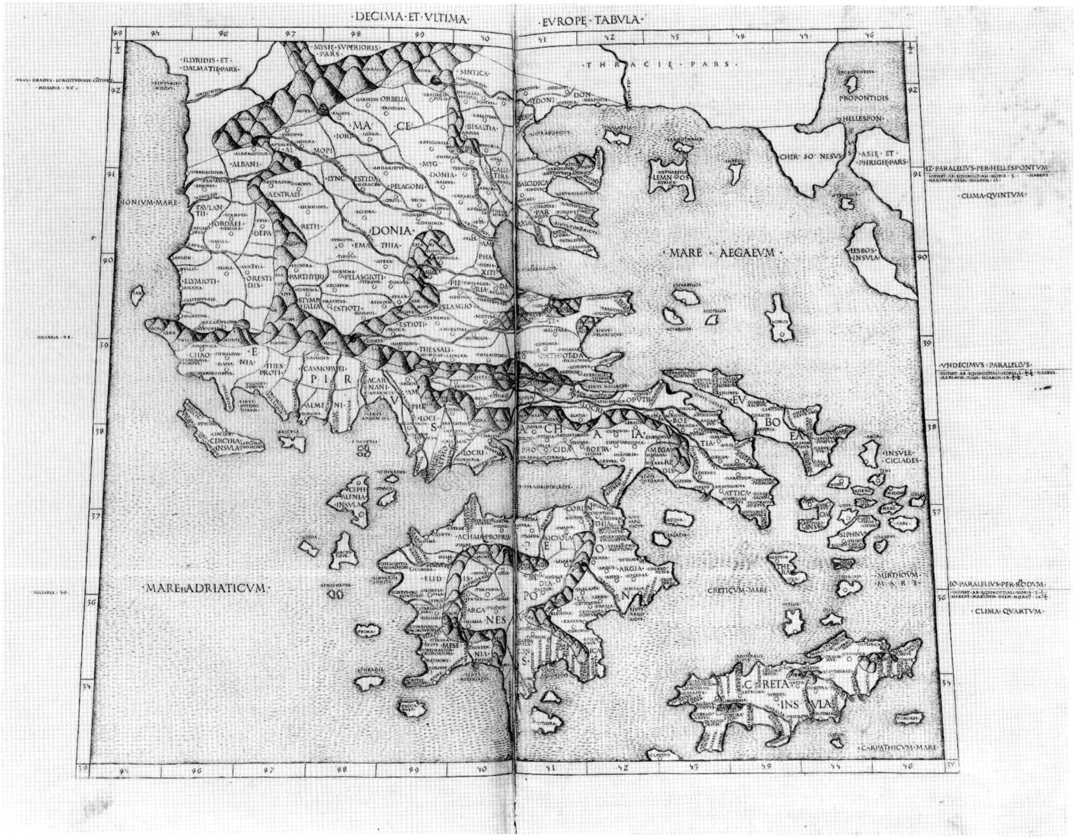

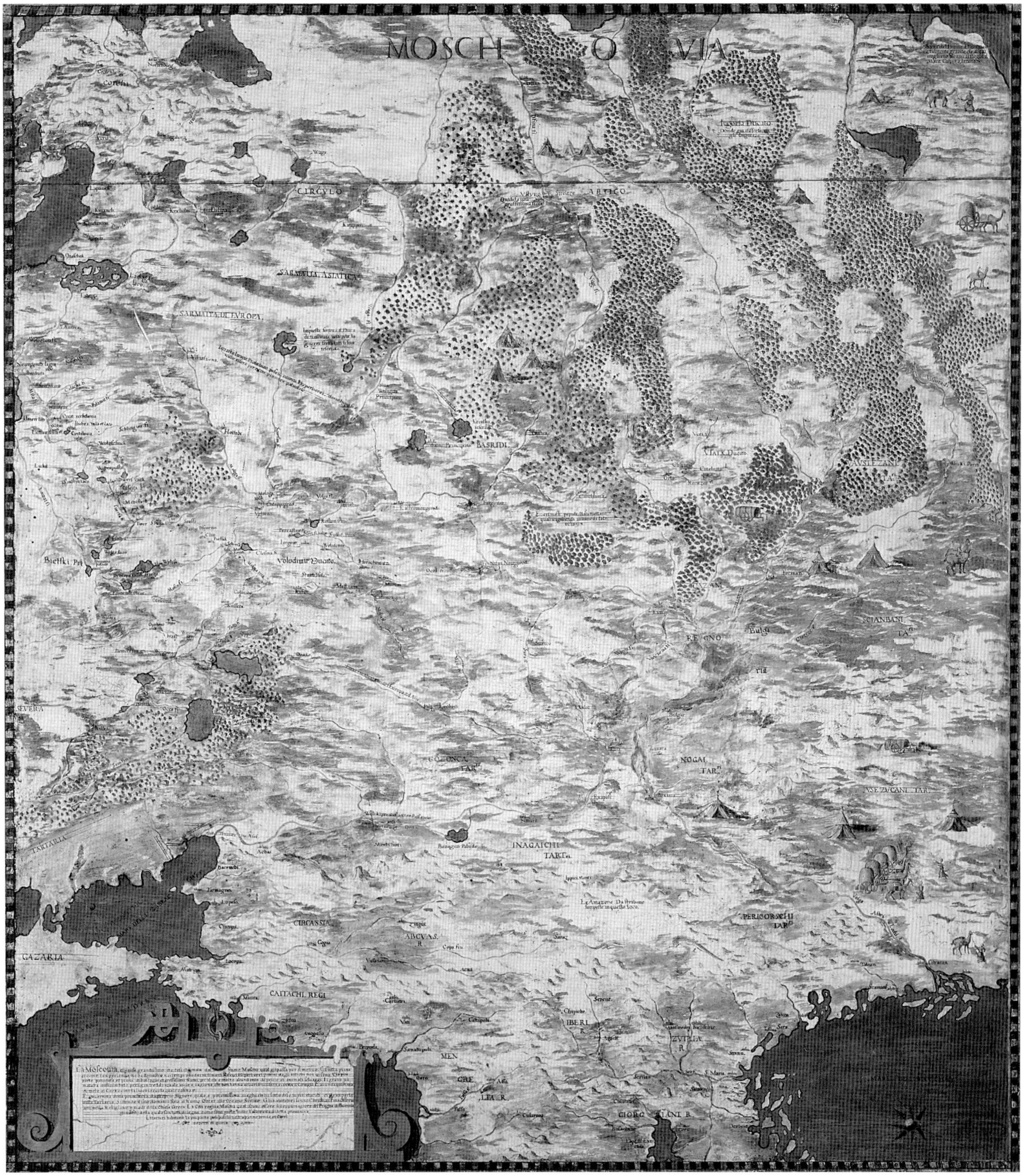

5.2 Egnazio Danti, *Moschovia* (Muscovy), not dated, panel. Florence, Palazzo Vecchio, Guardaroba Nuova jection, and his order. But Vasari might have considered the Florentine maps as the "Tables of Ptolemy" for a third, equally important, reason. Like Ptolemy's original 26 regional maps, they were geographical maps representing the world "with its general features" and were thus inherently different from chorographical maps, which instead illustrated partial views of the whole.

In the first chapter of *Geography*, Ptolemy defined geography as the graphic representation of the whole known world, and chorography as the graphic representation of a part of the whole.[3] Consequently, the geographer dealt with the quantitative values of positions and distances while the chorographer dealt with the qualitative details of individual places. The geographer made schematic drawings representing a coast with a line and a city with a point, while the chorographer, a skilled draftsman, portrayed the likeness of a place illusionistically. In order to better clarify his distinction between geographical and chorographical maps, Ptolemy introduced the comparisons between map-making and painting, between the body of the earth and the body of human beings. As the geographer painted the known world, so the painter painted the head of a person. As the chorographer represented particular places, so the painter represented the detail of an ear or an eye. Ptolemy's comparison between the activity of the map-maker and that of the painter was popular in Renaissance rhetoric and art, while his suggestive association between the body of the earth and the body of human beings kindled the imagination of many Renaissance authors who wrote metaphorically on the human body as a map or, conversely, on the map of the world as a body. But what needs to be stressed for the present discussion is that Ptolemy's comparison between the map of the whole world and the head of a person also introduced an additional meaning to his previous understanding of geography, since obviously the head is part of the human body. In fact explaining this comparison, Ptolemy defined geography as the graphic description of "the regions and their general features," an issue he further clarifies at the beginning of Book Eight, quoted here from Ruscelli's edition of 1561:

> After having made a competent and convenient description of the whole inhabited world in one single map so that we can look at it [in its entirety], it has to be explained the way to make brief descriptions by dividing the world, or the general map, into many maps in order to represent the known places clearly and with their accurate measure from one to the next. Therefore in the universal description it is better that, in order to keep the proportions of the parts of the world to each other, some parts, in which many places have to be located, would be enlarged, and other parts, in which there are no places to locate, would be reduced or would be left empty . . . Therefore it is not necessary that each map would be as large as the others, but it is sufficient that in each of them it would be kept the proportion that the places have to each other, just like we do when we paint only a head: we keep the proportion of those things that are in the head, or if we paint a hand, we keep the proportion of only those [things] that are in the hand. And when we do the head we do not look at the proportions of the hand, unless we are painting the entire body of a man. But as in the whole nothing prevents from making things bigger and smaller, or increasing or diminishing, similarly in the parts, when they are by themselves, nothing prevents from increasing or diminishing them according to the capacity of the maps in which they have to appear.[4]

The analysis of this passage in Book Eight in conjunction with Ptolemy's theoretical discussion at the very beginning of his manual confirms that his geographical maps were of two kinds. One was the graphic representation of the whole known world as unequivocally defined in Book One. The other kind was exemplified by the remaining 26 maps, representing regions of the world, or parts of the whole. According to Ptolemy's definition in Book One, these maps should have been regarded as chorographical maps, but his explanation in Book Eight leaves no doubt that they also fall under the definition of geographical maps. In fact, after having defined chorography at the beginning of his manual, Ptolemy never again used the word, and Renaissance editors of Ptolemy's text labeled his maps either cosmographical maps (*tabulae cosmographiae*) or geographical maps (*tabulae geographiae*), depending on the general title they gave to the book, but never chorographical maps (*tabulae chorographiae*).[5]

The maps of the Guardaroba Nuova are Ptolemaic geographical maps (Fig. 5.2). Like Ptolemaic maps, they represent large regions, are made according to one of the Ptolemaic projections, and include the grid of coordinates and the degrees of latitude and longitude. Moreover, both in the Ptolemaic maps and in the Florentine maps the interaction between word and image touches the very structure of the cartographic image. Words identify the elements of the map, like place names or toponyms, or explain the cartographic symbols, usually listed in a legend at the margin of the cartographic outline, and are structurally indispensable to the communication of the geographical content.

However, a closer look at the maps of the Guardaroba Nuova reveals many differences. Together with the grid of coordinates and the scale of representation, the Florentine maps include a plethora of qualitative details that did not appear on the Ptolemaic maps. Hills, mountains, forests, deserts, steppes, lakes and estuaries, cities, occasionally people, buildings, and animals proliferate on these maps. Such elements appear at a larger scale than the mapped territory and are represented in perspective rather than in orthographic view. Lengthy legends, cartouches, and inscriptions informed the viewer of the main geographic features of the mapped territory, the coordinates of its major cities, the main rivers and lakes, the mountain chains, and the flat lands. They described the goods and minerals produced in the mapped territory, its political divisions, the etymology of its place names, the ethnography and history of its people, their habits and beliefs, the look of their dwellings and their costumes, and the military and intellectual achievements of its population. Although it is not known what Ptolemy's original maps looked like, yet his text is quite clear about the kind of information that had to be included on the map itself. Introducing the regional maps in Book Two, he instructed the map-maker to: "leave out the great mass of reports about the characteristics of the peoples (unless perhaps some bit of current knowledge calls for a brief and worthwhile note)."[6] While Vasari called them the "Tables of Ptolemy," the Florentine maps were ostensibly different from the Ptolemaic maps. The difference involved not so much the cartographic accuracy of a place (this would hardly be surprising) but rather the very content of the geographical map.

The Florentine maps combined in one single image different systems of representation: the orthographic grid, the bird's-eye view, and the perspectival view. They also included an abundance of qualitative geographical details and different kinds of information – historical, geographical, zoological, botanical, and ethnographical. They described large regions of the world with the pictorial means Ptolemy would have associated with the chorographical description of small areas, and with the variety of historical and mythological details that would have been pertinent to narratives of local history. In short, the maps of the Guardaroba Nuova were both geographical and chorographical maps; they represented the mapped places as both mathematical sites and individual locales.

Such a comprehensive description of a place exemplified by the Florentine maps has traditionally been regarded as the heritage of the medieval world maps, which organized geographically and synthetically the encyclopedic knowledge of the world in words and images. A late example of this genre can be found in Fra Mauro's splendid world map made in Venice in the mid-fifteenth century. Although Fra Mauro knew Ptolemy's text and quoted it in his world map, he was utterly uninterested in Ptolemy's system of geographical coordinates, and instead filled his map with information pertaining to trading routes, spices, luxury goods, wonders, botany, zoology, ethnography, and religious beliefs and practices. Although there are some undoubted similarities between the medieval world maps and the Florentine maps, the interpretation of Renaissance maps as derived from medieval moralized cartography does not take into full account the fact that Renaissance map-makers and scholars seem to have entirely ignored the magnificent world maps illustrating the encyclopedias by Oronce and Beatus of Seville or the monumental maps exemplified by that of Fra Mauro. Rather they regarded their maps as resulting from the philological exegesis of the authoritative texts by Ptolemy and Strabo.[7]

Cosmography, Geography, and Chorography

Ptolemy's theoretical definitions of geography and chorography were amply commented upon in the Renaissance editions of his text as well as in the introduction to geographical publications of the period. In fact, Renaissance authors and commentators, including Egnazio Danti, scrutinized Ptolemy's definition of geography and chorography so closely that they realized that each term actually included two different concepts and consequently discussed them not only in relation to each other, as Ptolemy had done, but also in relation to cosmography and topography. Cosmography, which defined the main circles of the earth and corresponded to what might today be called mathematical geography, was not mentioned by Ptolemy in his *Geography*, although he did write about it extensively in his *Almagest.* Topography, literally the description of a *topos*, or a small place, usually corresponded to the view of cities, castles, or some other circumscribed locales. Modern scholars have focused on how accurately each Renaissance author understood Ptolemy's text, but have given far less attention to the way Renaissance interpretations of Ptolemy's geographical theory influenced the descriptive language of Renaissance maps. In fact, the internal structure of the maps of the Guardaroba Nuova and, more generally, of Renaissance printed maps was profoundly connected to the interpretations of Ptolemy's geographical theory provided by Renaissance humanists and map-makers who, while strongly inspired by Ptolemy, also proposed a very different notion of geography.[8]

As is often the case in the Renaissance, the meaning of cosmography, geography, chorography, and topography fluctuated from author to author: their oscillations affect even Ptolemy's revered text. The Greek scholar Chrysoloras wanted to use the title *Geographia* for his never-completed Latin translation – maintaining the title of the Greek manuscripts. His student Jacopo d'Angelo, who eventually completed the task, chose *Cosmographia* instead, grounding his choice on the authority of Pliny, who had called Ptolemy a cosmographer, and on the theoretical argument that the major circles of the earth were established through cosmography.[9] Others, including the humanist Guarino Veronese, claimed that the correct title of Ptolemy's text would have been chorography, since the manual dealt ultimately with the mapping of individual regions.[10] In 1482 Francesco Berlinghieri finally entitled *Geografia* his rendition of the Ptolemaic text in Italian verse.[11] The importance of Berlinghieri's publication has long been recognized: for the first time it included marvelous engraved maps; it added printed modern maps; and it testified to the relevance of Ottoman patronage in Renaissance Italy (one copy of the book was dedicated to Sultan Soliman the Magnificent). Usually, however, his geographical text is dismissed as a poetic but inaccurate rendition of the original text. Indeed Berlinghieri greatly expanded Ptolemy's notion of geography: from the mathematical description of the main geographical features that had been defined by Ptolemy, geography was transformed into the description of whatever the particular place contains (peoples, animals, plants, and rocks), including the historical and mythological events that happened in it. It may well be that Berlinghieri, a poet more than a map-maker, was a poor editor of Ptolemy's text, but he was a perceptive interpreter of his own culture, since his understanding of what a map should contain reflected and shaped the practice of Renaissance mapping.

The most illuminating evidence for understanding the meaning of cosmography, geography, and chorography in the sixteenth century comes from Peter Apian's *Cosmographicus liber* (Landshut, 1524), an influential geographical manual frequently republished in the sixteenth century that provided a verbal and graphic description of these concepts (Fig. 5.3). According to Apian, cosmography referred to the mathematical description of both the cosmos and the earth; it pertained to the four elements, and established the main circles of the earth by projecting the sphere of the cosmos onto the sphere of the earth. But cosmography also dealt with the earth's globe by enabling the mathematical location of each place of the earth in relation to the main circles of the terrestrial globe. According to Apian, this cosmographical description of a place, based on latitude and longitude, differed from a geographical map because the latter described places not mathematically but according to the categories of land, sea, river, and mountains. Apian clarified this double meaning of cosmography in an accompanying woodcut that included two separate diagrams, each corresponding to one of the different meanings of cosmography. On the left, the sphere of the cosmos and the sphere of the earth, seen from below by an imaginary viewer, were not represented

5.3 Peter Apian, *Liber cosmographicus*, Basel 1545, fol. 1 and 2. The University of Virginia Library

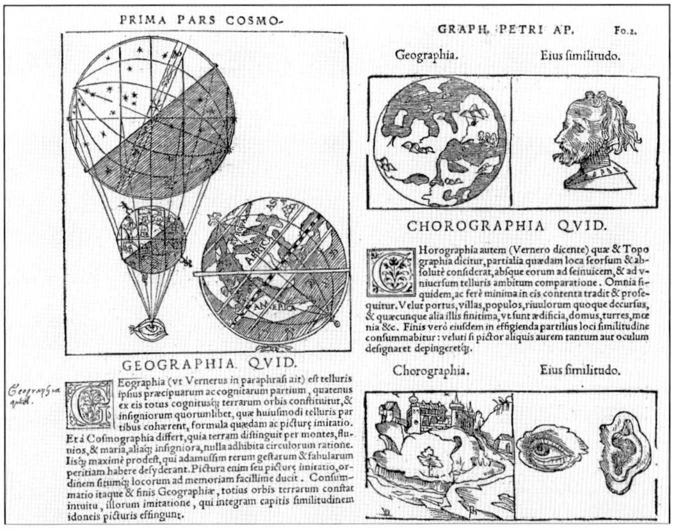

one inside the other, as the Ptolemaic cosmological system would dictate, but rather one above the other in order to visualize how the circles of the celestial sphere were projected onto the terrestrial globe. In the second diagram on the right, the main circles are marked on the terrestrial globe, suggesting visually how, "cosmography describes the earth by the circles of the heavens, beneath which it is, and not by mountains, seas, rivers nor other particulars, as does geography."[12] In this second meaning, Apian's cosmography as the mathematical description of places according to their coordinates corresponds to Ptolemy's geographical description of the whole known world. In accordance with this definition of cosmography, most Renaissance printed maps of the world were called maps of cosmography, not geography. Similarly, the Guardaroba Nuova in its entirety was regarded as a work of cosmography, one, according once again to Vasari, "greater and more perfect than any that has ever been done."[13]

If Apian's cosmography corresponded to Ptolemy's geography, the question arises of what geography was for Apian, and how it was different from chorography. Although Apian was obviously inspired by Ptolemy's comparison between painting and map-making, and by his suggestive association between the body of the earth and the body of human beings, he elaborated independently on the Ptolemaic theoretical distinction between geography and chorography. Geography was a map of a large region revealing oceans, lakes, rivers, islands, and mountains, and was compared to the head of a man. Chorography was a perspective view of a castle on a hill surrounded by a stream and was compared to the drawing of an eye and an ear. For both Ptolemy and Apian, the two kinds of maps differed in terms of extension of land represented, but Apian alone was explicit on their different systems of representation: the orthographic view was proper for geography while the perspective view was proper for chorography. More importantly, for Apian the geographical map included both a mathematical description of the main geographical features and a qualitative representation of other minute details, but was different from a chorographical map since the latter, "consider[s] particular places separately and apart, without consideration of or comparison with other similar places or the fullness of the earth."[14]

Apian's definitions were still very popular at the end of the sixteenth century and authors and map-makers such as Sebastian Münster, Giacomo Gastaldi, Girolamo Ruscelli, and Egnazio Danti adopted them almost verbatim.[15] Girolamo Ruscelli's commentary to his edition of Ptolemy's *Geography* (Venice, 1561) is particularly interesting, since he speculated not only on the meaning of geography and chorography, but also on the kind of projects map-makers of the late sixteenth century were pursuing. As a humanist, Ruscelli's explanation of Ptolemy's geographical theory was largely based on etymology. The Greek root of geography (*geo*), referred both to the whole earth (*terra*) as well as to the "region, or country, or province" (*regione, ó paese, ó provincia*), or even to an individual "land" (*terreno*). Thus the readers of Ptolemy's text should, "completely understand the true meaning of this word Geography, and not be mistaken when eventually they will find the said word earth (*terra*) used according to its different meanings in this book."[16] Etymology also explained the meaning of cosmography as both the description of the universe and of the ornaments of the terrestrial globe. As such, "cosmography is the discipline that does not consider the peculiar quantity or measurement of distances between places, but rather describes and narrates the nature and characteristics of countries and the things that are in them, the customs, people, and notable events that happened in the past."[17] According to Ruscelli, Strabo's

Geography rather than Ptolemy's text exemplified this kind of cosmographic description of the earth's ornaments.

A text written in the first century A.D. but completely ignored by the major geographers of imperial Rome and Alexandria such as Pomponius Mela, Pliny, Marinus of Tyra, and Ptolemy, Strabo's *Geography* was as revered as Ptolemy's in the Renaissance.[18] As Strabo himself explained in his lengthy introduction, his *Geography* was conceived as complementary to his historical writings, as a philosophical enterprise with the political and ethical mission of educating statesmen and commanders on the government of the world. Although Strabo discussed in his introduction the mathematical foundation of geography and the contribution of his predecessors, his book was entirely devoted to the verbal description of places recounting in detail the peculiarities of their terrain, waters, animals, and plants, the history of their inhabitants, and the marvels of their cities. For Strabo, geography was the study of human actions in space as much as history was the study of the same actions in time. The humanist Guarino Veronese, who translated Strabo's text into Latin in 1469, shared Strabo's view on geography and regarded the text as suitable reading for his aristocratic students, the future rulers from the Gonzaga, d'Este, and Montefeltro families. Strabo's encyclopedic description of the world was expressed exclusively in words. Even though he may have consulted maps during the compilation of his work or imagined his readers to be familiar with the map of the known world, he was not concerned with the visual description of the earth, let alone with its measurement. The medieval world maps attempted to organize spatially the encyclopedia of the world Strabo had so profusely described in words, but they could not serve as suitable models for Renaissance maps which, after the rediscovery of Ptolemy, also aspired to the mathematical representation of the world.

In fact, for Renaissance scholars and commentators, Ptolemy's mathematical geography had to be combined with Strabo's literary tradition. This synthesis was the way to interpret and improve the classical tradition. For instance, Ruscelli could not find any reasonable explanation as to why Ptolemy, who wrote after Strabo, did not mention Strabo among his sources, or why the geographer from Alexandria did not combine his mathematical geography of the earth with Strabo's description of its ornaments. This was, in Ruscelli's view, the scope of geography: to describe the entire globe of the earth together with its parts both mathematically and pictorially or, as Ruscelli put it, to combine Strabo with Ptolemy, adding "to Strabo's work, in which the names of provinces and places are reported and even repeated many times . . . the measurements in degrees [of latitude and longitude], which are reported in this book by Ptolemy together with the twenty-six maps that he made."[19]

This was also the way Egnazio Danti understood geography. In his book *Le scienze matematiche ridotte in tavole* (Bologna, 1577) written when he was a professor of mathematics at the University of Bologna but in fact reflecting his teaching at the Medici court, Danti divided the description of the earth into four branches, namely geography, hydrography, chorography, and topography. It is worth quoting Danti's definitions in full. Not only are they extremely relevant to the cartography of the Guardaroba Nuova but they also give an indication of the geographical knowledge at the Medici court:

> The description of the earth is dealt with in:
> GEOGRAPHY, which describes it [the earth] without taking in consideration the heavens as Strabo and Pomponius Mela did, who described it only with itinerary distances, in the same manner as Domenico Mario Negro and many others did. And especially Ptolemy,

whose book some think should be titled Cosmography since he describes the earth with the degrees of latitude and longitude; but those who consider that in fact he calculates the said degrees not in the heavens but rather on the equinoctial line and on the earth's meridians they will agree that his book should be title Geography since it describes only the earth, and not Cosmography, which instead describes the heaven.

HYDROGRAPHY, which describes the entire sea, taking from Geography the description of the maritime shores, in order to show the boundaries that the sea has with the earth both in continents and islands as well as in cliffs, as it can be seen in nautical charts.

CHOROGRAPHY, which in our language means the description of an individual place. An example is the chorography of the city of Perugia with the surrounding countryside made by my father Giulio Danti, in which the plan of the whole is drawn, and with all diligence the places are drawn after nature, with every street and house, trees, and other things, which are distinguishing features here. Thus one recognizes clearly that even though chorography can be practiced by a geometer, and is a science subaltern to geometry, it can, as a mechanical art, also be practiced by a simple painter, whereas geography can be apprehended only by a geometer, since it does not require painting at all, given that it describes a city with a dot, and a river or a sea coast with a straight or curved line.

TOPOGRAPHY, which also describes individual places, but since it does such description with words it does not need either Geometry or painting. In this manner Polybius described beautifully Lake Trasimeno before he narrated the defeat that Hannibal gave to the Romans. Similar is the description that Titus Livy made of Cannae in Apulia.[20]

The maps of the Guardaroba Nuova then were the embodiment of the Renaissance comprehensive geographical description of the earth. The Florentine maps were Ptolemy's maps since they described large regions of the world. They were also the geographical maps theorized by authors such as Apian, Gastaldi, Ruscelli, and Danti himself. They combined Ptolemy with Strabo, mathematical geography with descriptive geography, and orthographic projections with perspectival views. While maintaining the Ptolemaic grid of coordinates, they also included more qualitative descriptions of places in the form of lengthy legends, vivacious vignettes, and prolific inscriptions, which were not mere embellishments of the cartographic content but rather an integral part of the map. The Florentine maps reflected sixteenth-century toponymy and geographical knowledge, but they also illustrated the history of a territory and its inhabitants, integrating geographical descriptions with historical, mythological, botanical, and zoological details. The maps of the Guardaroba Nuova constructed a complex, encyclopedic image of a place. Their cultural value resided precisely in such syncretizing of different sources in terms of accuracy, chronology, and origin. For Danti and for many other Renaissance map-makers, a map was an encyclopedic description of the world projected onto an accurate geographical framework. Ultimately, it was a syncretistic image that integrated geography with encyclopedic knowledge.

* * *

Renaissance Maps and Classical Geography

At the core of the encyclopedic description of the earth proposed by the Florentine maps was the reconciliation of ancient and modern geography. Danti and other Renaissance map-makers regarded Ptolemy's system of coordinates of latitude and longitude as the backbone of cartography, but were also well aware of its shortcomings. For over a century, the sailing of unknown oceans, the discoveries of new lands, and the crossing of known but unexplored lands had been shaking the Ptolemaic world. Not only were the borders of the Ptolemaic world being extended, but also the knowledge of the old world was being revised and corrected. Yet, to discard Ptolemy's authority altogether was inconceivable. The geography of the ancient scholar had to be reconciled with the medieval heritage of Marco Polo and the reports of modern travelers. The modern map-maker had to extend and complete Ptolemy's work, not to replace it. The shape of the modern world could emerge only by combining modern measurements with classical texts, mathematical geography with humanistic philology.

Renaissance map-makers not only were skilled mathematicians, but also experts of ancient cultures, dealers in antiquities, philologists, or, at the very least, friends of accomplished scholars. The map-maker Giacomo Gastaldi completed his maps with lists of ancient and modern toponyms, possibly compiled with the help of his learned friend Giovan Battista Ramusio. Abraham Ortelius, who was close to humanists and dealers in antiquities in Antwerp, included descriptive legends and similar lists of ancient and modern toponyms in his maps; he even added an appendix of historical maps to his famous *Theatrum orbis terrarum.* Danti, who was well read in the classics and a friend of humanists at the Medici court, discussed ancient geography in the extensive legends of his maps. But it was the famous Gerard Mercator who most vividly articulated the modern cartographer's task of reconciling the geography of the ancients with the discoveries of modern times.

Mercator is mainly remembered as a cartographer and his name is connected to the projection he invented to represent as straight lines the rhombi needed for navigation. But he was also a scholar of philosophy and theology, a mathematician, a historian, an engraver, and a talented maker of globes and scientific instruments. On presenting his *Map of the World* (Duisburg, 1569) (Fig. 5.4) to the readers, a map considered to be one of the greatest cartographic achievements of all time, he declared that his aim was not only the accurate representation of the world, but also, "to show which are the parts of the universe that were known to the ancients and to what extent they knew them, in order that the limitations of ancient geography be not unknown and the honor which is due to past centuries be given to them."[21] It was in this map that Mercator, for the first time, used the projection still named after him today. In another legend of the same map he discussed the location of the Golden Chersonese, the region described by Ptolemy as rich in gold, and whose position in Renaissance maps was one of the most disputed issues, stating the dilemma of Renaissance cartographers even more precisely: "Should we, at the first incidental occasion, transpose, modify or discard the discoveries of the ancients, not only will we not improve but, by correcting a single error we will alter a hundred truths and, in the end, we will have an extremely confused mass of lands and names in which neither the parts will appear under their true names nor the names on their proper parts."[22]

In Mercator's view, which was shared by many others, including Danti, the strife over cartographic accuracy proceeded together with that of reconciling the geography of the ancients

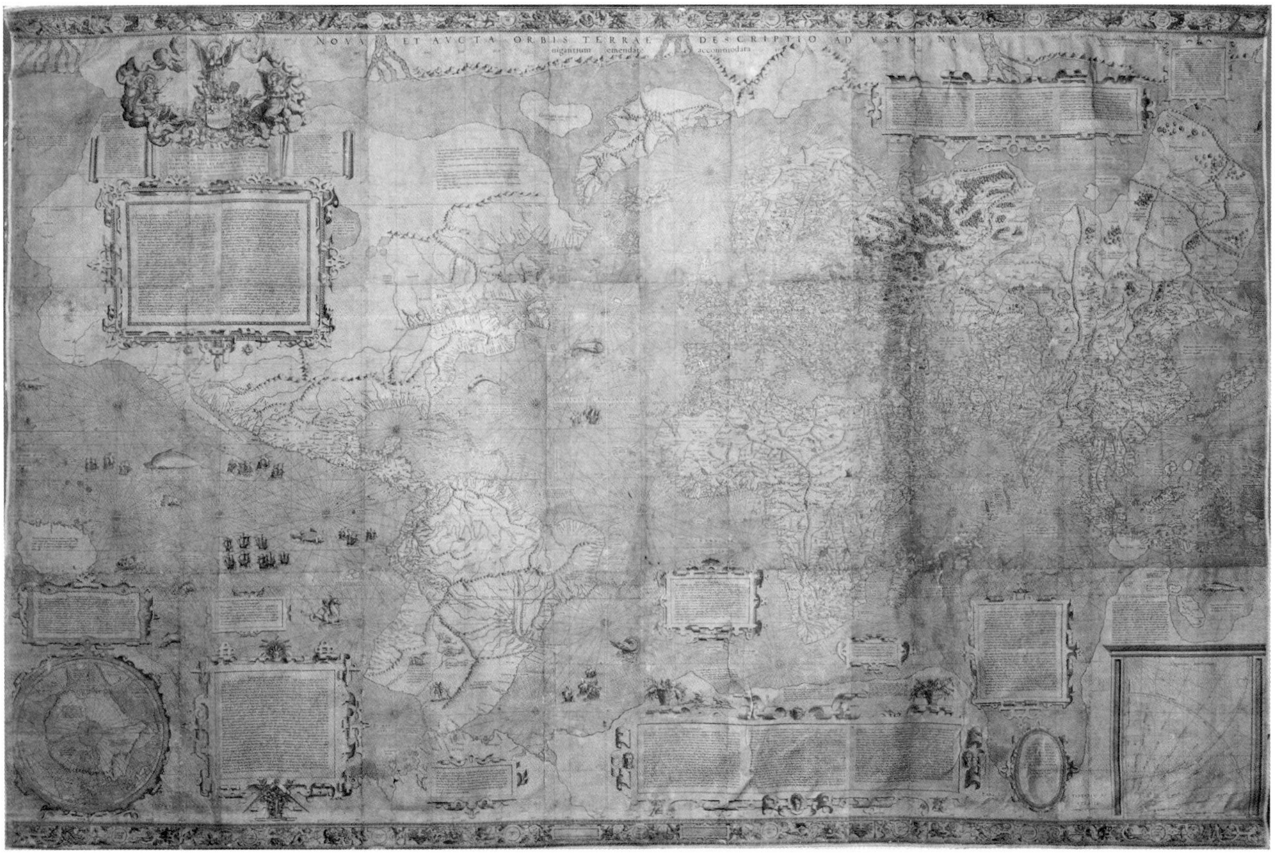

5.4 Gerard Mercator, *Nova et aucta orbis terrae descriptio ad usum navigatium* (Modern Map of the World for Navigators), 1569, engraving. Bibliothèque nationale de France

with the geography of modern times. He regarded maps as the product of both the humanistic analysis of ancient texts and the visual comparison of modern cartographic images, as images that had to represent the spatial relation of places but also their stratification in history. Ultimately, maps came to articulate, at times superbly, the interactions and tensions between the tradition of classical geography and the discoveries of the modern voyages. As a synthetic construct of modern observation, classical knowledge, and visual conventions, Renaissance maps shared the same epistemology of other Renaissance scientific illustrations, that is, they aspired to a record of the world informed both by the standards of cartography and by other non-mimetic forms of knowledge. Because of their encyclopedic meaning, Renaissance maps, magnificent images at the boundaries between art and science, could accommodate and visually reconcile the new and the traditional, the rational and the imaginary, the wondrous and the scientific. They became a privileged visual field on which naturalists, humanists, and artists joined the observation of the world with the humanistic exegesis of classical texts.

* * *

The Florentine Maps and Their Cartographic Sources

The maps of the Guardaroba Nuova were made for the exclusive view of Cosimo I and his court, but the comprehensive geographical description they proposed was a common language shared with maps made for a wide circulation. The pervasiveness of such an encompassing geographical description is best gauged through the analysis of the Florentine maps in relation to their cartographic sources, which were generally the most accomplished geographical prints produced north and south of the Alps. Such an investigation explains not only the cartography of Cosimo's map cycle, which, despite its historical significance, is known only sketchily, but it contributes more generally to the understanding of the mapping process for geographical prints and, in particular, emphasizes how deeply integrated modern geography was with its classical heritage in the mapping of the modern world.[23]

Danti and Bonsignori's working materials have not survived, but the analysis of the maps of the Guardaroba Nuova, made possible by their excellent state of conservation, shows that their favorite cartographic sources were the large wall maps made of multiple sheets, which were printed in large numbers in the 1550s and 1560s.[24] These wall maps were expensive in comparison to other prints, geographical or otherwise, and required a large house to be properly displayed, but they were greatly appreciated by learned patrons for their artistry and rich cartographic content. As masterpieces of copper engraving and woodcut these wall maps were characterized by refined graphic lines, a precise design, and clear lettering: all features that facilitated enormously the spatial understanding of the mapped territory. Furthermore, their large scale favored the inclusion of numerous geographical features, details, names, and inscriptions, so that the wall maps often offered the richest representation of an area available in print. Extremely popular, wall maps were hung in the residences of popes, rulers, cardinals, and aristocrats and, later, as the readership for maps expanded, also in the houses of scholars and merchants. It is worth noting, however, that while wall maps recorded the recent discoveries, the progress of colonial penetration, and the division of power in the new lands, they were never as informative as their manuscript counterparts or local maps, a fact that should warn the modern scholar not to apply the epistemology of the second to the first. What these wall maps surely provided was the integration of modern geography with other facets of Renaissance culture. This was a distinct advantage of printed maps over manuscript maps for the educated readers, who used them as visual aids to read the Bible and the classics, to estimate the daily reports on the European wars, to calculate the freight of merchandise, and to administer and defend the state.[25]

These popular large maps, representing large areas of the world on a unified scale and according to a single projection, greatly simplified Danti's and Bonsignori's cartographic work owing to their size, scale, and richness in geographical details. Instead of using multiple regional maps, Danti and Bonsignori selected one single wall map as their major cartographic source for each continent and divided it anew to fit the panels of the Guardaroba Nuova. They adopted the grid of coordinates of latitude and longitude of their cartographic sources, their geographical content, the legends and vignettes, which indeed offered a convenient, ready-made summary of the knowledge pertaining to the mapped territory. But Danti and Bonsignori enhanced the visual efficacy of their maps by painting them in tempera on fairly large panels rather than cutting them on wood or engraving them on copper plates. The bright colors, the vignettes, and the elaborate cartouches of the Florentine maps are the

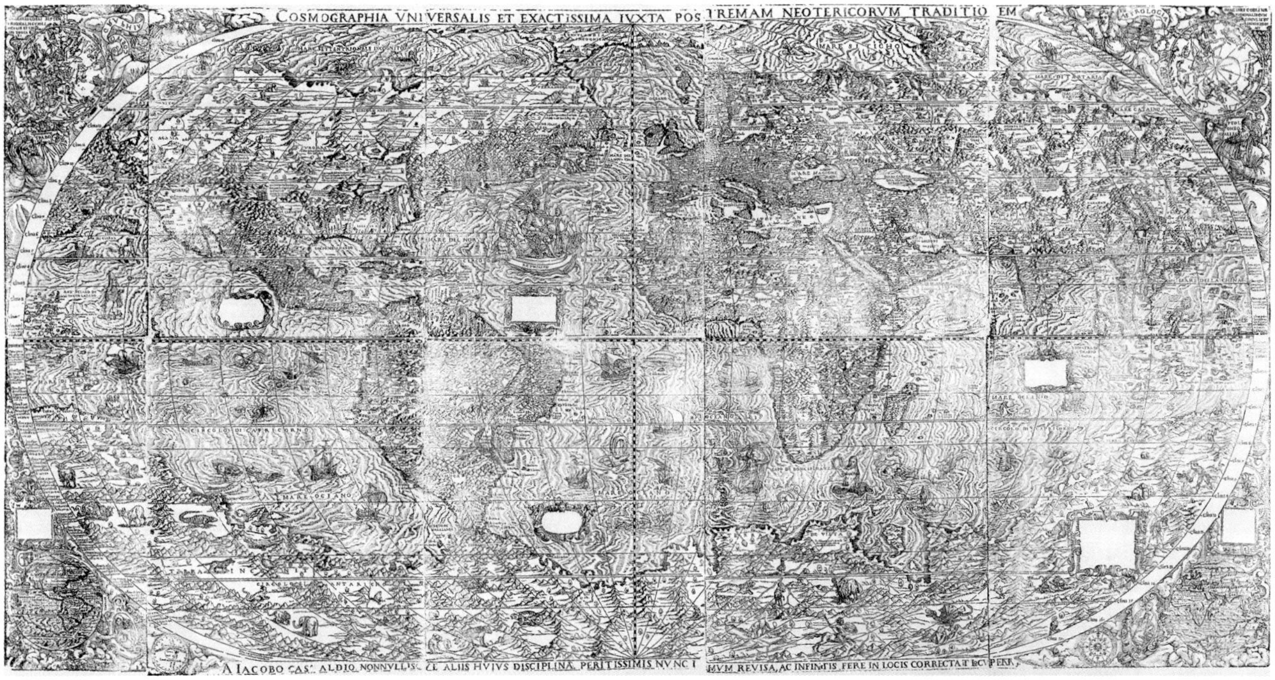

5.5 Giacomo Gastaldi, *Cosmographia universalis* (Map of the World), seventeenth-century reprint of the original map published in Venice in 1561. The British Library

ingredients of the best visualization of the differences between water and land, mountains and plains, forests and deserts – producing a map cycle that still fascinates the modern visitor, just as it must have delighted the Renaissance viewer. Although Danti and Bonsignori often used the same cartographic sources, their procedures were different: Danti, who started the cycle, was much more imaginative in combining different sources, while Bonsignori, who completed what Danti had started, was blindly imitative in following a single printed source for an entire continent, although he did show great inventiveness in the compilation of his legends. Indeed, Bonsignori's legends depart from Danti's as well as from those of most printed maps, including Ortelius's brief notes printed on the back of the maps in his *Theatrum orbis terrarum.* Instead of recounting the geographical, ethnographic, and cultural characteristics of the mapped territory, Bonsignori focused on the history of the mapped place with a very peculiar blending of biblical history and pagan mythology.

But it is Danti's mapping process rather than Bonsignori's that better exemplifies the originality of Renaissance mapping. At the beginning of his cartographic work, Danti selected one map as a guideline for the entire map cycle, the *Cosmographia universalis* (Venice, 1561) by Giacomo Gastaldi (Fig. 5.5), on whose work he relied heavily.[26] The Piedmontese Gastaldi had acquired both a solid reputation in Venice, where he had worked as a civic engineer for the republic, and a widespread fame for the maps he made for the Italian edition of Ptolemy's *Geography* (Venice, 1548) as well as for the maps he drew to illustrate the popular collection of travel reports assembled by his friend Giovan Battista Ramusio, *Delle Navigationi et Viaggi* (Venice, 1550–9). A tribute to his skills and fame is the fact that the Venetian Republic asked him to provide the sketches for the map murals to be painted in the Sala dello Scudo of its famous palace on the lagoon. His cartographic accomplishments, which included maps of Africa (1564) and Asia (1559–61), both widely used in the Guardaroba Nuova, culminated in the magnificent nine-sheet woodcut representing the earth's globe within an oval. In Gastaldi's oval map, the world, without major lateral distortions, was organized around the city of Venice, which was represented through its Lion and allegorical personification in the medallions at the bottom of the map. In the Guardaroba Nuova the centrality of Venice was replaced by the glorification of the Medici duke, but other elements of Gastaldi's world map were maintained. The two celestial maps at the top corners of the Gastaldi map (one each for the northern and southern hemispheres) and the two terrestrial globes at the bottom corners recall the *mezze palle* that welcomed the visitor at the entrance of the Guardaroba Nuova. The terrestrial globe and the armillary sphere held by the personification of cosmography and astronomy respectively in Gastaldi's map are the forerunners of the monumental globes originally planned for Cosimo's cabinet of art and curiosities. The portraits of Strabo and Ptolemy at the top of Gastaldi's map not only represent the two famous ancient geographers but also stand symbolically for mathematical and descriptive geography, the two branches of ancient geography that were reunited in Renaissance maps. Finally, Gastaldi's geographical outline of coasts and islands and the content of many of his legends reappeared word for word in Danti's Florentine maps.

Although Danti used Gastaldi's *Cosmographia universalis* as the main cartographic source for the Florentine maps, he also integrated it with the information provided by other famous wall maps pertaining to smaller areas. Gerard Mercator's *Map of Europe* (Duisburg, 1554) was the main source for the old continent; Abraham Ortelius's *Map of the World* (Antwerp, 1564) was used to represent America and as a source of legends and inscriptions throughout the Florentine maps; and Mercator's *Map of the World* (Duisburg, 1569) was used to fill miscellaneous gaps. Occasionally, Danti also used Portuguese and Spanish unpublished maps, nautical charts, and travel reports that Cosimo made available to him from the rich collection of geographical materials that the Medici had assembled for over a century.[27] Danti found useful the nautical charts made by Bartholomeo Velho in 1561, from which he copied the coastline of southern Africa, the East Indies, and South America.[28]

In one instance only did Danti refer specifically to his manuscript sources and he did so some twenty years after terminating the cartographic work on the Guardaroba Nuova. In his oft-cited letter to Abraham Ortelius of 1581, he recalled two books that he had received from Cosimo I:

> I had from His Excellency two books, one concerning the voyage from Lisbon to the Moluccas, written by a Portuguese pilot who made that voyage fourteen times, and he marked from place to place the height of the pole, and the character of the localities and their distances, with the depth of the sea, and other information concerning that course, [all] done with such care that if that course should at any time be lost, this book will be sufficient to bring it back to light. Another book does the same for the West Indies. Grand Duke Cosimo had the first work from the Cardinal Infant of Portugal; and the second from King Philip.[29]

Danti's informative description makes it possible to identify his source with a manuscript currently kept in the Biblioteca Nazionale, Florence, but once belonging to the Medici library, a copy of which he must have kept for himself after he left Florence. The manuscript contains the Italian translation of two travel reports, one Portuguese and the other Spanish, very rich in nautical instructions but very poor in verbal descriptions, documenting respectively, the Portuguese route from Lisbon to the East Indies, and the Spanish route from Seville to the West Indies. Dedicated to Cosimo, who commissioned the translation, and dated September 1562 (only a year before the planning of the map cycle), the manuscript was prepared by Bernardo Baroncelli, a professional sailor who had traveled extensively on the duke's behalf and later became a ducal administrator (*provveditore*) of the port of Livorno.[30] Cosimo had received this manuscript as a gift from a member of the Portuguese royal family, but it is still impossible to identify the Portuguese original, its author, or the circumstances of its production. The origins of the Spanish text also remain obscure, since it is still unclear whether Baroncelli translated it into Italian or wrote it himself on the basis of Spanish sources. Whatever the original texts were, the manuscript was crucial to Danti, who used it to make the Florentine maps and, later, around 1580, to paint a mural world map for Pope Gregory XIII in the Vatican Palace.[31]

Danti's working method involved the integrated use of different sources, both verbal and visual, and in many respects it was remarkably similar to that of Abraham Ortelius, who is generally credited with the making of the first modern atlas. Although there is no evidence that Danti and Ortelius knew each other or corresponded in the 1560s, they proceeded along similar paths in imagining the modern world as a series of maps organized according to Ptolemy's order. They both reduced their source maps (often the same ones) to a homogeneous size, the format of a large folio for Ortelius, the size of a wood panel for Danti. They submitted their sources to a similar process of comparing and collating, a process that aimed at the integration of geography with encyclopedic knowledge. Their respective map cycles, however, were sharply different in terms of circulation and methods of viewing. While Ortelius's collection of maps was bound in a book and made accessible to a large readership who could easily turn the pages in the space of a private library or study, Danti displayed his maps on the walls of a princely palace for the privileged audience of his patron and court. Their individual maps, however, shared the same cartographic language, a language that was crucial for integrating geographical knowledge with other facets of Renaissance culture. Both Danti's and Ortelius's maps combined different methods of representation and different kinds of information in one synthetic image, which defined the mapped territory both mathematically and pictorially.

In making his maps Danti did not follow any specific geographical order. Instead, he was attracted first by the areas that had not been mapped by Ptolemy, such as northern Europe, southern Africa, and Southeast Asia, and left to Bonsignori the better known parts of the world. However, Danti and Bonsignori must have known well the place of each of their maps within the map cycle, even though they did not execute them according to that order. Since the current display of the maps in the Guardaroba Nuova does not correspond to any historical geographical order, the following analysis of Danti and Bonsignori's maps adopts the geographical order proposed by Giacomo Gastaldi in his edition of Ptolemy's *Geography*. This contemporary book, which intercalated the Ptolemaic maps with the modern maps, offers the closest intellectual framework for interpreting the Guardaroba Nuova as a project aspiring to the emulation and completion of Ptolemy's mapping of the world.

The Expansion of Europe

Ptolemy had represented Europe, the first continent of his scheme, in ten maps, but in the Guardaroba Nuova the maps of the old continent were increased to fourteen in order to represent the northern regions, which "were little known to the ancients," as Danti himself reported in one of the legends.[32] Fascinated by the prospect of completing the project of the ancients by mapping Europe beyond the sixty-second degree of latitude, at which point the knowledge of the ancients had arrived, Danti started his map cycle with the representation of the British Isles and the Scandinavian peninsula. Later Bonsignori made the maps of central and southern Europe.[33]

Danti's main cartographic source was Gerard Mercator's *Map of Europe* (Duisburg, 1554), a magnificent fifteen-sheet wall map, now lost, which influential politicians throughout Europe listed among their possessions (Fig. 5.6).[34] Dedicated to Bishop Anthony Perrenot and possibly made with the purpose of mapping the Catholic presence in the religiously divided continent, Mercator's map was surely available in Venice, where a copyright request was submitted to the Republic in 1554, and from Venice it could easily have passed to other Italian cities. Its presence in Florence in the early 1560s is documented by Danti's maps *Thile Insula* and *Norvegia Gotiae* (Fig. 5.7), which closely follow Mercator's map both in the outline of the geographical areas and the content of the legends and inscriptions. Inspired by Mercator, Danti used his legends to comment on a variety of topics. The people in these areas, "are all Christian but follow their own creed." They are notable for their hunting and fishing knowledge, including a special technique for breaking the iced sea. The Island of Thule is renowned for its glacial weather, for the "pyramids in which the ancients from this island gathered with the Asians," pyramids which indeed are represented on the map, and for its mention in Seneca's tragedies (two verses from Seneca's tragedy *Medea* are quoted in the cartouche).

Danti's other maps of Europe, *Isole Britanniche* and *Livonia et Littuania* show instead his attempt to integrate Mercator's representation of the old continent with additional cartographic sources in the painstaking effort to offer a rich and homogeneous representation of the world. These maps exemplify remarkably the bricolage work of a Renaissance mapmaker. The cartographic content and the legends of Danti's map *Isole Britanniche* demonstrate that Danti skillfully combined Mercator's *Map of Europe* with the map of the British Isles made by George Lily, an English Catholic exile at the papal court, published in Rome in 1546 (Figs. 5.8 and 5.9). Danti selected Mercator's representation of Ireland, which indeed was the most detailed map of the island, available in print, especially of its west coast, but he departed significantly from it in the representation of England, Scotland, and Wales and followed instead Lily's map. Compiled from a variety of published and unpublished continental and British sources, Lily's map was then the most detailed representation of England and Scotland (Ireland was still shapeless). It was also a religious tool. Lily marked on his map the episcopal and archiepiscopal sees but excluded the controversial episcopal sees of Gloucester, Petersburg, and Oxford which had been elevated by the Anglican church against the papacy, thus mapping the Catholic presence in the Anglican land. The historian Paolo Giovio suggested the use of Lily's map alongside his book *Descriptio Britanniae* (Venice, 1548), and the map may have been bound together with Giovio's text. Whatever its religious meaning and uses, Lily's map enjoyed considerable success, and could have been easily acces-

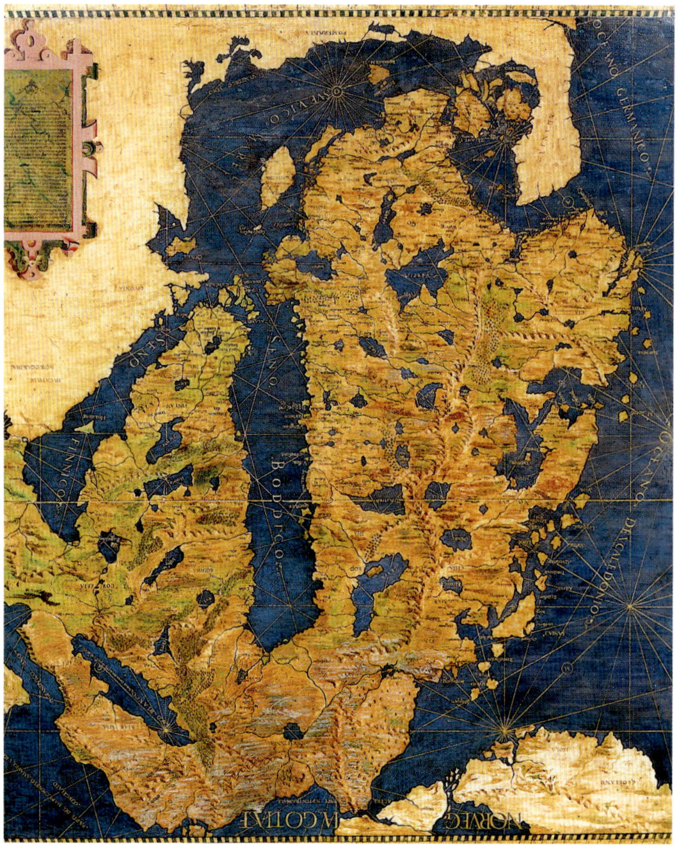

5.7 Egnazio Danti, *Norvegia Gotiae* (Map of the Scandinavian Peninsula) ca. 1565, panel. Florence, Palazzo Vecchio, Guardaroba Nuova

sible to Danti either in its original issue or, more likely, through one of its ten derivative maps published between 1558 and 1563. The orientation of Danti's map and the position of its legends suggest that Danti's most likely source was the derivative map issued by Fernando Bertelli and engraved by Paolo Forlani in Venice in 1562. Although Danti did not mark the Catholic and Protestant episcopal sees in his map, he did comment that the people there, "are all heretics and separated from the Catholic Church." He also reported that the British Isles, once poor in music and literature, were then prolific in culture and strong in arms, rich in gold, silver, and tin, in wool, meat, and imported wine. Scotland is particularly famous for a special stone that burns like coal, while in one of the northern islands is St. Patrick's well.[35]

Danti illustrated Russia, the vast land between Europe and Asia that Renaissance cartographers placed ambivalently in either continent, in two separate maps, *Livonia et Littuania*, which mapped the European side, and *Moschovia*, which instead illustrated its extension in Asia (Fig. 5.10). Even though Danti's maps of Muscovy mapped a land divided between two continents, they were conceived as a unity and based on the same cartographic source, Mercator's *Map of Europe*, from which Danti copied the coordinates of places, the outline of lakes and rivers, and the literal translation of the legends. Only one of Danti's legends commenting on the location of the Mountains of Hyperborea is not derived from Mercator, revealing Danti's receptiveness to new reports from Russia. News from Russia, a land still relatively unknown to sixteenth-century Europeans, fired the imagination of readers and map users, who fantasized about the richness of its products and the stellar gains to be derived from a possible northern trading route, as well as about the violent character of its inhabitants and their scandalous familial habits. In 1557 the English traveler Anthony Jenkinson traveled in Russia in search of a trading route from Moscow to Persia and China, a route which had long been a fixation of European merchants and an obsession for English and Dutch traders trying to catch up with the Portuguese and Spanish trade in the East Indies. In his travel report, published in 1560, Jenkinson considerably updated the European knowledge of Russia, which was reflected also in the map *Russia et Moschovia* (London, 1562), made to accompany his book (Fig. 5.11). Among Jenkinson's revisions was the location of the Mountains of Hyperborea, which Ptolemy, and

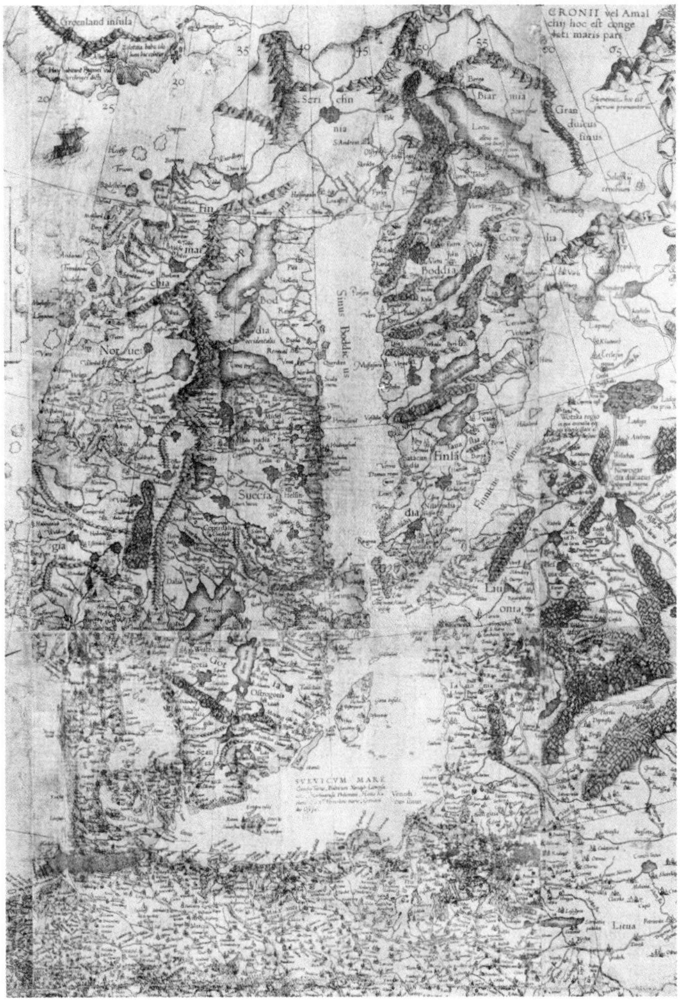

5.6 Gerard Mercator, Map of Europe, 1554, engraving. Modern facsimile edition of the copy once in the Library of the University of Breslau (destroyed). Yale University Library

Mercator after him, had placed east of Novgorod. Jenkinson reported that, on his trip to that area, instead of climbing the Ptolemaic mountains, he had to walk through an extensive flat land. Danti mapped a flat land in the area, adding the following inscription: "In this place Ptolemy located the Mountains of Hyperborea, but today it is known there are only marshlands." Danti may have had access to Jenkinson's map either in its original form or in a later Dutch reprint of 1568, or even from Ortelius's *Theatrum orbis terrarum*, which from 1570 onward included Jenkinson's map. It is according to Jenkinson's map that Danti corrected the delta of the Obi River in the North Sea, the course of the Volga and its estuary in the Caspian Sea, added the Pechora River that Mercator did not know about, and updated the outline of the Caspian Sea.[36]

Although in his maps of Muscovy Danti was able to dismiss the Mountains of Hyperborea as fictional, thanks to Jenkinson's firsthand report, he continued to be fascinated by other such classical features as the land of the Amazons, which he firmly located according to Strabo; or such fanciful news as the report that, "all the animals taken in this province [near the Lake Ladoga] soon become white." From Jenkinson's travel report Danti also took the ethnographic information pertaining to the people of Russia that reported that the lands were so marshy people could travel only in the winter when the wet lands were covered with ice and snow. These lands produced apples, wax, and sables, and had important trade cities such as Vilna. Their inhabitants, the Tartars, lived in tents and carriages, which Danti depicted skillfully in out-of-scale vignettes on his map. Tartar men could not have concubines, but their women could marry more than once and in fact were encouraged to have lovers which they called, "helpers to procreation" (*aiutatori alla generatione*). Danti was so puzzled by their habits that he could relate to them only through Aristippus's famous dictum, "honestum non natura sed consuetudinem constare," appropriately quoted on the map.

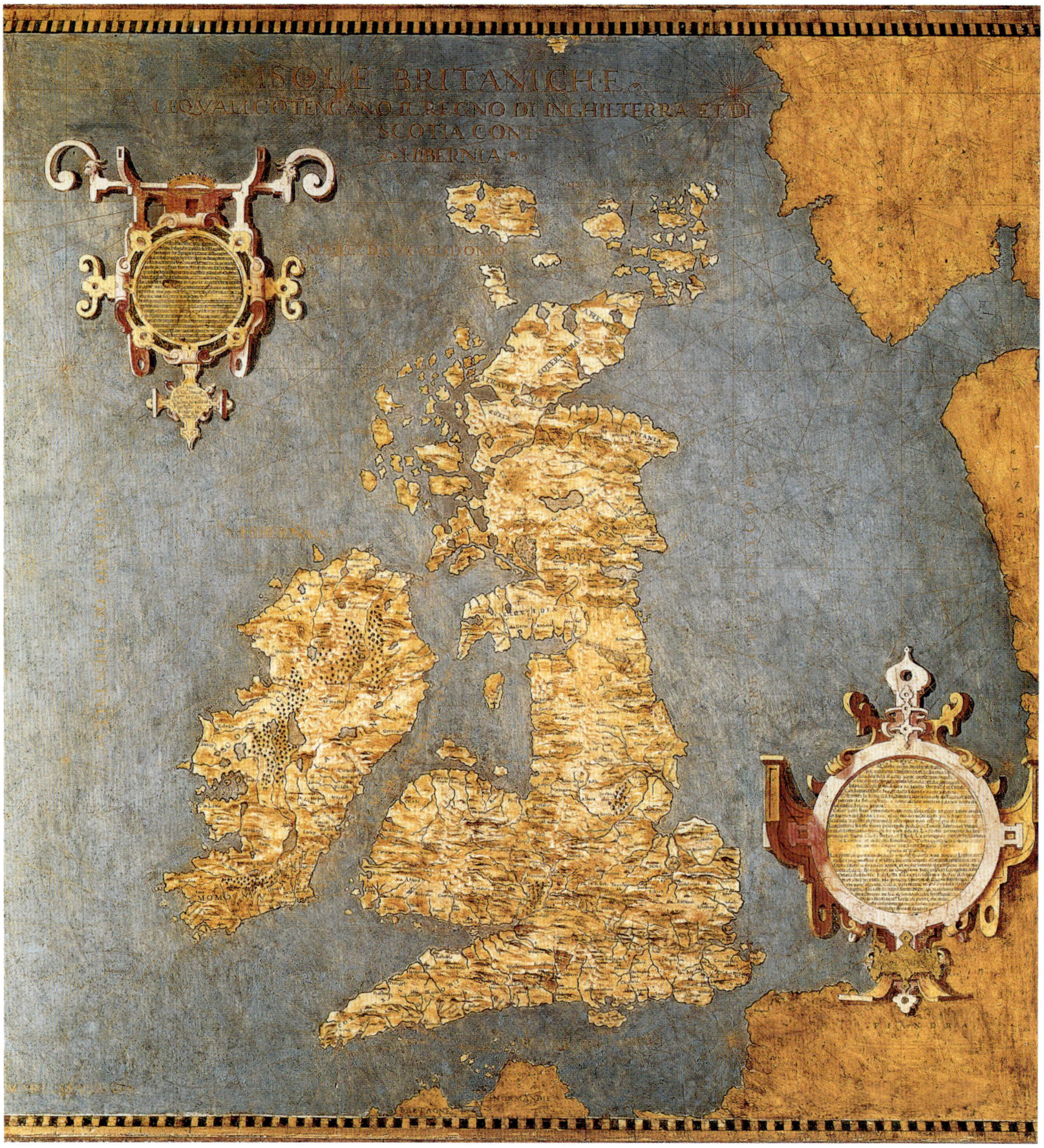

5.8 Egnazio Danti, *Isole Britanniche, le quali co(n)tengono il Regno di Inghilterra et di Scotia con l'Hibernia* (Map of the British Isles), 1565, panel. Florence, Palazzo Vecchio, Guardaroba Nuova

When Bonsignori took over the completion of the map cycle, he could count on the selection of maps assembled by Abraham Ortelius in his *Theatrum orbis terrarum*, whose maps he simply reproduced in tempera. Occasionally, however, Bonsignori used Ortelius's original cartographic sources rather than the reductions he had published in his map book, as in the case of his map *La Spagna* (1577), which was based on Carolus Clusius's *Map of Spain* (Antwerp, 1571) rather than on the smaller version of Clusius's map that Ortelius had printed in his map book.[37]

The Circumnavigation of Africa

According to Vasari's description, in the Guardaroba Nuova the continent of Africa had to be illustrated in eleven maps, a considerable increase to the mere four that Ptolemy had dedicated to the second continent of his scheme.[38] Like most Renaissance printed maps of Africa, Danti's and Bonsignori's maps offer an interesting case in the evaluation of how map-makers combined the conflicting authority of Ptolemy and modern Portuguese sailors in one single cartographic image. Ptolemy, who had limited knowledge of the continent south of the equator, believed Africa was attached to Asia and considered the Indian Ocean to be a large internal sea, a sort of Asian counterpart to the Mediterranean Sea. From the early fifteenth century the Portuguese started to explore the coasts of Africa. They sailed down the coast, around the Cape of Good Hope, along the eastern coast, and into the Indian Ocean. In a few decades they redesigned entirely the trading route from Europe to the East Indies. They cut out the Mediterranean and its ports, avoided the land routes and taxes across mainland Asia, sailed from Lisbon to Gibraltar, circumnavigated Africa, reached India and the Moluccas Islands by sea, brought spices and luxury goods directly to Lisbon and, from there, disseminated them into the markets of Europe. While successfully searching for a seabound route, the Portuguese also altered the map of the earth. They showed that Africa and Asia were separated, that the Indian Ocean was a vast sea full of islands, and that the coastal outline of Africa was completely different from that conceived by Ptolemy. Portuguese nautical charts redrew coastal

5.9 George Lily and Fernando Bertelli, *Britania insula quae duo regna continet Angliam et Scotiam cum Hibernia adiacente* (Map of the British Isles), 1562, engraving. The Newberry Library (Novacco Map 2F 5740.3 (4F128))

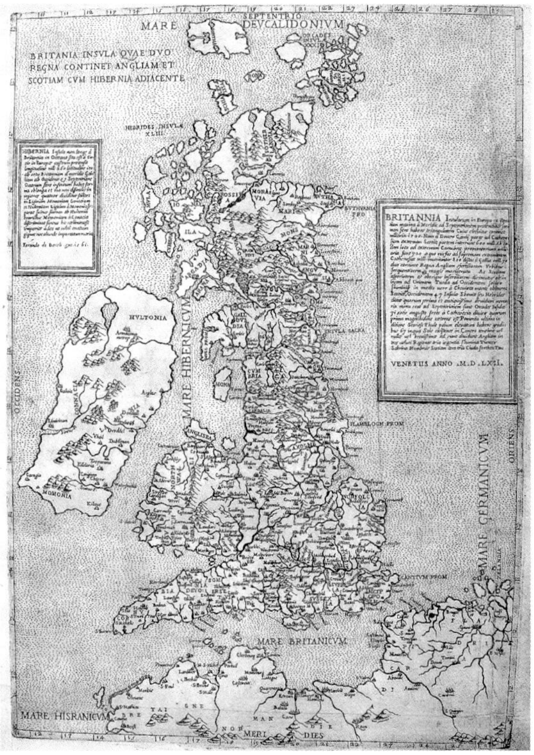

Africa and, owing to their prolonged contacts with Islamic sailors and traders, also introduced important innovations in the cartography of central Africa, an area not touched by their explorations.[39]

Danti trusted the depiction of Africa in the Portuguese nautical charts, although he was much more skeptical about their accuracy in the mapping of the Indian Ocean. For example, Danti's representation of the Island of Madagascar, which appears twice in the Florentine map cycle, was based on different sources, an issue he addressed in one of the legends: "If this part of the Island of San Lorenzo is somewhat different from the [same] part represented in the individual map [of the Island of San Lorenzo] the reason is that one is taken from a trusted Portuguese map while the other comes from different authors" (Fig. 5.12).[40] Danti's map *Isola di San Lorenzo* did not correspond to any printed map, but it was very close to the Portuguese navigational charts available in Florence at the time, particularly the nautical charts by Bartholomeu Velho (1561).[41] Velho's nautical charts revised the Ptolemaic representation of Africa both along the coasts and in the interior (Fig. 5.13). One of the many innovations Velho introduced pertained to the representation of the *Monti della Luna*, Ptolemy's fictional mountains from which, it was believed, the Nile originated, and the two large internal lakes that Ptolemy had imagined on the same parallel, but that Velho instead mapped with a north–south alignment. However, as it was typical for nautical charts, Velho did not include any legend discussing his revision of Ptolemaic features, forcing Danti to borrow the inscriptions for his maps from a famous printed source, Ortelius's *Map of the World* (Antwerp, 1564). In fact, printed maps of Africa rather than navigational charts combined the novelty of Portuguese discoveries with the authority of Ptolemy and Strabo.

5.10 Egnazio Danti, *Livonia et Lituania* (Livonia and Lithuania), not dated, panel. Florence, Palazzo Vecchio, Guardaroba Nuova

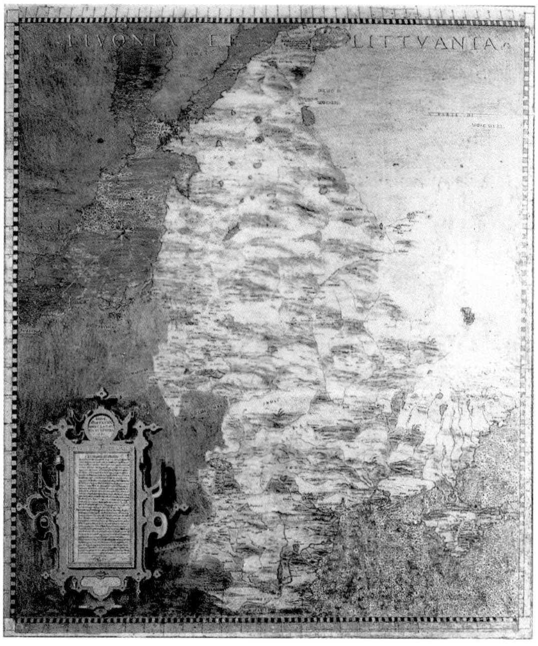

Exemplary of this kind of geographical print is Giacomo Gastaldi's *Africa* (Venice, 1564), the eight-sheet map Bonsignori selected as the cartographic source for his maps of Africa in the Guardaroba Nuova (Figs. 5.14 and 5.15).[42] The culmination of a lifelong cartographic work on the continent, Gastaldi's *Africa* was as successful as his *Cosmographia universalis*, making available to the general public the coastal outline of the continent that Portuguese sailors and cartographers had mapped. Like his Portuguese sources, Gastaldi removed the Ptolemaic *Monti della Luna* as the source of the Nile, replacing them with a large lake at the center of Africa. However, unable to escape fully the visual appearance of the Ptolemaic maps, he created a second lake in central Africa and placed it at the same latitude as the first one, just as Ptolemy had done. Finally, unlike his Portuguese sources, he completed his map with the list of correspondences between ancient and modern toponyms, a list absolutely irrelevant to sailors but of great interest to the general learned public who bought his prints. Gastaldi's *Africa* went through two new editions in its original large format and was published many times at reduced size. Even Abraham Ortelius, who in the first edition of his *Theatrum orbis terrarum* had produced a map of Africa more directly based on Portuguese sources, returned to the prototype in 1573, while Mercator never truly departed from it. It was Gastaldi's syncretistic reconciliation of modern Portuguese cartography and traditional Ptolemaic cartography that appealed to map users and map-makers of the following decades, rather than the more accurate, but historically mute, navigational charts. Similarly, the maps of Africa in the Gua-rdaroba Nuova followed Gastaldi's model in combining Portuguese nautical charts, sai-lors' travel reports, and Ptol-emy's authority.

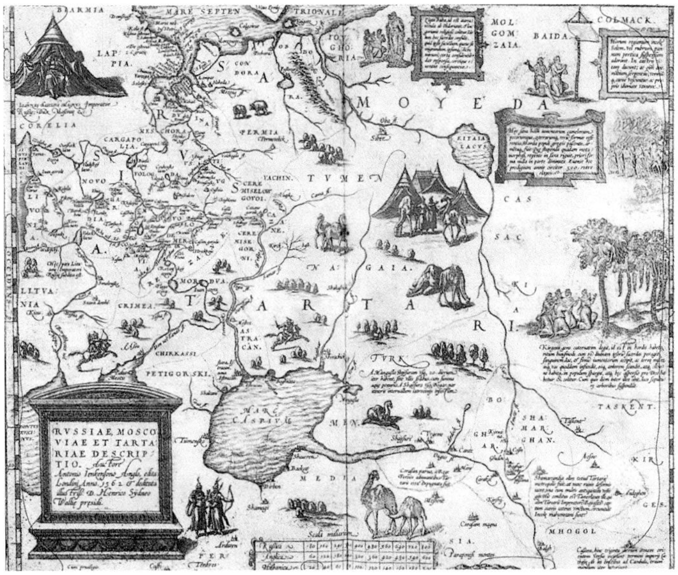

5.11 Anthony Jenkinson, *Russiae Moschoviae et Tartariae Descriptio* (Map of Russia, Muscovy, and Tartary), engraving, from Abraham Ortelius, *Theatrum orbis terrarum*, Antwerp 1570. The University of Virginia Library

The Reduction of Asia

The continuous negotiations, interplay, and tension between the geography of the ancients, the medieval heritage of Marco Polo, and the modern travel reports, typical of Renaissance geographical prints, are particularly acute in the mapping of Asia, a continent that Ptolemy had represented in twelve maps and that had been visited by Marco Polo in the thirteenth century and, more recently, by Portuguese sailors. The exaggerated extension of Asia handed down from the Ptolemaic tradition had been revised by modern sailors and also the separation between Asia and America had been slowly understood, but exactly how far apart Asia was from America remained to be established, while the direction of the Chinese coastline, the location of Japan, and the shape of the south eastern islands were still up for debate. The maps of Asia in the Guardaroba Nuova, like the printed maps from which they derived, reflected the uneven cartographic knowledge of the period and provide telling evidence on how map-makers faced their cartographic challenges.[43]

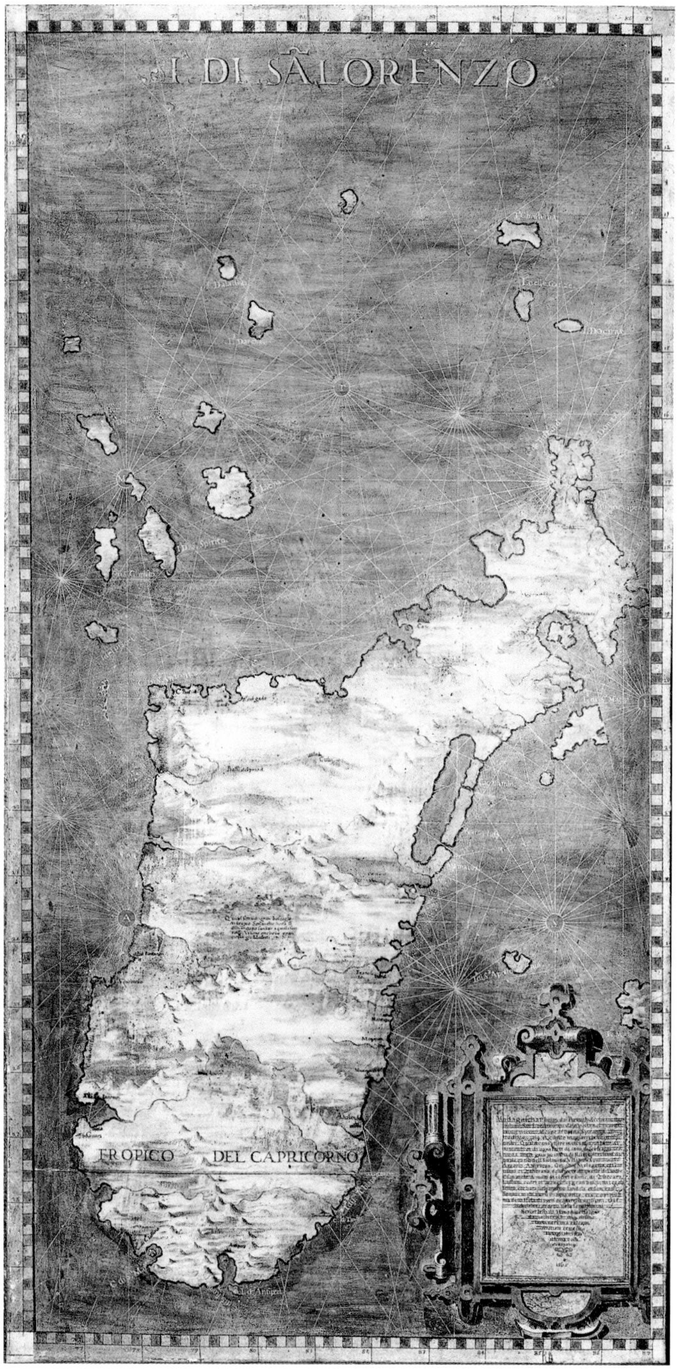

5.12 Egnazio Danti, *Isola di San Lorenzo* (Madagascar), 1565, panel. Florence, Palazzo Vecchio, Guardaroba Nuova

Danti made his maps of Asia over a period of a decade, but based them on the same cartographic source, Giacomo Gastaldi's maps of Asia (Venice, 1559–61), which he integrated with Gastaldi's *Cosmographia universalis* (1561) for the representation of the Indonesian islands south of the equator. Danti's choice of Gastaldi's maps of Asia as his main cartographic source is unsurprising, since these were the most updated descriptions of Asia available in print. Compiled from a multitude of published and unpublished sources that Gastaldi had obtained from his friend Ramusio, these magnificent copper engravings had been conceived in three distinct sections, each made of multiple sheets and published separately (two more sheets representing the Indonesian islands were added in 1565) (Fig. 5.16). Danti sliced the three source maps into thirteen panel maps and followed Gastaldi closely not only in the area represented – they both omit the northern areas of the continent – but also in the grid of coordinates of latitude and longitude, the geographical outline, the hydrography, horography, and toponyms of the mapped territories (Figs. 5.17, 5.18, 5.19, and 5.20). Danti, however, found that Gastaldi's maps of Asia failed to combine the mathematical geography of the continent with the qualitative description of its people and products. Gastaldi had included a legend in each of his maps of Asia explaining the geographic limits of each area; he had even complemented the maps with a separate sheet listing the correspondences between ancient and modern toponymy; but it was only for the third map of Asia that he had provided a series of brief inscriptions marking the location of unusual and precious products, which indeed Danti copied in the legend of his map *Indostan fuori al Gange*. Generally, however, Gastaldi's verbal descriptions

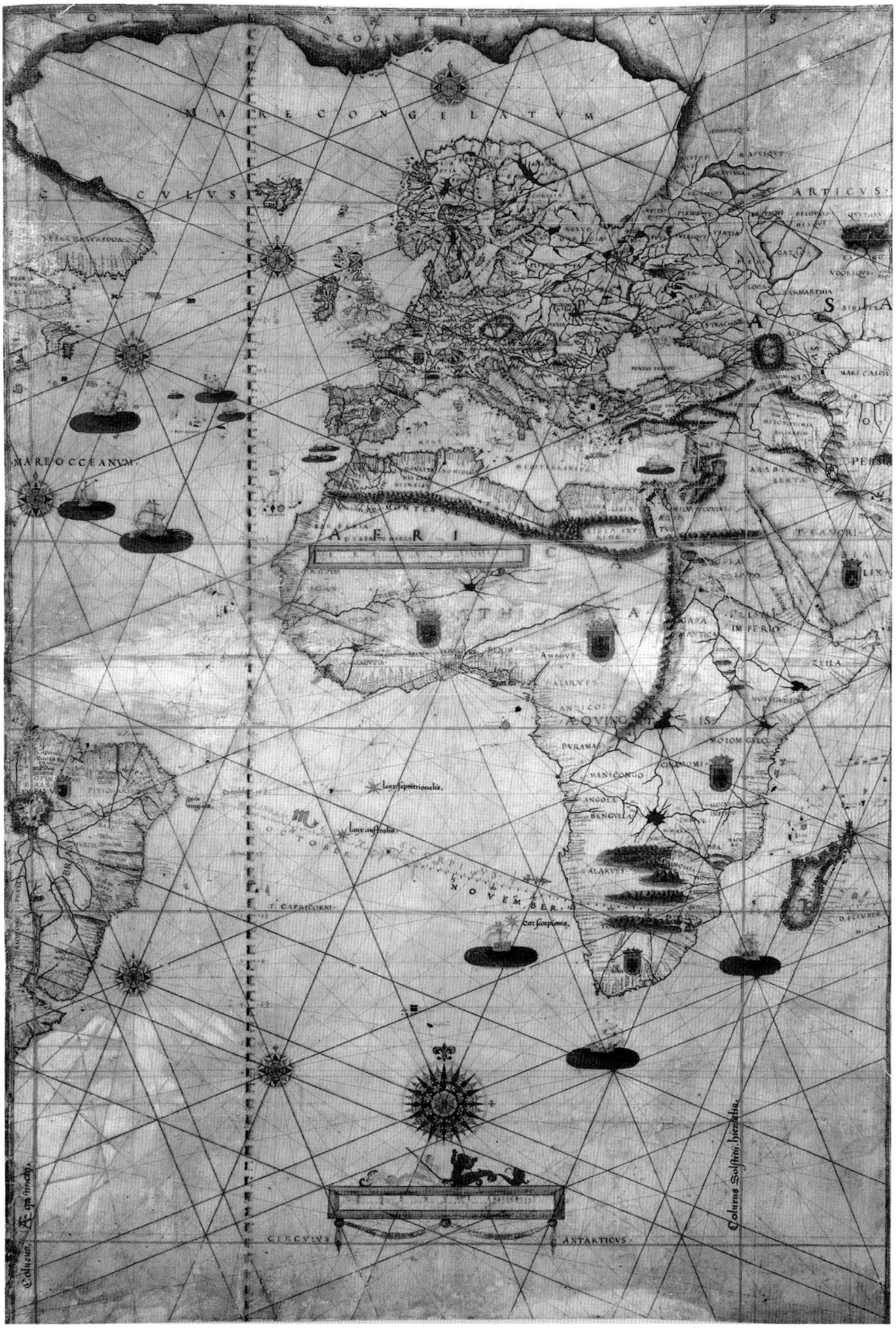

5.13 Bartolomeu Velho, Africa, 1561, parchment. Florence, Museo di Storia della Scienza

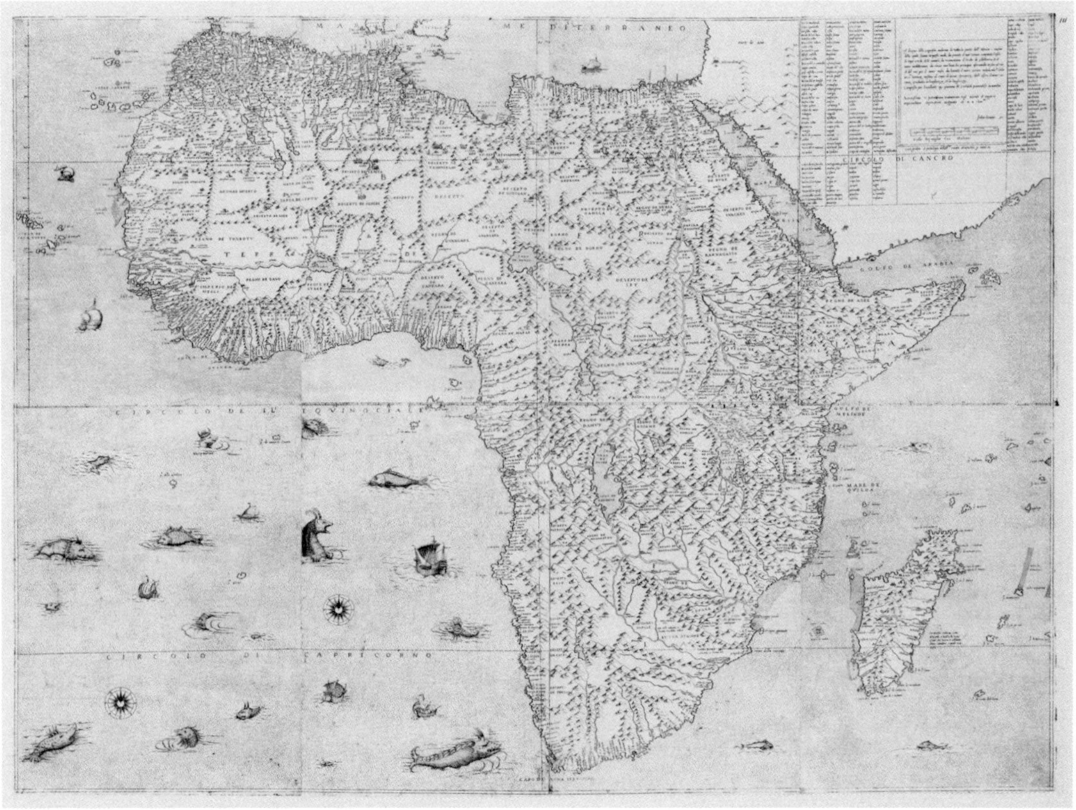

5.14 Giacomo Gastaldi, *Il disegno della geografia moderna de tutta la parte dell'Africa* (Map of Africa), 1564, engraving. Öffentliche Bibliothek, Universität Basel

were limited in content and topics, a fact quite uncharacteristic for this successful map-maker and utterly unsuitable for the encyclopedic description of Danti's maps.[44]

As Danti was making his panel maps of Asia, another influential map of Asia appeared, Ortelius's *Asiae orbis partium maximae nova descriptio* (*New description of Asia, the largest part of the earth*) (Antwerp, 1567) (Fig. 5.21).[45] Although it claimed to be a novel description of Asia, Ortelius's map very closely followed Gastaldi's, but it complemented its geographical outline with many legends, cartouches, and inscriptions. Characteristically, Ortelius's legends not only described the geographical features of the mapped territories, but they also systematically articulated the correspondences between Ptolemy's geography, his regions, islands, cities, and mountains, and the travel reports of modern voyagers. Danti fully endorsed Ortelius's determination to reconcile ancient and modern geography. He repeated Ortelius's legends word for word, translated them into Italian, and thus added the dimension of time to his maps of Asia. The content of Ortelius's legends and the fact that Danti found them so appropriate that he copied them verbatim are further indications that Renaissance map-makers regarded the discussion of ancient geography as an integral part of the mapping of the modern world. Particularly pertinent in this respect is Ortelius's legend discussing the correspondence between the islands of Ceylon (*Zeilan*) and Sumatra (*Tabrobana*)

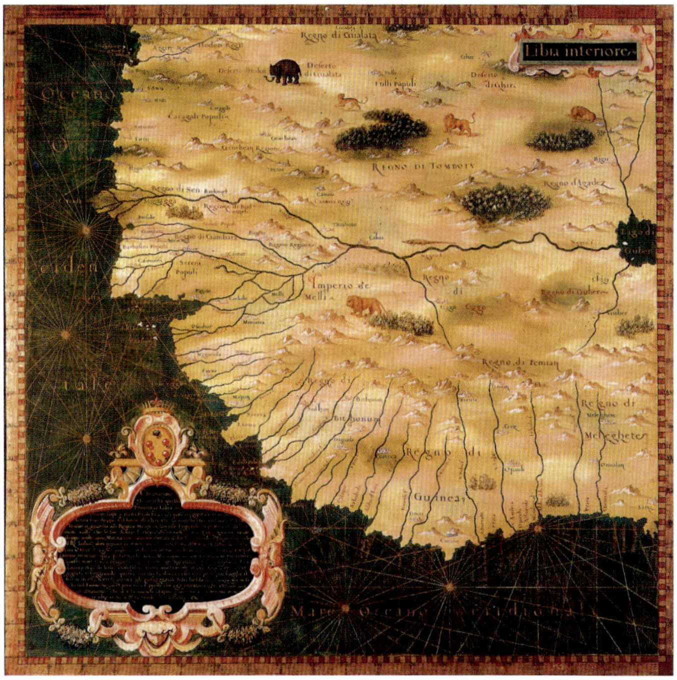

5.15 Stefano Bonsignori, *Libia interiore* (Map of the Western Coast of Africa), 1580, panel. Florence, Palazzo Vecchio, Guardaroba Nuova

with Ptolemy's Golden Chersonese, a legend that Danti translated into Italian in his Florentine map of India (Fig. 5.17). Like Ortelius, Danti identified the Golden Chersonese with the island of Sumatra, rejecting the opinion of other cartographers, namely Mercator, who identified it with the Chersonese Malacca, the peninsula of Malaysia (Fig. 5.4). Ortelius's argument, which Danti repeated, is a fascinating mixture of cartography, philology, and rudimentary geology. The longitude of Sumatra, he claimed, corresponds to the longitude Ptolemy assigned to the Golden Chersonese not to the Chersonese Malacca. Additionally, the Ptolemaic name implied that the peninsula was full of gold while in Malacca, "there is no gold apart from that which is imported." Finally, he reinforced his identification of the peninsula of the Golden Chersonese with the island of Sumatra by arguing that the island was originally a peninsula, which "was separated from the continent by the force of the ocean after Ptolemy's time."[46] To locate correctly the legendary Golden Chersonese was one of the obligations of Renaissance cartography, and this was information that map users expected to find in a geographical print.

In other maps of Asia Danti acknowledged explicitly the syncretism of his cartographic sources and the difficulties in achieving a consistent mapping of the modern world. While the wish for consistency was obviously in Danti's mind, so too was the awareness of its elusiveness. What is worth noting, however, is not the presence of inconsistency, but rather the way in which Danti dealt with the conflicting evidence of modern maps. Often unable to choose between different representations of the same area, Danti exposed the contradictions in the legends of his maps, without even attempting to solve the problems they raised. For instance, in the legend for the map of China and Japan Danti discussed the Chinese coastline and the location of Japan, two of the most debated areas among Renaissance map-makers (Fig. 5.19):

> Great difference exists among geographers on the location of this coast of China and of the Island of Japan, which have been placed in this map according to the opinion of the most modern authors, although it seems that this coastline from Cape Litampo to Paugiu had to be represented going from south-east to north-west rather than from north-east to south-west as it is represented here. And the Island of Japan ought to go from south to north rather than from east to west. But since in the large globe (with the approval of the Most Serene Grand Duke of Tuscany Cosimo) this has been represented according to

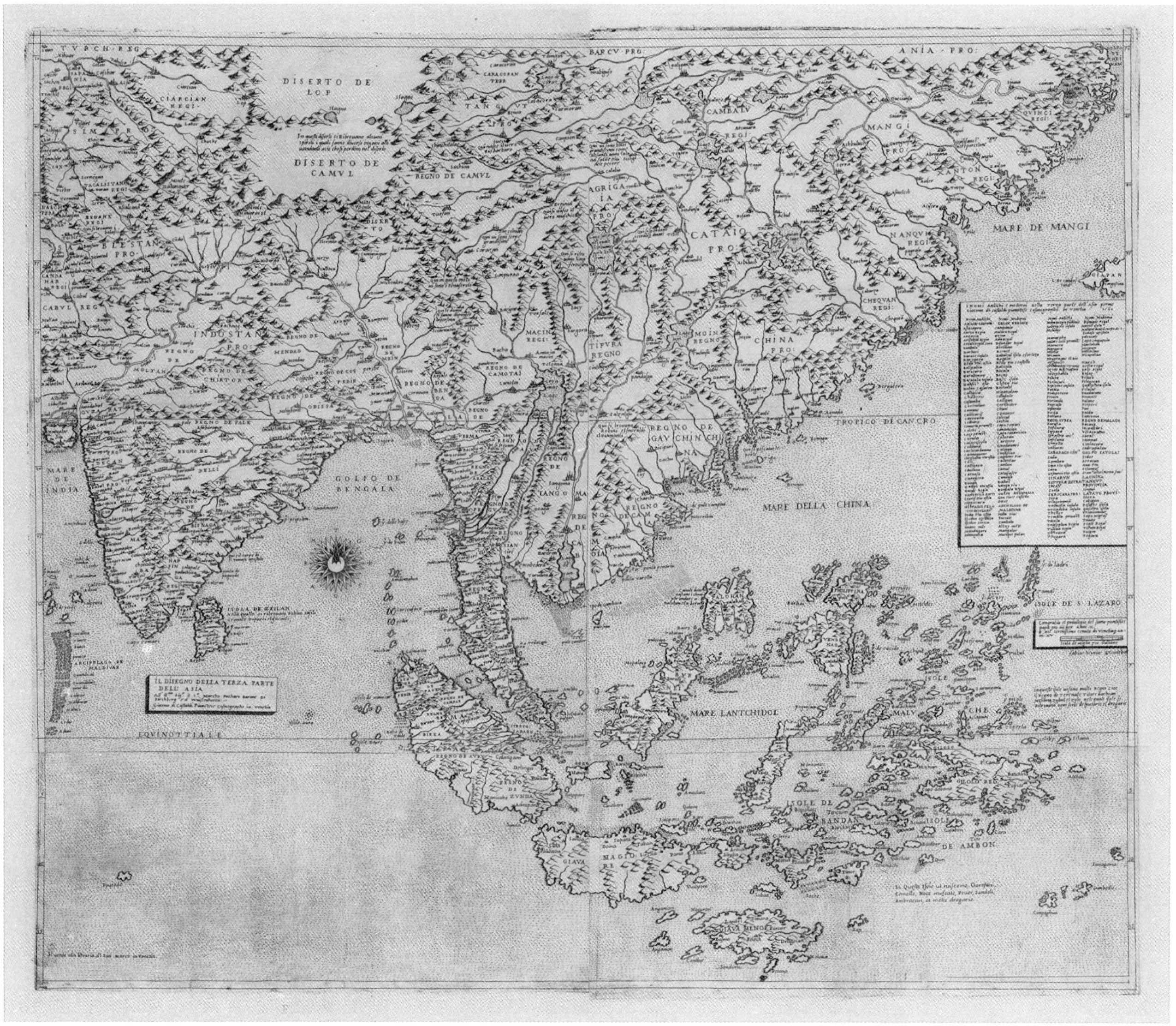

5.16 Giacomo Gastaldi, *Il disegno della terza parte dell'Asia* (Map of East and South-east Asia), 1561, engraving. Öffentliche Bibliothek, Universität Basel

Marco Polo's description and according to the location of Juan do Barros, who explicitly says that the most easterly cape of the discovered Indies is Cape Litampo, which is near the city of Nimbo, after which it is also named Cape Ninsu. Without mentioning that the coast of China running from south-east to north-west appears in an old world map similar to the one made earlier in Venice by Marco Polo, who wrote on that province many things, which although they were not believed then, it was later found out that they were all true.[47]

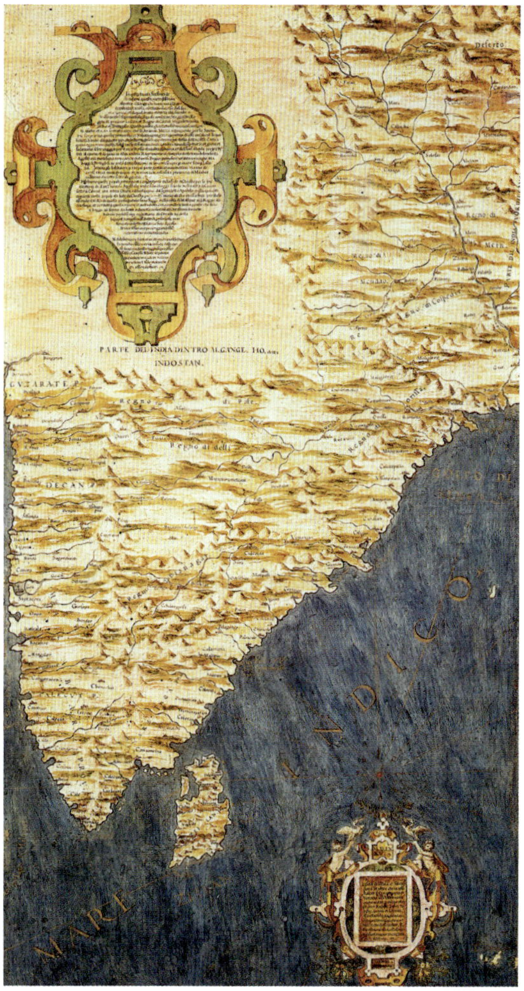

5.17 Egnazio Danti, *Parte del India Dentro al Gange Hoggi detta Indostan* (Map of India), 1575, panel. Florence, Palazzo Vecchio, Guardaroba Nuova

Following Gastaldi and Ortelius, Danti represented the northern part of China (the coastline from Cape Litampo to Paugiu) extending eastward, but in the map's legend he acknowledged that other cartographers had represented the same portion of the Chinese coast extending in the opposite direction, westward. Significantly Danti noted that the evidence supporting the representation of the area he did not choose in his panel map was strong: Marco Polo wrote about it; a medieval world map "similar to one made earlier in Venice by Marco Polo," which probably Danti had the chance to see, depicted Polo's description; and the Portuguese traveler Juan do Barros had confirmed in the early sixteenth century that the Cape Litampo was the most easterly part of Asia. Danti himself had accepted this representation of northern China in his terrestrial globe completed in 1569, adding that Cosimo I supported and encouraged this cartographic choice. But mapping the same region on the cupboards' doors Danti chose to disregard the consistency with his own globe located in the Guardaroba Nuova, and mapped instead the alternative representation of the Chinese coast. Not only did he represent the northern coastline of China extending westward, but he also placed Japan perpendicular to the coastline of China. Danti's representation of Japan followed closely Gastaldi's map of Asia of 1561, but Danti felt compelled to acknowledge that other cartographers had placed Japan along the coast of China. Such a different representation of Japan was based on Portuguese nautical charts, including Bartolomeu Velho's, which was available to Danti. Also Ortelius based his representation of Japan in his map of Asia of 1567 on similar Portuguese sources (Fig. 5.21), and Mercator followed Ortelius in his world map of 1569 (Fig. 5.4). Another cartographic problem that Danti could not resolve was the representation of the connection between Asia and America, another major challenge for Renaissance cartographers. Danti could not decide whether water or land separated the two continents and he presented both options as possible. He accepted the Strait of Anian, the sea passage dividing Asia from America, which had been first introduced by Gastaldi in his world map of 1561 and later accepted by Mercator in 1569, but he chose to leave the

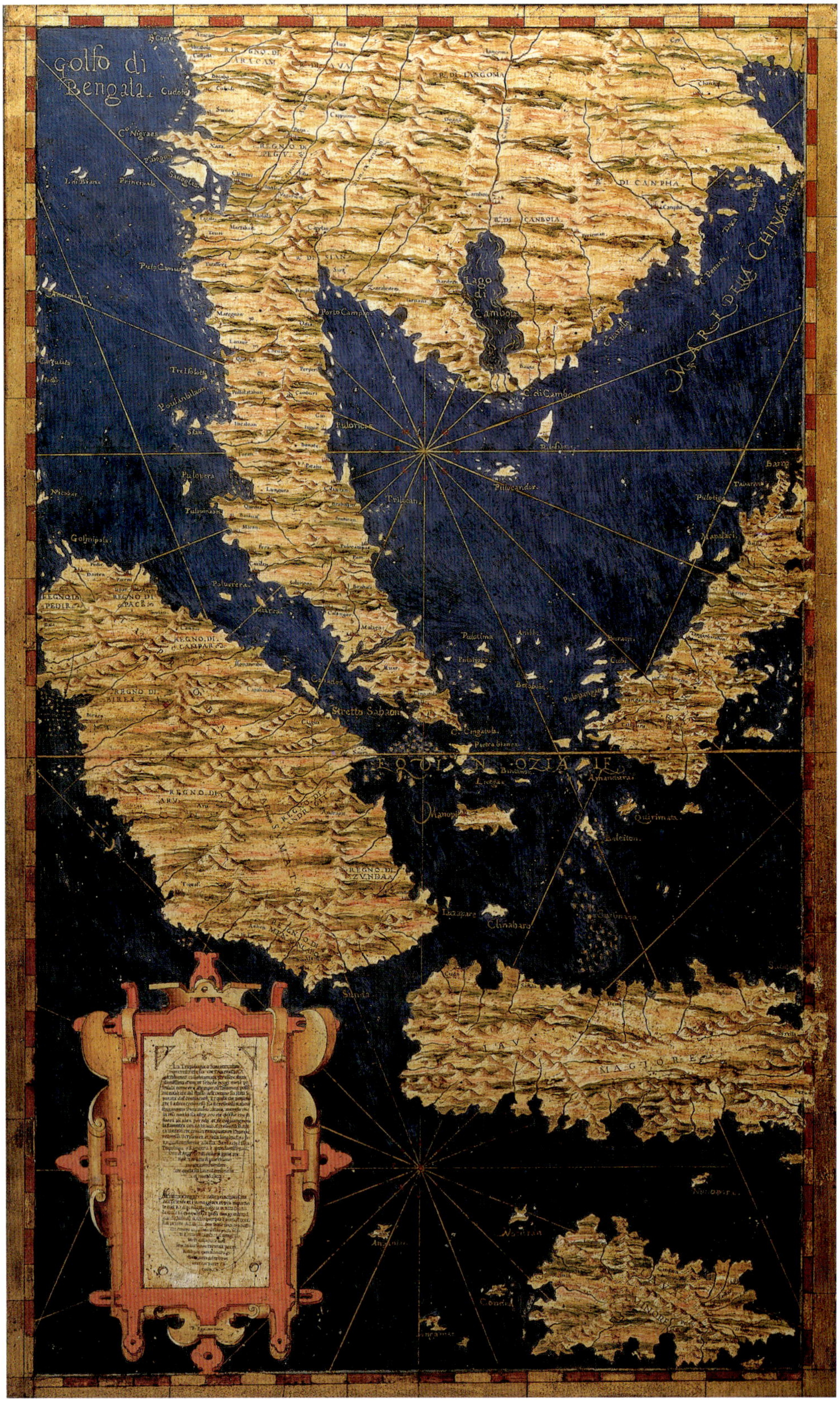

5.18 Egnazio Danti, Map of Indochina and Indonesia, 1573, panel. Florence, Palazzo Vecchio, Guardaroba Nuova

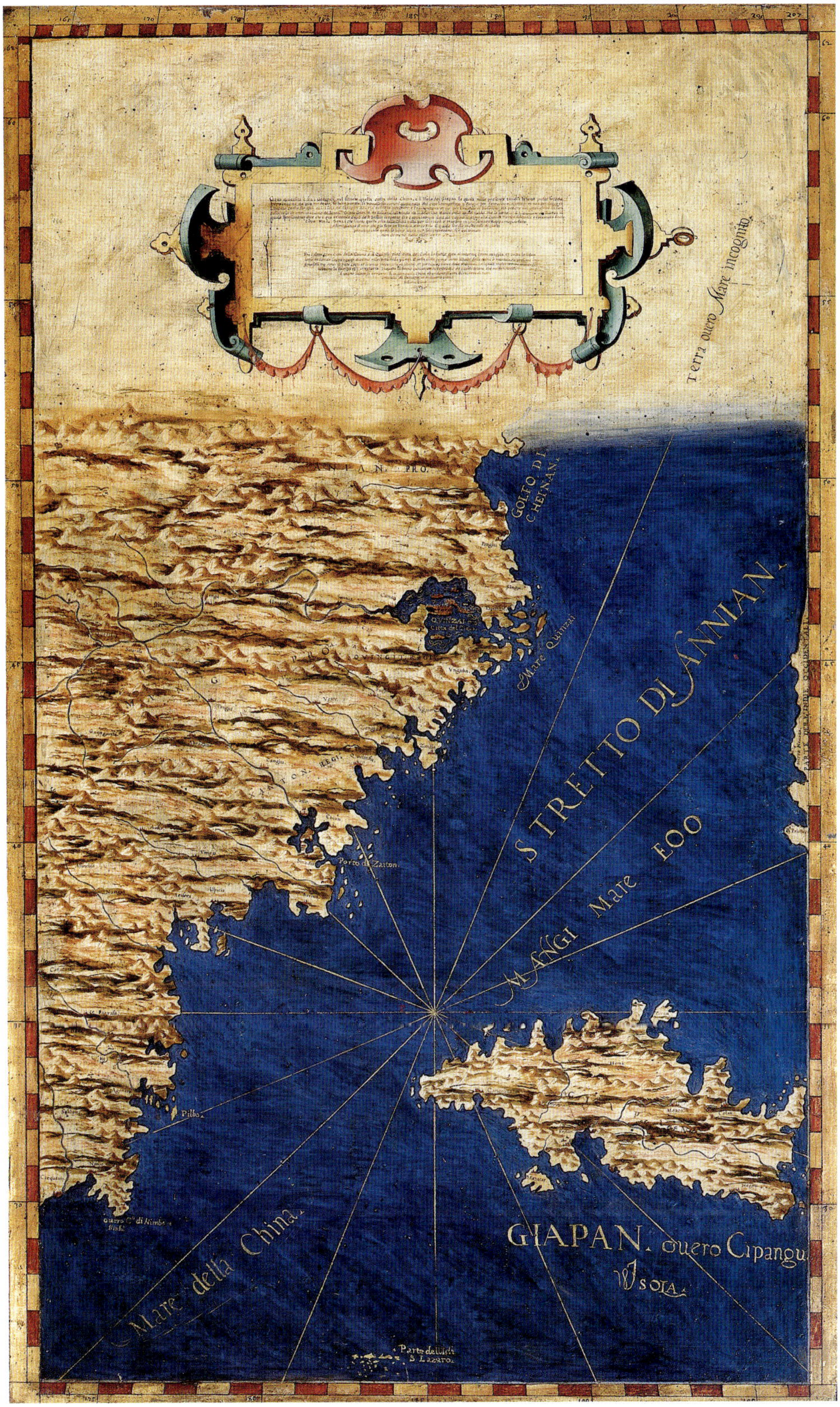

5.19 Egnazio Danti, Map of China and Japan, not dated, panel. Florence, Palazzo Vecchio, Guardaroba Nuova

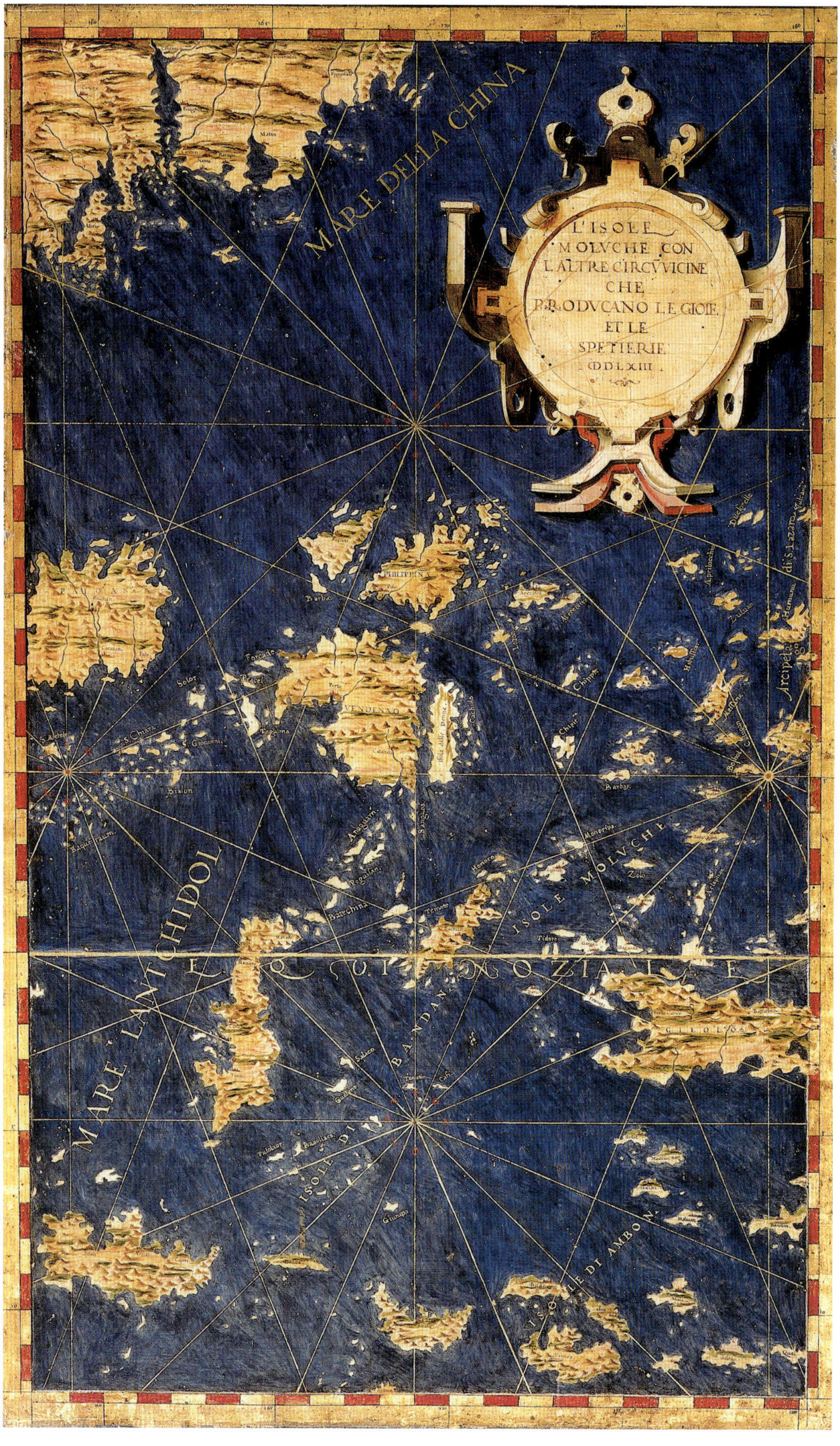

5.20 Egnazio Danti, *L'Isole Moluche con l'altre circonvicine che producano le gioie et le spetierie* (Map of the Moluccas Islands), 1563, panel. Florence, Palazzo Vecchio, Guardaroba Nuova

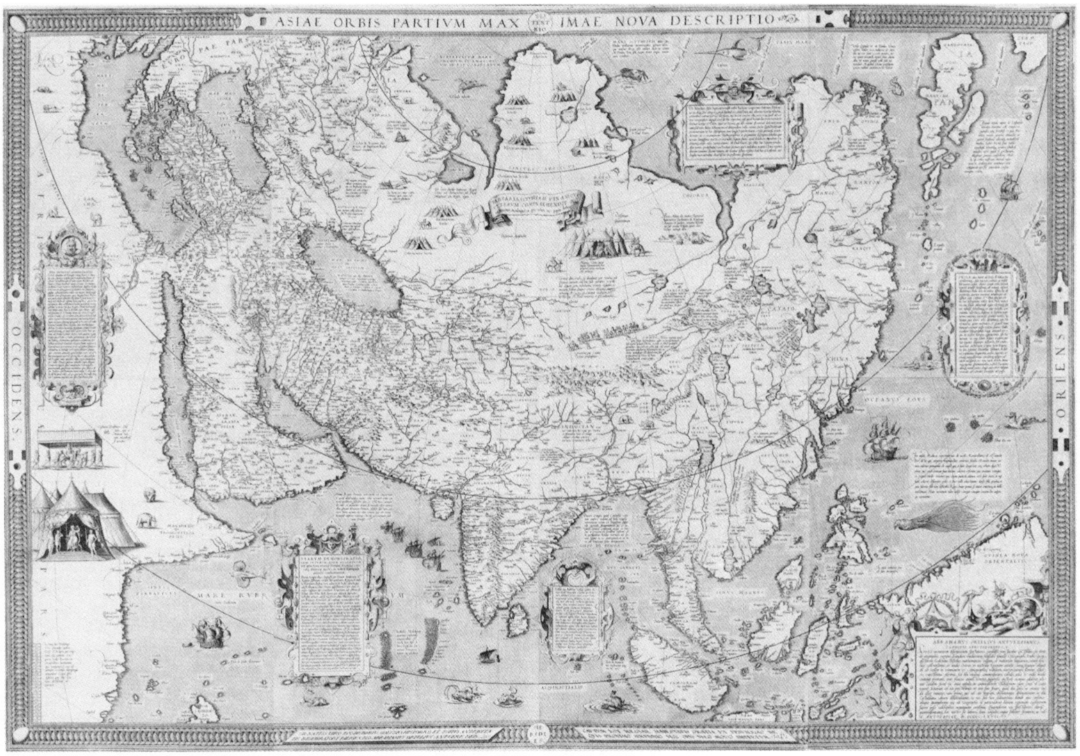

5.21 Abraham Ortelius, *Asiae orbis partium maximae nova descriptio* (Map of Asia), 1567, engraving. Öffentliche Bibliothek, Universität Basel

issue open with the noncommittal inscription, "Unknown Land or Sea" (*Terra overo Mare incognito*).[48]

The cartographic sources of Danti's maps of Asia are found in the famous large maps printed in the 1560s, but he never missed an opportunity to gather updated information. In 1569, Danti wrote to Bernardo Neri, who was on his way to Lisbon, asking him, "to get a copy of the cosmography of China that Joan do Barros said he had translated from Chinese into Portuguese, and to find out if the said Signore has any map of that country and province of China, Mangi, and Cathai, and especially any information on the great city of Quinsai. Your Lordship should try to get a copy because I know for sure that there is no other thing that would bring more joy to His Excellency the Duke than this, since no good information of the above mentioned places exist. I am not talking about the ordinary navigational charts, because of these the duke has many, but of maps that record the lands of the said country and the above mentioned provinces of India."[49] Indeed, in some instances he succeeded in consulting Portuguese manuscript maps and travel reports, over which he exercised critical judgment. Although the Portuguese had the most updated knowledge of the East Indies, the accuracy of their maps was a matter of debate, to which Danti referred in his map of Hindustan: "In describing the maritime shores of this map I did not follow the measurement of nautical charts of the Portuguese because they are false, indeed made false by them in order to prove that the Moluccas Islands are included in their territory, while they lay outside it."[50]

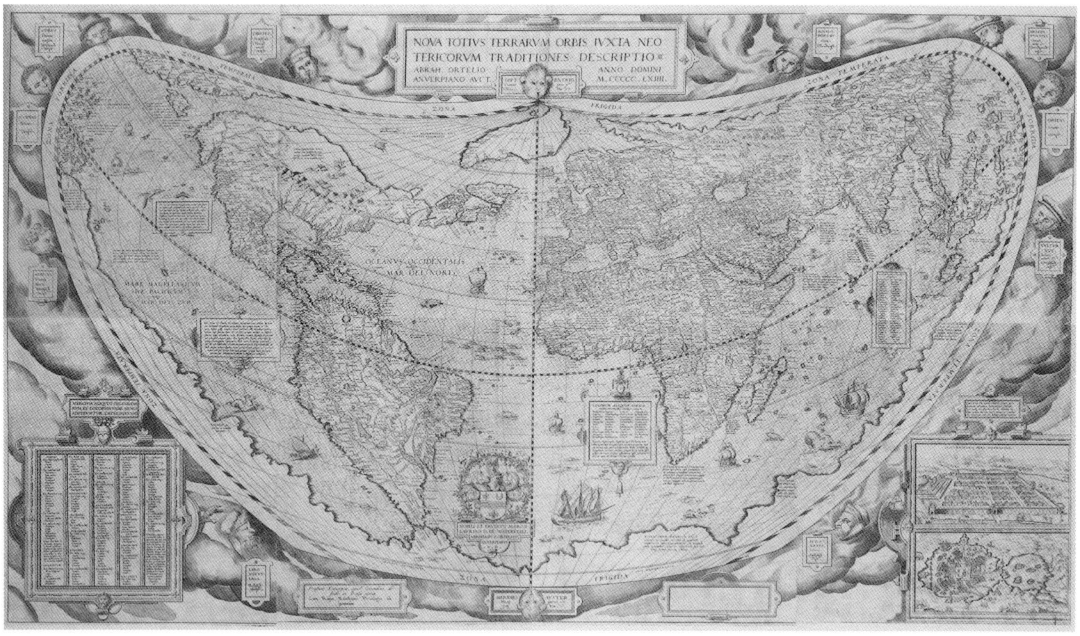

5.22 Abraham Ortelius, *Nova totius terrarum orbis iuxta neotericorum traditiones descriptio* (Map of the World), 1564, engraving. Öffentliche Bibliothek, Universität Basel

The misleading record of the latitude and longitude of the southeast islands was at the core of the disputes between the Spaniards and the Portuguese on the rule of the Moluccas Islands. Although Pope Alexander VI had divided the West Indies between the Spanish and Portuguese kingdoms with the Treaty of Tordesillas in 1494, the possession of the Moluccas was still debated for decades, during which time famous cosmographers of both sides excelled in producing false maps to support the conflicting territorial and political claims of their royal patrons. Danti's discriminating use of Portuguese nautical charts shows that while he regarded them as fully reliable in the mapping of Africa, he still questioned their accuracy in the latitude and longitude of East and Southeast Asia.

Like Gastaldi, Danti did not represent the land and sea north of the fiftieth degree of latitude, since the knowledge of these areas was still very rudimentary and relied primarily on Marco Polo's medieval travel report. Later Bonsignori represented the northern regions of Asia, which Danti had omitted, in two maps entirely based on Mercator's world map of 1569.

America

By the time Danti and Bonsignori made their maps of America, not only was the cartography of the new continent well advanced but it had also reached the public at large through the geographical prints that disseminated the maps the Spaniards and Portuguese had treated as classified material at the beginning of the century. Indeed, the shape of America emerged rapidly in the sixteenth century together with information on its peoples, their civilizations, as well as the progress of their conversion to Christianity and the appropriation of their resources. Paradoxically, however, the mapping of the new world was a far simpler matter than the mapping of the other continents since its shape did not have to be matched with ancient accounts.

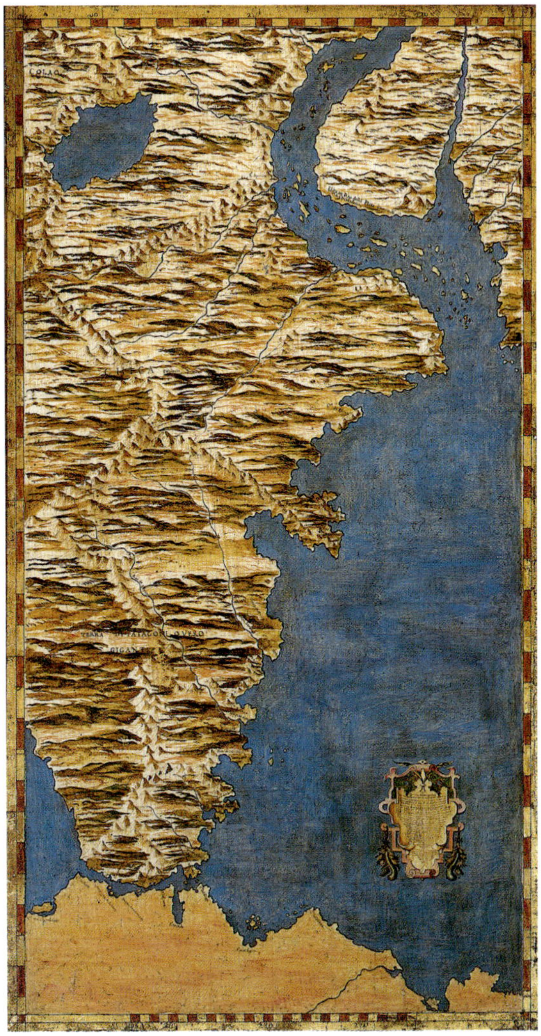

5.23 Egnazio Danti, Map of South America and Strait of Magellan, not dated, panel. Florence, Palazzo Vecchio, Guardaroba Nuova

In the Guardaroba Nuova, America was illustrated with maps that matched in numbers those of the other continents and that, according to Danti and Bonsignori, were based on unpublished cartographic sources.[51] For instance, Danti himself reported in the legend of his map of New Spain that the coastal outlines were "based on Castilian nautical charts, and the remaining inland was mapped according to the travel reports by Cortez and others that have been there and from other local maps" (Fig. 4.13).[52] But Danti's unpublished sources turned out to be Gastaldi's and Ortelius's world maps, two maps that he used extensively in the entire cycle of the Guardaroba Nuova (Figs. 5.22, 5.23 and 5.24).[53] Danti's claim to have used unpublished sources pertains only to the main islands, including Cuba, which were in fact mapped after unpublished Spanish nautical charts (Fig. 5.25). Also the Tierra del Fuego and the Strait of Magellan did not follow Gastaldi and Ortelius's maps but rather Portuguese navigation charts, such as the ones by Velho, or perhaps Mercator's *Map of the World* of 1569, which in turn was based on Portuguese sources.[54]

Many maps of America do not have legends, an indication that Danti left these maps incomplete when he abruptly left Florence in 1575. The few existing legends are derived from Hernando Cortez's letters. They document the recent voyages, the names of the early explorers, the urban and architectural characteristics of the main cities, such as Cuzco and Mexico City, the peculiarities of the inhabitants and products of the new world. In a rare instance of pride in his monastic order, Danti also emphasizes the Dominican contribution to the

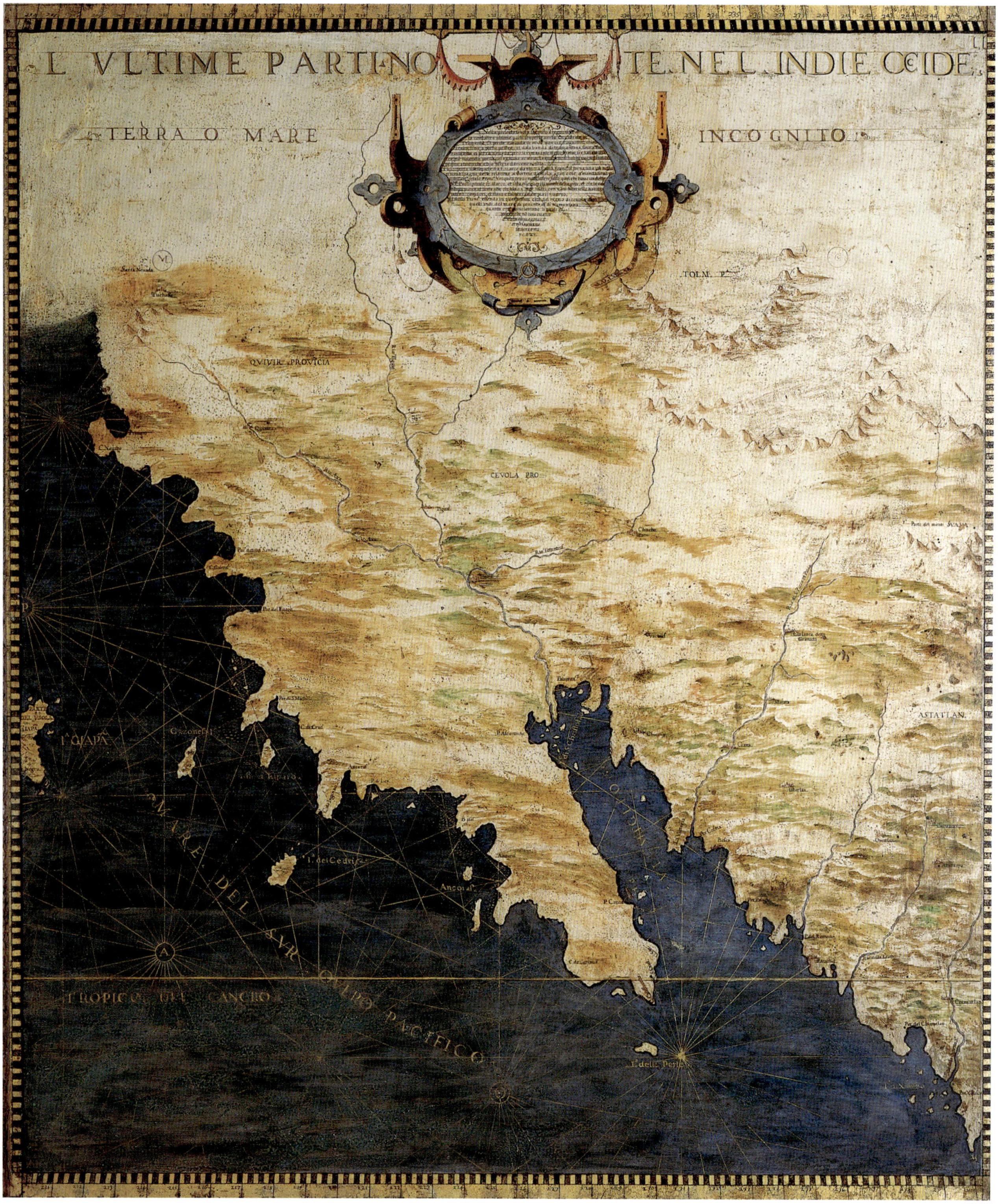

5.24 Egnazio Danti, *Le ultime parti nel Indie occidentali* (Map of California), May 1564, panel. Florence, Palazzo Vecchio, Guardaroba Nuova

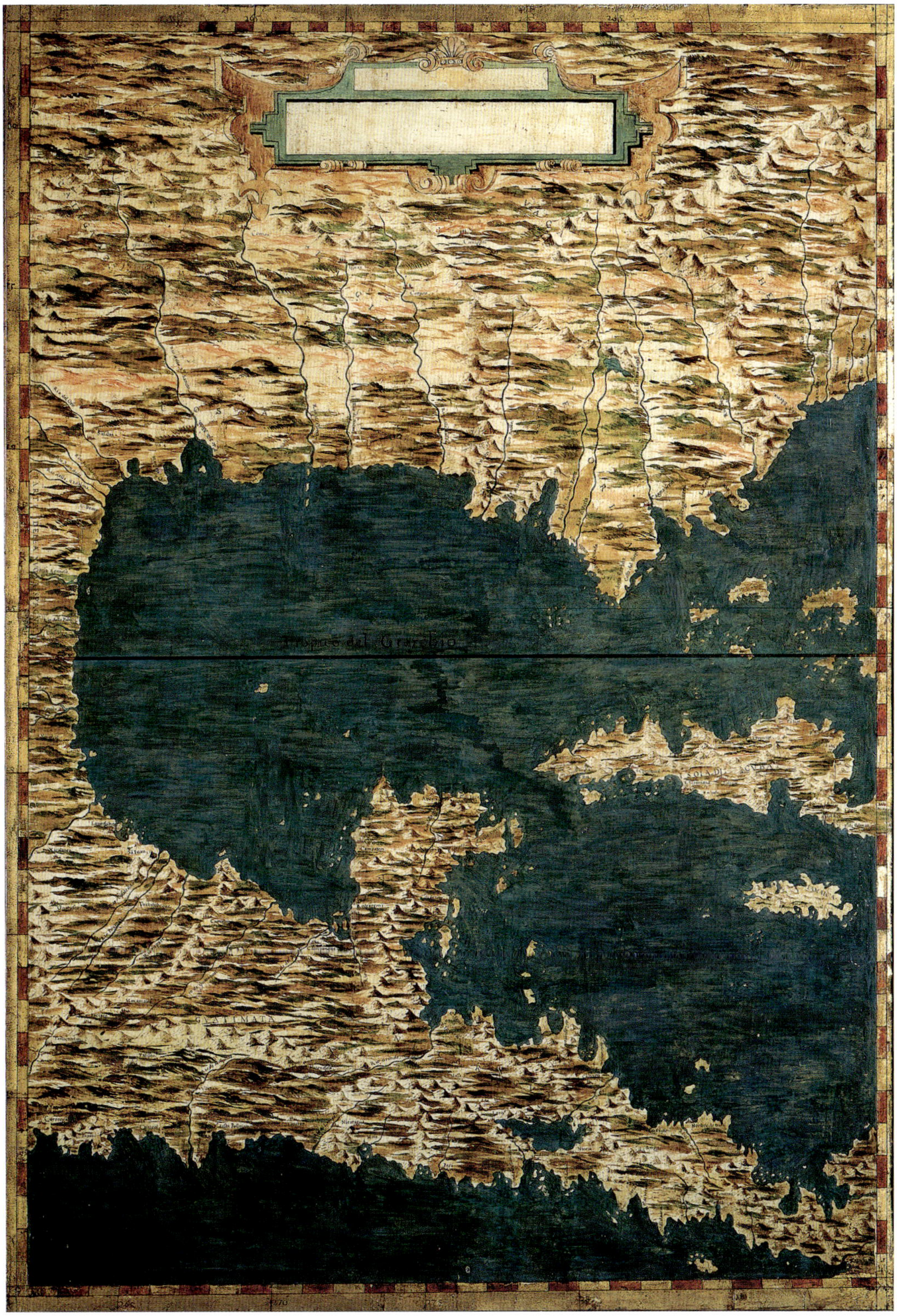

5.25 Egnazio Danti, Map of the Mexican Gulf, not dated, panel. Florence, Palazzo Vecchio, Guardaroba Nuova

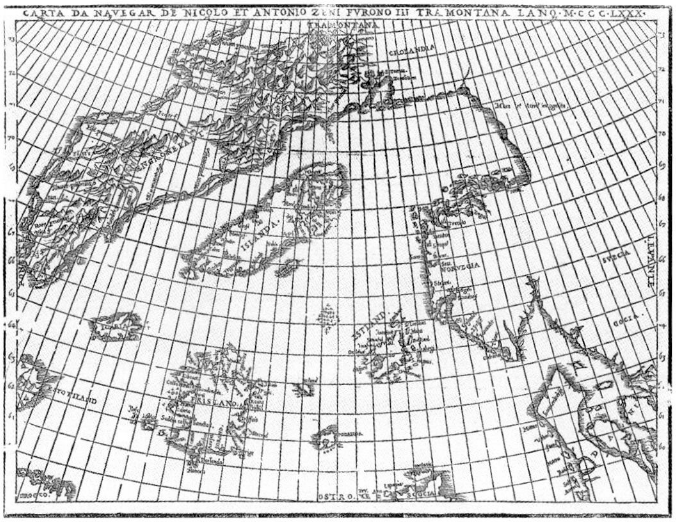

5.26 *Carta da navegar de Nicolo et Antonio Zeni* (Zeno Map), engraving, from Caterino Zeno, *Dei Commentarii*, Venice 1558. The University of Virginia Library

conversion of the inhabitants of America. In the legend of the Antilles Danti explained that the Island of San Dominico owed its name to the Dominican friar who first celebrated mass there. In another legend, Danti mentioned, "friar Alfonso of [the order of] Saint Dominic" (*frate Alfonso di San Domenico*), a Dominican born in Mexico from a Mexican father, who told Danti many things about the new world and even corrected for him the errors in the printed account by Cortez. However, six maps out of the planned fourteen were never executed and large parts of the new world were not mapped, especially the eastern and central areas of North America, perhaps an indication that updated Spanish sources were not readily available at the Medici court.

Greenland and the North Pole

The mapping of Greenland and the North Pole was regarded as part of the mapping of the new world, and their maps were often included together with those of America in the Renaissance editions of Ptolemy's *Geography*. In fact, the need for an accurate description of the north pole was so urgent and the absence of cartographic sources so painful that few mapmakers were able to resist the fascination of the fictional maps of these northern regions produced throughout the sixteenth century.

Unquestionably, the most famous forged map of Greenland was the so called Zeno Map (Venice, 1558), a map published by Nicoló Zeno to illustrate the imaginary voyages of his ancestors, the brothers Nicoló and Antonio Zeno in the 1380s (Fig. 5.26). Based on fictional letters that Nicoló Zeno claimed to have accidentally found in his family house, the voyages of these Venetian brothers make for one of the most fascinating readings of the period and, thanks to the accompanying map, also for a very instructive journey in the mapping of the modern world. The fascination with this Renaissance fabrication was irresistible. Ruscelli included the Zeno Map in his edition of Ptolemy's *Geography*. Ortelius assured its proper diffusion by including it in his world map of 1564, Mercator provided his authoritative support by inserting Zeno's Greenland in his world map of 1569, and Danti included it among his maps for the Guardaroba Nuova (Fig. 5.27). The nomenclature and the cartographical details of Danti's map show that he consulted the original Zeno Map, which would have been easily accessible together with its accompanying travel report. With his characteristic independence from earlier sources, however, he departed from the original Zeno Map by detaching Greenland from northern Europe, a solution Ruscelli had already adopted in 1561.[55]

The Renaissance knowledge of the North Pole was also based on a fictional map, the stereographic projection of those lands that Mercator had inserted in the left corner at the bottom of his famous world map of 1569 (Fig. 5.4). Inserted in a circle, Mercator's North Pole is symmetrically divided into four parts by the four rivers which "run to North with such speed that the ships which enter it cannot in any way go back, pulled by an internal depth."[56] Mercator's North Pole was divided into four panels in the Guardaroba Nuova, although these maps were not part of the original plan since Vasari did not mention them. Their authorship by Danti or Bonsignori is questionable, and they might have been added in the late sixteenth century by Antonio Santucci, who also restored Danti's monumental terrestrial globe.

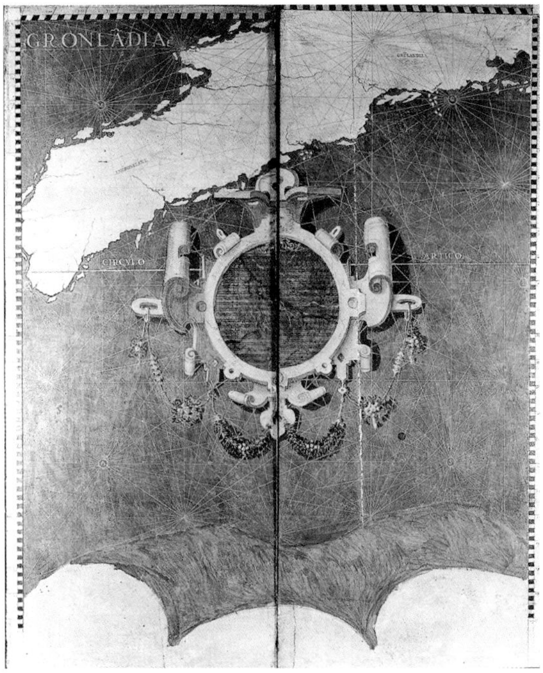

5.27 Egnazio Danti, *Gronlandia* (Map of Greenland), not dated, panel. Florence, Palazzo Vecchio, Guardaroba Nuova

The maps displayed along the walls of the Guardaroba Nuova represented the world piecemeal, region by region. The integration of the parts into a coherent whole was left to the viewer with some familiarity with the geography of the modern world and was facilitated by the monumental terrestrial globe placed in the middle of the room. The visual and geographical interactions between the panel maps and the monumental globe was stressed by Vasari who wrote that on the globe "there will be seen reflected all the tables that are right round in the pictures of the cupboards, and they will have a counter mark in the globe wherewith to find them with ease."[57] The globe must have been a magnificent piece that can only be imagined today. Together with its counterpart, the celestial globe that was planned but never executed, it would have offered a display of the descriptive possibilities offered by Renaissance mapping. In the very room that Vasari called a work of cosmography, the description of the cosmos appeared as a three-dimensional model of the universe, as a globe with the diagram of the earth's circles, as a mathematical mapping of its regions, as a description of its ornaments, and as a collection of its products.

For its very size, the erudition it embodied, and the mathematical accuracy it aspired to, the crafting of this map cycle was a daring project of science and art, imagination and scholarship, the worthy enterprise of a sophisticated ruler who regarded the cosmos as his own ornament.

As a masterpiece of cartography and as an exceptional cabinet of art and curiosities that organized artifacts, marvels, and exotica geographically, the Guardaroba Nuova is fundamental for understanding the expanded boundaries of political representation at the Medici court. With his impeccable sensitivity to the power of art as a vehicle to convey political and

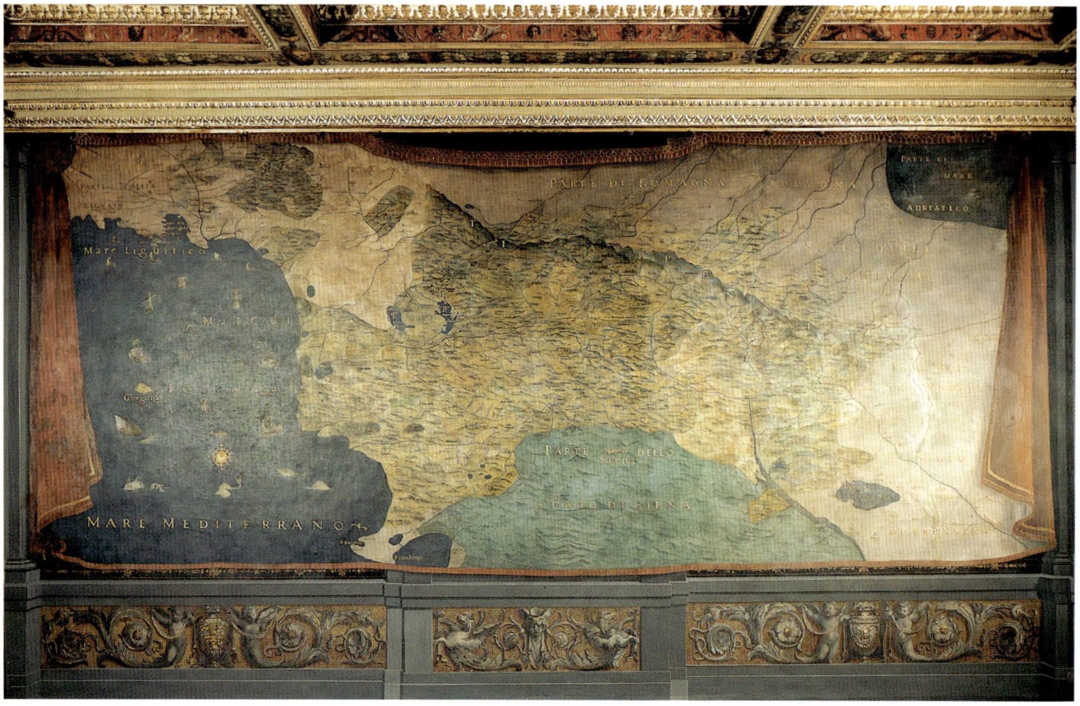

5.28 Stefano Bonsignori, *The Old Dominion of Florence*, ca. 1587, fresco. Florence, Uffizi Palace, Sala della Cosmografia

intellectual prestige, Cosimo had a task force of courtiers concentrated on the creation of his personal imagery, an imagery whose novelty and sophistication would make him stand out among fellow monarchs. At the core of this expanded dimension of political representation was the combination of modern cartography with traditional imagery of power and authority. At Charles v's imperial court, the orb of ancient heritage had already assumed the geographical features of the earth's globe, but only at Cosimo I's court could the earth's globe become inextricably intertwined with the persona of the ruler through his very name. The search for novel imagery made possible the use of maps of the world as a system for the organization of a collection of artifacts, and, consequently, the creation of an exceptional cabinet of art and curiosities. While collecting was an integral component of royal imagery and the display of dynastic collections in appositely built and decorated rooms a major concern of European rulers, only the Medici duke conceived a special cabinet for the display of his exotica, a theater of the world that indeed attempted the integration of the exotica within the geography of the world.

No historical source reveals why the Guardaroba Nuova was never completed, but it is easy to imagine that its imagery, which had been designed as a three-dimensional rendition of the motto "Cosmos is Cosimo's Ornament", was drastically reduced in significance once the new Medici duke carried another name. What remained significant for future grand dukes, however, were the two driving forces that made the Guardaroba Nuova possible: the attention to the display of the family collections and the study of the universe.

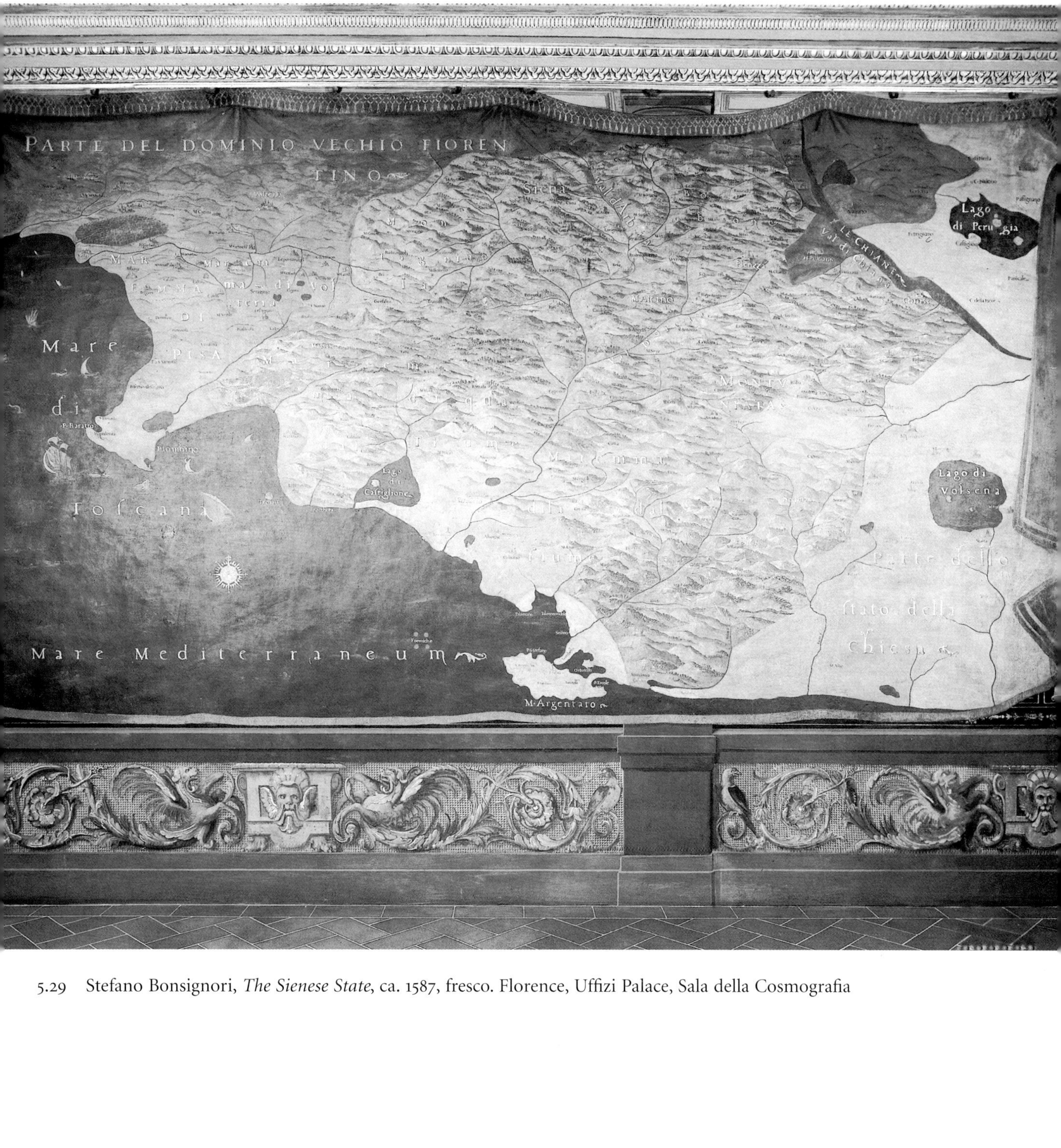

5.29 Stefano Bonsignori, *The Sienese State*, ca. 1587, fresco. Florence, Uffizi Palace, Sala della Cosmografia

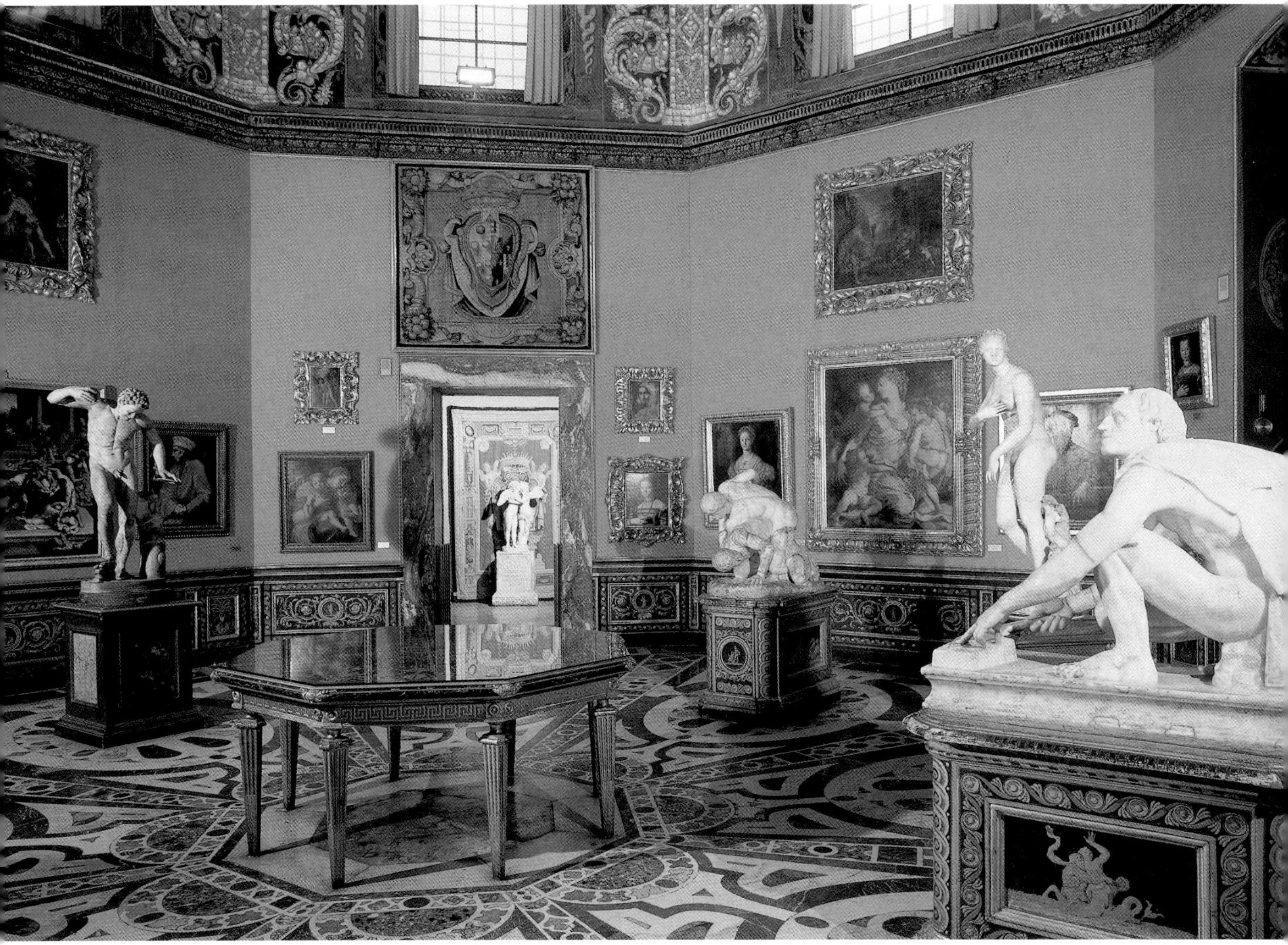

5.30 Tribuna. Florence, Uffizi Palace

The dismantling of the Guardaroba Nuova coincided with the relocation of the family collections on the third floor of the Uffizi Palace, a project started by Francesco I and completed by his brother Ferdinando. The new grand duke Francesco found his own Studiolo and his father's studies as inherently unsuitable for the display of a collection in constant expansion: their invention seemed to him too rigid and unmalleable. It is plausible that a similar judgment applied also to the Guardaroba Nuova. The new arrangement in the Uffizi Palace presented the advantage of displaying the collections in an array of rooms appositely designed for different kinds of artifacts, and many objects from Cosimo's Guardaroba Nuova found proper display there, although not the maps. Cristofano dell'Altissimo's portraits of famous people were hung in the long corridor, together with other images of Medici family members. Danti's terrestrial globe was displayed in the Loggia della Cosmografia and later joined by the armillary sphere made by Santucci. In that loggia, the two monumental spheres were surrounded by maps, but the maps of the world of the Guardaroba Nuova were replaced by the maps of the Duchy of Tuscany and Siena, the actual territorial domain of the Medici grand dukes (Figs. 5.28 and 5.29). Lorenzo della Volpaia's armillary sphere was moved to the Sala delle Matematiche together with other scientific instruments, including those made by Danti. The arms, the exotic feather cloaks, and the Turkish ceremonial weapons were displayed in the armory. The ambitious frieze of plants and animals planned but never realized for the maps in the Guardaroba Nuova was eventually executed as the socle of the Tribuna, the jewel of the new organization (Fig. 5.30). The Tribuna contained on its ceiling a monumental anemoscope, the instrument for the measurement of the wind's direction that Egnazio Danti had reinvented after ancient models. This special room abolished the distinction between small and big objects and their differentiation according to material that was so crucial in the earlier collector's studies. But it maintained the interplay between the nature-made and the man-made, the carved and the uncarved, the exotic and the familiar that informed its predecessors. Also the scheme of the four elements was preserved, although subsumed into a much looser program in the Tribuna, which was ultimately a cosmological room without maps.

Cosmography itself had flourished at the Medici court because of the duke's own name: as the name Cosimo had been an omen good enough to assure the rule of the young Medici as the legitimate successor of Cosimo the Elder so it was later justification of his patronage of cosmography. Indeed, the patronage of the investigation of the universe remained a defining component of Medici power even after Cosimo's death: it was another grand duke named Cosimo, Cosimo II, who brought Galileo to Florence.

The realization of the Guardaroba Nuova and the development of cosmography at the Medici court owed a great deal to the imagination of Cosimo I's cosmographer. Danti had created a map cycle that was unique for its cartographic accuracy, comprehensiveness, and monumentality, a map cycle that anticipated the very idea of the modern atlas. But Danti's position as court cosmographer involved much more than the making of maps and the teaching of mixed mathematics. It also entailed the cultivation of a network of connections with the ducal family, other courtiers, scholars, and artists. He had contacts with distinguished collectors and scholars who shared his passion for drawings, books, and manuscripts, such as Niccoló Gaddi, the learned aristocrat and passionate collector, and Giuliano de' Ricci, another noble with whom he also kept in contact after his departure from Florence.[58] He also looked for rare books for others, such as his Perugian friend Prospero Podiani, on whose behalf he went to meet Benedetto Varchi, and eventually assembled his own collection, unfortunately now lost, of printed books, manuscripts, and drawings.[59] Danti's circle of friends and acquaintance included many influential members of the court closely related to Cosimo. Vasari supervised his cartographic work for the Guardaroba Nuova. Vincenzo Borghini, Cosimo's iconographic adviser, had known Danti from the early days of the decoration of the Guardaroba Nuova, and kept in contact with him after his departure from Florence. He owned many of Danti's books, and suggested to Vasari the inclusion of the detailed description of the Guardaroba Nuova in the 1568 edition of his *Lives*, the very description from which this investigation started.[60] Giovan Battista Strozzi, an influential poet and writer of the Florentine Academy (he later became its director), shared with Danti an interest in translating ancient authors into the vernacular and publicly praised Danti in his writings. Piero di Gherardo Capponi, a member of the distinguished old Florentine family and one of Cosimo's private counselors, also wrote a sonnet to celebrate Danti and his scientific pursuit.[61]

The ducal cosmographer was among the courtiers invited to attend the major courtly events, such as the performance of the comedy *La coronatione di Saul* by G. M. Cecchi, organized by the confraternity of San Giovanni Evangelista and attended by the duke himself. He also attended the performance of the play *La Vedova* written by Giovan Battista Cini and staged in the ducal palace in 1569 to celebrate the visit of Archduke Charles of Austria, brother of Johanna of Austria, Francesco's wife. An acute observer of court life, as well as of the accomplishments of mixed mathematics, Danti left the most vivid account of the marvel and amazement aroused by the stage machinery (*periaktoi*) designed by Baldassarre Lanci of Urbino, a skilled engineer in the service of the duke, for the play written by Giovan Battista Cini, another influential member of Cosimo's court and the biographer of the duke. In fact, Danti personally knew both Lanci and Cini. The Urbino engineer shared with him an interest in mixed mathematics and may have provided Danti with the correct interpretation and use of his own perspective instrument. Cini shared with Danti an interest in ancient scientific instruments, and was the first patron to install an anemoscope by Danti in his country villa at San Lorenzo alle Rose, a locality five kilometers outside Florence in the direction of Porta Romana.[62]

Danti's high standing at the ducal court aroused concern among other courtiers as well as his fellow Dominicans, and he was challenged at least three times. The exact charges against him remain a mystery, although some allegations point to his misbehavior, a very common charge used then as now to dismiss unwanted rivals. The attacks came to nothing as long as Cosimo was in a position to defend his cosmographer and write letters to the Dominican authorities on his behalf; on one occasion he had to excuse his cosmographer to the pope himself.[63] Danti's standing at court changed dramatically after Cosimo's death. Grand Duke Francesco did not defend the Dominican but rather initiated the accusation. He asked, and obtained, Danti's immediate departure from Florence.[64] The maneuver must have been questionable, if Serafino Cavalli, the general father of the Dominican Order, felt compelled to take care of the objections he anticipated from Cardinal Ferdinando, a supporter of Danti who however was living in Rome far from the everyday intrigues of the Florentine court. Serafino Cavalli wrote Ferdinando that he was calling Danti in Bologna, "wishing from him information that he only can deliver."[65] Forced to leave the city within twenty-four hours, Danti left the Guardaroba Nuova incomplete.

But his skills as a polymath, a courtier, and an interpreter of his patron's ideals through maps, scientific instruments, and cosmography were soon to be put at the service of another patron in another court. In 1572, after only one day of conclave, the Bolognese Cardinal Ugo Boncompagni was elected pope. In honor of St. Gregory the Great, on whose feast day he had been elected cardinal several years before (1565), he took the name of Gregory XIII. Expert in the canon law that he had studied and taught at the University of Bologna, skilled in the politics of the papal court, which he had served in diplomatic missions at the European courts, and versed in ecclesiastical matters, he proposed to define at the Council of Trent, Gregory XIII lost no time in adjusting to his new position and immediately concentrated his efforts to get Rome ready for the Holy Year in 1575. Within the celebration of the Holy Year, he made clear his commitment to the use of cartography and mixed mathematics for the articulation of the ideals of the post-Tridentine church. He found in Danti an imaginative interpreter of his wish for a new sacred cartography. Gregory XIII brought Danti to the papal court in Rome and made him his papal cosmographer. Willing to please Gregory XIII as fully as he had pleased the Medici grand duke, Danti imagined for his new patron another exceptional room, a long corridor that mimicked the geography of the Italian peninsula. This time, however, this inventive map-maker imagined gigantic chorographical maps that aspired to the mathematical accuracy of geographical maps. And this time, he finished the project.

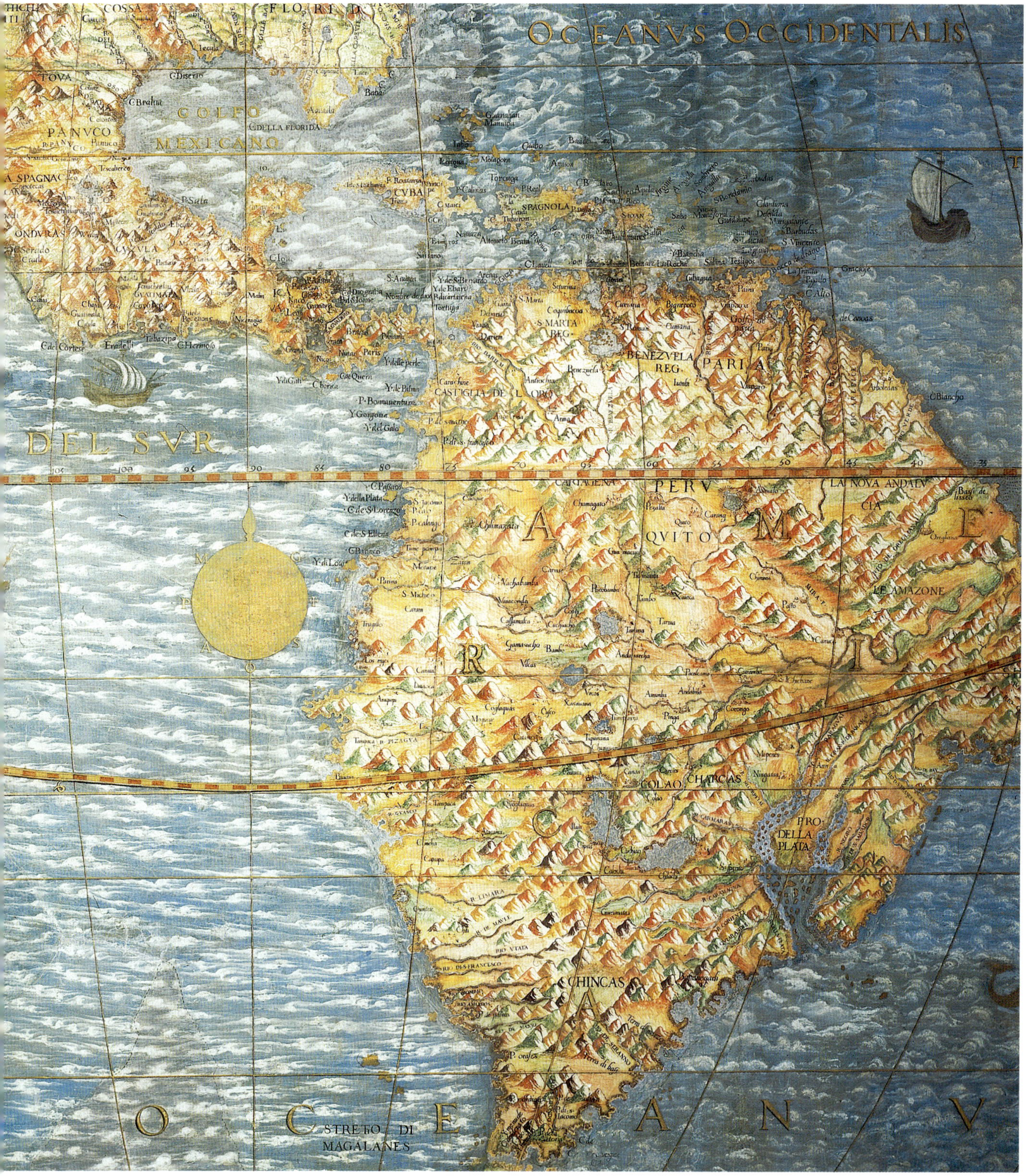
OCEANVS OCCIDENTALIS
GOLFO MEXICANO
PANVCO
CVBA
SPAGNOLA
DEL SVR
PERV
QVITO
A M E R I C A
LE AMAZONE
CHARCAS
COLAO
R. DE LA PLATA
CHINCAS
O C E A N V
STRETO DI MAGALANES

PART TWO

Maps as Sacred Art

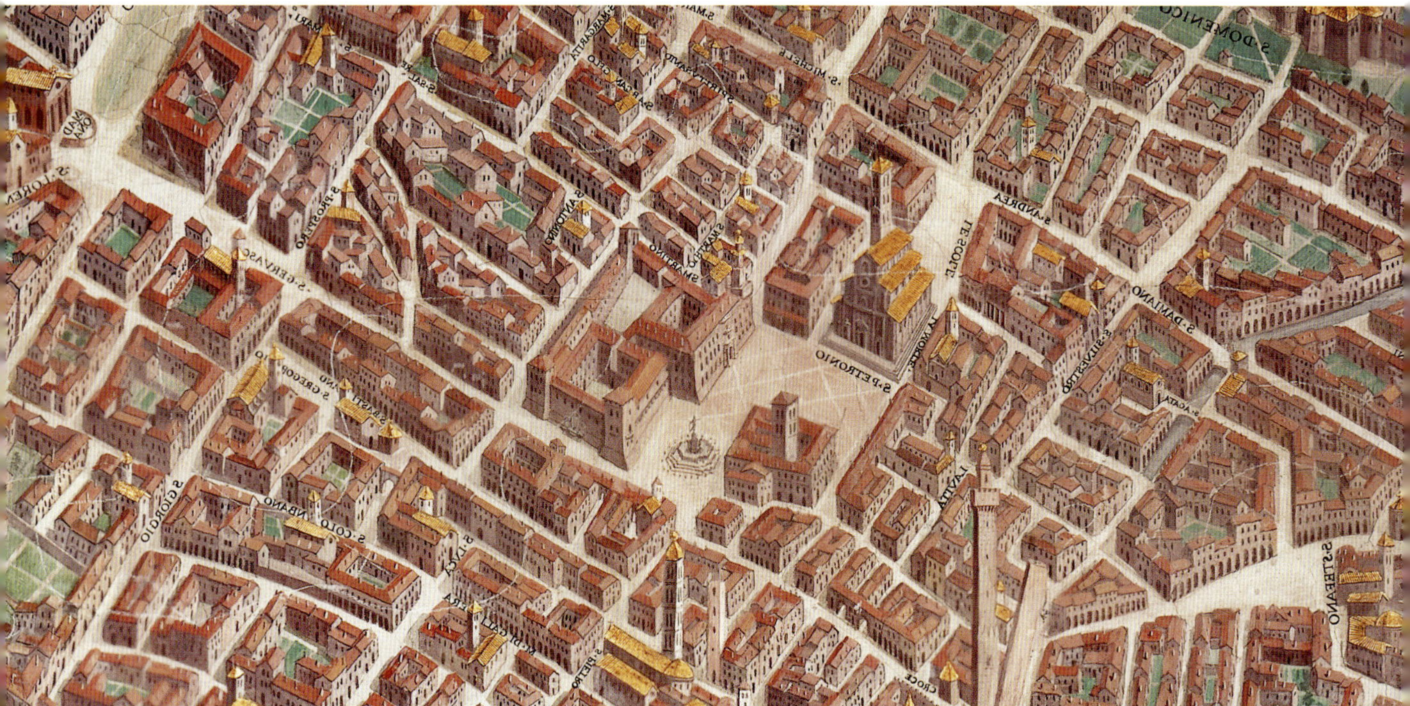
S·DOMENICO
S·PETRONIO
LE SCOLE
S·ANDREA
S·DAMIANO
S·STEFANO
S·GIORGIO
S·GERVASO
S·M·MAGGIORE
S·MARTINO
S·GIACOMO
S·CECILIA
S·GISMONDO
S·MICHELE

6

The Sala Bologna of Gregory XIII

The Holy Year 1575

On the evening of 24 December 1574, less than three years after he had been elected Pope, Gregory XIII Boncompagni led the most solemn procession of his pontificate. Together with patriarchs, archbishops, bishops, and ambassadors, he left the Vatican Palace, made his way through the multitude of pilgrims standing in front of the church of St. Peter, entered the Constantinian atrium of the old basilica, proceeded towards the *porta santa*, and received the ceremonial golden hammer to open the closed door. The moment was of great intensity, since the hammer broke and the pope slightly injured his finger, but the ceremonial act was completed, and the Holy Year 1575 inaugurated successfully. Giovan Battista de' Cavalieri made a print to celebrate the ceremony, which represented the atrium of St. Peter's, the Constantinian colonnade marked prominently *porticus constantinianus*, and Michelangelo's dome still under construction (Fig. 6.1). Cavalieri's print spread news of the great event throughout Europe, conveying one of Gregory XIII's dearest themes: the continuity between the new and the old basilica, between the church of Constantine's time and the church of the late Renaissance.[1]

As the first Jubilee year after the conclusion of the Council of Trent, the Holy Year 1575 was regarded as particularly important in the spreading of the spiritual renewal that the Catholic church had envisaged as the most effective way to face the challenges posed by the fragmented Christian world. Gregory XIII frantically prepared the city to meet the needs of the multitude of devotees he expected in Rome and even emanated a papal bull that, in the name of civic decor, forced Romans to restore their palaces along the main urban roads. He donated cash, buildings, and churches to the existing congregations for the assistance of the pilgrims, succeeded in opening the aqueduct of the Acqua Vergine, added new fountains in many squares, ordered bridges to be repaired, churches restored, national colleges founded, new altars consecrated, altarpieces painted, and sacred ornaments added everywhere. But Gregory XIII's major effort concentrated on the urban renovation to facilitate the procession to the seven basilicas, the procession revived by Filippo Neri a few years before which pil-

(facing page) Detail of fig. 6.5

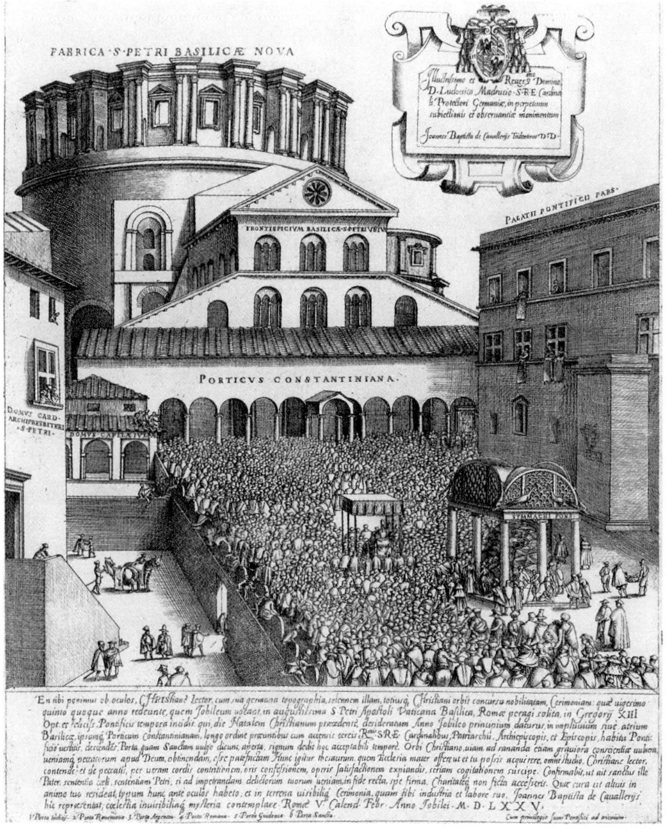

6.1 Giovan Battista de' Cavalieri, *The Opening of the Porta Santa in St. Peter's for the Holy Year*, 1575, engraving. London, British Museum, Department of Prints and Drawings

grims had to repeat for three consecutive days in order to obtain their plenary indulgence. At the center of the papal urban renovation and of the procession stood the church of San Giovanni in Laterano, the first official Christian church founded by Constantine as well as the cathedral of Rome. In another print celebrating the Boncompagni Holy Year, the eternal city, centered around San Giovanni in Laterano, was reduced to the seven basilicas of early Christianity, the city's Roman past limited to a couple of ruins engulfed in the basilicas, and its Renaissance urban fabric totally erased (Fig. 6.2). Even the urban roads, marked by processions of pilgrims moving from one church to the next, were limited to those connecting the basilicas, some of which Gregory XIII had recently built: the Via Gregoriana, appropriately named after the pope, which connected San Giovanni in Laterano to the church of Sta. Maria Maggiore; and the Via Ferratella that led to the church of San Sebastiano and provided a shortcut to St. Paul's. This was the post-Tridentine view of Rome that Gregory wished pilgrims would take back to their native cities.

The pope also refurbished the first Constantinian basilica. A new entrance from the main square was added and a new altar dedicated to the Eucharist was erected in its ambulacrum; based on a double *aedicola*, the design of this altar became the prototype of post-Tridentine altars throughout Europe. The nearby Baptistery was renovated, and a black marble basin, believed to be the very basin Pope Sylvester I used to baptize Constantine, was moved from the periphery to the center of the octagonal baptistery and extensively used during the Holy Year to baptize new converts. To the Scala Santa, the much revered flight of stairs made with marble from the steps of Pilate's palace where Christ had been brought on his way to Calvary, were added additional stairs so as to facilitate the flux of descending pilgrims.

Major projects were also undertaken in the Vatican Palace. Gregory XIII ordered frescos for stairwells, rooms, and apartments; completed the decoration of the Pauline Chapel, the papal private chapel in which Michelangelo had painted the martyrdom of Peter and Paul in the 1540s.[2] On 21 May 1573, he inaugurated the Sala Regia, the official audience room, which Pope Paul III had begun forty years earlier. Vasari, who had received special permission from the Medici grand duke to come to Rome and serve the pope, completed the Sala Regia in conformity to the already existing frescos that stressed the "submission due to the

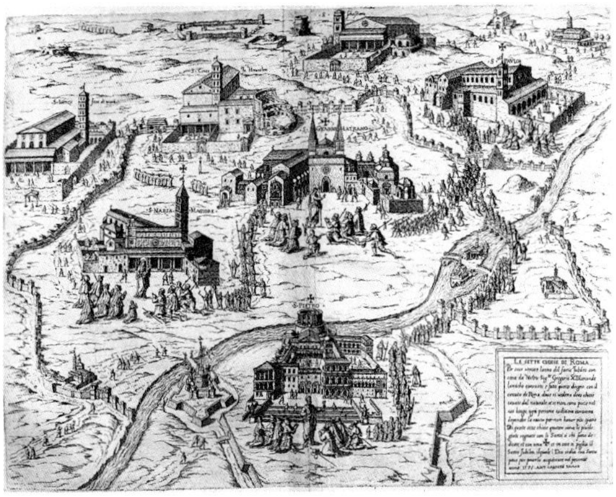

priesthood and the inferiority of the temporal rule."[3] He selected historical scenes representing kings and emperors rendering public obedience to the pope, with special attention to their potential connection to the reigning pope. But, as news of the massacre of the Huguenots reached Rome, Vasari revised his original iconography and invented three scenes depicting the cruel French events, which Gregory genuinely believed achieved the eradication of heresy from France. These three frescos outraged Protestant Europe.

6.2 Pietro de'Nobili, *Le sette chiese di Roma* (The Seven Churches of Rome), 1575, engraving. Rome, Biblioteca Hertziana

In order to accommodate dignitaries and rulers, Gregory also completed the new Palace of San Damaso, a large square building adorned with loggias, which were modeled after the adjacent loggia Bramante had built and Raphael painted earlier in the century (Fig. 6.3). Pius IV had started the palace and painted the upper floor of its west wing, the Terza Loggia, with the map cycle that had inspired Cosimo's Guardaroba Nuova. Gregory XIII continued the project and succeeded in having the north wing ready in time for the jubilee, but was forced to postpone the completion of the map cycle in the Terza Loggia to a later date (the map cycle of the Terza Loggia is discussed in chapter nine). In charge of the decoration of the new palace he put the Bolognese painter Lorenzo Sabatini. Sabatini not only modeled the iconography of the new loggias after the decoration Raphael invented for the adjacent wing, but he also adopted the organization of Raphael's famous workshop, which had proved effective in painting large surfaces in a short time. Rather than painting the loggias by himself

6.3 Matthijs Bril and Antonio Tempesta, *The Arrival of Gregory Nazianzus's Relics in St. Peter's*, 1582, fresco. Vatican, Terza Loggia

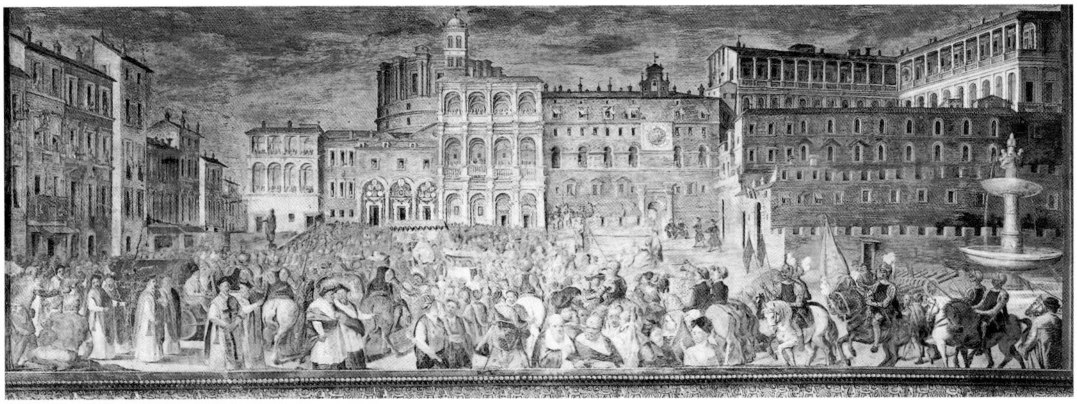

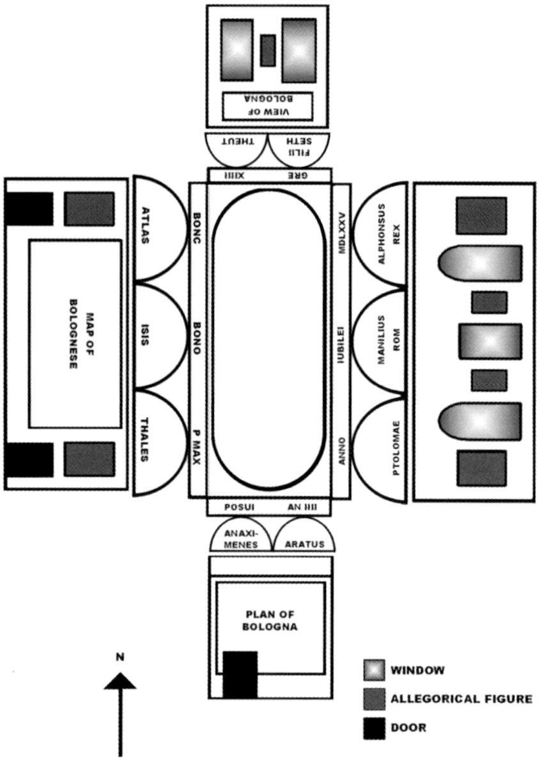

6.4 Reconstruction of the original layout of the Sala Bologna, Vatican (not to scale)

Sabatini supervised a small army of artists, each specialized in a different genre: grotesque, landscape, human figures, animals, plants, maps, and perspective and urban views. Such an organization of the artistic labor was to become characteristic of Gregory's papal commissions, one of the legacies he left to his successors.[4]

In the Palace of San Damaso, the pope also had built a new papal apartment that opened onto the Terza Loggia. For the dining room of his new quarters Gregory XIII planned a fully original, indeed personal, invention. The Sala Bologna, as this dining room was called, was a sunny hall with five large windows distributed on two walls that provided a stunning view of Rome and its countryside (Fig. 6.4).[5] Its decoration was entirely dedicated to the city of Bologna, the native city of the Boncompagni Pope and the second most important city in the Papal States. A dynamic center of scholarship and art since the middle ages and the seat of the oldest European university, the city acquired additional significance in the decades after the Council of Trent as a model of spiritual renewal thanks to its cardinal bishop, Gabriele Paleotti, who was also an old friend of the Boncompagni Pope. Indeed, with the election of Gregory XIII to the papacy, the religious, cultural, and artistic milieu of Bologna surged to a notable position at the papal court, a position made manifest in the Sala Bologna. But the iconography of this dining room, made up of maps, daring perspective views, and charts of the heavens, offers additional, unexpected insights into the Boncompagni Pope, his cultural interests and, especially, his vision of the usefulness of cartography and astronomy for the spirituality of the universal church.

The Sala Bologna

In the Sala Bologna, the city of Bologna was conspicuously celebrated through cartographic images: a monumental perspective plan of the city decorates the south wall (Fig. 6.5); a gigantic map of the county of Bologna embellishes the western wall (Fig. 6.6); and traces of a perspectival view of the city appear on the east wall between the windows, while allegorical figures filled the remaining space (today only one representing Peace is identifiable). As is typical in Renaissance maps, the mapping of Bologna and its territory was complemented by the representation of events of local history. The glory of the city and its university, where Ugo Boncompagni himself had studied and taught law before joining the clergy, was conflated with the glorification of the Boncompagni Pope: *Gregory IX Promulgating the Decretales* depicted the medieval pope with the likeness of the modern Gregory; and *Bonifacius VIII Confirming the Privileges to the University of Bologna* referred to the historic precedence of Gregory XIII's promulgation of the Decretals. Like Boniface VIII, Gregory XIII published a new edition of the official canon laws, entrusting the University of Bologna to their study and conservation.[6] These Decretals appeared only in 1582, but the frescos in the Sala Bologna suggest that Gregory planned their revision since the early days of his papacy.

The glorification of Bologna extended to the ceiling of Gregory XIII's dining room, which was not only painted by two Bolognese artists, Ottaviano Mascherino and Lorenzo Sabatini, but also included a stunning loggia in perspective, one of the specialties of Bolognese painting (Fig. 6.7). Designed and painted by Mascherino, this daring loggia in perspective was highly praised by contemporary viewers, especially by Egnazio Danti, who regarded it as exemplary of how architecture should be painted on an irregular vaulted ceiling, a concern of increasing importance for Roman and Bolognese artists experimenting with illusionism in painting. Mascherino's painted loggia housed ten ancient astronomers, painted by Sabatini, and sustained a magnificent sky map representing an unbroken panorama of the heavens designed, but not painted, by Vanosino. An artist not from Bologna but rather from the northern town of Varese, Vanosino was included in the artistic team for his expertise in adapting maps and globes to large walls, and had just completed a similar sky map for Cardinal Alessandro Farnese in his palace at Caprarola.[7]

The celebration of the hometown was common in papal iconography, although for Gregory XIII, a newcomer to Bolognese aristocracy, this traditional theme might have had additional urgency. The Boncompagni had entered the Bolognese aristocracy only a few decades before, when Cristoforo Boncompagni, a wealthy merchant and father of the future pope, married into an aristocratic family. However, the position of the Boncompagni within the Bolognese ruling elite remained unsecured and the luxurious palace that Cristoforo built near the cathedral had been highly criticized for being inappropriate in scale for the social standing and the financial resources of the family. What the family did not own in wealth, prestige, and land, Gregory acquired through his papal election, which made him the secular and spiritual ruler of Bologna and its surrounding territory. Undoubtedly he was the secular ruler of the entire Papal States and the spiritual ruler of the Catholic world but, in the early days of his papacy, what mattered to him was to assert his and his family's prestige within his native city of Bologna.

Gregory XIII's parochial vision of power that he expressed in the Sala Bologna comes into full focus once the room is compared with the Sala della Cosmografia that Cardinal Farnese had completed in 1574. These two rooms have often been compared because of their chronological proximity, the identical sky map depicted on their respective vaults, and for the map murals, even though Cardinal Farnese chose to represent maps of the world, while Gregory limited himself to the county of Bologna. Generally, it is noted that, in copying the sky map of the "gran cardinale," Gregory XIII eliminated the astrological references to Alessandro Farnese, such as Jupiter, the crown near Phaeton, and the fleur-de-lys. But in fact the Sala Bologna was not a mere copy of the Farnese Sala della Cosmografia, rather an ironic response.

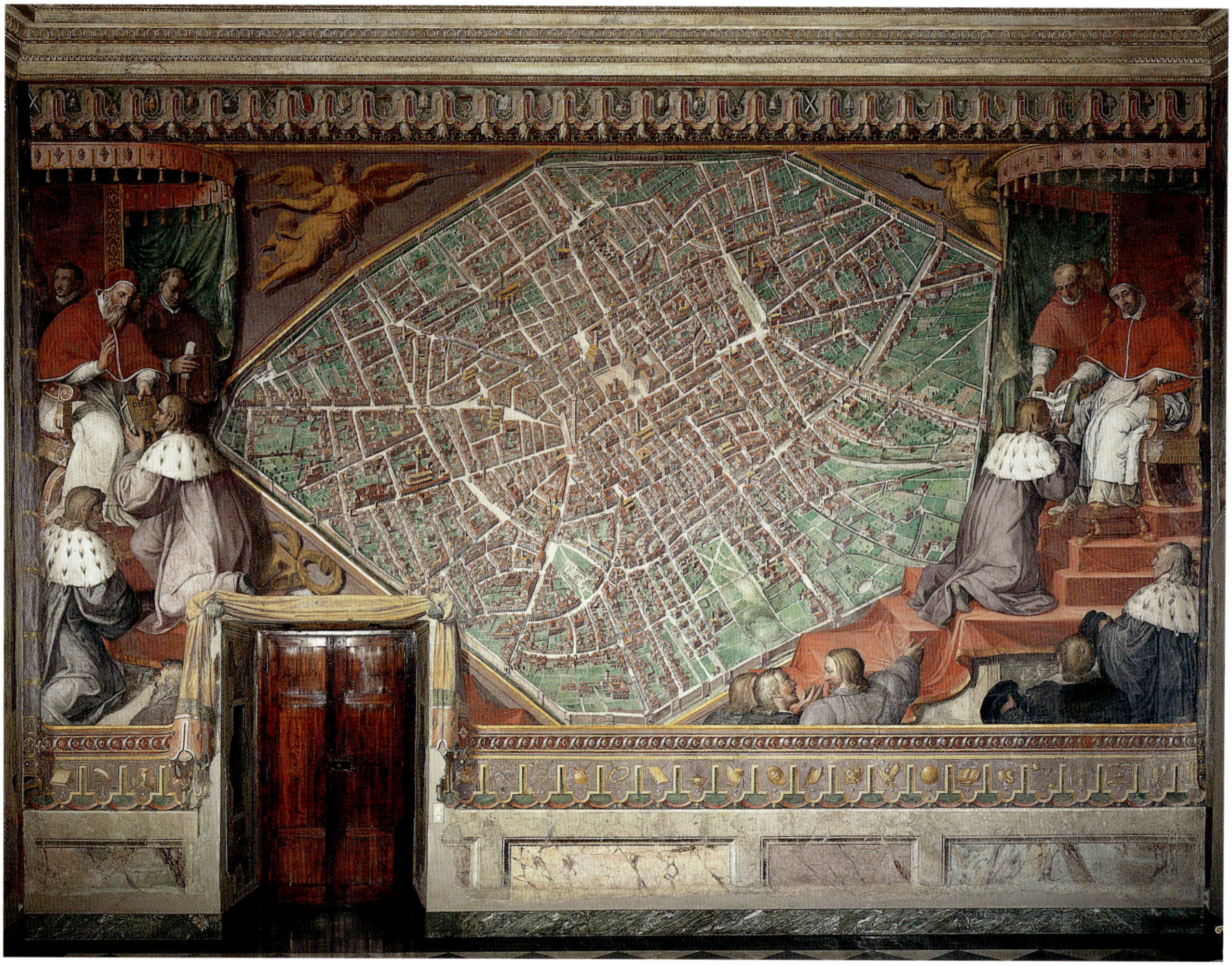

6.5 City View of Bologna, 1575, fresco. Vatican, Sala Bologna

In the language of astronomy and cartography, the Farnese room articulated the political and ecclesiastical ambition of the "gran cardinale," specifically his desire to be elected pope (Fig. 6.8).[8] The star map was a map of the destiny of the young Farnese, who was elected cardinal despite his right of primogeniture. The astrological frieze sustaining the sky map revealed that the cardinal's papal expectations were written in the stars. And the maps of the world depicted on the walls displayed the stage of his secular and spiritual power. The great expectations of Alessandro Farnese were no secret within his own circle, the papal court, and the European powers. What, surprisingly, has been underestimated so far is that Cardinal Farnese commissioned the iconography of the Sala della Cosmografia, emphasizing his destiny as cardinal, his future as pope, and his worldwide influence, right after he had lost yet another papal election. In 1572 the stern opposition of the King of Spain, Philip II, forced him to postpone his papal aspirations and Gregory XIII was elected instead. The defeated cardinal whom the stars destined to be the spiritual ruler of the world had to reaffirm his

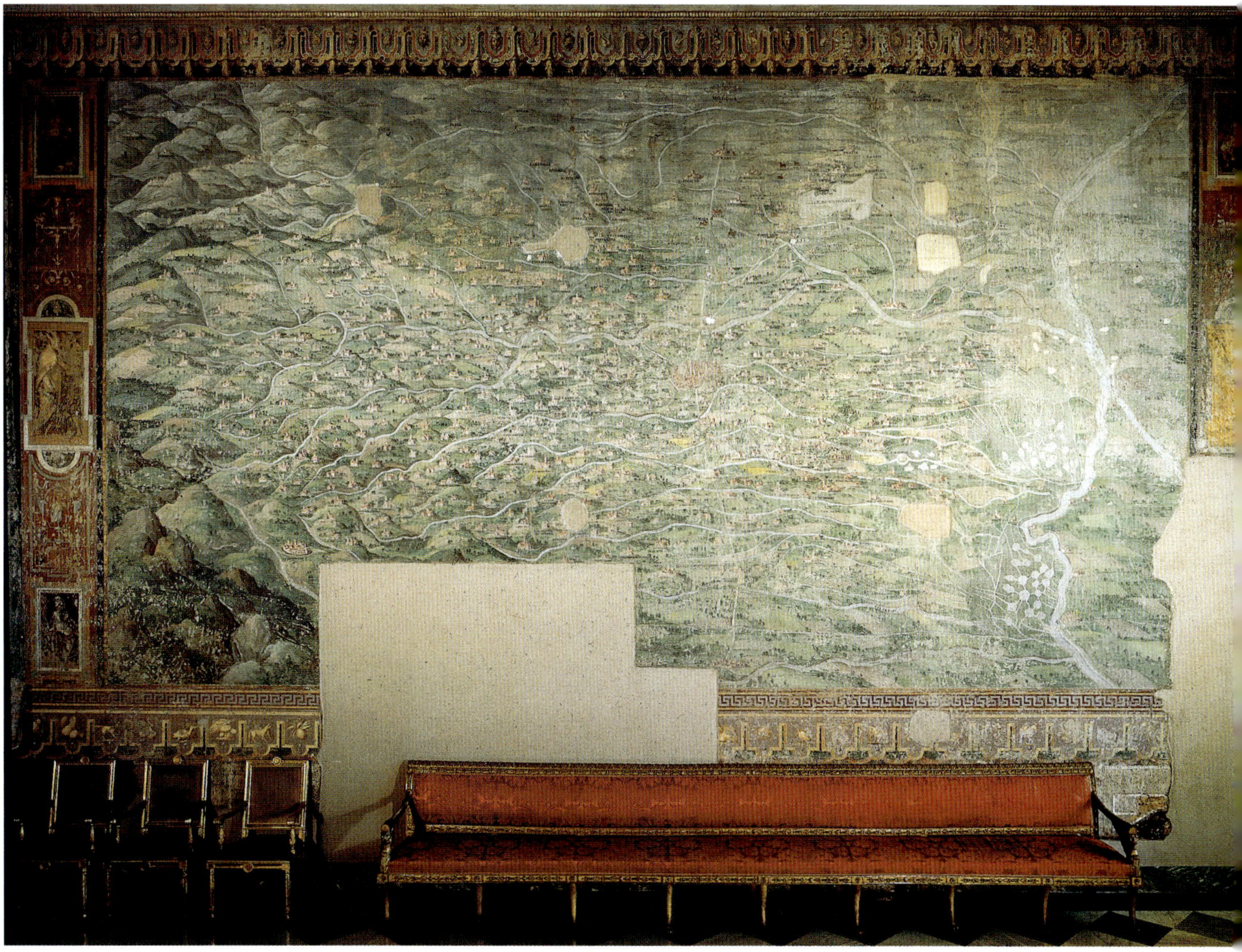

6.6 Map of the Territory of Bologna, 1575, fresco. Vatican, Sala Bologna

destiny against the contingency of politics. To claim his place at the top of the ecclesiastical hierarchy he used the cartographic language of papal decorations. Modeled after Pius IV's map cycle in the Terza Loggia, the map murals of the Farnese room expressed the universal aims of its patron, and generated the ironic response of the reigning pope. Using the same cartographic language and even the same sky map of the "gran cardinale," Gregory XIII confidently asserted his parochial view of power by replacing the maps of the world with the maps of his hometown.

Gregory's papal election was not written in the stars, nor did it come from his family's power and influence. His election was favored precisely because he was not a member of an influential political family, but rather a representative of the growing group of experienced clerical jurists and administrators. He had debated ecclesiastical matters at the Council of Trent and in papal tribunals and participated in political and diplomatic missions at many European courts. But when he became the spiritual leader of the Catholic world he felt he

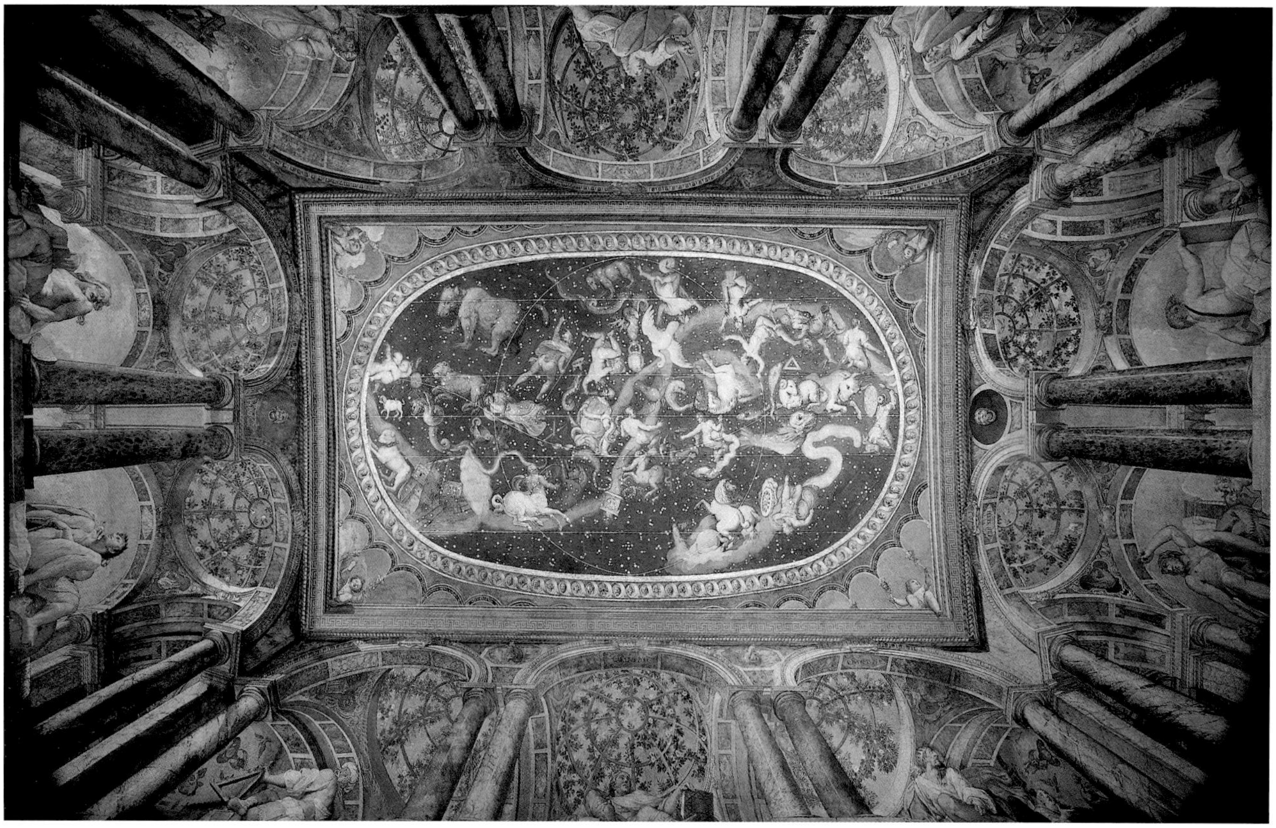

6.7 Sky Map, 1575, fresco ceiling. Vatican, Sala Bologna

had first to assert his and his family's power in the native city of Bologna. And added to this parochial view of power was the celebration of his own interests in cartography and astronomy, that is, in the mixed sciences of the sixteenth century that were revising the shape of the earth's globe and the structure of the universe.

Indeed, the images that Gregory selected as a vehicle to convey such a traditional theme of papal iconography as the celebration of the pope's birthplace transformed the Sala Bologna into an intellectual portrait of the Boncompagni pope. Documentary evidence shows that he was personally involved in the decoration of the Sala Bologna, especially in the making of the map murals. He chose the contents of the maps, selected the cartographer who was to make them, indicated the cartographic sources that were to be used, and imposed a very pressing schedule that made it impossible to achieve the cartographic richness he required, let alone to survey the city and the county anew.[9] Even though the papal aspiration to great cartographic accuracy had to be put aside because of the need to have the room ready for the Holy Year, the maps of the Sala Bologna were the most detailed cartographic representation of the city of Bologna and its territory available at the time, at least until Gregory himself embarked on the ambitious project of having the entire Papal States surveyed anew. Although no explicit evidence survives to indicate his direct involvement in the other parts of the iconography, still he is present everywhere in the Sala Bologna, a fact that can be better appreciated if different parts of the decoration – city view, map, painted architecture, and sky map – are considered not only as straightforward representations of a city, a region, a loggia, and the sky, but also as a visual synthesis of the mixed sciences of map-making, perspective, and astronomy, the so-called *scientiae mediae* that made it possible to transfer the abstract principle of Euclidean geometry into practice.

Vanosino's chart of the heavens was a standard one with no astrological references, as far as can be determined, to the bright career of the newly elected pope, and it stood simply as an illustration of the sky. It also signified visually and synthetically the skills and knowledge required to study astronomy: basic Euclidean geometry, astronomical observation, and mathe-

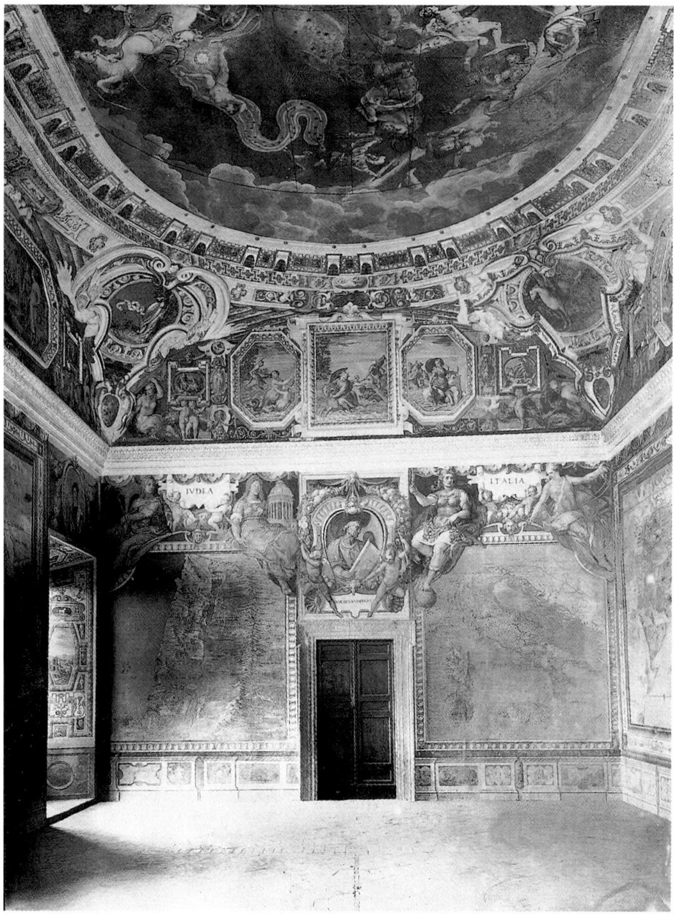

6.8 Sala della Cosmografia. Caprarola, Palazzo Farnese

matical computations. Similarly, Mascherino's loggia in perspective had the formal function of housing the ancient astronomers, whose observations and knowledge were basic to the making of the sky map painted above. But it was symbolic not of man's position at the center of the world but rather of the position of the *scientia media* of artistic perspective in relation to astronomy.[10] In other words, the same knowledge of Euclidean geometry was required to make a sky map and to paint a perspective view, or, for that matter, to survey a region, and make a map. Based on terrestrial and celestial maps and on perspective views, the Sala Bologna emphasized visually the interrelations between the most respected mixed sciences of the sixteenth century, and was ultimately a gigantic glorification of their usefulness, at least in civic life, if not in religious matters.

As far as is known, Gregory XIII never expressed in writing his interests in mixed mathematics, but his Sala Bologna, which was entirely decorated with the scientific images of cartography, artistic perspective, and astronomy, suggests more vividly than any written document his predilection for the *scientiae mediae.* Moreover, the iconography of the Sala Bologna raises the question whether, in this semi-official room of his palace, the pope was expressing only a personal predilection or, instead, was envisaging an expanded role of the *scientiae mediae* in the spirituality of the post-Tridentine church. After all, the Vatican Palace was not simply the official residence of a princely ruler but also the center of Catholic spirituality and, by necessity, the iconography of its halls, rooms, and corridors combined politics and religion, diplomacy and spirituality, classical knowledge and devotion. The iconography of Gregory's Sala Bologna requires an investigation into how the *scientiae mediae*, which were acquiring increasing importance in the explanation of the natural world, were regarded in terms of the spiritual renewal that Catholics advocated after the Council of Trent.

Paleotti and Bologna

The spiritual renewal that emerged from the Council of Trent was articulated in different projects, which were grounded on the contempt for earthly things, the rigid observation of the doctrine, the enforcement of discipline, the repression of heresy, the evangelization of the world, the exercise of individual contemplation, or the study of the natural world.[11] Building on the authority of Augustine and Thomas Aquinas, as well as on the achievements of Renaissance botany, zoology, geography, and astronomy, a group of reformed Catholics regarded the inquiry into the natural world as an integral part of the spiritual renewal they advocated and the contemplation of scientific images as a step toward the knowledge of the divine. Among the most convinced supporters of the study of scientific images as instruments of metaphysical contemplation was Gabriele Paleotti, Cardinal Bishop of Bologna and an influential figure at the Boncompagni court.

A native of Bologna too, Gabriele Paleotti shared a career path and interests with Gregory XIII.[12] Like the Boncompagni pope, he had studied and taught at the University of Bologna; was well versed in the politics, administration, and spirituality of the papal court; sat on papal committees in the 1560s; and distinguished himself at the last sessions of the Council of Trent. Paleotti and Gregory also shared a commitment to the spiritual reform of the Catholic church and a deep admiration for Charles Borromeo, Bishop of Milan, whom they both regarded as their spiritual guide. When Paleotti became the bishop of Bologna, he tried to transform his native town into a model of spirituality by following the reforms that Borromeo had introduced in Milan. The reform that Borromeo advocated and Paleotti endorsed was based on the mediation and dialogue between secular and ecclesiastical groups within the city, as well as the involvement of scholars in religious life, a vision that Gregory XIII must have shared since he always presented the Bologna of Gabriele Paleotti and the Milan of Charles Borromeo as models of Catholic reform. Their shared vision of spiritual renewal, however, did not prevent friction in the daily administration of power: the bishops resented the challenges that Gregory's legates posed to their ecclesiastical autonomy, while the pope wished to centralize the control of ecclesiastical matters in Rome.[13] These disputes, however, did not overshadow their shared interests, especially in the role they assigned to the knowledge of the natural world for the achievement of metaphysical contemplation.

Throughout his life, Paleotti cultivated personal interests in the natural sciences, mathematics, and the visual arts. He owned "four volumes in which he had pasted more than fifteen hundred plants, which he collected in Trento, Rome, Bologna, and elsewhere."[14] He planted a botanical garden in his episcopal palace, had a comprehensive knowledge of animals, and during his visits to the dioceses for religious purposes did not hesitate to stop along the way to collect rare specimens. As bishop, he made extensive use of maps. In 1572, he commissioned at least three large maps representing the city and county of Bologna and displayed them in the audience hall of his palace. These maps are now lost but, perhaps, they inspired the cartographic imagery of Gregory XIII's Sala Bologna. In 1581 Agostino Carracci dedicated a map of Bologna to the cardinal as a tribute to his interest in maps and commitment to the spiritual reform of the city (Fig. 6.9).[15] It has been suggested that Paleotti might have used maps during the visitations to his dioceses, a practice that his friend Charles Borromeo had introduced in Milan as an effective way to control and visualize the implementation of ecclesiastical reforms, but no conclusive evidence of this has emerged.[16]

6.9 Agostino Carracci, Map of Bologna, 1580, engraving. Bologna, Pinacoteca Nazionale, Gabinetto Disegni e Stampe

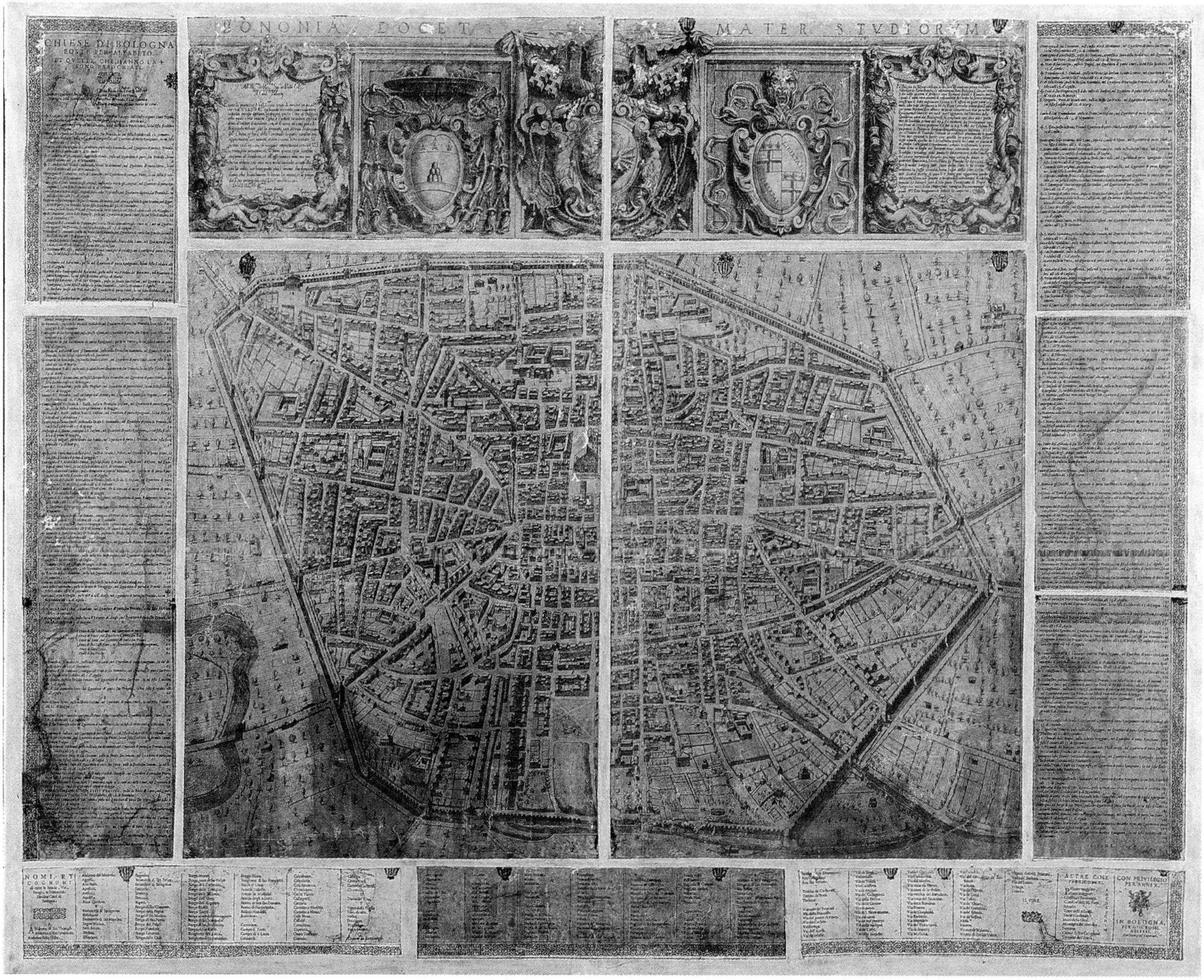

Paleotti also had close relationships with many Bolognese scholars. He regularly visited his friend Ulisse Aldrovandi in his museum, a trove of international fame full of maps, scientific instruments, dried specimens, drawings, and prints illustrating the world's flora and fauna. Paleotti had regular meetings with Carlo Varolio, professor of medicine and an expert in anatomy, and with Prospero Fontana, one of the city's leading artists whose paintings were known for their attention to nature. The historian Carlo Sigonio was also a close friend and started his work on ecclesiastical matters, including a history of Bolognese bishops, at Paleotti's request. Paleotti also knew Egnazio Danti who, expelled from Florence, had relocated in Bologna. To these trusted friends, Paleotti gave a draft of his *Discorso intorno alle imagini sacre e profane (Discourse on sacred and profane images)*, a treatise which, among other things, articulated his positive views on scientific illustrations in relation to metaphysics.

Written between 1578 and 1582 but never completed, Paleotti's *Discorso* was published in Bologna in 1582. Paleotti had conceived his treatise, which was dedicated to the uses and abuses of sacred images, as part of his duty for the spiritual reform of his diocese of Bologna. The debate over sacred images, briefly addressed by the early fathers of the church, had acquired new meaning and urgency after the Protestant ban on sacred images and their worship. The Council of Trent entrusted bishops with the supervision over the creation and use of sacred art in an attempt to channel the power of art in a positive way. Notoriously, the Tridentine decrees on sacred art were generic and vague, delegating to local bishops and Catholic writers on art the specification of individual cases. Indeed, after 1563 numerous Catholic treatises on the production, use, and function of sacred images appeared, among which Paleotti's treatise was the most comprehensive.[17] Generally read only in relation to sacred images and their abuses in the attempt to identify post-Tridentine views on art, Paleotti's treatise is in fact vital for the understanding of how reformed Catholics interpreted scientific illustrations, including maps, as sacred images.

Maps as Sacred Art

In his *Discorso*, Paleotti identified the abuses of sacred and profane art such as the depiction of heretical, apocryphal, or inaccurate subject matters, false gods, and lascivious figures, and discussed the general rules on how to control the production of sacred images so that they would conform to doctrine and facilitate the contemplation of the divine. His main theoretical assumption, based on the power of images, was that images have the capacity, as he said, "to delight, instruct, and move" the beholder. These three different functions of sacred images corresponded to three different levels of knowledge: sensual, rational, and spiritual. According to Paleotti, the ultimate end of a sacred image was to imitate the real world in order to persuade people to practice piety and move them toward knowledge of the divine. Paleotti's ideas on art, imitation, and religion were hardly original. In fact, they were firmly rooted in the Aristotelian aesthetics of imitation and in the orthodox Christian view of images as a universal language comprehensible to men and women, young and old, learned and ignorant.

But Paleotti's treatise contained, in addition, original ideas in relation to scientific illustrations, including maps. According to Paleotti, some profane images could function as sacred images, that is, they could ultimately contribute to metaphysical contemplation. Paleotti dedicated to these special profane images an entire chapter entitled, "Of profane images representing such diverse things as wars, landscapes, buildings, animals, trees, plants, and similar," clarifying that these useful images were: "geographical maps, [maps illustrating the] navigation to the Indies, descriptions of the sky and the stars, drawings of cities and countries, assaults to fortresses, great victories against infidels, images of animals, plants, and rocks, architectural things, buildings, drawings of fortifications, of drainage of rivers and valleys, in short all the subjects that are taught in the public academies and that St. Augustine (*De Doctrina Christiana*, II, 3r) wishes that not only should not be forbidden to a clerical doctor, but rather [regarded as] indispensable."[18] In another passage the good profane images are listed as, "representations of places, regions, provinces, kingdoms, and continents . . . images made and colored after nature of trees, plants, birds, fish, snakes, insects, rocks,"[19] that is, maps as well as images of plants and animals. These profane images, like sacred images, had to be painted properly, or as Paleotti said, "without vices, superstitions or other mistakes."[20] In other words, profane images had to imitate nature accurately. Such likeness could only be achieved through the observation and knowledge of nature and its mechanisms. Consequently, the delight a profane image provided was commensurate with the instruction it imparted.

Paleotti expressed a complete trust in scientific images that record the world of nature accurately, although the relation between images and knowledge had been a topic of contention from ancient times to the Renaissance.[21] Grounding his argument on the authority of Augustine, Paleotti presented the knowledge derived from scientific illustrations as useful and necessary for a devout Catholic. He also referred to the use of maps as effective tools for moral teaching, a point made already by ancient authorities:

> At times, beside their proper use, [these profane images] can also be useful for other purposes. For instance, in relation to geography, it is said (Aelian, *Varronis Historia*, III, 28) that Socrates wished to repress the haughtiness of Alcibiades, who was full of pride because of the richness and abundance of his possessions. Looking at a map of the world, he [Socrates] asked him [Alcibiades] to show him [on the map] the location of his land. Since Alcibiades could not find it marked anywhere, Socrates said: "Then there is no reason to rise for a thing that has no size and does not even appear on the earth's globe."[22]

Maps could teach the ephemeral nature of earthly things and indeed were widely used as symbols of the transience of life in Western art. But maps could also help artists to avoid abuses in the making of sacred images, especially what Paleotti called the "abuse of location" (*abuso di luogo*), a common mistake in art pertaining to geographical accuracy: "Another abuse would be that of location, an abuse made by the artist who put Rome in Asia, or Bethlehem outside Judaea . . . elephants and camels in the Alps of Italy, and firs and beeches in the valleys of Chioggia."[23]

It is clear that maps and scientific illustrations could easily fulfill the first two of the three functions that Paleotti attributed to sacred images: to delight and instruct. But it is less evident how they could also move the beholder. Paleotti unambiguously expressed his metaphysical attitude toward images of nature that could eventually play a crucial religious role. He wrote: "Profane images, which could bring us various advantages, should not be refused immediately but should instead be well considered and soundly judged so that, like bees that make honey from various flowers, we could take advantage of these images for religious purposes."[24] From this passage it can be inferred that scientific illustrations, for Paleotti, stirred and even moved the beholder to the contemplation of divine things. Within Paleotti's spirituality, one could ascend to a metaphysical plain through meditation on the mundane world and its images. And maps and scientific illustrations, along with a renewed religious movement in art, could serve as effective instruments in the propagation of the faith.

Furthermore, in a few passages of rare clarity in sixteenth-century artistic treatises, Paleotti discussed how viewers reacted differently to the same image depending on their education and knowledge. He limited the spiritual use of maps and scientific images to educated viewers who, because of their familiarity with the science of the map, could ascend to metaphysical speculation. Although, as a Catholic reformer, Paleotti was certainly concerned with the simplicity and clarity of sacred art in order to teach religious stories to the illiterate, he was also interested in the advantages that profane images provided for the educated reformer.

Whether devout and educated Catholics perceived and experienced maps as instruments for metaphysical contemplation, as Paleotti prescribed, is hard to tell. But, undoubtedly, Paleotti's *Discorso* gives a whole new dimension to Gregory XIII's Sala Bologna. It is through Paleotti's treatise that the scientific images of the Sala Bologna can be interpreted as images of religious renewal, indeed as sacred images. As large, beautiful, monumental frescos, these paintings delighted. As accurate images of the world, they instructed. And as both beautiful and scientific images, they moved the cultivated papal court to the contemplation of the divine order of things.

In this respect, the Sala Bologna appears as an all-encompassing portrait of Gregory's native city, its territory, art, and spirituality. While the perspective plan of Bologna represented the urban space, and the map of Bologna represented the city within its own territory and landscape, these maps also celebrated, together with the loggia in perspective and the sky map, the peculiar Bolognese interpretation of the post-Tridentine spiritual reform, an interpretation that envisaged the *scientiae mediae* at the service of the universal church. A manifesto of sorts that Gregory XIII made public to his dinner guests for the Holy Year, the Sala Bologna foretold what was to become one of the characteristic traits of his papacy: the use of scientific images, particularly maps, as a vehicle to articulate the ideals of the post-Tridentine church. Nowhere is this manifested more eloquently than in the Gallery of Maps, the long corridor along the Belvedere courtyard that he had decorated with forty monumental maps of Italy. Such a grandiose commission stemmed from the convergence of Paleotti's ideas elaborated in the cultural and spiritual milieu of Bologna, with Gregory XIII's preference for scientific illustrations to articulate the hopes of the Catholic church, and Egnazio Danti's expertise in the making of grandiose map cycles. In order to understand how the complex iconography of the Gallery of Maps came about, indeed how Danti was transformed from Cosimo I's cosmographer into the perceptive interpreter of Gregory XIII's vision of spiritual reform, Danti's activities in Gregory's native city are fundamental.It was here that the nexus of the Gallery of Maps began in the mid-1570s, when Danti and Paleotti met in Bologna.

* * *

Danti, Paleotti, and Bologna

After his abrupt departure from Florence, Danti relocated in the Dominican convent in Bologna. Although initially confined within the convent's walls and engaged in minor projects, he was soon able to resume his scientific interests and build a new network of friends and patrons. Danti installed an anemoscope and a meridian line in the Dominican convent. The anemoscope was one of the largest he ever built, but the meridian line was too small to continue the astronomical observation he had begun in Florence.[25] Nonetheless, it was very successful in convincing potential patrons of its usefulness in a monumental version. Giovanni Pepoli, president of the *fabbrica* of the church of San Petronio, authorized the installation of a monumental meridian line in the main Bolognese church, possibly in anticipation of its use in the evaluation of the proposal for the reform of the calendar, which the papal commission, instituted by Gregory XIII in 1575, was preparing. In place by April 1576, Danti used his meridian line to measure the tropical year.[26] A few months later, he was nominated professor of mathematics at the city's university with an annual salary of 200 libre.[27] Although in published works Danti appeared to be proud of this university position and of the recognition that came to him from the city, in private letters he resented the position as a setback after his previous appointment at the Medici court.[28] That university position, however, was instrumental in establishing his connection with the cultural, political, and religious elite of the city and, eventually, in making possible his later appointment as papal cosmographer.

Danti's classes at the University of Bologna were similar to his lectures at the Medici court: they started with the sphere of Sacrobosco, continued with Ptolemy's astronomy, Euclid, and the theory of the planets. Contrary to the practice at other universities, including the University of Pisa, but in line with Danti's ideas expressed in his *Trattato dell'uso et della fabbrica dell'astrolabio* (Florence, 1569), astrology did not receive much attention at Bologna, perhaps because it was covered in other classes of the medical curriculum. Danti's lectures, attended by about seventy or eighty students, were so successful that the Dominican authorities seriously considered using Danti's talent for the education of young Dominicans in the order's seminars. This project had to be abandoned because of the fierce protests of influential cardinals such as Alciati, Commendone, and Guastavillani (the latter was Gregory XIII's cardinal nephew) who wished that so excellent a teacher remain at the city's university.[29]

In Bologna, Danti also met members of the Boncompagni family, who interceded on his behalf to secure him prestigious tasks such as the Ufficio dei Lunari, the annual compilation of the *taccuino* reporting planetary movements, lunar phases, and astronomical and meteorological observations pertaining to the city. Also the Senate of the city used Danti's cartographic expertise to solve controversial territorial disputes; his surviving maps made for the city are stunning for the mastery of color and design that ultimately enhance the visual understanding of the disputed areas.[30] It is reasonable to suppose that Danti also knew some Bolognese painters, possibly Tommaso Laureti, whom he would meet again in Rome, and certainly Bartolomeo Passerotti who painted his portrait (Fig. 6.10). He befriended Ulisse Aldrovandi and visited his museum. Danti's signature is still kept in Aldrovandi's guest register: *Cosmographus Ignatius Danti*, while Antonio Gigante, who worked in Aldrovandi's museum, wrote a sonnet for one of Danti's treatises.[31]

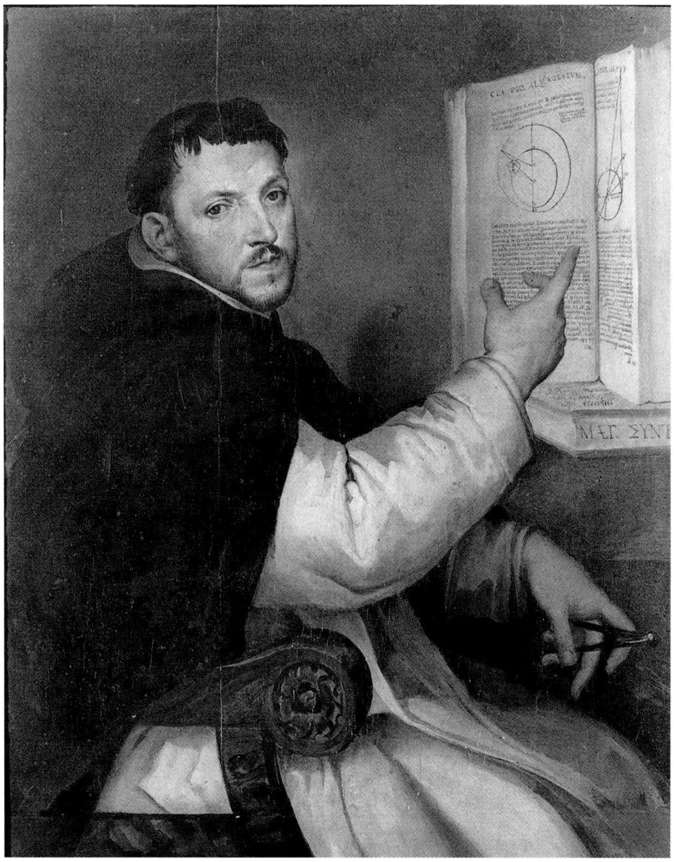

During his years in Bologna Danti became acquainted with Cardinal Paleotti. Their relationship has been entirely overlooked, although it was to have important consequences for both Danti's career and Paleotti's ideal of spiritual reform. In 1577, Danti greatly interested the curious and active cardinal in his scientific instruments, a fact he himself recounted in one of his treatises: "I set my mind to persuade the Most Illustrious Cardinal Paleotti to built such an instrument [an armillary sphere], a task that was easy to accomplish, since that Signore is knowledgeable of any kind of science and well disposed toward virtuous deeds."[32] The armillary sphere, similar to the one Danti had previously built on the façade of Sta Maria Novella in Florence, was meant to be placed in the courtyard of the episcopal palace, although it is unclear whether this project was ever realized. A few months later, Danti installed an anemoscope for Paleotti in the same palace; Paleotti's anemoscope was the largest Danti ever made; unlike others, it could mark the direction of the winds both indoors and outdoors; and it aroused the interest of Ulisse Aldrovandi, who described it in one of his manuscripts. To Paleotti Danti dedicated his Latin edition of the short treatise *Anemographia* (Bologna, 1578), which Pietro Antonio Cataldi, Danti's successor at the University of Bologna, later translated into Italian (Fig. 6.11).[33] It is further reported that in 1577 Danti was "working hard to complete other things for the Cardinal,"[34] while in February of the following year he received as a gift from Paleotti some relics that Danti himself had dug up in the capitular church of Santo Stefano in a public ceremony attended by numerous dignitaries, including Cardinal Paleotti.[35]

6.10 Bartolomeo Passerotti, *Portrait of Egnazio Danti*, ca. 1580, oil on canvas, Brest, Musée Municipal

Furthermore, Danti was one of the readers of Paleotti's *Discorso* before its publication in 1582. As is well known, Paleotti circulated a draft of his treatise to trusted friends, Aldrovandi, Sigonio, and Prospero Fontana, among others, who left their comments in a series of letters now kept in the Archivio Isolani, Bologna. Among those letters there are also a few by Danti. While the comments by Aldrovandi, Sigonio, and Fontana on Paleotti's treatise are well known to scholars, Danti's letters are virtually unknown.[36] They comment specifically on the genre of the grotesque, which Paleotti had condemned as inappropriate for post-Tridentine art because of its bizarre imagery. Danti, however, suggested reforming the genre of the grotesque by limiting its imagery to acceptable forms, a point particularly significant in relation to the reformed grotesques that were later painted in the Vatican Palace under Danti's supervision (Fig. 6.12). But Paleotti's treatise must have deeply interested Danti, particularly his ideas that scientific images could become instruments of faith. Danti himself echoed Paleotti's views in a treatise that he wrote in Bologna, in which he recalled the intermediary role of the mixed sciences between physics and metaphysics. Citing Augustine, Jerome, and Gregory Nazanzius, Danti argued that the mixed sciences "elevate the intellect, and sharpen [one's] wits to the contemplation of divine things."[37] Thus, they were indispensable to theologians seeking a correct interpretation of the Bible. Indeed Paleotti may have reminded Danti of the ultimate scope of the intense investigation of the universe that the Dominican had started in Florence.

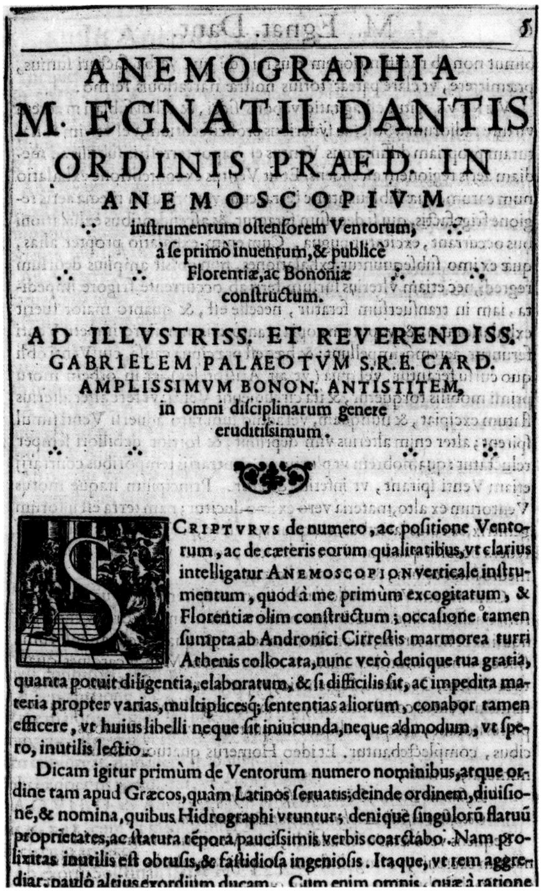
5

ANEMOGRAPHIA
M. EGNATII DANTIS
ORDINIS PRAED. IN
ANEMOSCOPIVM
instrumentum ostensorem Ventorum,
à se primò inuentum, & publicè
Florentiæ, ac Bononiæ
constructum.

AD ILLVSTRISS. ET REVERENDISS.
GABRIELEM PALAEOTVM S.R.E. CARD.
AMPLISSIMVM BONON. ANTISTITEM,
in omni disciplinarum genere
eruditissimum.

SCRIPTVRVS de numero, ac positione Ventorum, ac de cæteris eorum qualitatibus, vt clarius intelligatur ANEMOSCOPION verticale instrumentum, quod à me primùm excogitatum, & Florentiæ olim constructum; occasione tamen sumpta ab Andronici Cirrestis marmorea turri Athenis collocata, nunc verò denique tua gratia, quanta potuit diligentia, elaboratum, & si difficilis sit, ac impedita materia propter varias, multiplicesq; sententias aliorum, conabor tamen efficere, vt huius libelli neque sit iniucunda, neque admodum, vt spero, inutilis lectio.

Dicam igitur primùm de Ventorum numero nominibus, atque ordine tam apud Græcos, quàm Latinos seruatis; deinde ordinem, diuisioné, & nomina, quibus Hidrographi vtuntur; denique singulorũ flatuũ proprietates, ac statura tẽpora paucissimis verbis coarctabo. Nam prolixitas inutilis est obtusis, & fastidiosa ingeniosis. Itaque, vt rem aggrediar, paulò altius exordium ducam. Cum enim omnis, quæ à ratione

6.11 Egnazio Danti, *Anemographia*, Bologna 1578, Dedication to Cardinal Gabriele Paleotti. Florence, Biblioteca Nazionale Centrale, MS Magliabechiano 5.1. 301(2), fol. 5

Soon Danti's expertise came to the attention of papal administrators and, through them, of Gregory XIII himself, who commissioned from him the mapping of the entire Papal States for the purpose of administration. Not only was Danti's mapping of the Papal States so successful that it gained him the involvement in the Gallery of Maps, but it was also exceptionally well documented. His own treatises elucidate the procedures he used to make maps, while his correspondence, his maps, and, above all, his sketchbook with almost three hundred drawings documenting the last stage of the mapping process, offer invaluable insight into his actual work in the field. For its relevance to the Gallery of Maps as well as for its sheer richness Danti's cartographic work deserves our close scrutiny.

The Mapping of the Papal States

The mapping of the dominion was one of the basic functions of early modern states. European monarchs employed cartographers and draftsmen to walk around their state, measure their lands and cities, draw views of buildings and landscapes in order to produce an accurate and detailed representation of their lands. Rulers found these regional maps par-

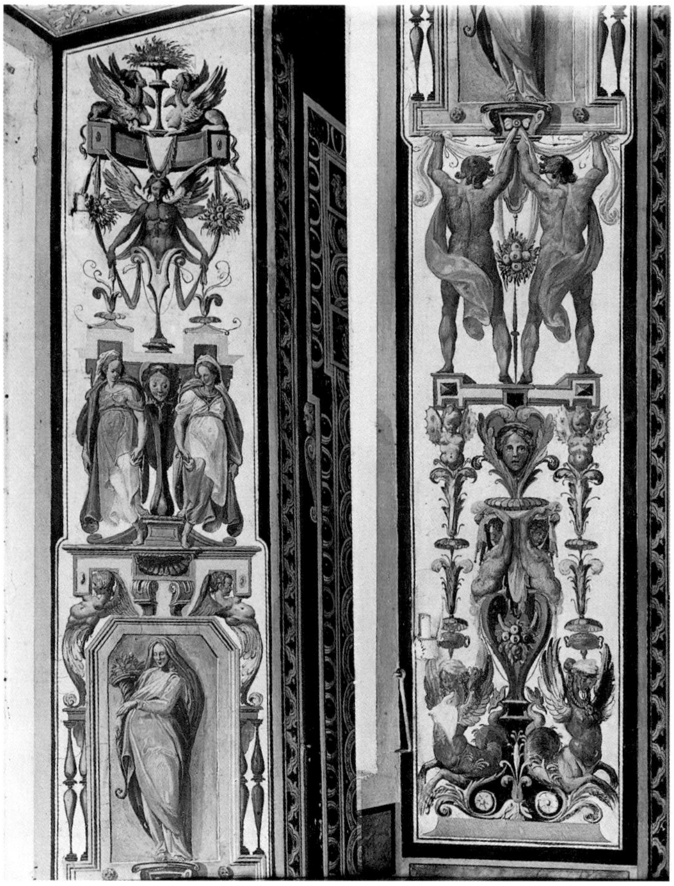

ticularly helpful to plan and monitor the collection and distribution of resources, the deployment of soldiers, and the resolution of border disputes. Like any other rulers, the popes were in need of such an effective tool for the administration and defense of their state, a tool which, however, became available only when a pope obsessed with cartography met a map-maker of exceptional talent.

In the summer of 1577, on the occasion of a trip to Perugia to visit his sick brother Girolamo, Danti met Giovanni Pietro Ghislieri, the papal governor of Perugia and a respected administrator at Gregory's court. Ghislieri commissioned Danti to survey the territory of Perugia and make a map. Danti did not fail to understand the potential of such a commission, and he wrote about it to Vincenzo Borghini, with whom he had kept in contact after his departure from Florence:

6.12 Grotesque, ca. 1580, fresco. Vatican, Gallery of Maps

I was asked there [in Perugia] by those gentlemen, and by the Governor, to make the chorography of that county, as indeed I did, drawing the plan in 28 days; I noted every distinguishing feature, and then, little by little, I portrayed everything in ink after nature, the mountains, major rivers, and remarkable buildings, and especially castles, of which there are two hundred and twenty-three, a section of the Flaminia road, and ten benefices of Malta, which produce about seven thousand scudi, and two of Santo Stefano with eleven other abbeys. I then noted for each place the quality of the air, waters, lands, what they mostly produced, the jurisdiction of the castles, whether they belonged to private lords, what the main activities of the inhabitants were, and other similar things. In the Palace in Perugia, with the same diligence, I then made a fifteen-foot mural painting drawn with the point of the brush, which showed the shadowed mountain and was later colored in watercolors, on which I marked in white all the main roads and with red lines all the divisions of the neighborhoods.

After hearing of the satisfaction the Perugians felt from this [chorography] that I made, Sig. Jacopo Boncompagni asked me, when he was here, whether I would like to survey the land of the whole Papal States; and we agreed that, when he would be in Rome, he would ask His Holiness.[38]

As far as can be ascertained, this was the first time Danti surveyed a large region of which he made a map. He surveyed the territory of Perugia in 28 days and made a map recording fortresses, fortified castles, villas, bridges, and water springs, among other features. It also marked the benefices that the Apostolic Chamber owned in the territory, the status of feudal properties, and the network of bishops' and archbishops' seats. In short, it represented the spatial distribution of both papal resources and ecclesiastical hierarchy. As a visual aid to the administration of the province, Danti's map was such a success that, immediately after it was completed in 1578, he was asked to have it painted as a mural in the governor's palace in Perugia. Danti's original maps of Perugia and its territory did not survive, but they are known through the monumental fresco painted later in the Gallery of Maps (Fig. 6.13) and through Mario Cartaro's engraved version, the *Descrittione del Territorio di Perugia* (Rome, 1580) (Fig. 6.14).[39]

The word about the usefulness of Danti's map must have spread among papal administrators, since a few months later Giacomo Boncompagni, son of Gregory XIII and then governor of the Papal States, extended Danti's commission to the mapping of the entire Papal States. In another letter, Danti shared with Borghini his anxiety concerning this challenging commission and his desire to make maps that would be even more detailed than the one he had made for Perugia.[40] Indeed, the mapping of the Papal States involved advanced skills in surveying techniques, instruments, and diplomacy. First of all, Danti had to inform the neighboring states of the new papal survey in anticipation of their objections concerning contested borders, a fact confirmed by his epistolary exchange with the Grand Duke of Tuscany, Francesco I.[41] He then planned the survey in detail, starting from the territory around Bologna. He began his fieldwork at the end of April 1578, "the first day after Easter," and completed it in sixty days with the help of three servants. In the first half of May he progressed westward from Bologna along the Via Aemilia, and then north towards the contested border with the Duchy of Ferrara. He returned to Bologna for Pentecost and left again on 21 May to survey the eastern region between the road to Ferrara and the Via Aemilia. During his trips through the Bolognese countryside he discovered a unused quarry, about which the senate of Bologna proudly reported to its ambassador in Rome (the quarry is recorded also in the Gallery of Maps). By the end of 1578 he had surveyed the whole region and made a map, which was sent to the Pope.[42] Danti's original chorography of the Bolognese area did not survive, but the content of his map is well known through its monumental version in the Gallery of Maps, the map murals *Bononiensis Ditio* (Fig. 6.15) and *Flaminia* (Fig. 6.16).[43]

Before being enlarged on the wall of the Gregorian corridor, Danti's Bolognese map was used as a visual commentary to the economic report that Giovanni Ghislieri, who had meanwhile been transferred from Perugia to Bologna, sent to Gregory XIII. Ghislieri wrote to the pope: "Your Holiness will easily have a perfect understanding of everything, if he reads this writing in conjunction with the drawing of this province, which I had Father Egnazio Danti make, and which I sent recently to your Holiness."[44] Like the report that each governor of the Papal States had to submit periodically, Ghislieri's described in detail the fortifications, bulwarks, roads, rivers, bridges, and mountains of the province under his jurisdiction, as well as the quality of the local products, the prowess of soldiers, and the sum of its income. At Gregory's request, Ghislieri and other governors had to report also in detail on the jurisdiction of each castle, town, village, monastery, and abbey, verifying the legitimacy of their feudal property. Lords who failed to pay their dues to the Apostolic Chamber had to return

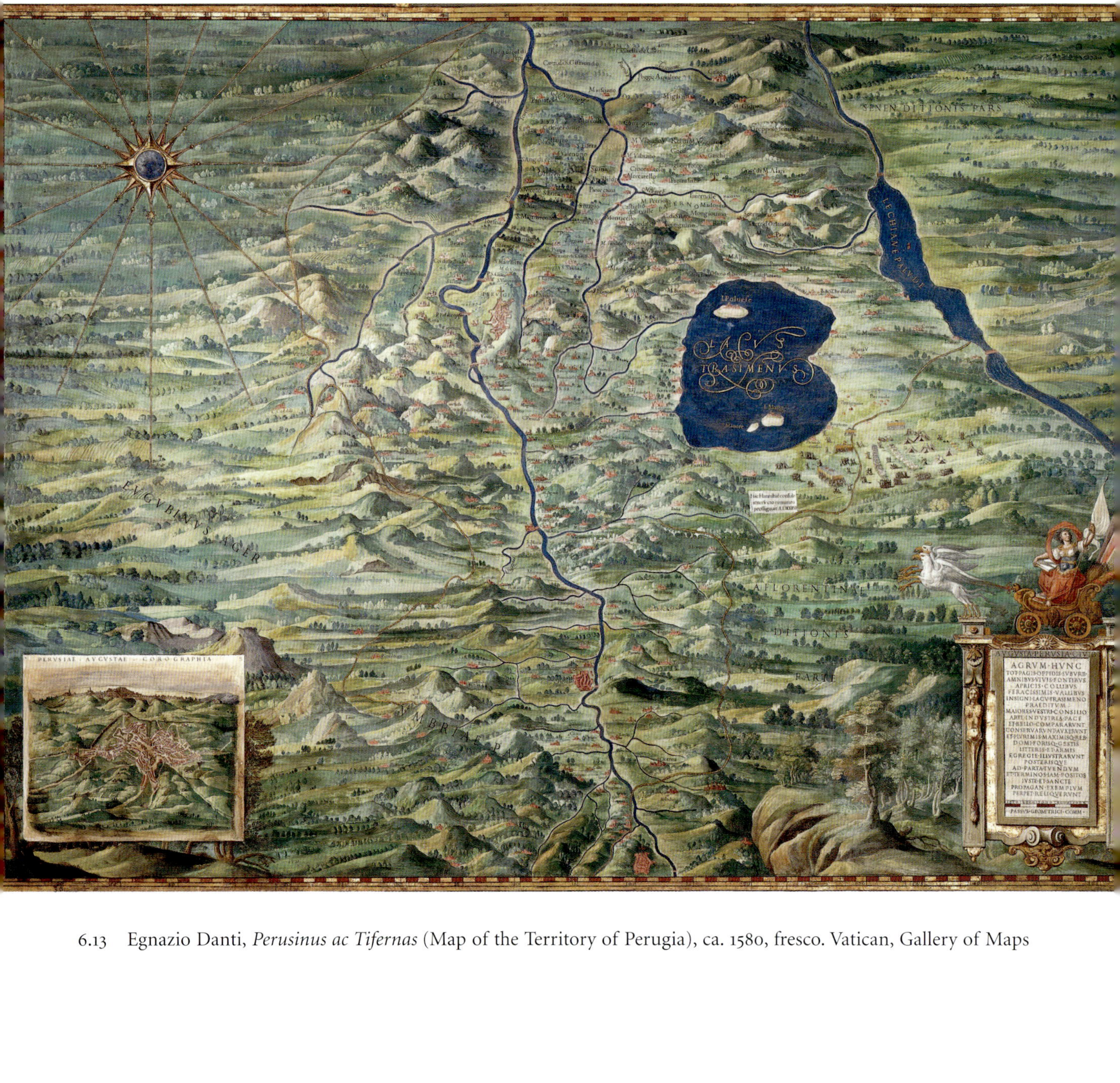

6.13 Egnazio Danti, *Perusinus ac Tifernas* (Map of the Territory of Perugia), ca. 1580, fresco. Vatican, Gallery of Maps

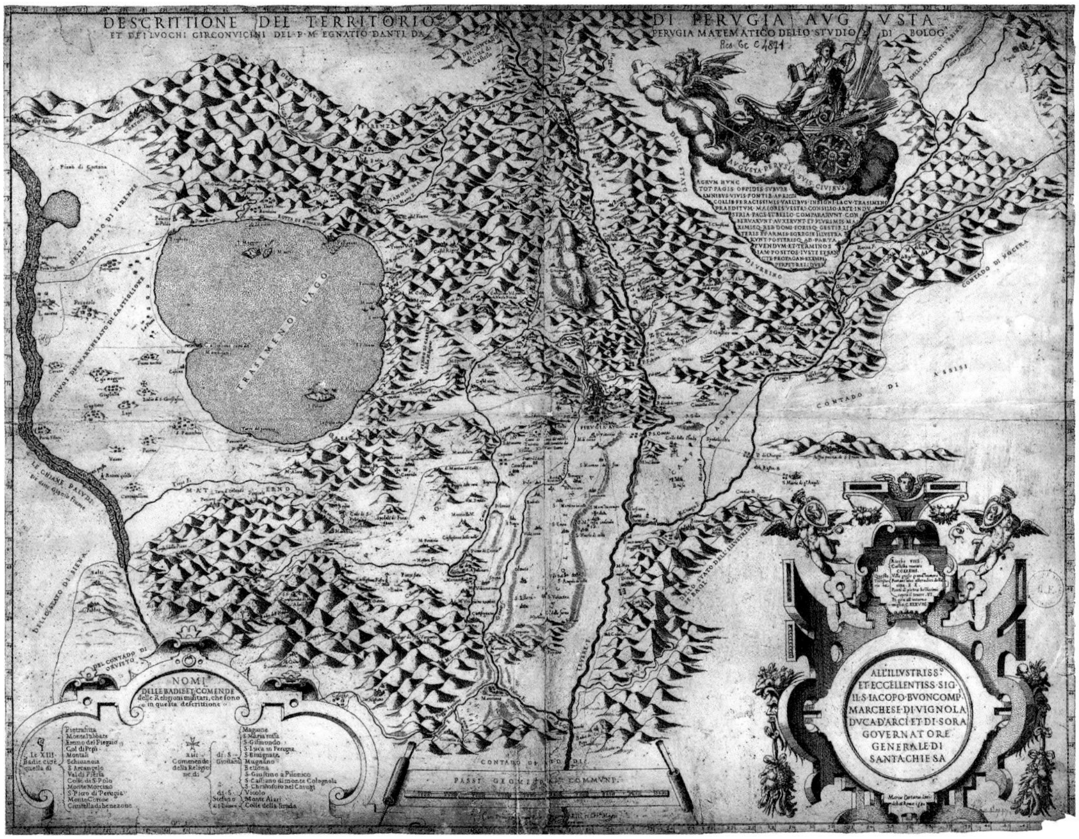

6.14 Mario Cartaro, *Descrittione del Territorio di Perugia Augusta* (Map of the Territory of Perugia), 1580, engraving. Bibliothèque nationale de France

their feudal property to the pope. Ghislieri's is the only report of this kind to use a map as a visual aid to his verbal description.

After he completed the chorography of the Bolognese area, Danti must have surveyed the remaining areas of the Papal States, and made their maps. These maps, now lost, would likely have been used as the cartographic source for the corresponding map murals in the Gallery of Maps: the maps of the Patrimony of St. Peter and Latium. Although Danti's maps of the Patrimony of St. Peter and Latium were repainted entirely by Lukas Holste in the 1630s, they are partially documented by other sources. In 1592, the cartographer Philip van Winghe executed a drawing of the map *Latium* at the request of Ortelius, who wanted to include it in one of the updated versions of his *Theatrum orbis terrarum* (Fig. 6.17). It is only the area around Orvieto that is recorded from the map of the Patrimony of St. Peter thanks to Mario Cartaro's engraving, *Urbis Veteris Antiquae Ditionis Descriptio* (Rome, 1583) (Fig. 6.18).[45]

The success and appeal of Danti's maps resided in the accuracy of his surveying techniques and in his ability as a draftsman. As he himself acknowledged in his *Trattato dell'uso et della fabbrica dell'astrolabio* (Florence, 1578), his surveys were made using methods based

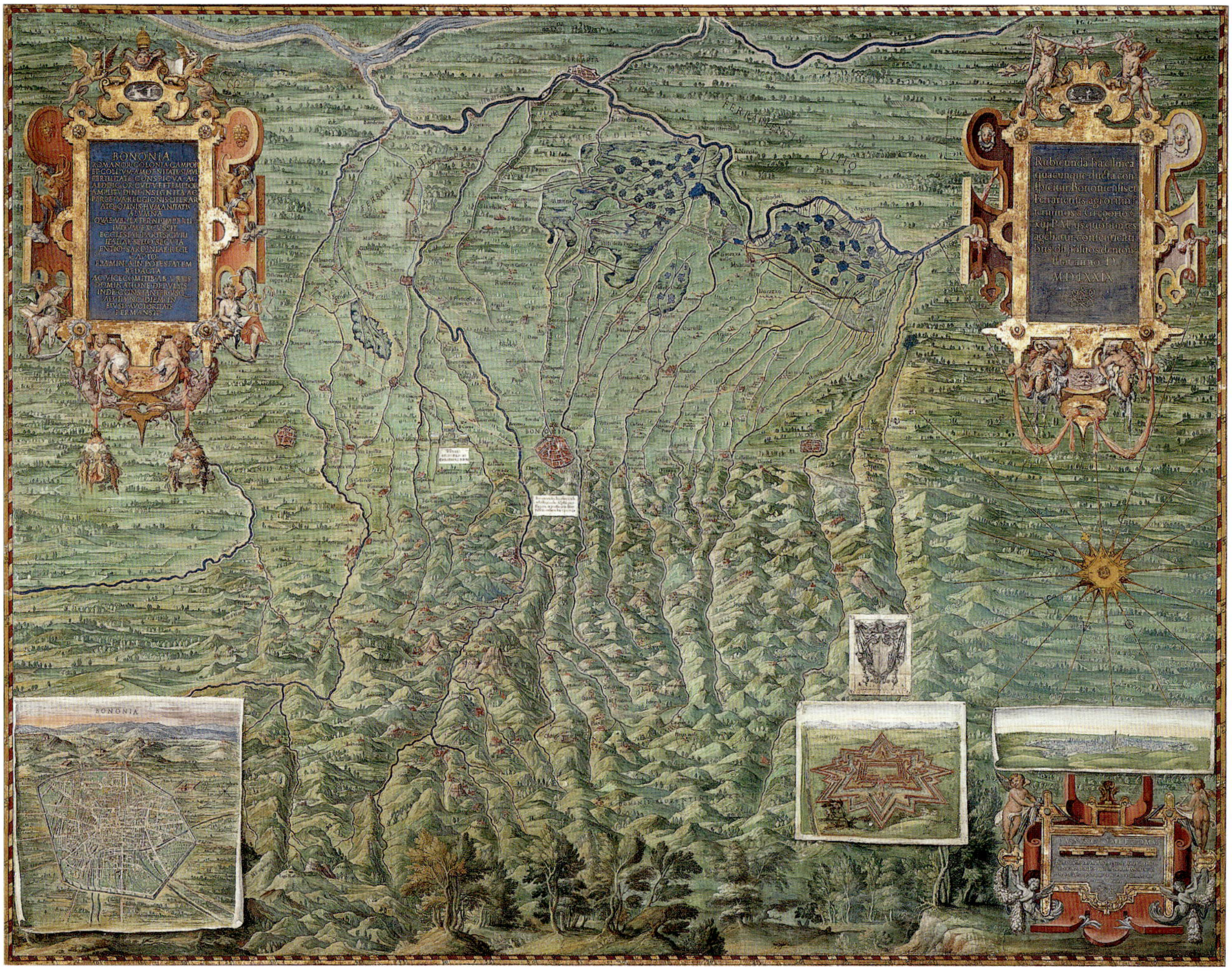

6.15 Egnazio Danti, *Bononiensis Ditio* (Map of the Territory of Bologna), ca. 1580, fresco. Vatican, Gallery of Maps

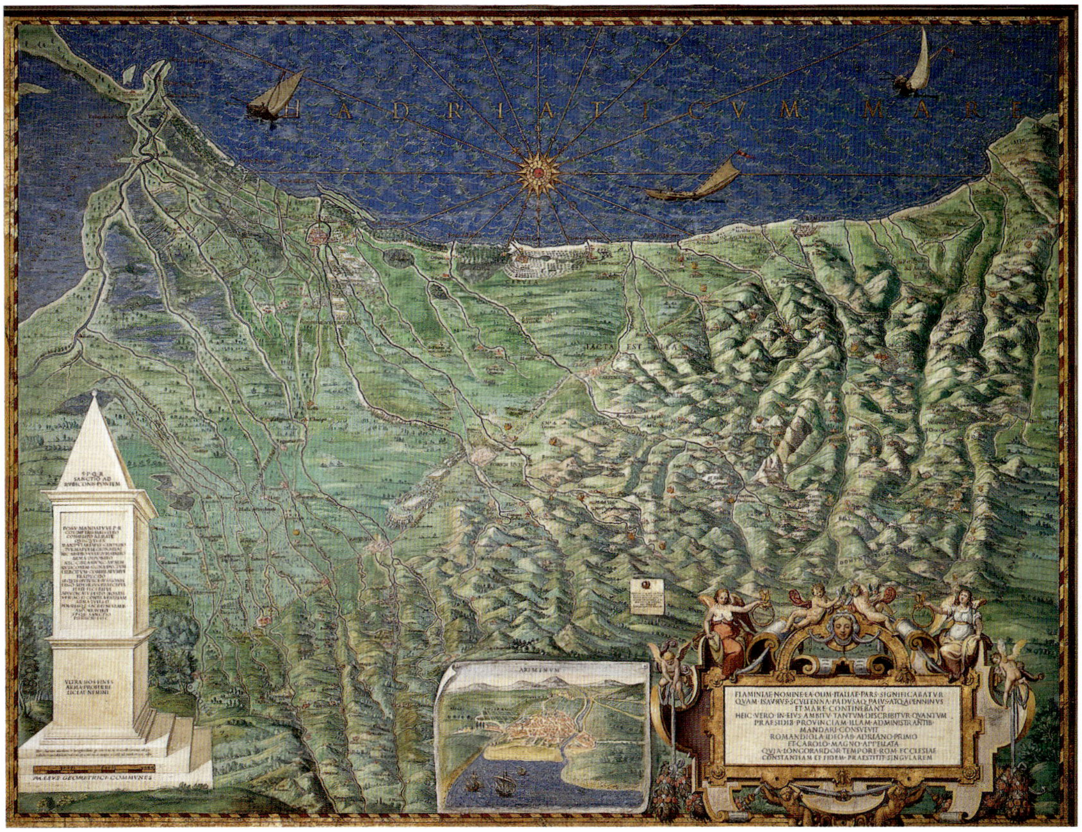

6.16 Egnazio Danti, *Flaminia*, ca. 1580, fresco. Vatican, Gallery of Maps

on triangles, such as those established by Peter Apian, Gemma Frisius, and Oronce Finé, and translated into Italian by Cosimo Bartoli.[46] In the same treatise, Danti also suggested adapting astronomical instruments such as the astrolabe and the quadrant for survey purposes. Despite Danti's assertion, however, the astrolabes and quadrants that he designed and had built (and which still survive) could hardly have been his surveying instruments. They are so rich and refined in material and so small in size that they could not have been meant for use in the field but rather as demonstration pieces for learned patrons.[47] In a later treatise, Danti stated that the instrument he used for surveying was the simple and inexpensive *radius latinus* (Fig. 6.19): "In the years 1577, 78, 79, 80, at the behest of the late Holy and Glorious Pope Gregory XIII, I surveyed most easily with this instrument [the *radius latinus*] a number of provinces, such as the entire Romagna, part of Umbria, of Latium, and of Sabina, and the whole territory of Perugia and Bologna; and extremely convenient was the fact that I could fold and unfold it without dismounting from my horse and that I could carry it while on horseback with no encumbrance."[48]

According to Danti, the *radius latinus* had numerous advantages since it included, "the radius of the Greeks, the cross staff of Spanish navigators (on which it depends), and the great *regula* of Ptolemy," and allowed one to survey two points simultaneously without

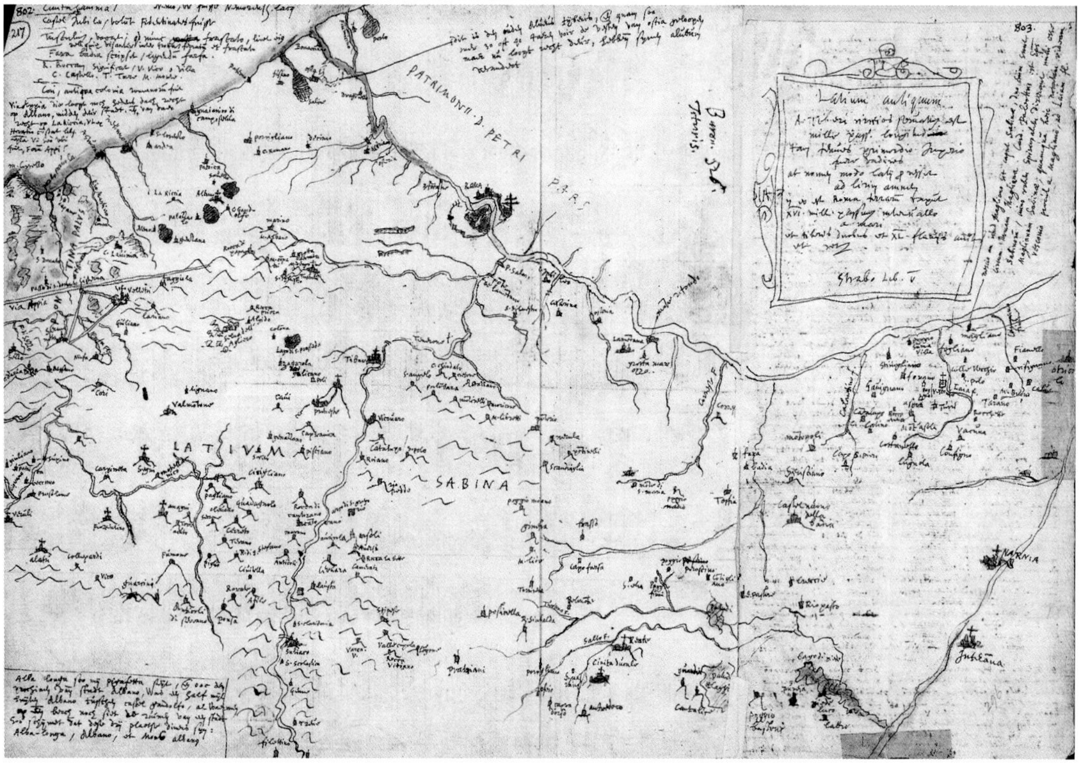

6.17 Philippe van Winghe, *Latium*, pen drawing. Leiden, Universiteitsbibliothek (BPL 2766)

moving the instrument.[49] Moreover, unlike other instruments, which are "difficult to build, often inaccurate, and require great diligence and time to place them correctly on the site where one has to do measurements with them," the *radius latinus* was very easy to fold, move around, level with the horizon, and adjust for measurements.[50] Although today one may question the accuracy of such an instrument with a mobile plumb line, which Danti used while on horseback, this fact does not diminish the historical significance of his attempt both to achieve greater accuracy and to speed up the surveying process.

This marvelous instrument was invented by Latino Orsini, a member of the famous Roman family and a skilled condottiere, who served under Giacomo Boncompagni in Gregory XIII's army. The instrument was of great help in calculating the trajectory of weapons, the height and distance of towers, bulwarks, and fortresses, as well as in drawing the plan of the enemy's camp. Latino Orsini wrote a short treatise on how to use and build the *radius latinus*, which was meant for his own use (*per suo memoriale*). Danti published Orsini's treatise, which he may have obtained from Orsini himself or from Giacomo Boncompagni, with his own explanatory notes in 1583 and in an augmented edition in 1586.[51]

In that treatise, Danti further explained that he also used the *radius latinus* to draw the map on paper, namely, to organize his list of measurements spatially. On a plane table

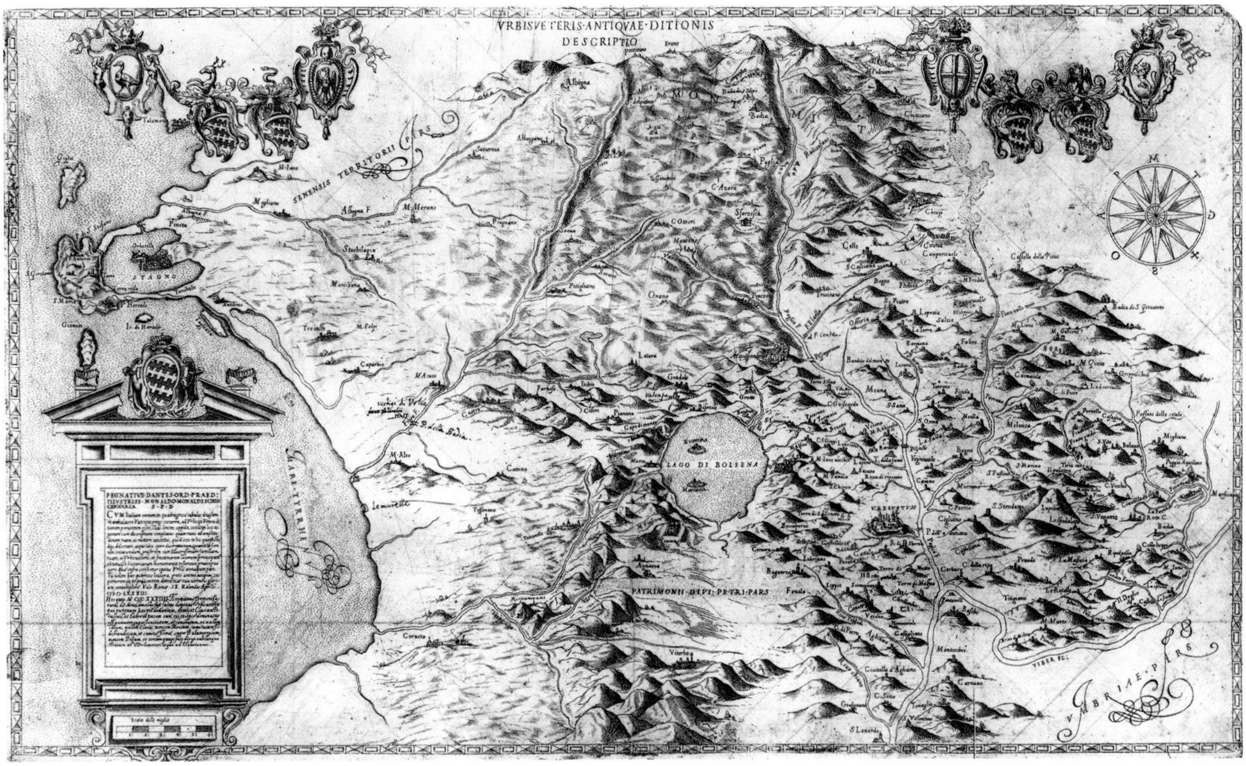

6.18 Mario Cartaro, *Urbis Veteris Antiquae Ditionis Descriptio* (Map of the Territory of Orvieto), 1583, engraving. Biblioteca Apostolica Vaticana (Stampe Barb. Lat. 9906(2), fol. 9)

(*tavolozza*), after drawing the distance between the two stations and having therefore established the map's scale, he positioned the instrument in line with the first survey station. He then reduced his measurements to scale and marked them on the paper. He repeated the same operation with the instrument in line with the second survey station. The intersections of the rays from the two stations marked the "real" position of a place on the map. At this point, Danti distinguished the surveying from the map-making as two different stages. But in the same *Trattato del radio latino*, he described in detail an instrument used by his father, Giulio Danti, to survey the city of Perugia (Fig. 6.20). Giulio's instrument was even simpler than the *radius latinus*, but it had this crucial advantage: "at the same time [as surveying] the map was drawn on paper."[52] According to Danti, the *radius latinus* could also be applied to the *tavolozza* during the survey.[53] In short, it is fairly certain that Danti used the *radius latinus* to survey and make maps. He may also have attached it to a plane table, as described here, in order to unify the separate stages of surveying and map-making.

Danti's surveying technique was commonplace at the end of the sixteenth century, and the instrument he favored, the *radius latinus*, was conceptually identical to other surveying instruments, although more convenient to use. What makes Danti's work so significant historically is the rich surviving documentation relating to his fieldwork. Although he held a prestigious position as court cosmographer and later as professor at the

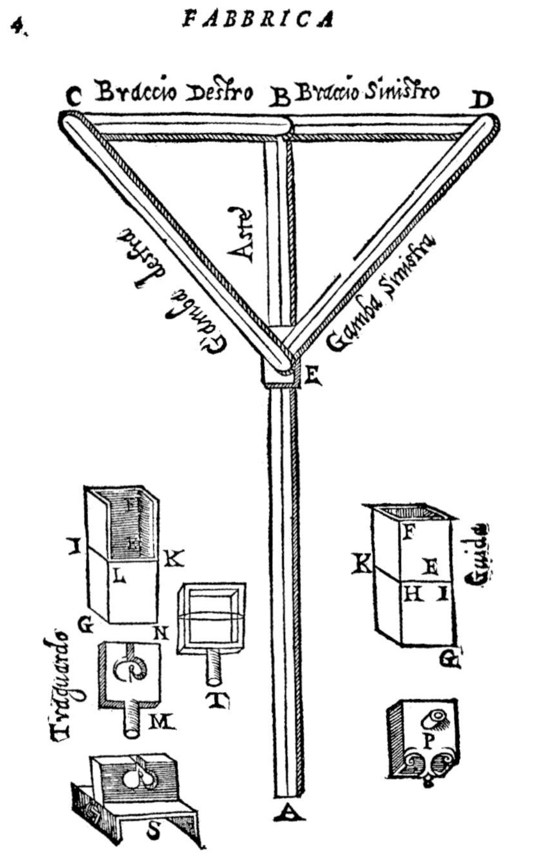

6.19 *Radius latinus*, woodcut, from Egnazio Danti, *Trattato del radio latino*, Rome 1586, p. 4. The Folger Shakespeare Library

University of Bologna, he did not disdain venturing out into the fields to survey regions by himself. A sketchbook, now in the Biblioteca dell'Archiginnasio in Bologna (MS Gozzadini 171), contains 292 drawings, part of Danti's work on the mapping of the Bolognese province.[54]

Page after page, numerous astonishing drawings "after nature" (*dal naturale*) provide a meticulous, visual account of the Bolognese countryside four centuries ago: its castles, palaces, villas, towers, and landscapes (Figs. 6.21 and 6.22). In these drawings Danti also recorded the distance and orientation of each place from two or three different survey stations, but made no astronomical measurements for coordinating longitude and latitude (Figs. 6.23 and 6.24). The manuscript also shows that Danti often returned to the same places, surveying them again and again in a painstaking effort to achieve an accuracy that his instruments could hardly deliver.[55] The main interest of the sketchbook, however, lies in the drawings *dal naturale*. They document Danti's last stage of the survey, recording, by different visual techniques, features that were no less essential than measurements to Danti's map. As he had explained in 1577 to his friend Vincenzo Borghini, after surveying an area he "depicted everything in ink after nature, mountains, principal rivers, distinguished buildings, and especially castles."[56] Particularly interesting are some drawings in which Danti effectively combined perspective and plan views in a single image, a representational technique that he adopted and refined in the Gallery of Maps.

Danti's multiple means of recording a place through both measurements and freehand sketching made it possible to make maps that not only represented the spatial relationship of places but also imparted the illusion of a countryside. Indeed, Danti conceived his regional maps as an encyclopedic description of a place projected onto an accurate geographical framework. He explained this approach in his *Delle osservationi de' viaggi* (*Observation on voyages*), a list of 26 subjects that a traveler should know when he is in "foreign regions in whatever part of the world, traveling by land, sea, or river . . . if he wants to benefit from his voyages."[57] These topics include such widely disparate matters as the ancient name, the borders, the distances from place to place, the manner, character, and language of its inhabitants, the entity and quality of its soldiers and fortifications, civic and religious buildings, the jurisdiction of every single castle and villa, and the number of episcopal and archiepiscopal seats. Danti's list asked essentially for information easily acquired by any curious and educated traveler and was, in fact, an index of information already amply provided by chorography books. However, Danti's last two points – concerning astronomical measurements and the drawing of a plan – would apply only to a traveler interested in map-making. At the end of the list, Danti recommended the reader to observe the following:

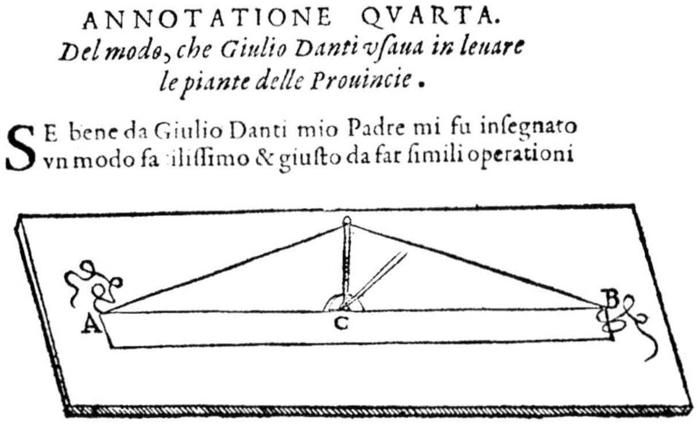
ANNOTATIONE QVARTA.
Del modo, che Giulio Danti vſaua in leuare le piante delle Prouincie.

SE bene da Giulio Danti mio Padre mi fu inſegnato vn modo fa iliſſimo & giuſto da far ſimili operationi

6.20 Giulio Danti's Surveying Instrument, woodcut, from Egnazio Danti, *Trattato del radio latino*, Rome 1586, p. 109. The Folger Shakespeare Library

> 25. Principally the latitude of the place, which is done by measuring the height of the pole, and the meridian height of the sun or of the stars, and from this one also deduces the greatest length or brevity of the days.
> 26. Where one can do so without raising suspicion, in detail the plan of the cities and the fortresses, and draw it, with the measurements of the curtains, bastions, cavaliers, and moats, with the nearby hills from where one observes if the place can be defeated, and if the land is such that it can be mined, if water can be drawn, if it can easily be besieged, and if one can flood it with some river.[58]

Danti's observation on voyages may thus be understood as a description of the multimedia format of regional cartography: to map a region, provide views of individual, notable buildings, report on its defense systems, and describe its inhabitants and peculiarities. The almost three hundred drawings of Danti's sketchbook strongly suggest that his list was not only a matter of cartographic theory but also of actual practice in the field.

Danti's knowledge of surveying and map-making, his meticulous registration of papal resources, feudal properties, and episcopal and archiepiscopal seats, as well as his ability to combine different systems of representation in one effective image, were all elements that together contributed to the great success of his mapping of the Papal States. Danti's maps appealed to Gregory XIII, who used them immediately for purposes of administration but who also understood that they could be used for more lofty reasons.

* * *

Gregory XIII made public his predilection for maps in the Sala Bologna, a predilection that indeed was confirmed in the decoration of other spaces in the Vatican Palace. His passion for maps was reinforced by his friend Paleotti, who not only viewed the contemplation of scientific images as a step towards the knowledge of the divine but also regarded the study of nature as a component of the post-Tridentine spiritual reform he advocated.

In the Sala Bologna, however, the pope had inserted the maps within a parochial view of

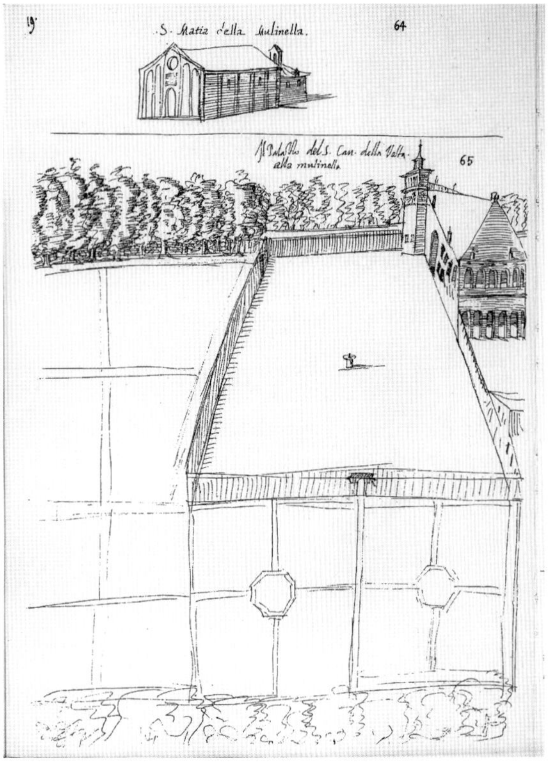

6.21 Egnazio Danti, The Palace of Signor Della Volta, ca. 1577, pen drawing. Bologna, Biblioteca Comunale dell'Archiginnasio, MS Gozzadini 171, drawing no. 65

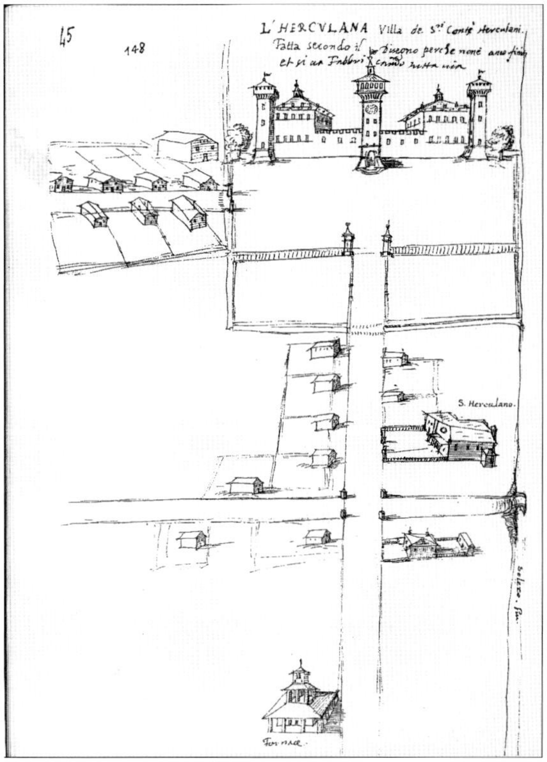

6.22 Egnazio Danti, Villa of Count Ercolani, Bologna, ca. 1577, pen drawing. Bologna, Biblioteca Comunale dell'Archiginnasio, MS Gozzadini 171, drawing no. 148

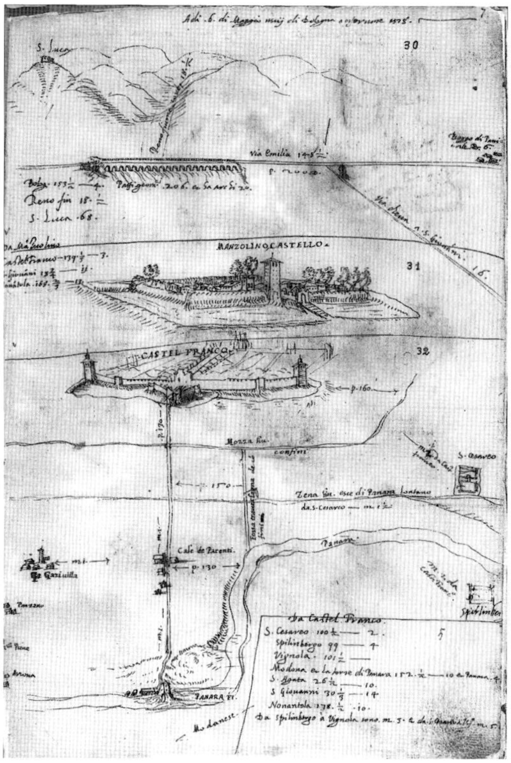

6.23 Egnazio Danti, The Town of Castelfranco, ca. 1577, pen drawing. Bologna, Biblioteca Comunale dell'Archiginnasio, MS Gozzadini 171, drawing no. 32

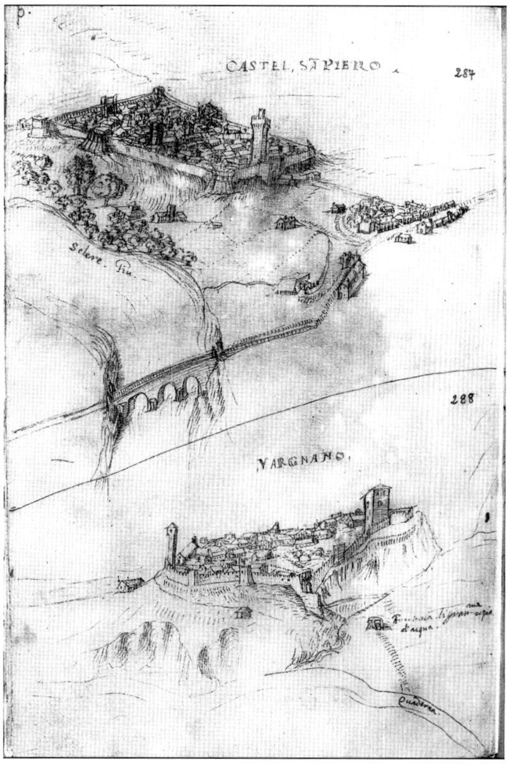

6.24 Egnazio Danti, The Villages of Castel San Pietro and Vargnano, ca. 1577, pen drawing. Bologna, Biblioteca Comunale dell'Archiginnasio, MS Gozzadini 171, drawing nos. 287 and 288

power, which did not correspond with the temporal power of the papacy, or with its role in Italy, let alone with its universal mission around the globe. As the years went by, Gregory rapidly expanded the horizon of his mapping projects. First, as a responsible ruler, he commissioned the mapping of the Papal States in order to control the economic, military, and spiritual resources of his dominion. Then, he turned to more daring and elusive mapping projects that had the scope of making visible his ideals for the post-Tridentine church. As he had commissioned to Danti the mapping of the Papal States, so to him he turned again when he wished to embellish the east wing of the Belvedere Courtyard with monumental maps. The convergence of Gregory XIII's cartographic passion with Danti's cartographic imagination transformed the long corridor into a fascinating, indoor walk along the Apennine spine of the Italian peninsula.

7

The Gallery of Maps

In his new corridor along the Belvedere Courtyard, Gregory XIII chose to represent Italy through a series of regional maps of the peninsula, a veritable cartographic achievement that, like Cosimo I's maps of the world, had no precedent in print or in paint. The cartographic representation of Italy that Gregory had painted on the walls of the papal residence was inspired by the classical notion of the peninsula described by authors such as Strabo and Pomponius Mela, a notion that was vividly revived in the antiquarian and chorographical tradition of the Renaissance. Like earlier chorographical representations of Italy, Gregory's Italy fused geography with local history but, unlike its predecessors, it presented the peninsula as the stage of church history. A marvelous example of Renaissance chorography in all its forms and meanings, the Gallery of Maps was a largely unsurpassed synthesis of modern geography and a papal interpretation of church history.

The Corridor

The Gallery of Maps, a corridor on the third floor of the Vatican Palace, was part of the Belvedere Courtyard, the area connecting the old Vatican Palace with the papal villa on the Belvedere Hill. In the early sixteenth century, the architect Bramante had planned the courtyard as a series of terraces at different levels flanked by corridors on both sides, an ambitious project that required the concerted efforts of many popes to be completed. By the time Gregory XIII became pope, the Belvedere Courtyard was close to completion, even though the east wing, the nearest part of the building to the Vatican gardens was one storey lower than the west wing. Renaissance popes often used the Belvedere Courtyard for important ceremonies. Gregory avoided staging there the theatrical performances that some of his predecessors had favored, but he liked to take walks along the slopes, and occasionally even gave audiences in the courtyard, as he did in July 1578 when, to the astonishment of his court and visitors, "he gave audience while walking indefatigably for two hours."[1] During the Holy Year 1575, Gregory started to use the Belvedere Courtyard for religious ceremonies. Appear-

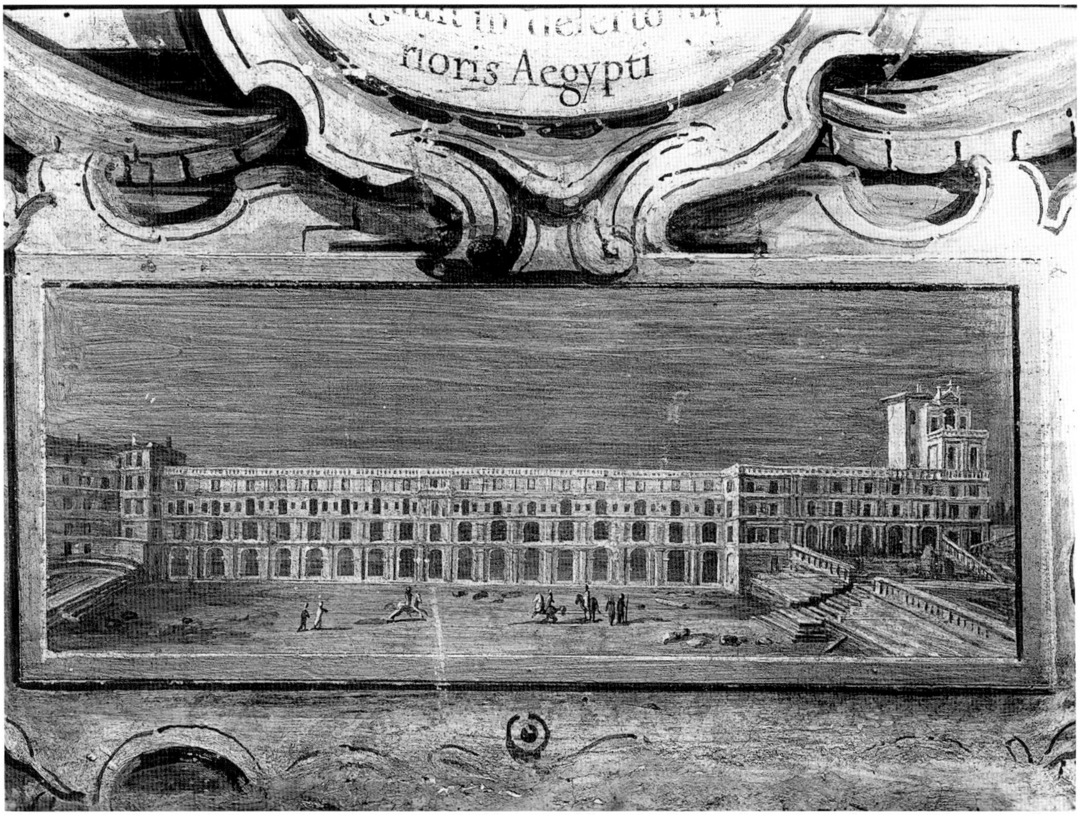

7.1 Mathijs Bril, *View of the Belvedere Courtyard*, 1581–1583, fresco. Vatican, Tower of the Winds

ing from one of the windows facing the main entrance to the courtyard, Bramante's Porta Rustica, he would bless the pilgrims gathered there. He must have liked this ceremony considerably, for he continued it throughout his pontificate.

The frequent use of the Belvedere made apparent the fact that its east corridor was inconveniently located one level below other rooms often used for meetings and audiences, such as the Sala di Costantino and Raphael's Stanze. In the autumn of 1578, Gregory XIII authorized the construction of an upper corridor on the east wing to provide a covered passage between Raphael's Stanze and the new papal apartment he was planning at the end of the same corridor.[2] Ottaviano Mascherino, who had become papal architect after his work in the Sala Bologna, designed the new papal apartment, its Tower of the Winds, and a magnificent spiral stairway in marble connecting the new rooms to the gardens, which Egnazio Danti later praised because it "turned out very beautifully."[3] Mascherino also built the corridor that was to become the Gallery of Maps, and constructed a terrace above it, so as to enjoy the walk along the Belvedere in mild weather (Fig. 7.1). For the middle of the corridor, which faced Bramante's Porta Rustica, Mascherino designed an imposing Serlian window with a balcony, which was perhaps used for the papal benediction. This window not only emphasized the direct connection between the Porta Rustica and the balcony, and

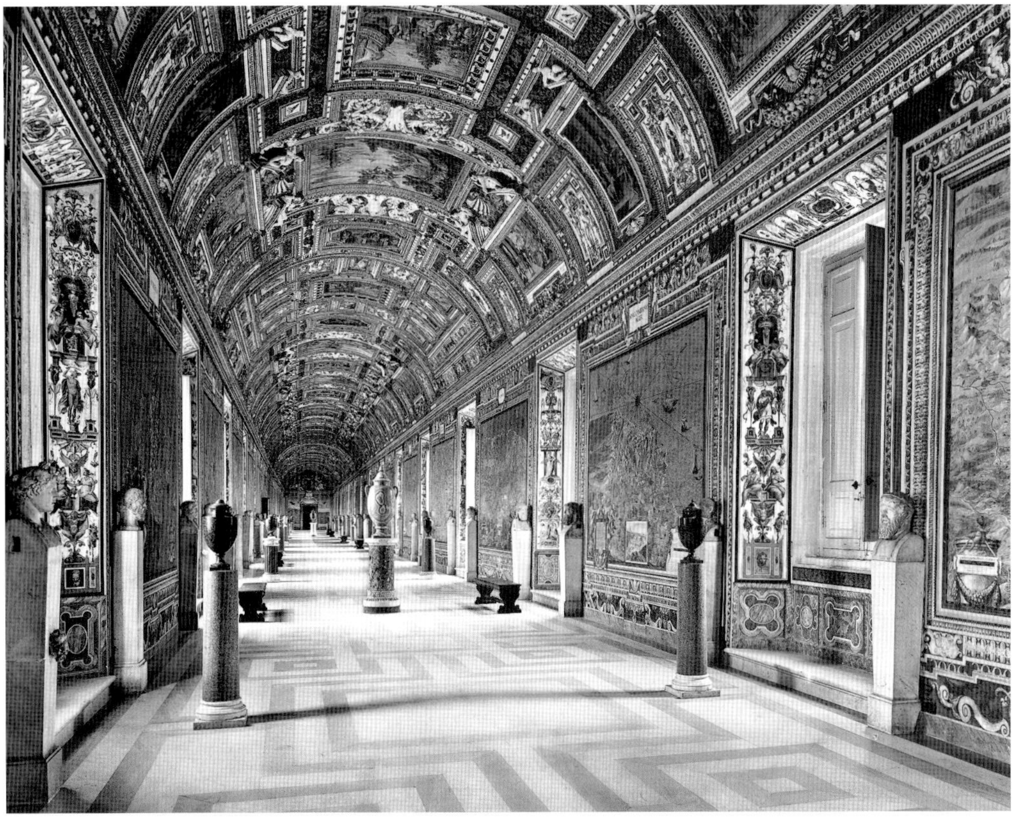

7.2 Gallery of Maps. Vatican

between the entering pilgrims and the pope giving the benediction, but it also recalled the similar Serlian window that Constantine had built in his palace in Constantinople.[4]

The construction of the corridor and of the papal apartment proceeded expeditiously due to the personal involvement of Gregory, who routinely visited the site during the construction, and postponed other projects to concentrate financial resources on the new corridor. By mid-1579 the corridor was ready for decoration. Two years later the frescos were completed.[5]

The decoration of the Boncompagni corridor elicited comments of marvel and admiration (Fig. 7.2). Vivid descriptions of the Gallery of Maps appeared in official diplomatic reports, in the *Avvisi di Roma*, in the diaries of many travelers, in poems dedicated to Gregory XIII, and in a poem entirely dedicated to its celebration, Giovanni Battista Stella's *Ambulatio Gregoriana* (*The Gregorian Walk*).[6] Forty magnificent, monumental maps, each about three by four meters, decorated the walls of the long corridor. With a profusion of bright colors, intense blues, and vivid greens, the maps represented the territories of the Italian peninsula, the main islands of the Mediterranean (Malta, Corfu, the Tremiti, and Elba), four important ports (Genoa, Venice, Ancona, and Civitavecchia), and the county of Avignon. The vaulted ceiling of the gallery was also filled with a multitude of scenes: 24 representing sacrificial subjects from the Old Testament, 51 illustrating miracles of saints, and over 100 personifications to fill the remaining space. An inscription on the northern portal explains the organizing principle of this unusual iconography (Fig. 7.3):

> Italy, the most noble region of the entire world: as it is divided by nature by the Apennines, similarly to this end the gallery is divided into two parts, on this side the one [part] bounded by the Alps and the Upper Sea, on this side the other [part] bounded by the Lower Sea; from the river Var as far as the furthest Bruttij and Sallentinians, with kingdoms, provinces, dominions, and islands arranged within their own borders, as they are now, the entirety [of Italy] is displayed in tables on both sides of the long hall. The vault shows the pious deeds of holy men in the places where they happened, each corresponding to the other. Gregory XIII, Highest Pontiff, in the year 1581 wished these things, begun by himself, to be finished with skill and splendor, not for his own advantage as much as for that of the Roman pontiffs, and so that utility not be absent from the delight and knowledge of things and places.[7]

Succinctly and effectively this inscription guided viewers in experiencing and interpreting the Boncompagni corridor. Visitors proceeded along the gallery as if they were walking along the Apennine ridge from north to south. As they were taking this imaginary walk, it seemed as though they looked out over the territories of Italy displayed on either side in accordance with their actual location along the peninsula: the regions of the Tyrrhenian Sea were represented on the east wall of the corridor, while the regions of the Adriatic Sea appeared on the west wall (Figs. 7.4 and 7.5). Indeed the space of this long corridor arranged the maps of Italy so effectively as to suggest their real, three-dimensional location in the world.[8]

The maps are arranged from the general to the particular: two maps of Italy – *Italia antiqua* and *Italia nova* – start the cycle at the south entrance (toward the old Vatican Palace) (Figs. 7.6 and 7.7); the regional maps of the peninsula follow. The maps adopt the conventional colors of Renaissance cartography, but the effect of chromatic richness and splendor is enhanced by their sheer size. Furthermore, each map is provided with a scale, windrose, coordinates of latitude and longitude, and with at least one legend that briefly explains the peculiarities and the history of the territory and its inhabitants. Historical vignettes are depicted on the topography of the maps, and landscape features are at the bottom of most maps (Figs. 7.8 and 7.9). There are also perspective and plan views of cities and other important places.

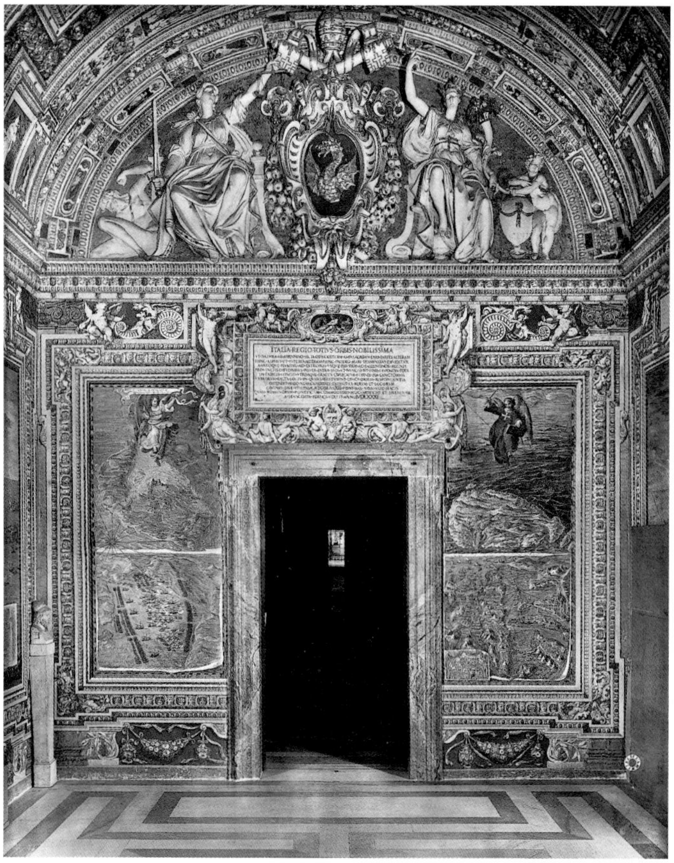

7.3 North Entrance, Gallery of Maps. Vatican

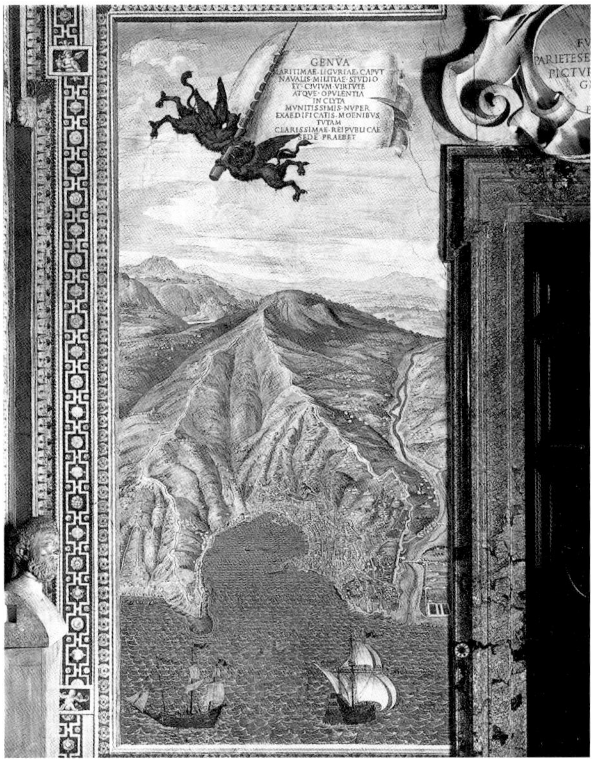

7.4 Egnazio Danti, View of Genoa, ca. 1580, fresco. Vatican, Gallery of Maps

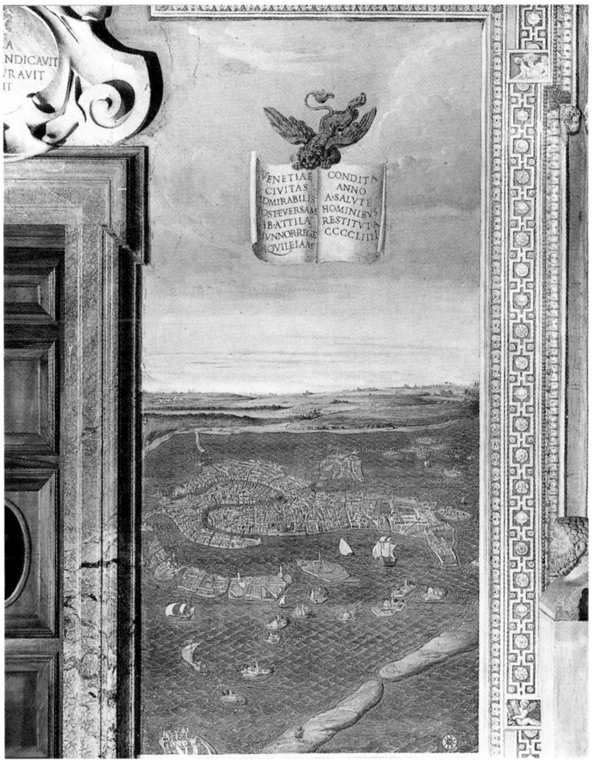

7.5 Egnazio Danti, View of Venice, ca. 1580, fresco. Vatican, Gallery of Maps

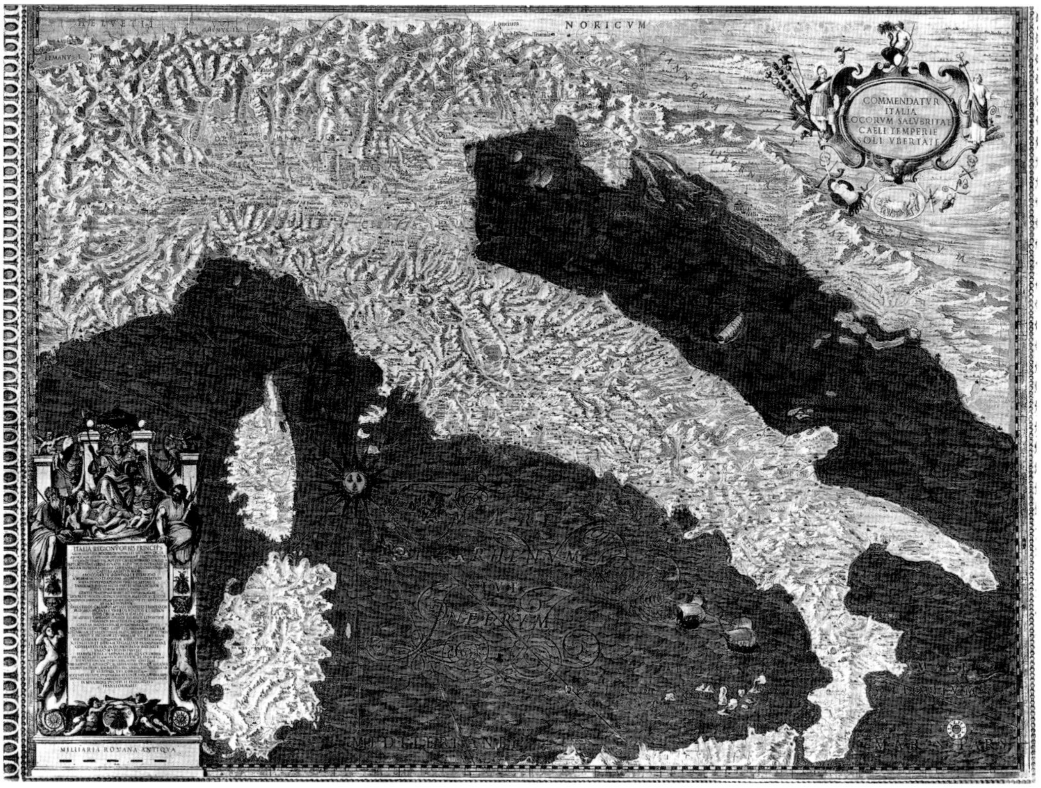

7.6 Lukas Holste, *Italia antiqua* (Map of Ancient Italy), ca. 1630, fresco. Vatican, Gallery of Maps

On the ceiling, but closely related to the maps, as the inscription also explains, are depicted pious deeds by saints that took place in the territories charted below. These start with those of Constantine, whose life is represented in five episodes connected to the two general maps of Italy, suggesting explicitly the continuity between the church of Constantine's age and the church of the late sixteenth century. From the Constantinian age, church history unfolds gloriously, and the association of geographical site with historical event is evident. For instance, *The Miracle of the Blood of St. Januarius in Naples* relates to the region of Campania (Figs. 7.10 and 7.11); *The Exposition of the Holy Shroud*, which took place in Turin, corresponds to the map of Piedmont; *St. Ambrose Refusing Emperor Theodosius Admittance to Church* illustrates the Duchy of Milan; *The Archangel Michael Appearing on Monte Gargano* illustrates Apulia.[9]

The artists involved in this huge project are well known, thanks to a manuscript note in the Boncompagni archive. Among the documents gathered by Giacomo Boncompagni in order to write his father's life, this note succinctly, but accurately, listed the accomplishments of the Boncompagni pope in art and architecture. On the Gallery of Maps it explained: "Ottaviano Mascherino was the architect of the great Gallery. The painting of the cosmographical maps was done following the plan and design of the most Reverend Father Egnazio Danti, Bishop of Alatri. The paintings of the vault were planned by Girolamo Muziano, but

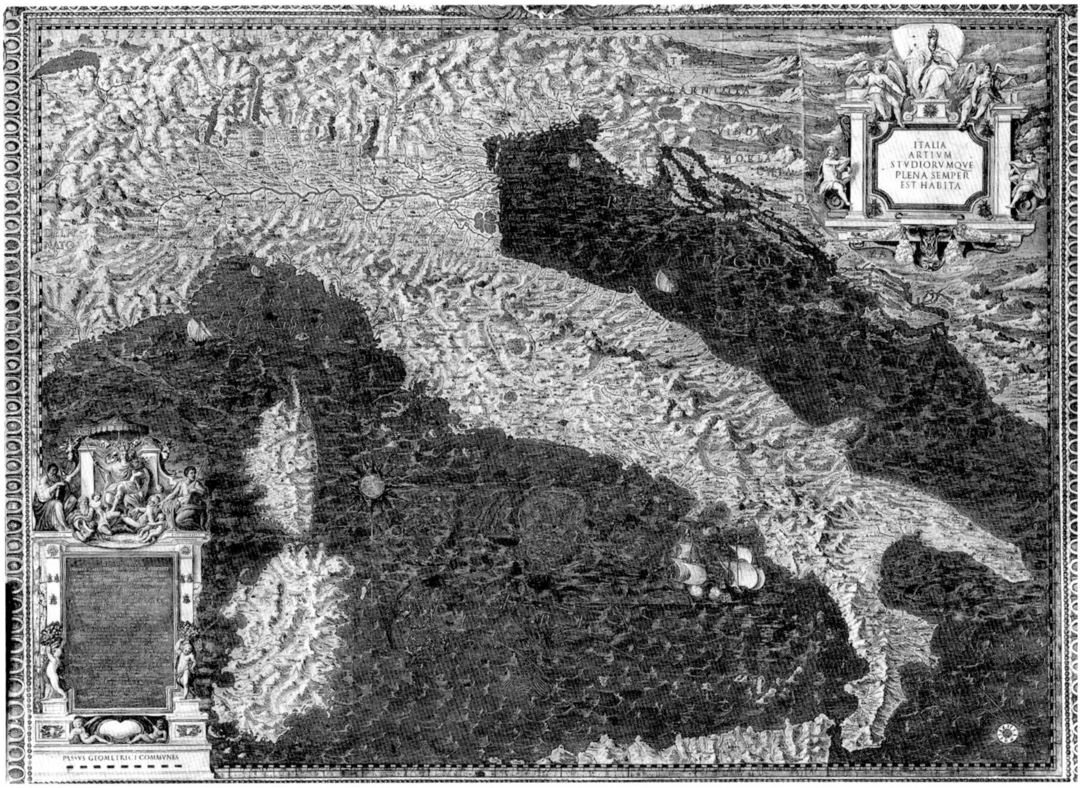

7.7 Lukas Holste, *Italia nova* (Map of Modern Italy) ca. 1630, fresco. Vatican, Gallery of Maps

designed by Cesare Nebbia of Orvieto."[10] According to this document, Danti designed and planned the maps, a task that is not difficult to assign him in consideration of his previous cartographic work at Cosimo I's court, while the painters Girolamo Muziano and Cesare Nebbia were responsible for the partition and design of the ceiling respectively. However, the nature of their artistic collaboration requires some further explanation.[11] Muziano must have secured the commission for the Gallery of Maps, since Gregory had already entrusted important commissions to him, including the design and supervision of the work on the mosaics in the vaults of the Gregorian Chapel in St. Peter's. But, absorbed by the mosaics of the Gregorian Chapel, Muziano could not supervise the Gallery of Maps daily, nor could he draw the preparatory drawings, let alone paint the walls. He simply provided the partition of the huge vaulted ceiling into smaller areas, and perhaps the inspiration for the great attention paid to landscape details in the maps and in the scenes of the vault. Nebbia, who had been Muziano's pupil many years before and had become his trusted collaborator, was responsible for the invention and design of every single scene, a fact confirmed by eleven preparatory drawings relating to as many ceiling scenes that still survive (Figs. 7.12 and 7.13).

Nebbia invented the ceiling compositions but did not paint them all. This task was left to a small army of artists, which the art critic Giovanni Baglione, generally well informed on

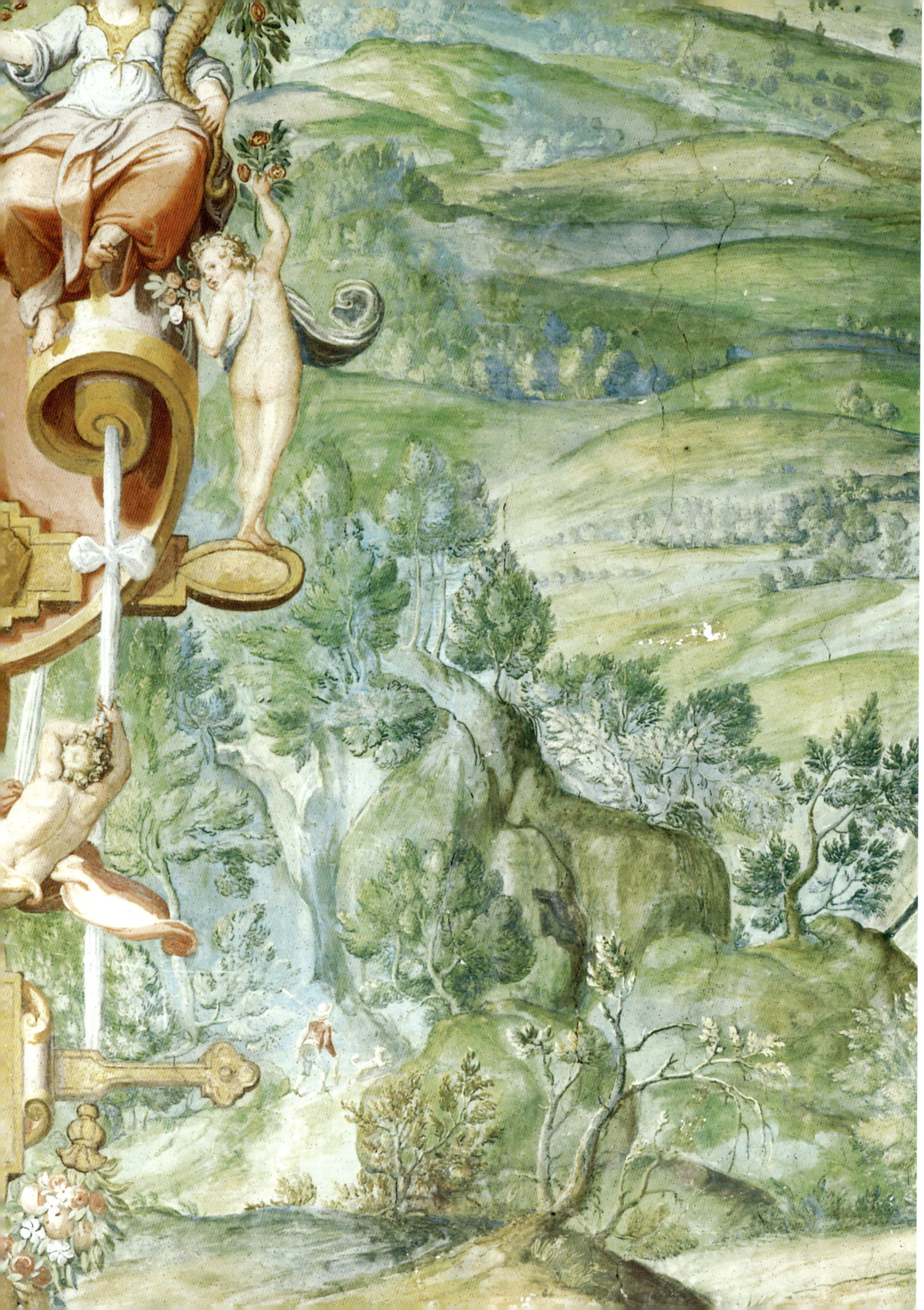

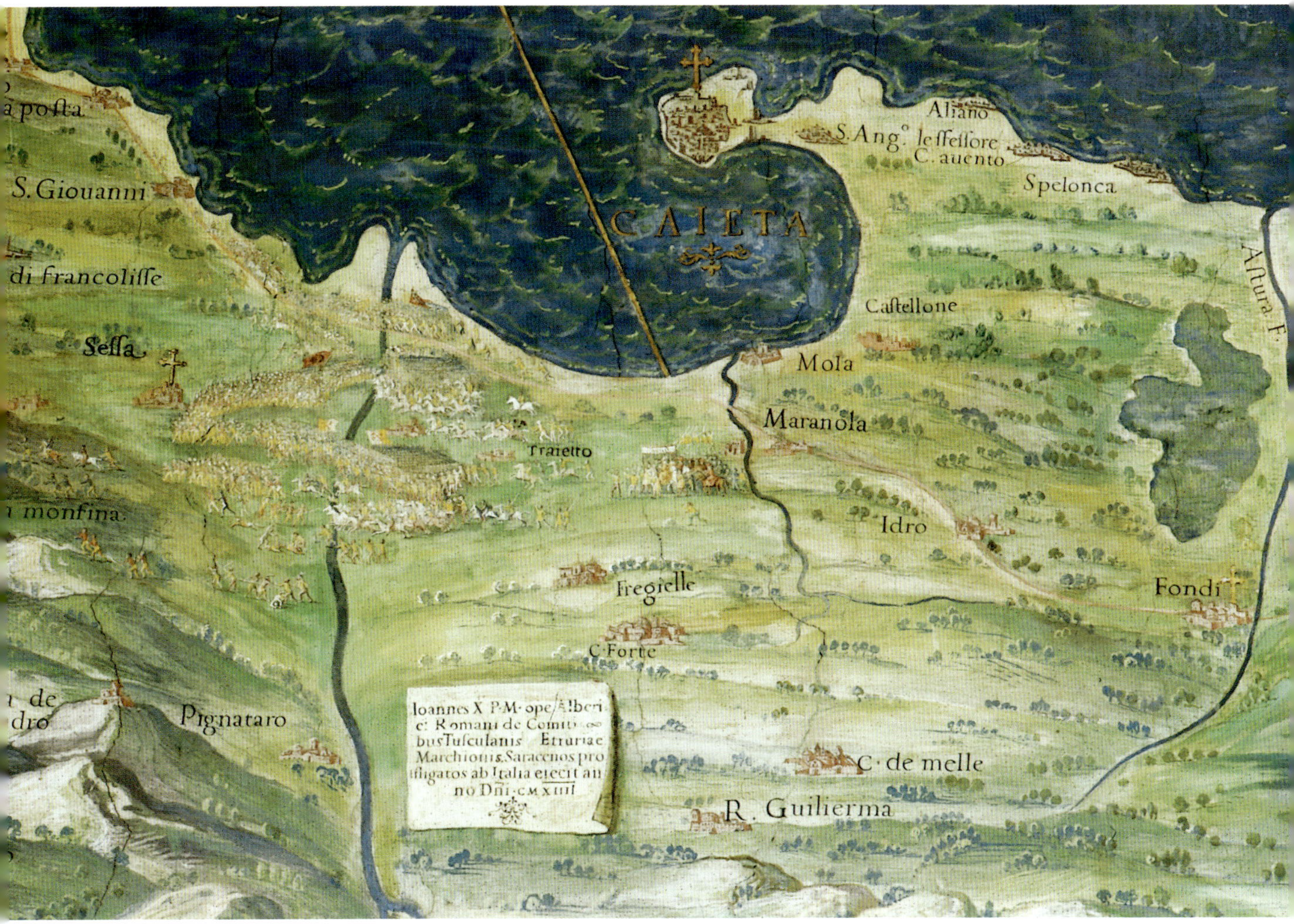

7.9 (*above*) Detail of fig. 7.10: *Pope John x Defeating the Saracens near the River Garigliano*

7.8 (*facing page*) Detail of fig. 7.10: *Landscape*

Gregory XIII's artistic patronage, listed.[12] Documentary and circumstantial evidence, however, demonstrates that the artistic team mentioned by Baglione was a smaller group that included the expert painter of grotesques Marco da Faenza,[13] the landscape specialist Matteo da Siena and his assistant Giovanni Battista Lombardelli,[14] the talented landscape painter Mathijs Bril,[15] the versatile Antonio Tempesta,[16] the experienced Donato da Formello and Paris Nogari,[17] and a group of younger and lesser known artists, who started their successful career in the Gallery of Maps, such as Baldassarre Croce, Giacomo Stella, and Jacopo Semenza.[18] Many efforts have been made to distinguish the contributions of individual artists who worked in this immense corridor, but questions of attribution should not overshadow the importance of the Gallery of Maps in terms of style, iconography, and organization of artistic labor. Nebbia evidently concentrated his efforts on achieving a stylistic unity from the work of many different artists, each with his own specialty. This conspicuous team of artists translated the cartoons of the maps into fresco, painted boats, grotesques, historical scenes, sacrifices, tablets, personifications, and numerous winged dragons, the heraldic symbol of the Boncompagni pope. Considering the speed of execution, the artists must have worked simultaneously in different sections of the corridor, probably starting at the south entrance. For obvious practical reasons, painting began in the vault and proceeded to the walls. Such an organization of artistic labor based on a large team of artists, each specializ-

7.10 Egnazio Danti, *Campania* (Map of Campania), ca. 1580, fresco. Vatican, Gallery of Maps

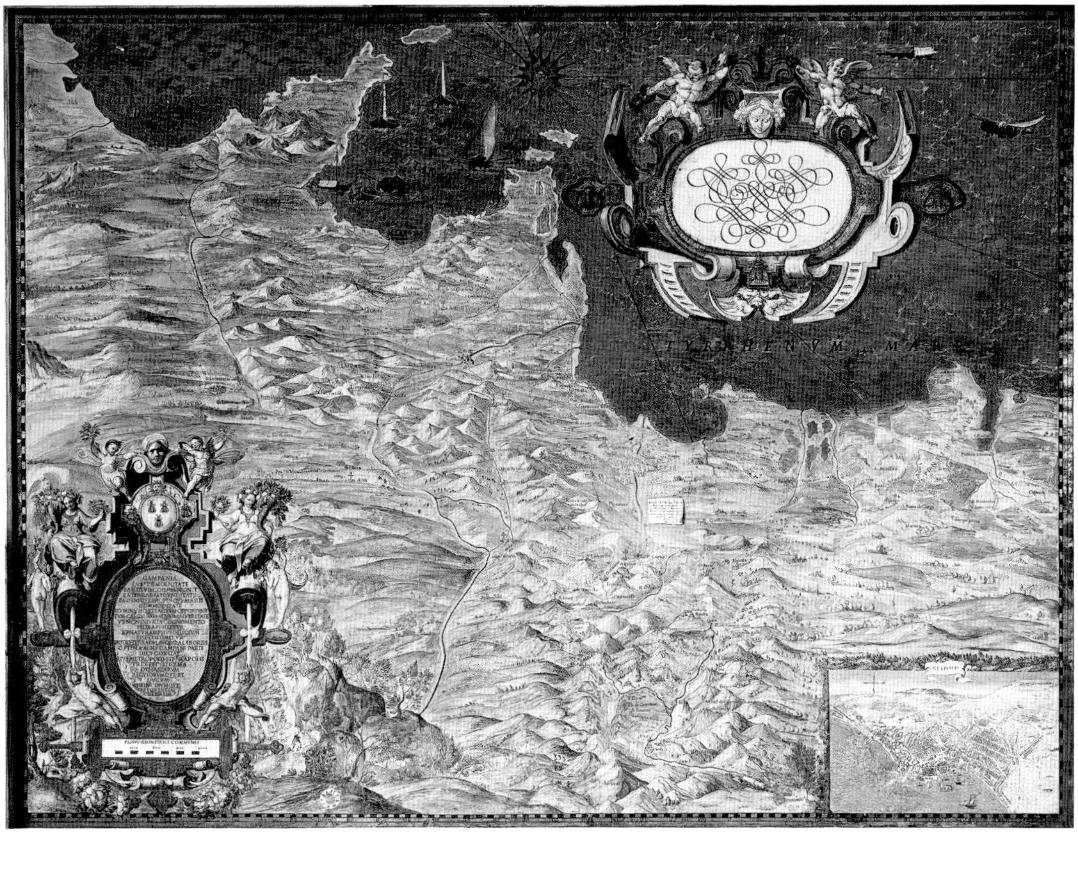

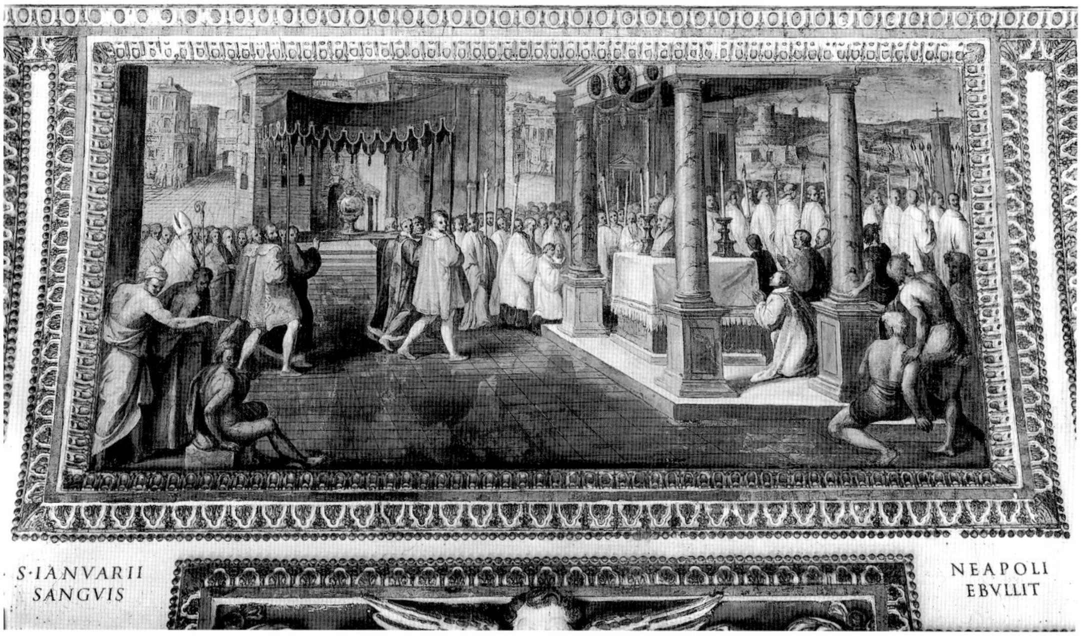

7.11 *The Miracle of the Blood of St. Januarius in Naples*, ca. 1580, ceiling fresco. Vatican, Gallery of Maps

ing in a genre, and dependent on the extensive use and reuse of the same workshop's cartoons, made the completion of these huge frescos possible in about two years.

But among the many scenes and maps that make up the Gallery of Maps, some general characteristics can be observed that suggest an overall plan executed by different hands. First, there is the predominance given to landscape both in the maps and in the vault scenes, which must be attributed to Muziano. Coming from Brescia and trained in Venetian painting, Muziano had acquired fame in Rome for his devotional paintings that represented saints in gorgeous landscapes. Equally important are the many scenes illustrating processions of small figures unfolding in accurate urban views and landscapes, a theme favored by the Boncompagni pope. The high quality of some of these processions suggests that the painters Matthijs Bril and Antonio Tempesta surely had a leading role within the artistic team and indeed may have started their successful collaboration *a quattro mani* in the Gregorian corridor.[19] Bril may have been responsible for the best landscapes, such as the complex and minute views with saints, hermits, hunters, and shepherds, at the bottom of the maps or in the vault scenes. Tempesta, who had been in Rome since 1576 and had built a reputation, "for his paintings of cavalcades, hunts, and battles,"[20] was perhaps responsible for the numerous narratives illustrating battles, processions, parades in scenographic landscapes, and for the miniature battles on the maps.

This large team of artists was called to paint an original iconography, a complex, exceptional system of interrelated images. In a radical invention, Italy and its regions were represented not only through modern maps rather than traditional allegorical personifications, but they were also connected with episodes of church history and scenes of biblical sacrifice.

The Invention of the Iconography

Individual elements of the iconography of the Gallery of Maps – maps, episodes of church history, and biblical sacrifices – had an honorable history in other parts of the Vatican Palace. In the Sala Regia the representation of church history focused on the subordination of the temporal rule to the papal spiritual rule through scenes of territorial donations from sovereigns to past popes; many of these scenes were repeated in the Gallery of Maps. In the Sala di Constantino, the large room that Gregory XIII used for mass, sermons, and other large meetings of his court, including the wedding of his son, celebrated the importance of Constantine for the triumphant church of the Renaissance, a theme further emphasized in the Gallery of Maps. In Pius IV's Terza Loggia, the maps of Europe were connected with religious scenes on the vault, a connection between wall and ceiling repeated also in the Gregorian corridor. Biblical sacrifices were represented in many places in the Vatican Palace, most notably in Michelangelo's Sistine Ceiling and in Raphael's Loggias. But the synthesis of these different iconographic traditions in one single, all-encompassing program was a unique accomplishment of the Gregorian corridor.

The representation of Italy in forty colossal maps had never been attempted before and unequivocally bespeaks the papal intention of producing an unprecedented cartographic enterprise. But the Italy depicted in the Vatican Gallery corresponded neither to the ancient geographical area described by Strabo and Ptolemy (both portrayed in the Gallery of Maps), nor to any past or future political partition. It represented, instead, a utopian construction of a place under the ecclesiastical authority of the papacy. This ideal, imaginary Italy included the regions of the peninsula and the main islands (Sicily, Sardinia, and Corsica). But it also included the military bulwarks against the infidels north and south of the peninsula: the island of Corfu, the gateway to the east, was depicted together with the Battle of Lepanto (1571), the major victory of the Holy League against the Turks (Fig. 7.3); the island of Malta, defended by the Order of the Knights of St. John, was represented together with the Siege of Valletta by the Turks (1565), one of the greatest events in the history of the Mediterranean. The county of Avignon, which had never been included in any chorography of Italy, was added because, as the inscription related, "although the ancient city of Avignon and the Comtat Venaissin, its capital Carpentras, and the other towns and cities are not strictly speaking part of Italy, they still belong to the church of Rome, and therefore are shown here."[21]

The regional divisions diverged from past partitions but served the interests of its papal patron. The papal states were represented in ten maps, while the much more extensive Duchy of Milan, then under Spanish domination, occupied only one map. Similarly, the territory of the Republic of Venice was represented in a single map, although the city of Venice was the subject of a spectacular city view at the very entrance to the corridor, a fact that the Venetian ambassador in Rome proudly reported to the Republic's senate. In a perfect convergence of papal iconography and celebration of Gregory XIII, the map illustrating the native land of the Boncompagni pope appeared at the very center of the corridor, in exact correspondence with the Serlian window; its related fresco on the vault, also at the very center of the corridor, apparently celebrated Gregory XIII so conspicuously that a later pope felt compelled to repaint it.

The maps of this notional, imaginary Italy were conceptually and spatially linked with episodes of church history, a fact recorded by the corridor's chief inscription: "The vault shows the pious deeds of holy men in the places where they happened." Consequently, church history was not recounted according to chronology, a choice that would have fit perfectly along the length of the corridor, but rather according to geography. In a radical revision of conventional iconography, the episodes of church history jump freely from ancient times to the late sixteenth century, from St. Paul's shipwreck, which illustrates the island of Malta, to *The Exposition of the Holy Shroud in Turin*, an event that took place in Turin in 1578 and is related to the map of Piedmont. Like the maps of Italy, the episodes of church history were represented according to a north–south sequence that violates chronology but respects geography.

The intended association of geographical site with historical event is explicit. At times, however, a region is illustrated by an event taken from the life of a famous local saint, even though the event itself took place elsewhere. For example, the episode depicted in *St. Francis Appearing to St. Anthony of Padua* took place at Arles in southern France but it refers to the map of Veneto, where the city of Padua, a major center for the saint's cult, was represented. Other regions are illustrated not by saints' lives but rather by the depiction of their relics, the devotion to which was one of Gregory's dearest concerns. For instance, *The Translation of St. John the Baptist's Ashes to Genoa* is related to the map of Liguria, since the Genoese brought the saint's ashes to their city after the First Crusade; *The Manna Coming From St. Andrew's Body* refers to the map of the region of Salerno, since it also mapped the city of Amalfi, where a famous relic of St. Andrew was kept. Occasionally, the connection between maps and church history is less evident, as in the case of the map of Sardinia, for which nothing better was found than the illustration of the story *Pope Symmachus Sending Aid to the Exiled Bishops.* This quite obscure episode of the life of Pope Symmachus, a native of Sardinia, however, resonated with Gregory XIII's claim to the nomination of bishops in the island, a claim the Spanish king contested.

Within this geographical framework, the scenes of church history had been carefully selected to respond to the most controversial issues raised by Protestants, notably the political role of the Roman church, the foundation of holy buildings in early Christianity, the historical truth of saints' lives and miracles, and the veneration of relics.[22] That the iconography of the Gallery of Maps is imbued with post-Tridentine meaning should come as no surprise given its immediate context. The rationale for selecting the episodes of church history was, first, the geographical site of the event and, second, its relevance to current religious debates.

On the gallery ceiling, the church narratives alternate with sacrificial scenes taken from the books of Genesis, Exodus, and Leviticus (Fig. 7.14).[23] Thematically and chromatically (for they are ochre monochromes) isolated from the main iconography, these biblical sacrifices pass unmentioned in the corridor's inscriptions and in later descriptions of the Gallery of Maps, as if they were marginal or self-explanatory. These scenes of biblical sacrifice referred unmistakably to the Eucharist, specifically to its sacrificial nature, which formed a major issue of contention with the Protestants. However, the reason why these biblical sacrifices were connected with the maps of Italy has never been explained.

In the Gallery of Maps, the ordering principle of the entire iconography is the geography of Italy, an undisputed fact acknowledged in the corridor's inscription but persistently ignored in modern interpretations of the Gregorian corridor. Exemplary of the little consideration that the geographical order has received is the discussion pertaining to Giovan

Battista Romanelli's *Feed My Sheep*, the only vault scene that does not conform with the geographical order of the Gallery of Maps and which indeed had been repainted in the 1640s. According to the Gospels (John 21:1–17), Jesus imparted to Peter the care of his flock ("Feed my sheep" were Jesus' words) on the shore of the sea of Tiberias, a place that is not represented in the Gallery of Maps. An iconography traditionally associated with the unity of the flock and the pastoral mission of Christ's representative on earth, the *Feed My Sheep* has been regarded as appropriate to the message of the Gallery of Maps and thus identified as the iconography that Gregory XIII originally chose. It is more likely, however, that the original scene would have conformed with the geographical order of the Gallery of Maps and thus illustrated an episode of church history that took place in (or near) Bologna, since Gregory's native land is in close proximity. The location of this scene near the map of Bologna and at the very center of the vault strongly suggests that it must have included an explicit celebration of the Boncompagni pope, perhaps such a conspicuous one that Pope Urban VIII decided to remove it altogether. Although it is impossible to ascertain the subject Gregory selected for the central fresco of the Gallery of Maps, in the Sala Bologna he chose the deeds of an earlier pope named Gregory to conflate the celebration of his own papacy and of his homeland.[24]

The proper consideration of the geographical order adopted in the Gallery of Maps can also help to clarify the identity of the inventor of this unique iconography made up of maps, scenes of church history, and biblical sacrifices. Numerous names have been proposed as the iconographic adviser of the Gallery of Maps, such as Cesare Baronio, Guglielmo Sirleto, Fulvio Orsini, and Egnazio Danti himself. But little consideration has been given to the fact that, apart from Danti, the others, while being prominent scholars, were not experts in cartography. Baronio was the leading expert in church history at the papal court. In the years that the Gallery of Maps was being decorated, he was lecturing on church history in the

7.12 Cesare Nebbia, *Stopping the Lava on Aetna By Holding St. Agata's Veil*, ca. 1580, pen on paper. Location Unknown

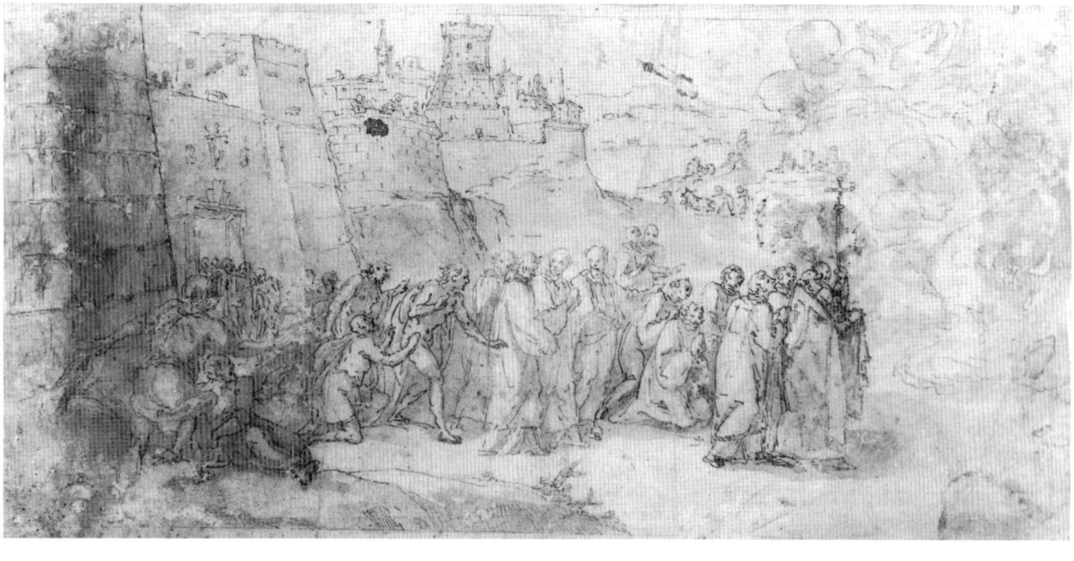

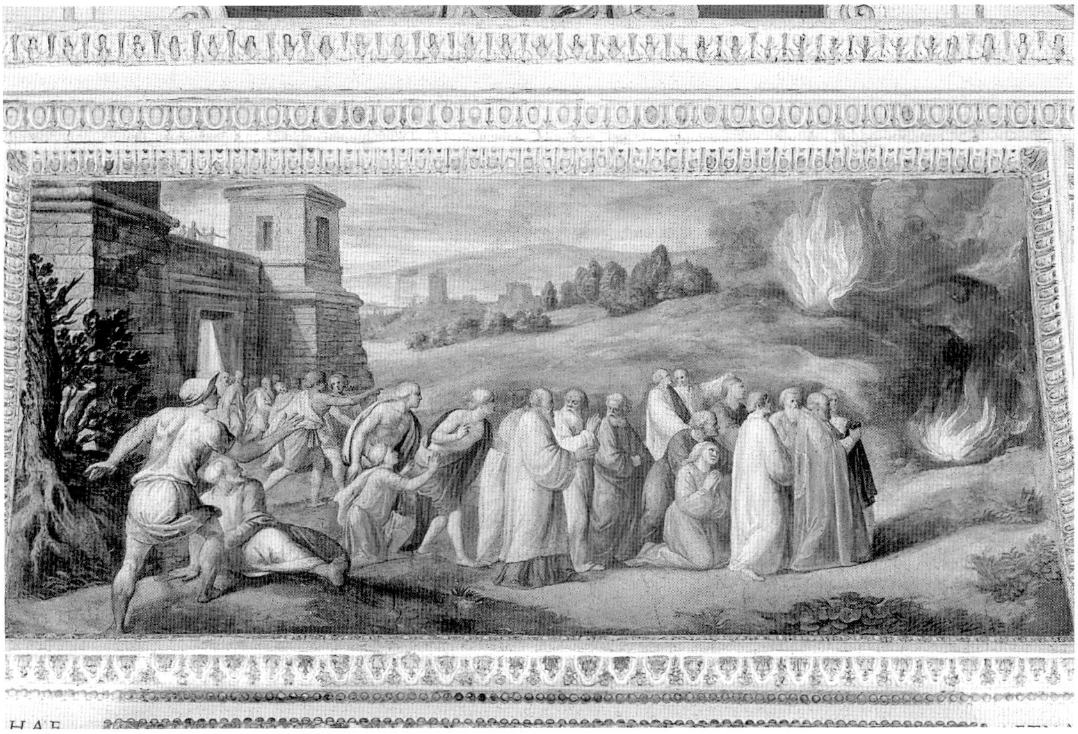

7.13 *Stopping the Lava on Aetna By Holding St. Agata's Veil*, ca. 1580, ceiling fresco. Vatican, Gallery of Maps

Oratory of San Filippo Neri and started to write his *Annales ecclesiastici*, the monumental official history of the church that would be completed many decades later. Later he devised religious iconographies but, as far as is known, he had no interest or expertise in cartography. His views on church history, however, were very present in the Gallery of Maps. Sirleto, the cardinal librarian of the Vatican library and one of the most erudite and influential figures at the Gregorian court, could have had a role in the Gallery of Maps, as he had in most papal affairs. Certainly he knew Danti well. But he had no knowledge of the cartography that is seen in the Gallery of Maps. It has been suggested that Fulvio Orsini, librarian of Cardinal Alessandro Farnese and one of the most erudite people in Rome, was involved in the program, but solely on the basis of a Greek inscription in the Gallery of Maps, a language that many mastered at the papal court, even Danti. As for Danti, his authorship for the map murals has never been questioned, but his role in the invention of the iconography is disputed, since it is not supported by definitive documentary evidence. However, the very invention of the Gregorian corridor, based on the geographical order of the maps to which everything else is subordinated, bespeaks Danti's pervasive role, compelling a reevaluation of the surviving evidence.

The note from the Boncompagni archive quoted above cited Nebbia as the supervisor of the artistic workshop, but the well-informed Baglione unequivocally wrote that Danti "apportioned" (*divisava*) the tasks among the artists, suggesting that the polymath rather than Nebbia was in charge of the supervision of the artistic team.[25] While it is documented that Nebbia took such a supervising role in later decorative campaigns in the Vatican Palace and elsewhere, there is no record of his expertise on cartography. Danti's cartographic expertise, on the other hand, is well known, as well as his involvement in later decorative campaigns at the papal court. In a note Danti himself wrote listing his activities as papal cosmographer, a title he obtained in 1580, he did not mention his scientific activities, such as the treatises he wrote, the meridian line he built, or the reform of the calendar he undersigned, but rather wrote about the rooms he supervised for the Boncompagni pope in the Vatican Palace: "for Pope Gregory of blessed memory I did the Gallery, the Loggia de' Venti, the rooms near the Sala degli Svizzeri, and the Sala dei Palafrenieri together with the other large loggia above the office of the Camera, in front of the Sala Bologna."[26] For these rooms Danti invented allegorical personifications, designed the architectural backgrounds, selected scenes and themes, chose the artists who were to paint them, usually from the artistic team of the Gallery of Maps, and distributed and supervised the artistic labor; in short, he was responsible for these decorative cycles from beginning to end.[27] It is also documented that for the Sala della Meridiana in the Tower of the Winds he not only selected the themes referring to the effects of the winds, providing pertinent literary references and suggesting their representation in the room according to the direction in which they blow, but he also offered the interpretation of those themes in relation to the concerns of the post-Tridentine papacy.[28] For instance, he explained that the winds menacing the ships of Saints Peter and Paul represented respectively the eastern and northern heresy menacing the Catholic church. Indeed, the supervision of Gregory XIII's artistic commissions was a defining component of Danti's position as papal cosmographer.

Moreover, the famous letter that Danti wrote to Abraham Ortelius in 1580, while the Gregorian corridor was under decoration, is such a detailed description of its iconography that it is worth quoting in full:

> Having divided Italy along the Apennines, I placed on one side of the Gallery that part which is washed by the Ligurian and Tuscan Seas, and on the other side that which is girt by the Adriatic Sea and the Alps, further dividing it according to the states and the prefectures of the governments into forty parts, the Gallery being divided into forty panels of such size that it required sixty-four sheets of royal paper for each panel to make the cartoons. I am now reducing the whole thing into a book, where the parts of Italy will be as many as forty-eight, and after [these] there will be eighty stories with figures, which are painted in the vault of the Gallery above each panel, representing some particular miracle which had happened in that province. I wanted to inform you about it so that, if you wish to engrave this book for me, you will let me know, for I would try to convince the pope to give us some assistance. After each table [map] there will be some particular and new annotation with the inscriptions which I made in that Gallery, and they have been all revised by Signor Moreto, Signor Bargeo and by Frizzolio . . . Additionally, I want to mention to you that I collected as many as a hundred pieces of cities and other great districts of Italy portrayed in plan after nature, which will be with the same book.[29]

Usually considered only in relation to Danti's authorship of the Vatican maps, this letter yields a good deal of information. First, it establishes that it was Danti's idea to divide the representation of the peninsula along the spine of the Apennines and to place the maps of Italy on the side walls of the long corridor. This seems further confirmed by the fact that the first sentence of Danti's letter follows the inscription over the northern portal almost word for word. Second, the letter identifies Danti as the author of the cartouches of the maps. Although the cartouches were strongly influenced by contemporary geography books, they were undoubtedly Danti's production, so much so that he felt compelled to obtain the editorial advice of three eminent scholars: Antoine Muret, professor of Latin at the University of Rome and well known for his polemics in favor of Latin against the vernacular; Pietro degli Angeli Barga, the private secretary of Cardinal Ferdinando de' Medici, Danti's patron since the Florentine period; and Lorenzo Frizzolio, a Roman erudite. Finally, and most important, the letter gives a detailed account of a map book inspired by the Gallery of Maps that Danti was planning to have printed.

This volume was evidently a collection of 48 maps of Italy, an augmented version of the Gallery of Maps, which contained only 40. From Danti's description, the book would have followed the model established by Abraham Ortelius's *Theatrum orbis terrarum* which had been adopted in many national atlases: the maps would have been accompanied by "some particular and new annotation" and by the gallery's inscriptions, both printed after the corresponding map. The book would have also included around a hundred city plans and/or views, further augmenting those in the Gallery of Maps.[30] Danti's book would have made available to the public at large the cartography of Italy, the same kind of national cartography that the monarchs of Europe were ordering for their own states and that were published in beautiful volumes or large broad-sheet maps.

But, by closely following the iconography of the Gallery of Maps, the proposed book diverged considerably from standard geography books. Danti's book was to have included also, "eighty stories with figures, which are painted in the vault of the Gallery above each panel, representing some particular miracle which had happened in that province." The placement of the religious narratives in the printed book remains unclear. It is unclear whether the word "after" (*apresso*) in the letter means "after the series of maps" or "after each pertinent map." Either way, for the Dominican, the maps were to have been as important as the episodes of church history. It was precisely this unexplored connection between maps and church history that Danti wished to be brought to public attention.

Danti's project for a book was never realized, possibly because of the high cost, or even because of the absence of a publishing house in Rome sufficiently equipped to print such an ambitious project.[31] Indeed, Danti asked Ortelius for technical help with the engraving (*intagliare questo libro*), but there is no evidence that the contact between Danti and Ortelius went any further, although the Dutch cartographer was still showing interest in Danti's Gallery of Maps a few years later.

As Danti was behind the book project inspired by the Gallery of Maps, so he was behind the invention of the iconography for the Boncompagni corridor. The papal court abounded with experts on the minutiae of church history, with exegetes of biblical sacrifices, and theologians of the Eucharist, scholars who could, and indeed did, help Danti in the realization of the Gallery of Maps. But the very idea of having a long gallery mimicking the Italian peninsula, this imaginative adaptation of the geography of Italy to the three-dimensional space of the corridor, could come only from an imaginative map-maker who had been trained in the visual arts. As an expert map-maker and a talented draftsman, Danti could best interpret Gregory XIII's wishes: to fill a cartographic void by mapping the regions of Italy, and to assert the role of Italy for the post-Tridentine papacy by representing church history unfolding on Italian soil.

The Chorography of Italy

Danti regarded the Gallery of Maps as a chorography of Italy, indeed as an exemplary case of the many images that were called chorographical in the Renaissance. Based on the combination of two Greek words, the noun *chora* (place), and the verb *grapho* (to describe), chorography indicates the description of a place. In ancient philosophy, particularly in Plato, *chora* was a philosophical concept that mediated between the absolute entities of time and space, and had a fundamental cognitive function in experiencing the world.[32] Indeed ancient philosophers discussed chorography in conjunction with chronology, the description of time. Ptolemy was the first author to define chorography in relation to geography, the description of the world: geography represented large portions of the terrestrial globe while chorography represented smaller areas. Ptolemy also distinguished the activities of the geographer from that of the chorographer and gave instructions on how to make geographical maps but not chorographical maps, which he regarded as pertaining to painting.

As previously discussed in relation to Cosimo I's maps of the world, Renaissance scholars and map-makers commented on Ptolemy's definitions of geography and chorography, acknowledging that chorography, as a mode of representation, could not be defined in absolute terms but only in relation to other representational modes. Apian visualized the relations between cosmography, geography, and chorography in a famous woodcut (Fig. 5.3). Gastaldi listed the subjects of geographical and chorographical maps in the popular booklet accompanying his world map of 1561. Ruscelli explained that the synoptic view of the world represented in geographical maps had to be complemented by the separate and enlarged representation of its individual regions as provided in chorographical maps. Similarly, the Renaissance editions of Ptolemy's *Geography* titled "chorography" the maps that represented on an enlarged scale an area that had been represented on a smaller scale elsewhere.[33]

Because of its intrinsic relative nature, Renaissance chorographical maps could represent a region, a county, a city, a castle, or an individual building. Typical Renaissance chorographies were maps of regions, such as Girolamo Bellarmato's *Chorographia Tusciae* (Rome, 1536), or views and plans of cities such as those assembled in Guillaume Gueroult's *Epitome de la corographie d'Europe* (Lyons, 1553), and Georg Braun and Franz Hogenberg's *Civitates orbis terrarum*, or local maps such as Cristoforo Sorte's maps of the Veneto, and Marc'Antonio Pasi's maps of the Duchy of Ferrara.

The Gallery of Maps best illustrates the different forms of Renaissance chorography as well as their relationship to each other.[34] The corridor includes: thirty-six regional maps of Italy, which are chorographical maps in relation to the general maps of Italy represented at the beginning of the corridor (Figs. 7.6 and 7.7); four large perspective views of the maritime cities of Venice, Genoa, Ancona, and Civitavecchia, which are chorographical views in relation to the regional maps of Venetia, Liguria, Latium, and the Marches (Figs. 7.4 and 7.5); four monumental views of the islands of Elba, Malta, Corfu, and the Tremiti, which are chorographical images in relation to the entire peninsula; and innumerable views and plans of cities, villages, fortresses, buildings, and monasteries that are chorographical images in relation to the individual map on which they appear.

Danti, who invented this multi-faced chorography of Italy, further distinguished chorography from what he called topography, or verbal chorography. Elaborating on Ptolemy's definitions, he regarded chorography as the visual description of a place through geometry and painting, and topography as the verbal description of the same place. For Danti, a model of chorographical description was the map of Perugia made by his father Giulio, while exemplary topographical descriptions were Polybius's description of Lake Trasimeno and Livy's description of Cannae in Apulia, descriptions that the ancient historians had added to their narration of the Punic Wars and that Danti had represented as historical vignettes in the Vatican maps (Figs. 7.16 and 7.17).[35] Although Danti's definitions were not accepted unanimously, they are instrumental for understanding that the Gallery of Maps was conceived as the visual counterpart of the verbal chorographical descriptions of the peninsula written by ancient and modern authors.

Italy had been described extensively in words in the ancient geographical texts by Strabo and Pomponius Mela, texts that inspired the erudite books written by humanists such as Flavio Biondo, Raffaello Maffei, and Leandro Alberti.[36] These books resulted from the antiquarian tradition of the fifteenth and sixteenth century and did not include a single map. Their verbal descriptions of the peninsula had no equivalence in maps. In Flavio Biondo's *Roma restaurata et Italia illustrata* (Rome, 1474), the historical identification of Rome as the *caput mundi* extended from the ancient city to the entire Italian peninsula. Biondo's opening chapters described the continuity of classical Rome into Christian Rome, and the merging of the ancient ruins into the Christian churches. But, in his final chapters, it is the entire peninsula that becomes the "most important part of the entire world" (*parte principale di tutto il mondo*), a definition of Italy inspired by Pliny and indeed repeated in the Gallery of Maps.[37] Literally, for Biondo the restoration of Rome also meant bringing fame to Italy, *Roma restaurata et Italia illustrata.* Raffaele Maffei (better known as Raffaele da Volterra) followed a similar strategy in his *Commentariorum urbanorum* (Rome, 1506) by describing Rome and Italy as the center of the world. Flavio Biondo and Raffaele Maffei were depicted in the Gallery of Maps, in the cartouche of the map of modern Italy (Fig. 7.7); as modern chorographers, they proudly face their ancient predecessors, Ptolemy and Strabo, who were represented in the cartouche of the map of ancient Italy (Fig. 7.6).

The chorographic description of Italy achieved an unprecedented richness of detail in Leandro Alberti's *Descrittione di*

7.14 *Adam's Sacrifice of Praise*, ca. 1580, ceiling fresco. Vatican, Gallery of Maps

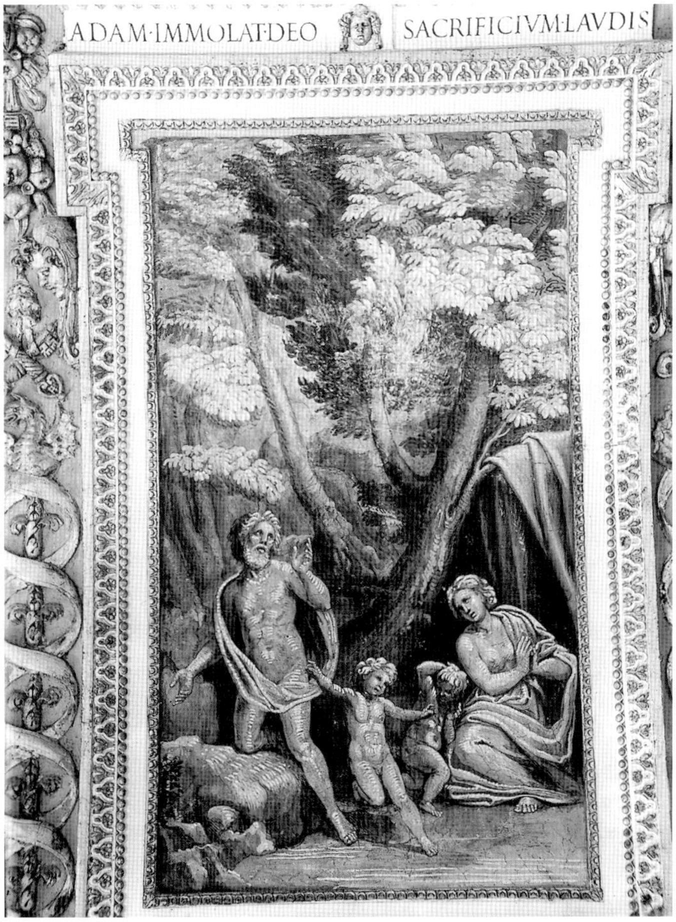

tutta l'Italia & isole pertinenti ad essa (Bologna, 1550), a popular text published in twelve augmented versions in less than a century, including twice in Latin. The 1577 Venice edition, which is the closest chronologically to the Gallery of Maps, consists of over five hundred pages and includes an appendix on islands. Although roughly modeled on its predecessors, Alberti's book is distinguished by the different order he adopted in the description of the regions of Italy. Rather than moving from north to south, Alberti pretended to circumnavigate the peninsula. Thus his journey starts in the northern regions of the Tyrrhenian Sea, in Liguria, and proceeds south along the coast of Tuscany; it then moves inland, to the Duchy of Spoleto and Basilicata, and returns later to the coast of Calabria. Alberti's circumnavigation continues south and, as it reaches the bottom of the peninsula, it turns around the boot of Italy, and proceeds from south to north, passing through Apulia, Abruzzi, the Marches, Romagna, Lombardy, Friuli, and Istria. The journey around Italy concludes at the top of the Adriatic Sea with a lively description of Venice. Following the tradition of the books of islands, the *isolari*, Alberti treated the islands separately.[38]

As Alberti moved around Italy, he reported about the etymologies of peoples, rivers, mountains, and towns documented by classical authors. For each region, he listed the most eminent peoples and their accomplishments, instructed by what ancient geographers reported about a place and what ancient historians wrote on the events that took place there. He reported extensively on the barbarian invasions, on the Punic Wars and Hannibal's deeds, on Attila, Alberic, Brennus, and Totila, on the occupation by the Lombards and their defeat by Pippin, and on Charlemagne's descent into Italy and his coronation by the pope. He described famous battles in detail, showing a definite preference for the most gruesome and devastating battles. In addition, as a Dominican, Alberti was very sensitive to the deeds of clergymen and of religious orders on Italian soil, particularly the deeds of members of the Order of Preachers, to which he belonged. He cited the location of important monasteries, sanctuaries, and churches, as well as the cult of local saints, their miracles, and the local celebrations on their feast days. His criteria for differentiating a city from a town or a village bespeak his religious belief. Instead of distinguishing urban areas according to the number of their inhabitants, he adopted the administrative division of the Roman church, which, to use his own words, "define as a city a place that has a bishopric."[39]

Alberti's attention to church history, the lives of saints, their miracles and cults made his chorography of Italy very different from Biondo and Maffei's, who were concerned primarily with the surviving vestiges of the classical world rather than with documenting the expansion of Christianity. Moreover, Alberti's attention to the achievements of the Christian church on Italian soil was meant to correct the description of Italy that Sebastian Münster had provided in his *Cosmographia universalis.* Rich with geographical, historical, and anecdotal details, and profusely illustrated by maps, Münster's text also contained an abundance of devastating comments on the corruption of the Roman Church and its representatives throughout the peninsula, information that the papal court censured in its Italian and Latin editions.

Alberti's *Descrittione di tutta l'Italia* did not include a single map, even though the full understanding of his chorography of Italy could be achieved only through the complementary reading of the text along with a map of Italy or, even better, with a set of regional maps of Italy. Maps of Italy could be easily bought in urban centers in the second half of the sixteenth century, and it is safe to assume that many readers of Alberti's text did use them.

Sets of regional maps of Italy were also readily available in modern editions of Ptolemy's *Geography* and, from 1570, in Abraham Ortelius's *Theatrum orbis terrarum*. These maps could also be assembled by individual readers from the scattered production of broad-sheet maps. There was still, however, the need to integrate Alberti's chorography of Italy with a coherent set of regional maps in a single publication.

The antiquarian Jacopo Strada planned to publish an augmented edition of Alberti's *Descrittione di tutta l'Italia* which, for the first time, would have been amply illustrated by maps and other pertinent images. Owing to his extensive contacts in Florence and Rome, on whom he relied to secure collectibles for the emperor, he was able to gather many maps to illustrate Alberti's text. Strada described his ambitious project in a letter to Guglielmo Gonzaga, Duke of Mantua, and to the editors Doetichum and Plantin, hoping to involve them as patron and publisher respectively:

> The things that I wanted him [Leandro Alberti] to have included are the following: First are the things pertaining to cosmography, namely that for every region the map would have been placed in front of its description, and that there would have been also well-portrayed illustrations after nature of all the pertinent cities in that region; and additionally, that next to each city there would have been illustrations of the most prominent features, such as ancient and modern buildings with accurate plans and elevations with their measurements written beside them. Then [he should have included] an accurate portrayal of the ancient marbles with their inscriptions and decorative elements surrounding them as they actually appear. Then a list of all the learned men with their names and all the works they have written, published and unpublished. Finally, a portrayal of all the coats of arms, that is to say, of each prince of each city, beginning with the first [coat of arms] of that house and proceeding forward to portray that of the current, living prince. After these he would have finished with a portrayal [of the coats of arms] of all the nobility of each city.
>
> Now I am working on the aforementioned order that the friar [Leandro Alberti] left out, and I am already advanced to such a point that the section on Italy has grown by more than two thirds and it will result in three big volumes.[40]

The project came to nothing, although Strada claimed he had collected a considerable number of maps and other images. Strada's purpose in publishing such a book was strictly antiquarian, but it is important to note that he followed the geographical structure of Alberti's book to describe the treasures and rulers of Italy. City views, ancient reliefs, coins, rulers and princes, genealogical trees, coats of arms, famous people, buildings, and works of art: all these things and people had to be organized according to place.

Similarly, the eclectic scholar Girolamo Ruscelli designed four volumes of maps of the world, one volume of which was to be dedicated entirely to Italy. Ruscelli's project was probably inspired by his edition of Ptolemy's *Geography* and by the desire to update the old text with accurate modern maps of the world. He publicly announced his project in two of his most popular books, his edition of Ptolemy's *Geography* and the *Lettere di principi* (Venice, 1562) a book that circulated widely in Renaissance courts:

> And since actually the most proper, fair, correct, and useful way to make such a description [of the world] is by means of tables [maps] or illustrations, the quality and quantity

of which can be filled in with words, I intend each of the four above mentioned volumes to have as many tables [maps] and illustrations as possible, always beginning with the general, and proceeding to the particular and to the even more detailed maps of the provinces, so that each volume will contain at least a hundred tables, while in Ptolemy's mere twenty-six [maps] the whole earth is loosely sketched rather than described in detail.[41]

Although publicly announced, Ruscelli's and Strada's projects were not realized. It is probable that these projects were known at the papal court and that Danti knew of them both. These earlier projects would have provided Italy with the chorographical description that other European nations were rapidly acquiring.

The idea of the Gallery of Maps then emerges as the grandiose fulfillment of earlier, unsuccessful projects, the first and unique attempt to provide Alberti's *Descrittione di tutta l'Italia* with the needed corpus of regional maps that would integrate history, geography, and religion. The influence of Alberti's chorographical text on Danti's Gallery of Maps is pervasive. Danti composed his inscriptions and legends from Alberti's text, selected his historical vignettes from Alberti's narration, and, above all, adopted Alberti's circumnavigation of the peninsula. In fact, although the maps are ordered in the corridor in such a way as to mimic a walk along the peninsula, they can be appreciated individually only according to the sequence of Alberti's text. It is this second sequence that seems to have been ingrained in Renaissance viewers and that I have adopted in Appendix B (in which the historical vignettes of the maps and the historical scenes of the vault are connected with specific passages of Alberti's text).[42]

Danti's tasks in the Gallery of Maps were numerous. He had to gather the best cartographic sources, compare and make them uniform in terms of scale and accuracy, and divide them according to the space available in the corridor. Occasionally Danti used printed collections of maps, such as Braun and Hogenberg's *Civitates orbis terrarum* for city views and plans, or Ortelius's *Theatrum orbis terrarum* for toponyms and geographical features; often, he selected one printed map for each region; generally, however, he followed the cartographic practice that he had developed in Florence, and instead of relying entirely on his cartographic sources, he combined, amended, and integrated them with each other. For instance, he used Bellarmato's *Chorographia Tusciae* for his map of Tuscany, and Gastaldi's Map of Sicily (1545) for his own map of the island. But his map of Venetia is the imaginative combination of ten different chorographical maps, while the map *Forum Iulii* is an original collage of four different printed maps.[43]

Many of Danti's chorographical maps are also based on manuscript sources, some of which are still unidentifiable today. Gregory XIII was personally involved in facilitating the gathering of manuscript chorographical maps and instructed his ambassadors to obtain classified maps from local rulers. A steady supply arrived in Rome in the late 1570s and early 1580s. Stefano Ghebellino had sent a map of Avignon in 1574, which was used for the map *Avenionesis Ditio*. In 1577, Rocco Cappellino, a military engineer working for the Spanish in Sardinia, sent manuscript maps and verbal descriptions of Sardinia to Rome, which were used in the corresponding map in the Gallery of Maps. In 1579, a papal administrator sent to Rome, "reports, models, and drawings" (*relattioni, modelli, et disegni*) for the map *Corsica*. The following year Francesco Bolognetti, the papal ambassador in Venice, sent "a roll with the description of the Tremiti that Father Egnazio Danti wanted for the work in the Gallery."[44] Filippo Pigafetta volunteered a drawing of the city of Vicenza, which was used for the city plan in the map of Venetia. Alfonse II d'Este must have sent a copy of the large chorography of his duchy that Marc'Antonio Pasi had completed in 1580. The duke of Urbino made available the chorography of his territory made by Giovan Battista Clarici. A Spanish viceroy sent maps of southern Italy through the Duke of Sora, Gregory XIII's son, maps that, however, have not yet been identified. Unknown cartographic and verbal sources, which possibly reached Rome through the diplomatic channels of the papal court, were also used for the maps of Liguria, Umbria, the Marches, the Duchy of Parma and Piacenza, and the Duchy of Milan, for the eight maps representing southern Italy, and for the depiction of the Boncompagni Duchy of Sora on the map of Campania.[45]

Additionally, in order to augment the geographical content of his chorography of Italy, particularly the location and names of small urban centers and villages, Danti extensively interviewed natives of various cities living in Rome, checking their oral knowledge against verbal chorographies, especially Alberti's *Descrittione di tutta l'Italia.* Danti himself commented on his cartographic work in the inscription attached to the map *Sallentyna Hydrunti Terra* (Fig. 7.16):

> Since it was decided to compile this chorography of Italy from those authors who had described numerous terrestrial and maritime places of Italy (having observed their differences in latitude and longitude) and from the various and very dubious accounts of those who had traveled in the particular places, it should seem strange to no one if less-known towns do not correspond here exactly to their position. However, we took care that the latitude and longitude of the more famous places correspond exactly (as much as this is pertinent to chorography). Father Egnazio Danti of the Order of Preachers wanted this to be mentioned.[46]

Although this cartouche is usually discussed in relation to Danti's authorship of the Vatican maps, it is important also for other reasons. Not only did it define the Vatican map cycle as a "chorography of Italy," but it also mentioned astronomical accuracy, which had previously been the domain of geography, as a component of chorography. Danti himself admitted that he had to resort to printed and manuscript maps from other authors and to the astronomical measurements of other travelers. It is clear from his inscription that he did his best to position the most important places correctly (*adamussim*), but there is little or no evidence to reveal to what extent he accepted or rejected the cartographic sources he consulted.[47] Danti could rely only on other people's astronomical measurements for the major cities of the peninsula. Even for the Papal States, which he had himself surveyed, there is no evidence that he systematically took astronomical

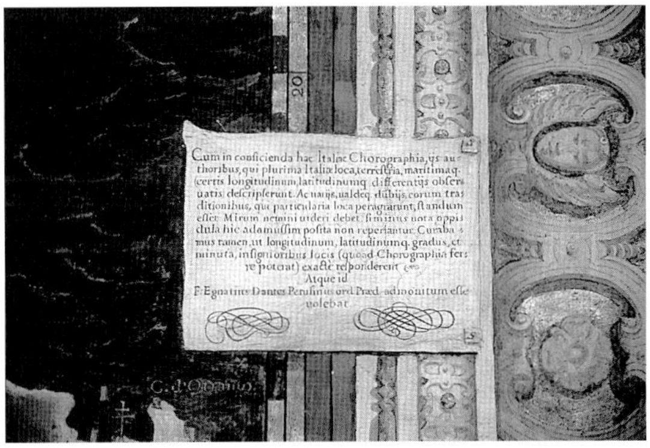

7.15 Cartouche, c.1580, fresco. Vatican, Gallery of Maps

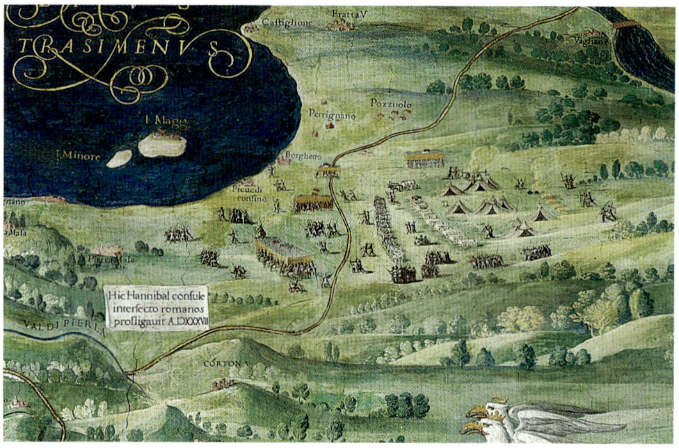

7.16 Detail of fig. 6.13. *Hannibal's Battle at Lake Trasimeno*

measurements for small places.[48] Generally, the Gallery of Maps is full of cartographic incongruities, which Danti, with the typical approach he had already developed in the Guardaroba Nuova, exposed, but did not solve. For instance, in regard to the exact location of the city of Milan he commented in the appropriate cartouche: "I instructed that [the artists] place here the latitude of Milan, which was measured by a very expert man and seems to be confirmed by others, although it differs slightly from that of the universal description [the general map of Italy]."[49] The monumental scale of the Vatican maps dramatically emphasized the lack of accuracy in astronomical measurements, and the speed of execution did not allow time for the consistent and balanced correction of astronomical positions for even major places.

But the Gallery of Maps was a huge cartographic accomplishment that had eluded others, a chorography of Italy that included many more toponyms and geographical features than any previous endeavor. Resulting from the gathering of local maps to Rome and from Danti's systematic work in collating this rich, yet scattered and mostly unpublished, assemblage of local maps, the Gallery of Maps was as much an achievement of chorography as it was a triumph of papal diplomacy.

Chorographical Maps

The Gallery of Maps is notable as the first atlas of Italy and as a visual complement to the verbal description of the peninsula, but it is equally famous for the cartographic language of the individual maps. Monumental in size, stunning in polychromy, and rich in geographical information, these maps enhanced the descriptive mode of Renaissance chorography by including both the qualitative description of a locale and its position within the Ptolemaic grid.

Like other Renaissance maps, the Vatican maps combined different systems of representation. The whole map was represented in orthographic projection, but other elements were painted in linear perspective or in bird's-eye view. At the edges of each map were painted perspective and plan views of cities, buildings or fortresses. Historical vignettes, represented in perspectival views and at a scale larger than the map itself, were placed on the topography of the maps (Figs 7.16 and 7.17). Landscape features, also represented in perspective and at enlarged scale, were painted at the bottom of most maps (Fig. 7.8). The Vatican maps also included other cartographic features that were common in geographical maps but unusual in chorographical maps: a frame registering the graduation of latitude and longitude in degrees and minutes; a cartouche providing the scale of the map in *passi communi*; and an inscription listing the latitude and longitude of the most important places; often also a windrose was included. By including the Ptolemaic grid, the scale in *passi communi*, and the astronomical location of the most important places, these chorographical maps aspired to the mathematical accuracy that was usually regarded as the domain of geographical maps.[50]

In this respect, the Vatican maps differed from common chorographical maps, which usually limited themselves to a more pictorial description of a locale, but conformed to the interpretation of chorography that Danti himself stated in the cartouche attached to the map *Sallentyna Hydrunti Terra*. In that cartouche Danti revised the Ptolemaic definition of chorography that other Renaissance authors had followed and proposed instead the making of chorographical maps that would represent individual locations as both a place and a site, as a landscape filled with nature and history as well as a fixed point in the cartographic grid. Danti accomplished his vision of an all-encompassing representation of a place in the maps he made for Gregory XIII.

Arguably, the most conspicuous elements of Danti's chorographical maps are the landscape features, the historical vignettes, and the scenes of church history connected to them. These elements, often regarded as decorative and external to the chorography of Italy, are instead defining components of Danti's cartographic language.

The landscape features are functional, giving visual access to the mapped territory (Fig. 7.17). Standing in front of one of these large chorographical maps, viewers enter the space of the map from the bottom, that is, from the landscapes represented in perspective. As they inspect the map visually by moving their eyes along the paths unfolding into the mapped territory, they walk mentally through the territory. As viewers look towards the upper part of the map, the center of the projection recedes until it reaches infinity, and the representation in perspective is slowly transformed into a plan view. Traveling over the map, the historical vignettes and the orography bring the viewer back into the depths of a three-dimensional space represented illusionistically (Fig. 7.8). In these maps, the different systems of representation coexist without contradiction; in fact, they complement each other in providing a complex, multifaceted image of a place.

The historical vignettes and the vault's scenes, on the other hand, deserve further

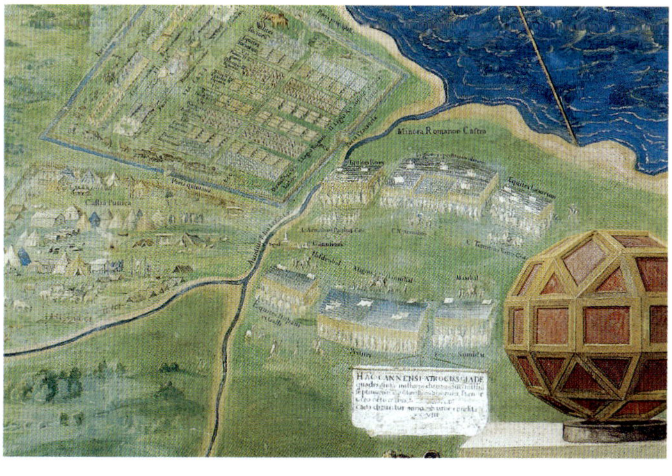

7.17 Detail of fig. 7.24. *Hannibal's Battle at Cannae*

investigation. Generally regarded as different in nature, language, and function, the historical vignettes and the vault's scenes are intrinsically similar in conception and function: they both illustrate the past of the mapped territory through the representation of complementary aspects of its history. But in order to appreciate this, the Vatican maps have to be considered simultaneously from the perspective of Renaissance chorography and of sixteenth-century church history.

The Historical Vignettes

Many maps include historical vignettes. Depicted at a much larger scale than the general map, these vignettes represent historical events in the places where they happened. They range from Roman history to very recent events, from the battles between the Lucanians and the Samnites in southern Italy in 330 B.C. to the battles between the French and the Spanish in Piedmont in 1544. Whoever selected the historical events to be represented on the Vatican maps did not attempt to give a comprehensive history of Italy but rather focused selectively on specific events. Great emphasis was given to Hannibal and the Punic Wars, to the medieval barbarian invasions (the Huns and Attila, the Lombards and Desiderius, the Saracens), to the French and Spanish battles in Renaissance Italy, and to the papal participation in those battles. Moreover, these highly selected events were not arranged chronologically but rather geographically. Therefore, events very distant in time but very near geographically are depicted side by side on the same map (Fig. 7.18). For instance, in the map of the Duchy of Milan, the battle between Hannibal and Cornelius Scipio that took place at Vigevano in 218 B.C. is depicted near the French siege of the city of Pavia, an event that took place in 1528 (Fig. 7.19). In the same map, Charlemagne's defeat of the Lombards is also represented, a battle that happened near Piacenza in 774, and Marcellus's defeat of the Insubrians, a story referring to the Gauls who founded Milan and occupied the region until Marcellus established Roman rule in northern Italy in 222 B.C. In the map of the Duchy of Parma and Piacenza, Hannibal's defeat of Scipio's army, which took place near Trebbia in 218 B.C., is represented close to the battle of Fornovo, in which the French army of Charles VIII confronted the army of the Italian League in 1495. In the map of the Bolognese area, Octavianus, Lepidus, and Marcus Antonius are shown near Bologna, where they gathered in 43 B.C. to form the second triumvirate; nearby are depicted the medieval inhabitants of Bologna, who are about to expel the Visconti rulers from the city in 1361 (Fig. 7.20). The map *Flaminia* is illustrated with Julius Caesar marching along the Rubicon River in 49 B.C., while the French army and the papal and Spanish soldiers face each other near Ravenna in 1512 (Figs. 7.21 and 7.22). In the map *Umbria*, Hannibal fails at Spoleto in 217 B.C., while Pope Leo III is rescued by Duke Winigis in 799.[51]

Typically, local history was depicted at length in chorography books, although in these books, unlike in history books, the chronological sequence of events was abandoned in favor of a geographical sequence. Therefore, the Punic Wars were not reported as they unfolded chronologically from the Battle of Pavia to the battle near Venosa but rather they were reported in fragments, each event linked to the place where it happened. For instance, in Alberti's *Descrittione di tutta l'Italia*, which was based on the imaginary circumnavigation of Italy, Hannibal's battles, fought on the Tyrrhenian side of the peninsula, were reported

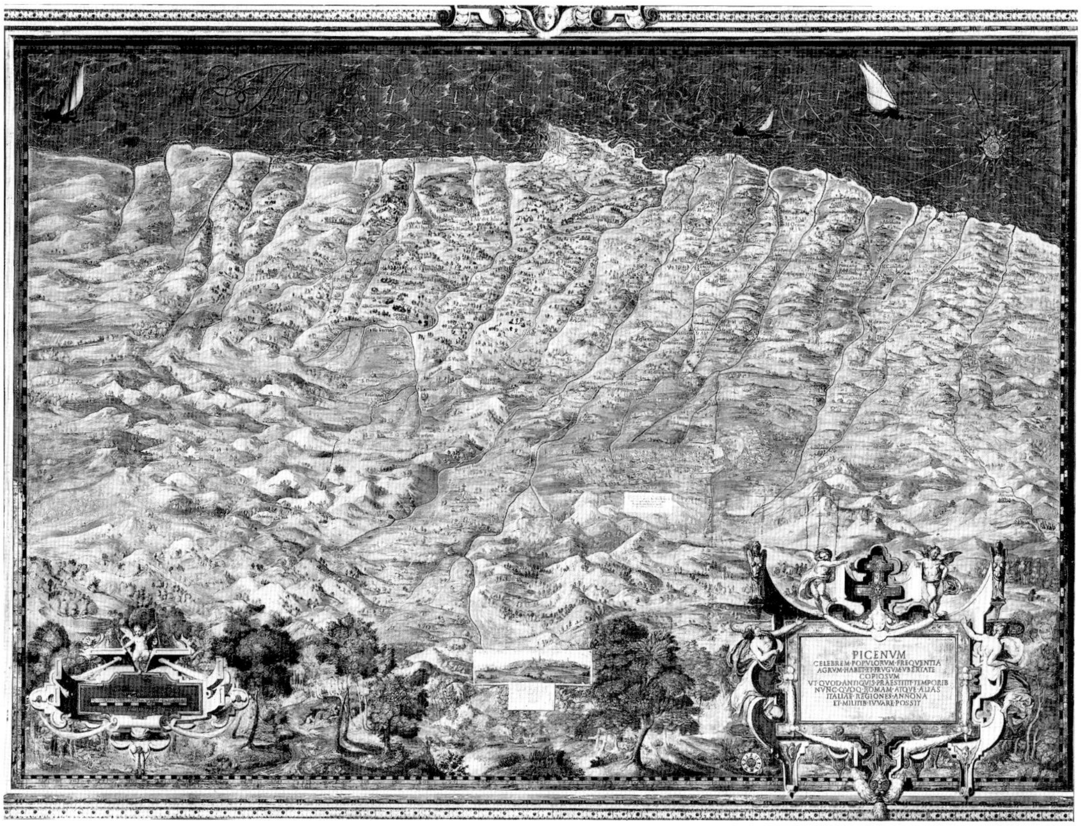

7.18 Egnazio Danti, *Picenum* (Map of Piceno), ca. 1580, fresco. Vatican, Gallery of Maps

before those which took place on the Adriatic side; and the battles fought in northern Italy were recounted before those that took place in southern Italy. Similarly, in the Gallery of Maps, which followed Alberti's imaginary circumnavigation, Hannibal's battle at Lake Trasimeno, represented on the map of the county of Perugia, was the first episode of the Punic Wars to appear in the map cycle, although it was not the first chronologically (Fig. 7.16); the battle at Pavia, the earliest episode of the Punic Wars, was actually the last to appear in the corridor, since it was included in the map of the Duchy of Milan, one of the last maps of the cycle.[52]

The selection of the historical vignettes depicted on the Vatican maps has struck historians as an odd and unrepresentative sequence of events.[53] While sometimes the historical vignettes portrayed important events in Roman history (Hannibal's battle at Cannae or Julius Caesar crossing the Rubicon), in most cases the episodes depicted were quite marginal to the universal history of the church or to the changing of powers in Italy. However, if these historical vignettes are considered not from the modern perspective of Italian history but rather from the parochial perspectives of individual places, the criteria for their selection become apparent. The historical vignettes primarily illustrated the participation of a locale, a region, a city or an area in historical events. For instance, the location of the formation of

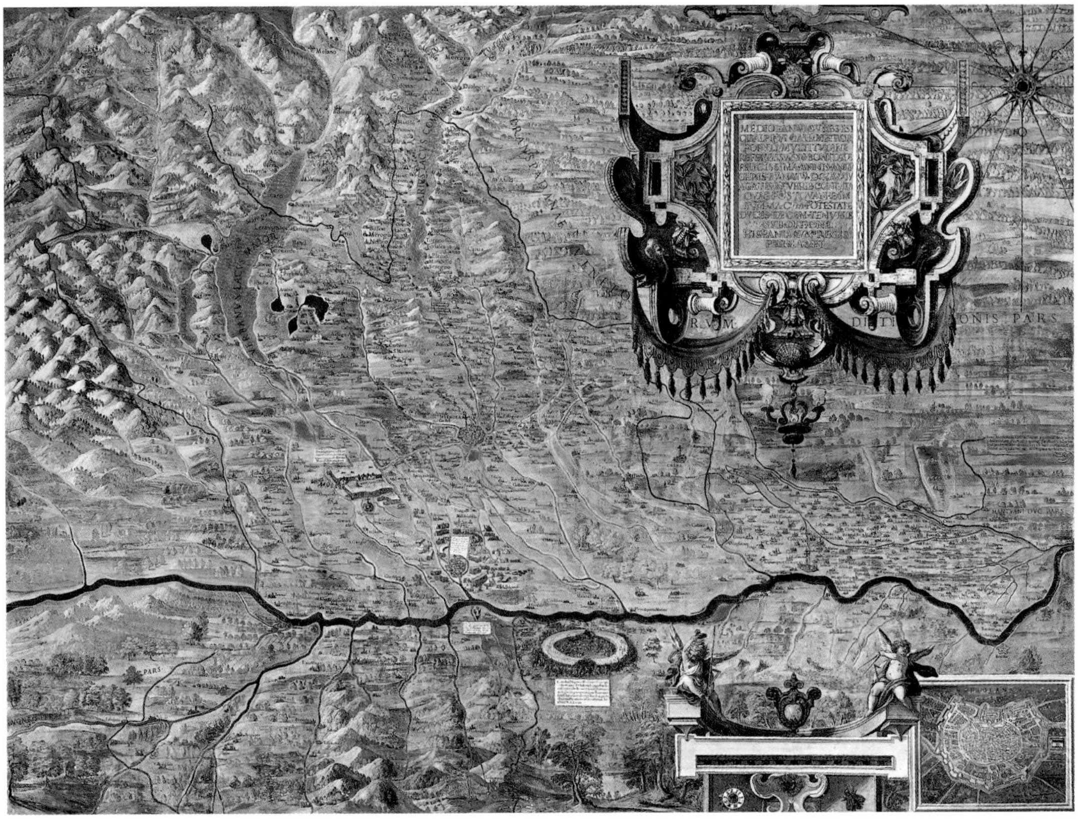

7.19 Egnazio Danti, *Mediolanensis Ducatus* (Map of the Territory of Milan) ca. 1580, fresco. Gallery of Maps, Vatican Palace

the second triumvirate near Bologna is an incidental detail in Roman history, but it becomes a crucial fact if the story to be told is not the history of the Roman Republic but rather the history of the Bolognese countryside (Fig. 7.20).

Alberti's *Descrittione di tutta l'Italia*, which was so important in the planning of the Gallery of Maps, also provided a ready-made selection of events to be represented in historical vignettes, as well as the explanation of their importance. For instance, the location of the formation of the second triumvirate was for Alberti an essential event in establishing the fame of Bologna, his native city, in Roman history. Similarly, the death of Alexander the Molossian, King of Epirus, by the Brutij near Pandosia is marginal to the history of ancient Rome, but in sixteenth-century chorography was very significant. Alberti mentioned the death of this obscure king three times, including in a page-length description, and related it to the etymology of the word Brutij, the original name of the people later called Calabri, and to the location of the ancient city of Pandosia. The battle Pope John x fought against the Saracens near the Garigliano River in 915 is of significance today as one of the many battles between the papal forces and the infidels, but Alberti considered it the defining moment after which Naples recognized the spiritual and temporal authority of the Roman

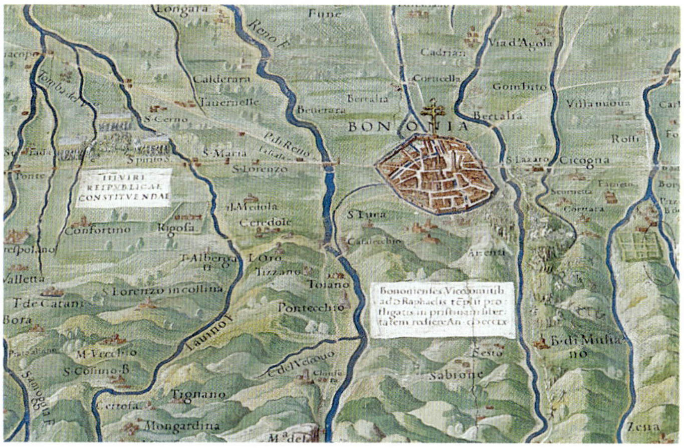

7.20 Detail of fig. 6.15. *The Bolognese Defeating the Visconti at San Ruffillo* and *The Formation of the Second Triumvirate*

pontiff. The Battle of San Ruffillo between the Bolognese and the Visconti demonstrated the historic dependence of the city on the pope, a reading provided by Alberti and reinstated prominently in the legend of the map *Bononinesis Ditio* in the Gallery of Maps (Fig. 7.20).[54]

The conception of history according to geography rather than chronology that informed Alberti's *Descrittione di tutta l'Italia* is crucial to the Gallery of Maps. Indeed, Danti's chorographical maps were the visual rendition of Alberti's text. What Alberti narrated verbally, Danti translated into cartographic images; the lengthy historical descriptions of the former became the vivacious vignettes of the maps. As Alberti's historical descriptions were an integral component of his chorography of Italy, so the historical vignettes of the maps were a constituent part of Danti's chorography of Italy.

The Vault Scenes

The maps of Italy are also connected to the scenes depicted on the vault, which illustrate events that happened in the territory mapped below. Although the connection between the mural maps and the figurative vault scenes was clearly stated in the corridor's main inscription, it is usually overlooked by modern viewers. Overwhelmed by the profusion of bright images, colored maps, and beautiful stucco, they generally ignore the relationship between wall and ceiling.[55] But these connection would have been obvious to the visitors to the papal court, who would have readily understood it as rooted in the Catholic version of church history. Indeed, the writing of the official history of the church was a topic passionately debated at Gregory XIII's court.

First, it is vital to clarify that, from the perspective of post-Tridentine Catholic historians, the events depicted on the ceiling of the Gallery of Maps were all regarded as episodes of church history. Contemporary sources labeled the vault scenes variously. Danti called the vault scenes, "stories with figures representing some notable miracles that happened in that province" (*storie di figure rappresentando qualche segnalato miracolo occorso in quella provincia*). The corridor's inscription, instead, described them as, "pious deeds of saintly men" (*pia sanctorum virorum facta*).[56] Danti was more interested in distinguishing the lan-

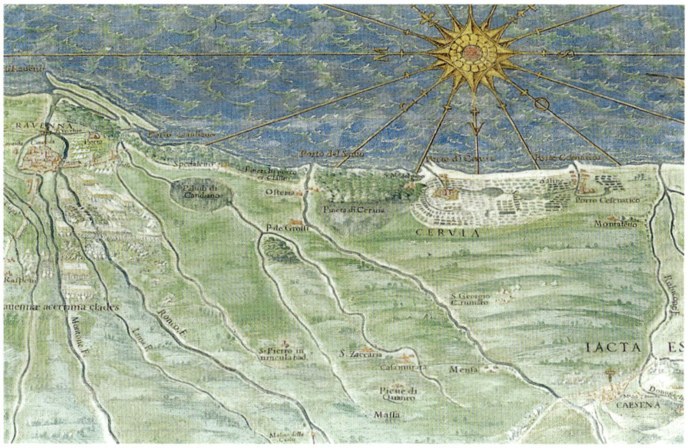

7.21 Detail of fig. 6.16. *The Battle Between the French Army and the Holy League at Ravenna*

guage of the vault scenes, which include figures, from the language of the maps, which instead are nonfigurative, although they include people in the historical vignettes. He also stressed the miraculous aspect of the deeds, that is, the fact that these events, unexplainable by the laws of nature, were held to be supernatural in origin, indeed an act, literally a wonder (*miraculum*) made possible by divine intervention. The main inscription in the Gallery of Maps, on the other hand, stressed the pious scope of these deeds rather than their miraculous nature, although it did mention that the pious deeds were performed by saints. Cesare Baronio included the majority of the vault scenes in his *Annales ecclesiastici*, the official history of the church, about which he lectured in Rome during the planning and execution of the Gallery of Maps.[57] For Baronio and other members of the papal court, the history of the church was made up of political and spiritual history, of human actions and divine intervention, of deeds and miracles. A non-anachronistic interpretation of the Gallery of Maps should be informed with Baronio's historical perspective, a perspective in which the actions of saints and the miracles of God were regarded as integral as wars and treaties to the history of the affirmation, diffusion, and defense of Christianity. Indeed if the vault scenes are regarded as episodes of church history, their relevance to the papal chorography of Italy becomes transparent and their inclusion in Alberti's *Descrittione di tutta l'Italia* fully justified.

In a few instances the connections between map and vault are particularly strong, since they are reinforced by cross-reference between the figurative scenes and the maps. For instance, the relationship between the map of Abruzzi, which included also the hermitage of the future Celestine V on Mount Maiella, and the vault scene *The Hermit St. Peter of Morrone Elected Pope*, is further stressed by mentioning the saint in the map's legend. The scene *Countess Matilda Donating Her Possessions to the Roman Church*, represented on the vault in connection with the map of the Patrimony of St. Peter, is mentioned also in the map's legend. The miracle of *St. Michael the Archangel Appearing on Monte Gargano*, represented on the vault in connection with the map of Apulia, is also explicitly mentioned in the legend of the same map (Figs. 7.23 and 7.24). The historic meeting between Pope Leo I and Attila the Hun near Mantua is represented on the vault in connection with the map of the Duchy of Mantua, but it is also illustrated in one of the historical vignettes on the map. The

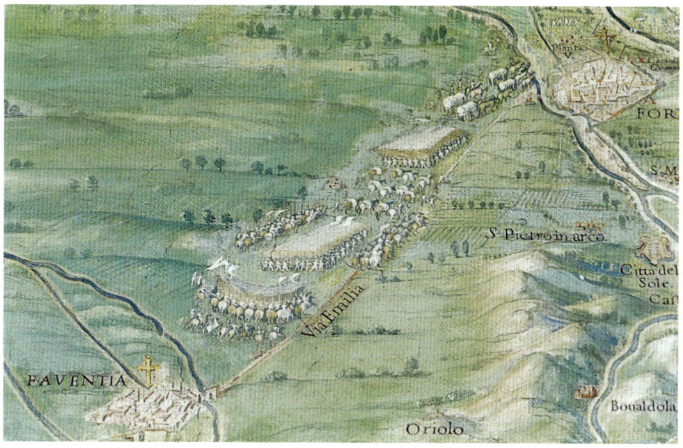

7.22 Detail of fig. 6.16. *Julius Caesar Crossing the River Rubicon*

House of Loreto, one of the most venerated Marian sites, is marked on the map *Picenum*, and illustrated in a separate view on the same map; additionally, its miraculous translation is represented on the vault in correspondence with the map *Picenum*.

But the maps and the episodes of church history are connected even more deeply, a fact that can only be appreciated fully when they are evaluated according to the perspective of the historians who worked for the Boncompagni Pope, Carlo Sigonio and Cesare Baronio. Sigonio, whom Gregory XIII appointed to write the official history of the church in 1578, only a few months before work on the Gallery of Maps had begun, was a professor at the University of Bologna and a close friend of Cardinal Paleotti, whose treatise on sacred images he had read before its publication. He had achieved international repute for his rigorous scholarship on the Roman Empire and late antiquity.[58] In his two influential books on medieval Italy, *De regno Italiae* (Venice, 1574) and *De occidentali imperio* (Bologna, 1578), he regarded the barbarian invasions of the Italian peninsula, many of which are depicted on the Vatican maps, as positive events in their outcome. While from a modern perspective they marked the decline of Roman civilization, from Sigonio's perspective they were vital to the spread of Christianity, making it possible the recognition of the temporal and spiritual power of the Roman popes by the barbarian kings. Although Sigonio was unable to complete his history of the church because of the opposition from conservatives within the papal court, his views on the Barbarian invasions were shared by other influential members of the papal court, including Cardinal Sirleto and Cesare Baronio, who eventually succeeded to write the official history of the church.

In line with Sigonio's and Baronio's view of history, the historical account offered in the Gallery of Maps is not simply an opposition between the grim human affairs represented in the vignettes on the maps and the perfect saintly history depicted in the vault. Rather it is an account that interprets late antiquity, the barbarian invasions, and the Renaissance struggles as an integrated providential sequence of events, all contributing to the affirmation of papal supremacy. Within such a history, the provinces of the Roman Empire were transformed into the dioceses of the Christian church, a fact hinted at in the legends and personifications accompanying the general maps of Italy. In the map of ancient Italy the personification of Italy, a woman holding a sword and a cornucopia seated under a baldachin

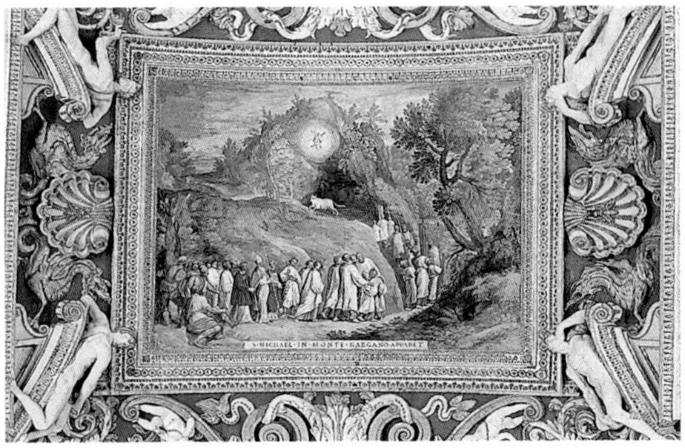

7.23 *St. Michael the Archangel Appearing on Monte Gargano*, ca. 1580, ceiling fresco. Vatican, Gallery of Maps

adorned with the keys of St. Peter, is flanked by Strabo and Ptolemy and by the rivers Tiber and the Arno (Fig. 7.6). The accompanying legend of the same map lists Augustus's division of the peninsula into eighteen regions; Constantine's partition of the peninsula and its islands into sixteen provinces; and the changes that, "the Goths, Lombards, and Greeks [brought to] the ancient partition of Italy with the new dominions of the duchy and the exarchate."[59] In the facing map representing modern Italy, the personification of Italy is flanked by Flavio Biondo and Raffaele Maffei and by the rivers Po and Adige, while the accompanying legend relates how Italy "receives honor from the primacy of the Holy Roman Church, now led by the servant of God's servants, Pope Gregory XIII, who built and decorated these buildings, the Patriarchs of Venetia and Aquileia, twenty-nine archbishops, and two hundred and fifty-three bishops" (Fig. 7.7). In other words, these two general maps of Italy, together with the portraits of ancient and modern geographers, the personifications of rivers, and their inscriptions, presented the uninterrupted historical continuity between the Roman Empire, the age of Constantine, the barbarian invasions, and the papacy. Gregory's courtiers and guests, well versed in this papal version of history, would not have failed to interpret the historical vignettes on the maps and the scenes on the vault as part of the same history that, from ancient times to the Renaissance, made possible the affirmation of papal supremacy.

The unique contribution that the Gallery of Maps offered to the chorography of Italy was precisely the integration of miracles and pious deeds with secular and antiquarian history in the description of the notable features of a place. Even though the episodes of church history were separated from the maps, they were an inextricable part of the map cycle, linked conceptually and not just by proximity. Similar to the historical vignettes on the maps, the criteria for the selection of the vault's scenes were dictated by the maps, and their purpose was to illustrate the participation of individual places in church history. Only in this papal chorography of Italy, made up of geography, history, and miracles, was the peninsula portrayed as the privileged theater for the affirmation of Christianity.

* * *

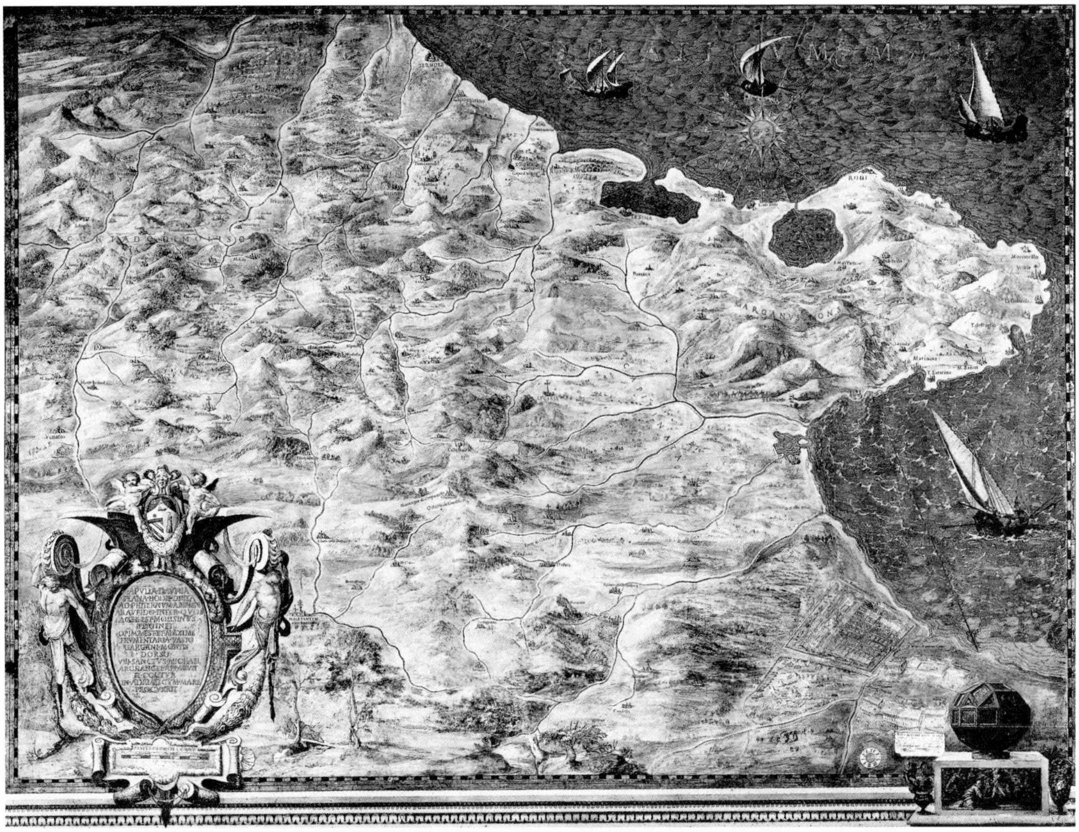

7.24 Egnazio Danti, *Apulia* (Map of Apulia), ca. 1580, fresco. Vatican, Gallery of Maps

The Atlas of Modern Italy

The maps of the Gallery of Maps should not be considered individually as historical maps, nor collectively as a historical atlas of Italy, even though they contain historical vignettes and are connected with episodes of church history.[60] Their purpose was not the understanding of Italian history, but rather the representation of modern Italy. The modernity of the Vatican project comes into better focus if we compare the Vatican maps with maps that were regarded as historical maps during the Renaissance.

By the end of the sixteenth century, historical maps had developed into a popular genre. Originally they were printed as single-sheet maps, illustrating the Holy Land, the Mediterranean, and Egypt, but soon they were also collected in historic atlases. Abraham Ortelius added an appendix of historical maps, significantly entitled *Parergon, sive veteris geographiae aliquot tabulae* (*Addition, or some maps of ancient geography*), to his popular *Theatrum orbis terrarum.* Playing on the double meaning of the word "parergon," which literally means "addition" but which had also come to denote landscape painting in the Renaissance, Ortelius's *Parergon* was both an addition to his atlas of the modern world as well as a land-

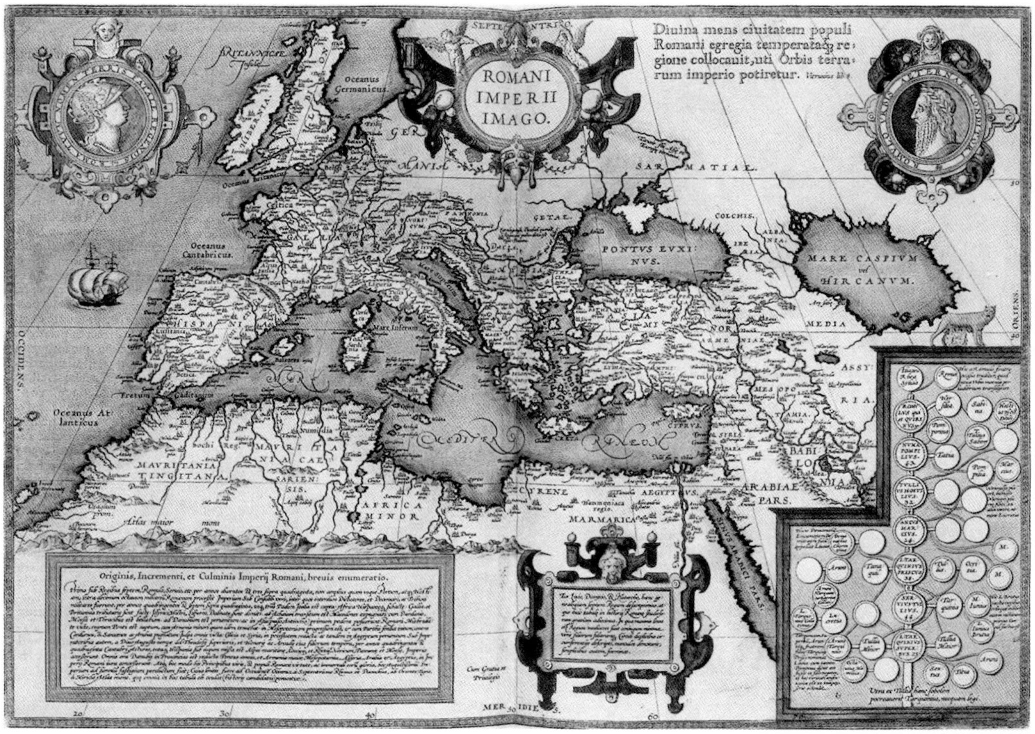

7.25 Abraham Ortelius, *Romani Imperii Imago* (Map of the Roman Empire), from Ortelius, *Theatrum Orbis Terrarum*, Antwerp 1579. John Carter Brown Library, Brown University

scape for historical events.[61] In its first edition in 1579, *Parergon* included only three historical maps, one representing ancient Greece, another illustrating the Roman Empire at the time of Alexander the Great (Fig. 7.25), and the third representing the travels of St. Paul from the Holy Land to Rome (Fig. 7.26), but in the subsequent editions it increased in size and scope. In its final rendition of 1598, Ortelius's historical appendix included maps on the ancient history of the Celtic lands, the British Isles, Greece, Latium, northern Africa, Gaul as conquered by Caesar, the Roman Empire (Fig. 7.25), the Navigation of Aeneas, the Travels of Abraham, the Route of the Exodus, the view of Mount Olympus and of the Temple in Thessalia, and the plan of Philip II's Escorial. Ortelius himself clarified the scope of his *Parergon* in his address to the reader published on the frontispiece of the historical appendix:

> Dear Reader, after our description of the world you have here the following maps which I have designed for the benefit of those interested in ancient history, sacred and profane. I have decided to publish these in a separate book because they did not seem to be relevant to the maps in our *Theatrum*, where I intended to show only present-day locations. However I gave into my friends' demands and have put them here at the end of our work as a Parergon. Greetings, and be well disposed to our efforts.[62]

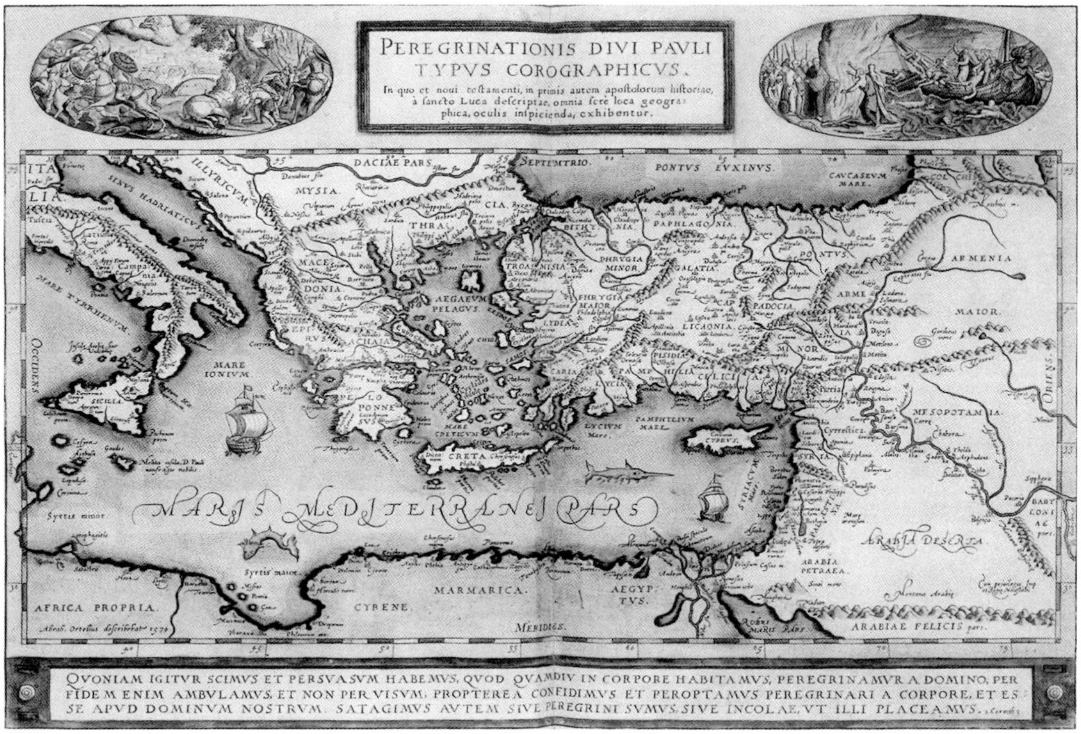

7.26 Abraham Ortelius, *Peregrinationis divi Pauli Typus Chorographicus* (Chorography of St. Paul's Peregrinations) from Ortelius, *Theatrum Orbis Terrarum*, Antwerp 1579. Folger Shakespeare Library

Ortelius regarded his historical maps, which represented the stage upon which past events took place, as intrinsically different from the modern maps of his *Theatrum*, which, instead, represented the modern world (Fig. 7.26).

In fact, in historical maps the land was mapped according to the most advanced knowledge of the place, but the toponyms were the old ones. Moreover, the maps included large excerpts from the text: historical, religious, or biblical, which they were meant to illustrate. Often they also included historical vignettes documenting the historical narrative as it unfolded in its proper place, or at the borders of the pertinent land.

The Vatican maps, however, were conceived as modern maps.[63] They represented Italy as it looked in the late sixteenth century. The cities were those that existed in 1580, the place names were the modern ones, and the borders those of the late sixteenth century. The historical vignettes that populated the maps did not transform them into historical maps but rather contributed to the general chorographical description of the mapped territory. The Vatican chorographical maps were not meant to be read as a single text, although obviously knowledge of Roman, medieval, and ecclesiastical history was required to understand them fully. Had Gregory wished to have a historical atlas of Italy, he would have selected a different model, which had been circulating in individual maps for many decades and which Ortelius made available with his *Parergon*. The Vatican maps should be considered for what they were meant to be: superb chorographical descriptions made possible by the fusion of

Danti's cartography with Baronio's history, modern maps in which word and image, history and geography, past and present, cartography and religion blended exquisitely.

The Visitors

As a marvelous visual synthesis of geography and church history, the Gallery of Maps was a much desired site for illustrious visitors to the Roman court, educated travelers, and cartographers. The first enthusiastic visitor was Gregory XIII himself, who rarely lost an opportunity to take a walk in his marvelous corridor. The gallery provided him with distraction from ruling the Catholic world. It was a place in which he could take the physical exercise he always enjoyed. And he used the Gallery of Maps for reviewing other projects which, at his request, Danti had recorded on the maps. For instance, as Danti himself tells us, while the architect Giovanni Fontana was restoring the Claudian port at Ostia, Gregory asked Danti to survey the ancient ruins and to compare them with Etienne Du Perac's reconstruction of the ancient port, which had been printed in 1575. Danti reported that the pope ordered him, "to draw the plans of all the ruins that today remain there, to draw their elevations, and to paint [the harbor] precisely, as indeed I did, in the Gallery which I have created for His Holiness in his palace in the Vatican, so that he can nonetheless see it before his eyes, and plan how he could restore such a noble and wonderful work to its original use."[64] Danti painted his survey of the ancient port on the map *Tremiti* and Du Perac's reconstruction on the map *Ilva* (these and *Italia antiqua* are the only historical maps in the Gregorian corridor).

Later popes also enjoyed the Gallery of Maps and regarded it as more than a fixed decoration. They modified maps and inscriptions in order to register the changing political situation of the peninsula and to update its geography. Clement VIII amended the legend of the map of the Duchy of Ferrara to record the annexation of the duchy to the Papal States under his papacy, a proud gesture repeated by Urban VIII, who recorded the acquisition of the Duchy of Urbino which he had secured for the papacy. Urban attached great significance to the Gallery of Maps. He had the terrace above the corridor walled to prevent water seepage and, in 1631, he commissioned Lukas Holste, the Vatican librarian and an erudite cartographer, to update completely the geographic content of the maps, though this project was only partially executed.[65]

As a gesture of great appreciation and distinction, selected visitors were admitted to enjoy a walk in the Gregorian corridor. Gregory XIII entertained the Grand Master of the Knights of Malta with a visit to the corridor during his prolonged stay in Rome to solve crucial disagreements over the succession in his order.[66] The Marquis d'Alincourt, who was waiting for Cardinal d'Este while he attended the meeting of the Consistory on 22 November 1584, was admitted into the spacious corridor as a sign of distinction, while everybody else awaited in a crowded hall.[67] Travelers, whose visits are documented in the Gallery of Maps from the later Renaissance to the eighteenth century, left accurate descriptions in their diaries. Charles de Brosses, who visited the corridor in 1739 or 1740, used Danti's map of Tuscany to verify the accuracy of his own map representing the defeat of Catiline near Pistoia. Johann Caspar Goethe, father of the more famous Wolfang and interested in ancient ruins, was allowed to study the Roman consular roads and Danti's plan of the Claudian port at Ostia.[68] Famous cartographers eagerly sought permission to copy the maps. Abraham Ortelius, who had previously been informed of the Gallery of Maps by Danti himself and was in constant search of new maps for his *Theatrum*, asked Filippo Pigafetta and Philip van Winghe to report to him on Danti's map *Latium*.[69] The two cartographers expressed a devastating judgment on Danti's cartographic work, echoed years later by Giovanni Antonio Magini, a polymath who was himself compiling a printed atlas of Italy.[70] It is not surprising that the Vatican maps looked obsolete only a few years after their completion. They had been executed very rapidly, not allowing time for the consistent and balanced correction of astronomical positions of even major places and thus contained geographical incongruities and cartographic errors. But for many decades to come the most detailed chorography of Italy was available only on the walls of the papal residence, where the geography of the peninsula blended with the papal post-Tridentine vision of church history.

The original iconography of the Gallery of Maps derived from the convergence of the interests of Gregory XIII, who selected a map cycle that filled a void in Renaissance cartography, and Danti, who imaginatively adapted the geography of the peninsula to the architectural space of the Vatican corridor. Both Gregory XIII and Danti envisaged the Gallery of Maps within the chorographical tradition of the Renaissance, indeed as the visual counterpart of the verbal descriptions by Biondo, Maffei, and Alberti. But both patron and mapmaker also aspired to novel achievements. For his part, Danti imagined his chorographical maps of modern Italy as an innovative synthesis that combined the quantitative and qualitative description of a locale, that fused the Ptolemaic grid with landscape views, modern cartography with local history. Gregory XIII, on the other hand, imbued the chorography of modern Italy with the Catholic version of church history.

But as the first atlas of modern Italy, the Gallery of Maps also poses different questions. As a cartographic enterprise, it looks strikingly similar to the contemporary national atlas that European monarchs sponsored to represent their own states – atlases that expressed literally and symbolically the ruler's control over his lands. Obviously, however, the pope was not the ruler of Italy, owning only a portion of the peninsula, and it is therefore legitimate to wonder whether, through the language of cartography that he favored, Gregory XIII was announcing his plans for the political conquest of the peninsula. Or whether, instead, his magniloquent representation of Italy through maps articulated a different state of affairs for the post-Tridentine church, a state of affairs within which Italy held a privileged place.

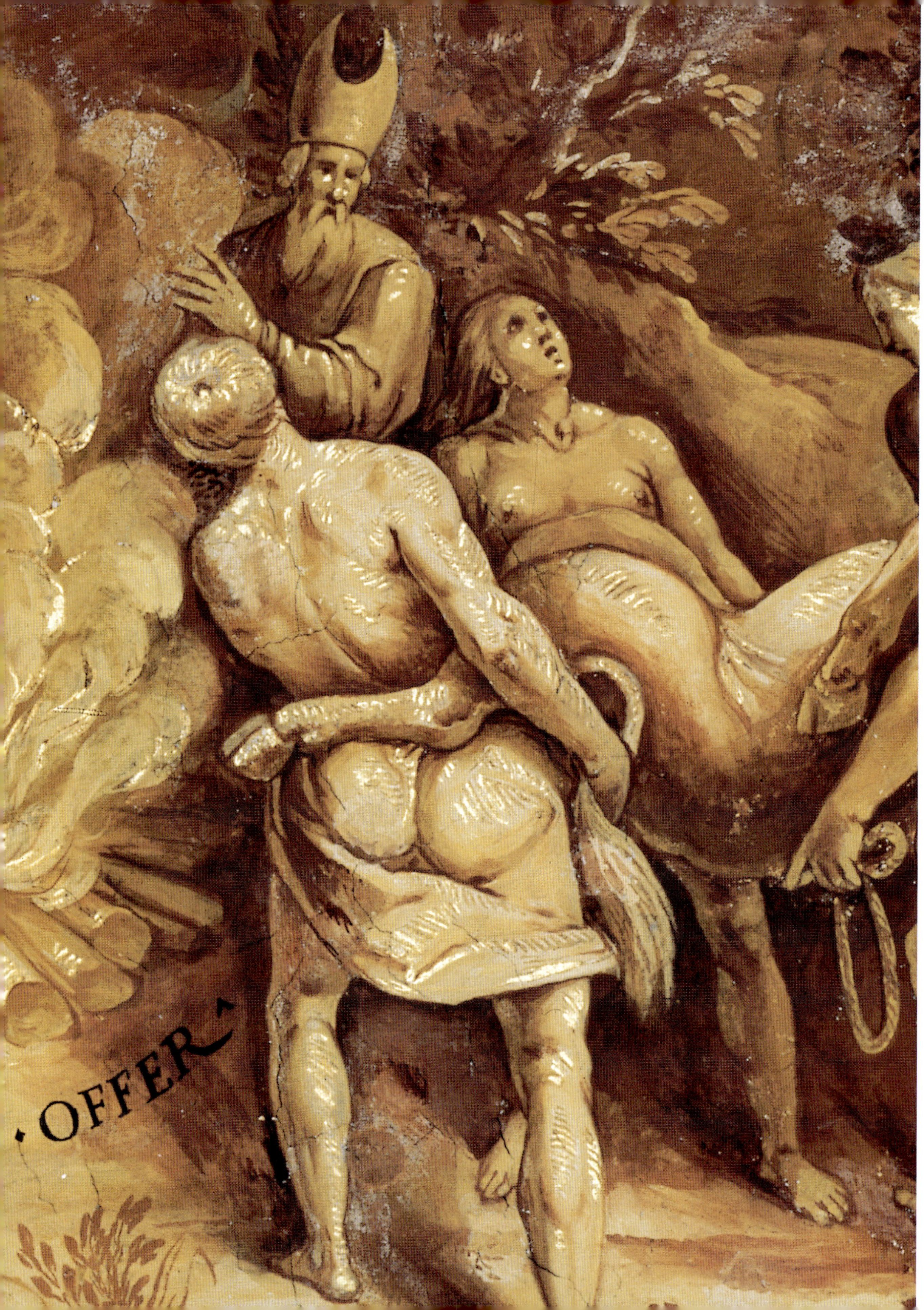
OFFER

8

The Primacy of Italy

The artistic and cartographic language of the Gallery of Maps suggests, at the very least, that Gregory XIII regarded Italy, its cities, and islands, as having a privileged place within the universal church over which his spiritual power extended. Indeed, Gregory and other post-Tridentine popes knew well that their universal mission could be accomplished more fully in Italy than in the rest of the world.[1] After the Council of Trent, the rulers of the Italian peninsula proved more receptive in accommodating papal requests for ecclesiastical reforms within their own states and, as a result, the Roman Church was able to establish an ecclesiastical authority in Italy that was more wide-ranging than in other parts of the Catholic world. The exercise of papal authority in the peninsula was far from uniform, being subject to local adjustments, different governments, and local situations; while the diplomatic relationships between the popes and the Italian rulers alternated success and support with disappointment and obstruction. But overall the post-Tridentine papacy enjoyed a primacy in Italy that remained unchanged in the expanded Christian empire of the following centuries.

In Gregory XIII's view, the primacy of Italy stemmed from the classical tradition that placed Italy at the center of world history, from his own peculiar interpretation of the peninsula as the land that Constantine had originally donated to the Roman popes, and from the fact that ecclesiastical revenues were largely collected in the peninsula rather than elsewhere in the Catholic world. It is this papal primacy in ecclesiastical matters that Gregory so triumphantly celebrated in the Gallery of Maps, not the political conquest of the peninsula. The ecclesiastical supremacy that the Gallery of Maps proclaimed is confirmed by the juxtaposition of the maps of Italy with the biblical sacrifices. These biblical sacrifices referred not only to the sacrificial nature of the Eucharist, which Protestant reformers had rejected, but also to the institution of the tithe – the means to support the Christian popes and their ministers – which Catholic rulers were supposed to offer spontaneously to the pope.

* * *

(*facing page*) Detail of fig. 8.16

Italy and Constantine

The literary topos of the centrality of Italy in world history originated in antiquity, was appropriated by the early Christian church, and enjoyed enormous success in the humanistic Renaissance, especially in the chorographical descriptions of the peninsula by such writers as Biondo, Maffei, and Alberti, whose books were fundamental to the conception of the Gallery of Maps. In accordance with this Renaissance tradition, the centrality of Italy was emphatically restated in three inscriptions in the Gallery of Maps, in which Italy was defined as: "The most noble region of the whole world" ("Regio totius orbis nobilissima"); "The principal region of the world" ("Regio num orbis primaria"); or "The nurturer of all the lands" ("Omnium terrarum alumna").[2] Another inscription (today erased) in the map *Italia antiqua* quoted Pliny describing Italy as a homeland given by God: "Italy is a land which is at once the foster child and the parent of all other lands, chosen by the providence of God to make heaven itself more glorious, to unite scattered empires, to make manners gentle, and to become throughout the world the single fatherland of all the peoples."[3] Using the words of this ancient narrator, which were altered only in a minor, yet crucial, detail – the plural gods (*deorum*) was changed to the singular God (*dei*) – the Gallery of Maps adapted them to the reality and aspirations of Gregory XIII's post-Tridentine papacy.[4]

Moreover, the primacy of Italy was linked with another favored Renaissance topos: the continuity between the pagan past and the Christian present, proclaimed in the Gallery of Maps by the juxtaposition of the map of ancient Italy (*Italia antiqua*) with the map of modern Italy (*Italia nova*). The legends of these maps (see pp. 172–5) listed the imperial partition of the peninsula and its later transformation into a network of bishops' and archbishops' seats (Figs. 8.1 and 8.2). The history of the peninsula presented in the Gallery of Maps consciously ignored different accounts that questioned Italy's centrality

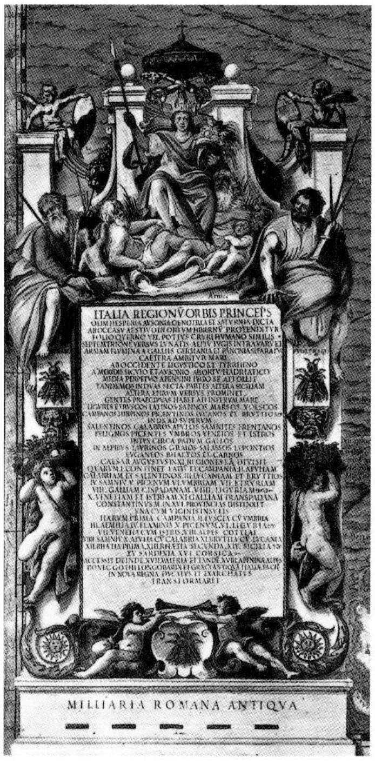

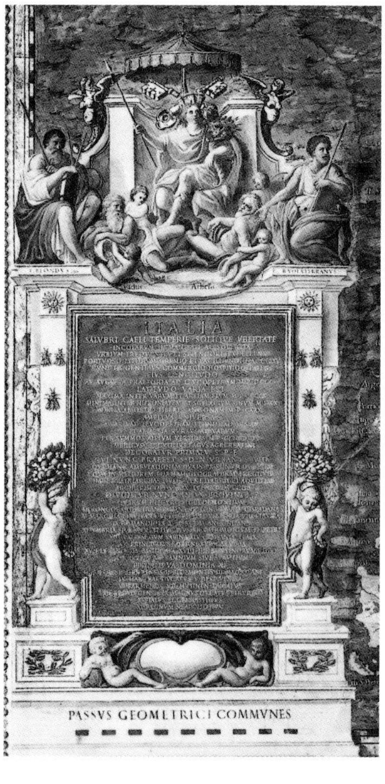

8.1 (*right*) Detail of fig. 7.6: Cartouche

8.2 (*far right*) Detail of fig. 7.7: Cartouche

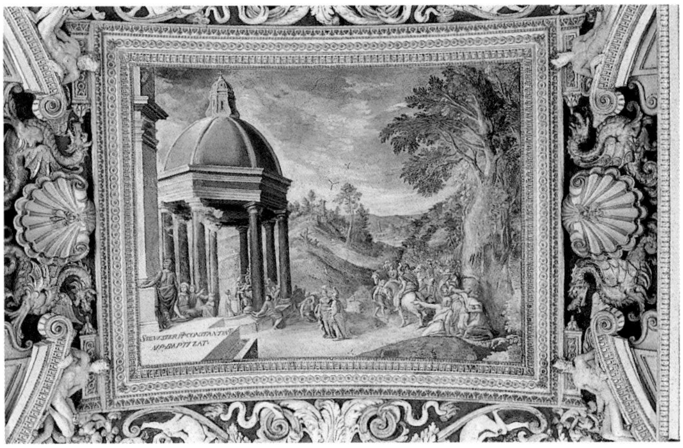

8.3 *St. Sylvester Baptizing Emperor Constantine*, ca. 1580, ceiling fresco. Vatican, Gallery of Maps

in world history, most notably Sebastian Münster's *Cosmographia universalis*.[5] In that polemical and informative book that European readers loved to possess, Münster presented sixteenth-century papal Italy as the negation, not the fulfillment of the early days of Christianity. But in the Gallery of Maps the reality of modern Italy is linked to the glory of the Constantinian age.

The maps of ancient and modern Italy are connected with five vault scenes illustrating episodes from the life of Emperor Constantine, the only figure represented so extensively in the Gallery of Maps. These scenes, which took place in Rome, are: *St. Sylvester Baptizing Emperor Constantine* (Fig. 8.3), *Emperor Constantine Building the Basilica of St. Peter* (Fig. 8.4), *Emperor Constantine Building the Basilica of St. Paul* (Fig. 8.5), *Emperor Constantine Holding the Reins of Sylvester's Horse* (Fig. 8.6), and *Constantine's Vision of the Cross Before the Battle at the Milvian Bridge* (Fig. 8.7).[6] Unsurprisingly, the continuity between ancient and modern Italy was marked with the conversion of the Roman emperor to Christianity, his subordination to the pope, the actual construction of the Roman basilicas, and the imperial commitment to the military defense of Christianity. But the Gallery of Maps included an uncharacteristic connection between Constantine and Italy, which must have reflected Gregory XIII's view on Constantine's posthumous role in the post-Tridentine papacy.

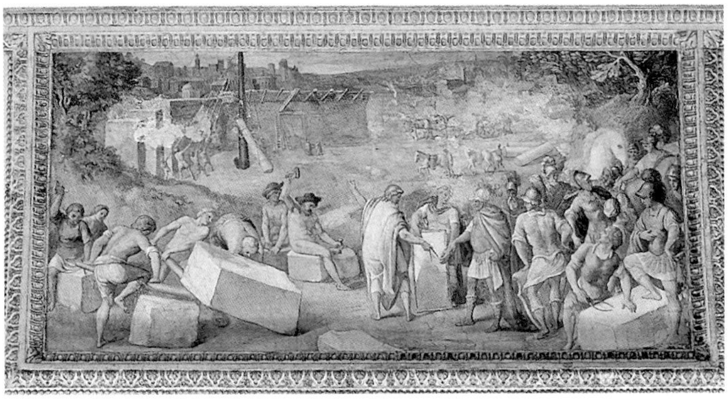

8.4 *Emperor Constantine Building the Basilica of St. Peter's*, ca. 1580, ceiling fresco. Vatican, Gallery of Maps

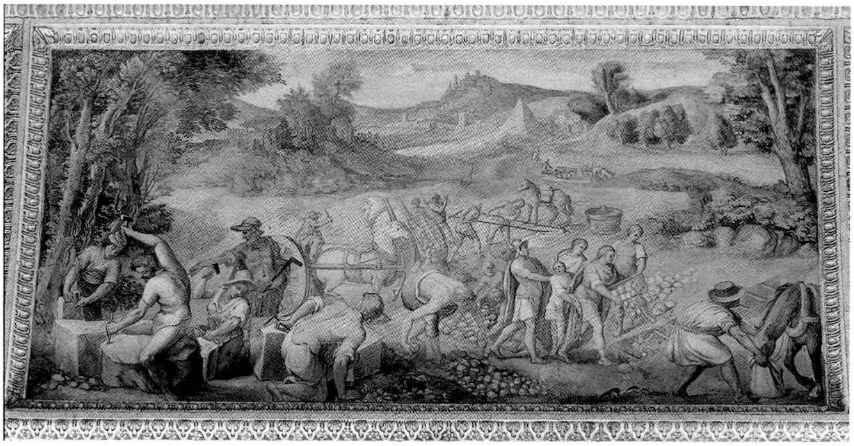

Like earlier popes, Gregory revered Constantine and, throughout his papacy, linked his own actions to those of the emperor whom he admired.[7] He restored the Constantinian buildings in Rome: the Lateran Baptistery, where Pope Sylvester had baptized Constantine (the baptistery features prominently in the Gallery of Maps); the Lateran Palace, which, as explicitly mentioned in the famous donation, Constantine gave to Sylvester; and the church of San Giovanni in Laterano, founded by the emperor, which Gregory XIII had put at the center of his urban renovation for the Jubilee of 1575. Even the Gregorian reform of the calendar that Gregory had promulgated in 1582 was a tribute to Constantine, since it established the date of the vernal equinox at the time of Constantine rather than at the time of the pagan Julius Caesar. The pope intervened also in the Sala di Constantino in the Vatican Palace, replacing the original wood ceiling with a high vault painted by Tommaso Laureti, an artist he brought from Bologna to Rome specifically for this project. It is the ceiling of the Sala di Constantino that reveals explicitly the meaning of the connection between Constantine and Italy proposed in the Gallery of Maps.

The ceiling of the Sala di Constantino is justly renowned as a tour de force of illusionistic perspective (Fig. 8.8). In an empty basilica a statue of Mercury, symbol of the pagan idols, is crushed to the ground and replaced by the crucifix. Around the central fresco are the eight provinces of Italy described by Strabo and the islands of Sicily and Corsica. Represented in pairs through allegorical personifications (Liguria and Tuscany, Latium and Campania, Lucania and Apulia, Piceno and Venetia, Sicily and Corsica), they hold cartouches stating the attributes of each region. In the ceiling there are also putti *dal sotto in su* holding the regalia that Constantine gave to Sylvester as symbols of the papal temporal rule over the donated territories: the tiara, crown, purple robe, and scepter. The representation of the regions of Italy was strictly connected to the Donation of Constantine, the forged document that popes had used since the ninth century to demonstrate their supremacy over temporal rulers. This fact is explained in a note written by an anonymous yet well-informed member of Gregory XIII's court who described the ceiling of the Sala di Constantino: "[the painter Tommaso Laureti] thought of painting deeds of the same emperor, particularly those he did in honor and benefit of the Holy Church. Having seen in one of the walls in the same hall that the donation of Italy made by Constantine to Sylvester and to his successors had been

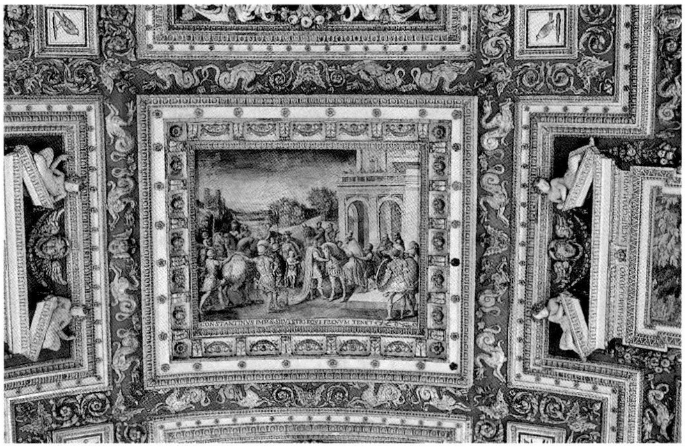

8.5 *(facing page) Emperor Constantine Building the Basilica of St. Paul*, ca. 1580, ceiling fresco. Vatican, Gallery of Maps

8.6 *Emperor Constantine Holding the Reins of St. Sylvester's Horse*, ca. 1580, ceiling fresco. Vatican, Gallery of Maps

represented through a small, not very legible figure, Tommaso thought of painting the same Italy divided in eight provinces according to Strabo's order so that this donation would be more intelligible."[8] The anonymous writer attributed the invention of the iconography to Laureti himself and connected it with the wall fresco painted by Giulio Romano and his workshop in the 1520s depicting *The Donation of Constantine* (Fig. 8.9). Giulio Romano had placed the scene inside St. Peter's and represented the act of the donation symbolically, through the exchange of a gold statuette of Rome between the emperor and the pope. But, in the late sixteenth century, Giulio Romano's representation of the donation was regarded as confusing since it could validate two contradictory interpretations: according to one interpretation, the donation was limited to the city of Rome, but according to another it pertained to Italy, and thus the statuette of Rome was to be interpreted as a metonym for the entire peninsula. In either case, Giulio Romano's fresco demanded a clarification pertaining to the geographical extent of Constantine's donation. Considering that the Sala di Constantino was part of the official halls of the Vatican Palace, it has to be assumed that Tommaso Laureti interpreted Gregory XIII's view in representing the Italian regions as part of the imperial donation.

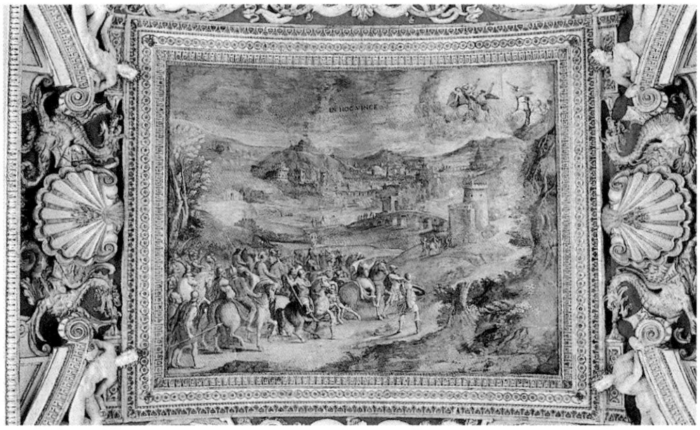

8.7 *Constantine's Vision of the Cross Before the Battle at the Milvian Bridge*, ca. 1580, ceiling fresco. Vatican, Gallery of Maps

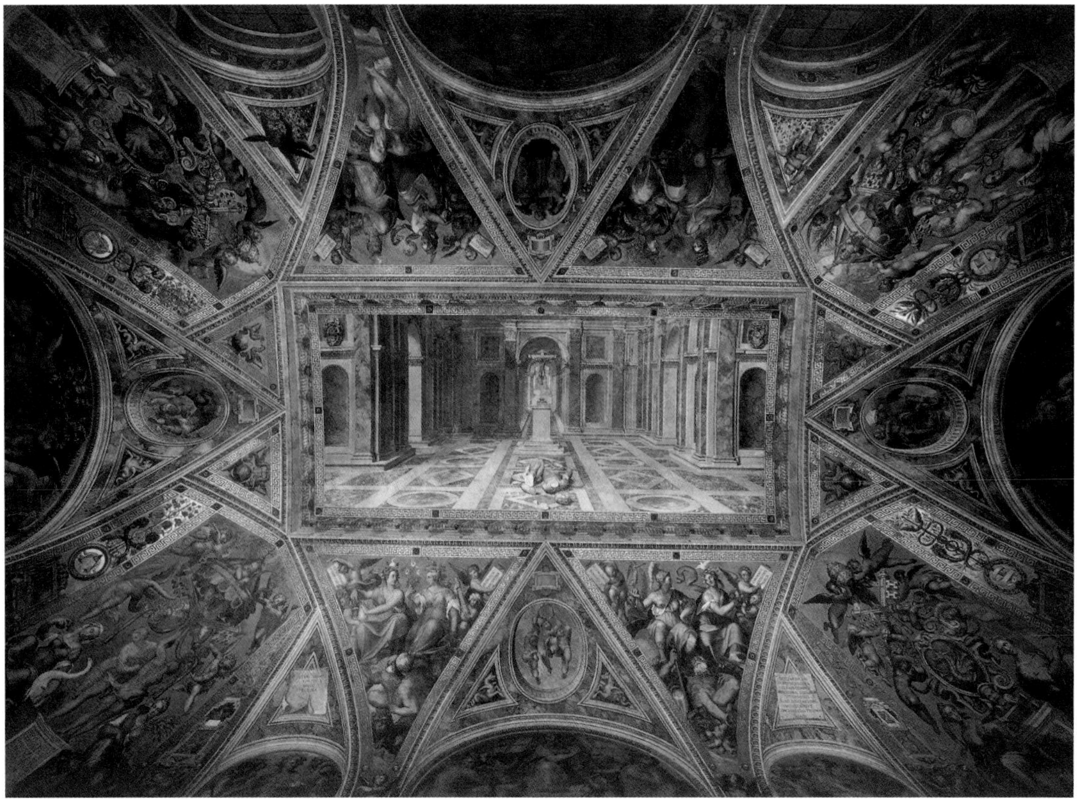

8.8 Tommaso Laureti, *The Triumph of Christianity over Paganism*, 1585–1586, ceiling fresco. Vatican, Sala di Costantino

The donation of Constantine was such an integral component of papal supremacy that it did not disappear completely from papal iconography, even after the humanist Lorenzo Valla had successfully proved it to be fake in 1440, and even after Martin Luther had attacked it vigorously. In the Sala di Costantino it occupied an entire wall, but in other cases it was either included as a detail in different episodes of the emperor's life, or it was alluded to by representing less controversial episodes explicitly mentioned in the contested document. The *officium stratoris*, the episode that represented Constantine as the groom to Sylvester, was regarded as the most compelling scene to reiterate the theme of papal supremacy, and indeed it was included in the Gallery of Maps (Fig. 8.6). At the core of the debate on the authenticity of the document lay the geographical extension of the lands that Constantine donated to the emperor. Although the donation of Constantine was interpreted, at the Boncompagni court, as the donation of the Italian peninsula, the geographical extension of the donation itself was an issue far from settled.

In the forged document the objects of the donation were listed as, "our palace [the imperial Lateran Palace], the city of Rome, and all the provinces, places, and cities of Italy, and the western regions."[9] While the Lateran palace, the city of Rome, and the cities of Italy never posed interpretative problems, the meaning of the "western regions" was ambiguous.

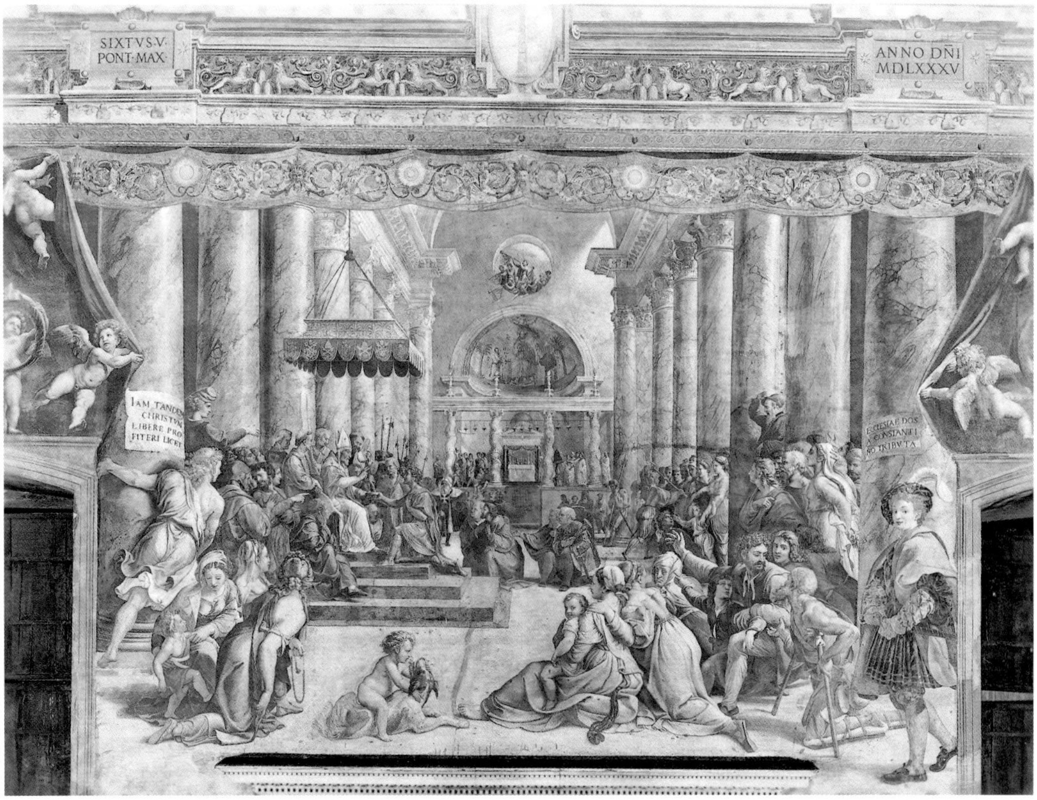

8.9 Giulio Romano, *The Donation of Constantine*, 1519–1524, fresco. Vatican, Sala di Costantino

Modern scholars seem to agree on the fact that the ninth-century author of the document identified the "western regions" as the northern regions of peninsular Italy (Lombardy, Venetia, Istria, and the adjacent islands), that is, the lands that were dominated then by the Frankish kings. Medieval jurists, however, tended to limit the "western regions" to the Patrimony of St. Peter, that is, the territories over which the pope had direct temporal authority. The fifteenth-century humanist Lorenzo Valla interpreted them as referring to the regions of Europe (Spain, France, Germany) and thus considered them, together with the language of the document, as crucial in disproving the text's authenticity since the emperor had no jurisdiction over these lands. At the Boncompagni court, the "western regions" were understood as referring to regions of the peninsula, a fact suggested by Laureti's ceiling, the anonymous writer's note from the Boncompagni archive, and the Gallery of Maps.

In the Gallery of Maps the contested donation was referred to indirectly, through the depiction of the *officium stratoris* on the vault, and explicitly, through the cartographic representation of the regions that were regarded at the Boncompagni court as the very object of the donation. For Gregory XIII, the maps of Italy represented not only the classical geographical unity that Strabo and Pliny described, but also the "western regions" Constantine donated to the papacy.

Furthermore, throughout the sixteenth century, the discussion of the Donation of Constantine became part of the fierce scholarly debates pertaining to the chronology of the emperor's life – debates that pitted Protestant against Catholic historians as well as different groups of Catholic scholars. The Gregorian corridor appears to be deeply informed by these debates that animated the papal court in the 1570s and 1580s.

Constantine and Catholic Historiography

In the late sixteenth century, the study of Constantine's life centered on the chronology recorded by Eusebius, one of the emperor's close contemporaries and author of the oldest document on the emperor's life. As an early source on Constantine, Eusebius's text held an authoritative position among humanists of any religious belief, although it posed considerable historiographical problems to orthodox Catholic historians since it eliminated the historical circumstances of the donation, namely the meeting between Constantine and Pope Sylvester in Rome and the emperor's baptism in the eternal city. According to Eusebius's account, Constantine founded the basilica of San Giovanni in Laterano not during the papacy of Sylvester, but rather under his predecessor Miltiades; he started to defend the Christians only after his victory at the Milvian Bridge in 312, a victory that evidently the emperor obtained without the help the Cross; and he was baptized not in Rome but on his deathbed in Constantinople.[10]

Reformed historians were quick to embrace Eusebius's chronology since it confirmed the forgery of the Donation and further ridiculed the Catholic version of papal supremacy. But Catholic historians, who wished to sustain papal supremacy and, at the same time, provide a plausible reconstruction of historical events, had to find new imaginative ways to argue their case, especially after a group of Protestant historians had written their version of church history, the Magdeburg *Centuriae* published in Basel in the 1560s. The value of the *Centuriae* lay in its ability to apply the methodologies of humanistic philology, biblical criticism, and legal scholarship to the study of church history. As the humanists of earlier generations had done for secular history, these reformed historians analyzed church history contextually, replacing the medieval *simplex narratio gestorum*, usually organized chronologically, with more compelling historical reconstructions. In this new historical writing, papal supremacy did not result from divine intervention or imperial donation, but rather emerged from the medieval struggles between the German emperors and the rulers of Italy.[11]

Also Catholic historians, trained in humanistic scholarship and in the rigor of philological analysis, were well disposed to accept Eusebius's history of Constantine's life and reject the Donation of Constantine, although they were not ready to reject papal supremacy, for which, instead, they offered alternative historical evidence. Among them, the most prominent was certainly Carlo Sigonio, who accepted Eusebius's version of Constantine's life and unequivocally regarded the Donation of Constantine as a fable "believed by pious men."[12] However, Sigonio convincingly argued that the later donations made by the Frankish kings to the popes in the eighth and ninth century and by Matilda of Canossa in the eleventh century were much more significant than Constantine's to the formation of papal supremacy. For Sigonio, the Papal States and the Roman primacy grew through the conflicts between the Kingdom of Italy and the German emperors, in which the papacy always sided with the Italian rulers. Because the history of medieval Italy was so inextricably intertwined with church history, Sigonio's books were scrutinized by the Roman Congregation of the Index during Pius v's papacy. In 1571 ultra-orthodox members of the papal court demanded from Sigonio a radical revision of his version of Constantine's life, including the recognition of the imperial donation. The dispute between Sigonio and these uncompromising members of the papal court ran for many months and was finally resolved after the intervention of Cardinal Paleotti and Sigonio's partial capitulation. In *De occidentali imperio* (Bologna, 1578), the historian was allowed to follow Eusebius in asserting that Constantine donated the Lateran to Miltiades and founded the church of San Giovanni in Laterano under this pope. He had, however, to agree to save the historical circumstances of Constantine's donation, writing that the emperor was cured and baptized by Sylvester in 315. As for the document of the donation itself, Sigonio was able to isolate this section of the book visually and conceptually from the rigorous scholarship of his work: instead of publishing the "original" text, he printed the letter Pope Hadrian i had written in 785 summarizing the content of the Constantinian edict; and to further isolate the document he had this section of his book printed in a different, oversize type.

Sigonio's views on the Donation of Constantine and on medieval Italy were shared by many other Catholic scholars and clergymen, including by influential members of the papal court. Cardinal Paleotti wrote to members of the papal court to support Sigonio's publications. Agostino Valier, bishop of Verona and a committed reformer, defended Sigonio's orthodoxy, proposing a pro-Roman and pro-Italian interpretation of his historical work. Cesare Baronio, whose adherence to the papal view of history few doubted, called the Donation of Constantine a forgery in a private letter, although in his official *Annales ecclesiastici* took a more moderate approach, and shared Sigonio's view on the vital importance of the medieval donations in the formation of the Papal States. Even Cardinal Sirleto, who strenuously defended the Donation of Constantine, had been convinced of its relative insignificance in comparison to the medieval ones. Indeed, the later donations by Pepin, Charlemagne, and Peter ii of Aragon had been depicted in the Sala Regia as examples of papal supremacy, and many reappeared also in the Gallery of Maps. Although no document survives to reveal Gregory xiii's view on the Donation of Constantine or on medieval history, circumstantial evidence suggests that he must have thought of Sigonio's historiography as an effective weapon against the Magdeburg *Centuriae*.

In 1578, to the astonishment of many at his papal court, Gregory appointed Carlo Sigonio to write the official Catholic version of church history. In the attempt to connect the historical research of early Christianity with the reform of the post-Tridentine church, Gregory wished to turn the tools of the Magdeburg *Centuriae*, namely humanistic philology and contextual analysis, to the advantage of Catholic historiography. Indeed, Sigonio was appointed precisely because of his conception of history, especially his abilities in reconstructing the historical context of events through the use of maps and the geographical description of places.[13] Sigonio never finished the book on church history that had been commissioned from him, since the Roman Committee of the Index of Forbidden Books blocked it. Even Paleotti's support was not sufficient to guarantee the papal commission to Sigonio for the writing of the official church history, which instead was later given to Cesare Baronio. Baronio was well versed in humanistic philology, had access to vital documents in the Vatican Library, and shared many of Sigonio's views on medieval history, including the importance of the historical context. But Baronio's strength as the official historian of the papacy resided in his ability to bend historical rigor to the needs of papal propaganda.

The Gallery of Maps seems informed by Sigonio's view of history. The events of church history depicted in the vault were contextualized in their most minute detail: buildings, clothes, and places were rendered accurately as they appeared at the time when the event took place. Geographical accuracy was an integral part of historical rigor. Moreover, the life of Constantine, which was so crucial to papal supremacy, was represented according to Sigonio's *De occidentali imperio*, that is, it reflected the compromise between Eusebius's life of the emperor and the papal propaganda of the Donation of Constantine. In the Gallery of Maps, the baptism by Sylvester in the Lateran Baptistery, an episode not recounted in Eusebius but that Sigonio was forced to include in his tract, was followed by the representation of the honors the emperor bestowed on the pope in the *officium stratoris*. Instead of the foundation of the church of San Giovanni in Laterano, which according to Sigonio Constantine had founded under Pope Miltiades, we see in the Gallery of Maps the building of the basilicas of St. Peter and St. Paul. According to the letter of Pope Hadrian I, which Sigonio had printed in his *De occidentali imperio*, the images of these two saints played a role in Constantine's conversion. The battle at the Milvian Bridge, which according to Eusebius happened before Constantine's conversion, concludes rather than begins Constantine's Roman sojourn depicted on the vault. Finally, the donation was not represented directly, but indirectly through the maps of Italy.

By interpreting the Donation of Constantine as the donation of Italy, Gregory XIII was not claiming political rule over the entire peninsula, but rather giving historical and religious foundation to the primacy of Italy within the universal church. Medieval popes had used the Donation of Constantine to claim their temporal power in Italy, but their post-Tridentine successors knew that the Donation of Constantine could not be adapted to the political divisions of sixteenth-century Italy: the Papal States were limited to central Italy, Spain dominated in the north and the south, and a number of smaller independent states and republics proliferated elsewhere. The political partition of Italy was an accomplished fact, and the post-Tridentine popes, including Gregory XIII, were the committed guardians of this precarious balance of powers, not claimants to the temporal dominion over the entire peninsula.[14] What post-Tridentine popes could claim on the Italian peninsula was greater ecclesiastical autonomy than in the rest of the Catholic world to order apostolic visitations, nominate bishops and archbishops, establish ecclesiastical tribunals, and collect ecclesiastical revenues. It is precisely this primacy that the Gallery of Maps triumphantly celebrates. The interpretation of the Gallery of Maps as a celebration of papal primacy in Italy is confirmed by the very presence of the scenes of biblical sacrifices in the vault of the Gregorian gallery. Their depiction in relation to the maps of Italy has thus far eluded interpretation.

The Biblical Sacrifices

On the vault of the Gallery of Maps, intertwined with the episodes of church history, is a series of twenty-four ochre monochromes. The series illustrates the biblical sacrifices performed by Adam, Abel and Cain, Noah, Abraham, Jacob, and Moses, and it includes a unique sequence of fourteen scenes illustrating the sacrificial rituals of the Levite priests, a subject rarely represented in western art. Traditionally regarded as prototypes of the sacrifice of the Eucharist, these Old Testament sacrifices had become crucial in post-Tridentine doctrine to demonstrate the sacrificial nature of the Eucharist, a doctrine that the Protestant reformers had rejected. Certainly, the biblical sacrifices had been included in the Gregorian corridor to reassert the doctrine that Protestants had questioned. But they also referred to the institution of the sacramental tithe, a topic of primary concern for Catholic rulers rather than Protestant reformers. First analyzed within the history of biblical sacrifices and, second, in relation to papal finances, the Gregorian series of biblical sacrifices suggests that the Gallery of Maps represented not only the region described by ancient geographers or the land donated by Constantine to the papacy, but also the specific geographical limits of papal ecclesiastical authority.[15]

The series of biblical sacrifices opens with a scene representing Adam and Eve contemplating the sky while their children Cain and Abel play nearby (Fig. 7.14). The inscription below the fresco is a direct quotation from the Vulgate, Psalm 50:14: "Adam offers to God the sacrifice of thanksgiving" ("Adam immolat deo sacrificium laudis"), explaining that Adam is offering his praise. Adam's sacrifice of praise has rarely been represented in Western art, the potentials of its visual rendition being rather limited in contrast to other more dramatic events of his life. But it is a crucial story in establishing the figurative sacrifice of praise as the foundation of the literal biblical sacrifice. In Psalm 50 the sacrifice of praise stands above that of produce and livestock; the Letter to the Hebrews defined it as, "the tribute of lips which acknowledges his name" (Hebrews 13:15). Later Thomas Aquinas recognized the sacrifice of praise as the principal condition that inspires and animates the external offering without which even the most perfect offering has no worth. In other words, the sacrifice of praise, the first biblical sacrifice on the vault of the Gallery of Maps, puts into a historical and spiritual perspective the offerings of produce and livestock represented in the scenes that follow.[16]

Adam's offering of praise is followed by the bloody and bloodless sacrifices offered by Abel and Cain respectively (Genesis 4:3–5). This scene too was rarely represented in Western art, although it is quite significant: Cain killed his brother out of jealousy over God's greater appreciation of Abel's sacrifice of livestock above Cain's offering of produce.[17] Its inclusion in the sacrifice series in the Gallery of Maps must be related to the need to demonstrate the contemporaneity of the two kinds of sacrifices as well as the higher status of the bloody over the bloodless offering, two notions the Protestant reformers contested.

The scenes that follow illustrate bloody sacrifices. Some of these sacrifices are common in Western art: Noah offering animals and birds, one for each species, on the altar he built after the deluge (Genesis 8:20–1), and Abraham sacrificing his own firstborn (Genesis 22:10–12). Others are quite rare. Jacob is the protagonist of five scenes of sacrifice that are very important not only in the history of biblical sacrifice, but also in establishing the connection between the offerings and the institution of the tithe. The first scene illustrates Jacob's dream, which notably is not a sacrifice, but the inscription below the fresco, "This is no other than the house of God" ("Erit mihi dominus in deum") (Genesis 28:17), points unequivocally to the vow Jacob made upon his awakening: if God assured him a safe journey, he would erect an altar on the stone on which he had rested his head and he would offer to him the tithe of his produce and gain. In the following scenes Jacob builds altars in places where God had spoken to him, and he also makes offerings. He pours oil on the altar he erected at Shechem in Canaan after he had made peace with Esau (Genesis 33:20). He builds another altar at Bethel, the very site of the dream of the ladder and of the institution of the tithe (Genesis 35:3–7). He pours oil on the altar he built at Bethel, on the site where God had given him his new name of Israel and predicted a multitude of descendants and the ownership of Abraham's land (Genesis 35:11–15). And he offers animals at Beersheba just before entering into Egypt with his people (Genesis 46:1). The next scene illustrates Moses and the Burning Bush, a subject apparently unrelated to the sacrifice series. But, as in Jacob's Dream, in this scene too the inscription below reads, "Moses made offerings to God on this mountain" ("Moses immolabis deo super montem istum") (Exodus 3:12), referring to the sacrifice of animals Moses made, after he had left Egypt, on the same mountain where God had spoken to him through the burning bush.

At first sight, the biblical stories of the vault seem to conform to Raphael's painted Bible in the Vatican Loggias, which also represented the deeds of Adam, Abel, Cain, Noah, Abraham, Jacob, and Moses.[18] But a closer examination reveals a different emphasis on the same biblical narratives in the two painted series. Unlike in Raphael's Loggias, in the Gallery of Maps these scenes from Genesis and Exodus document the early history of biblical sacrifices, many of which were performed by the royal priests Jacob and Moses. They also document the early instances of altar building, that is, the marking of a sacred place for the offering, and they establish the connection between the sacrifice and the tithe.

The difference in emphasis between the biblical scenes of the Gallery of Maps and Raphael's earlier precedent is particularly evident in the remaining fourteen scenes of the Boncompagni corridor. While Raphael had represented the deeds of the Jewish kings Joshua, David, and Solomon, the artists of the Gallery of Maps illustrated the sacrificial rituals performed by the Levites, the only priests from the tribe of Levi and the house of Aaron who, after Moses, could perform sacrifices. These rituals were codified in Leviticus, the third book of the Pentateuch generally dedicated to matters of concern to the priesthood which indeed inspired the depiction of sin offerings and sacrifices to restore legal purity depicted in the Gallery of Maps. This series is particularly significant in relation to the maps of Italy. It places great emphasis on the communal offerings, that is, offerings made by the priest in the name of the entire community in atonement for inadvertent transgression of the law (Leviticus 4), the very communal offerings that the Protestant reformers had abolished and the Council of Trent reaffirmed. Communal offerings are illustrated in a sequence of seven scenes. First the priest brings the sacramental bullock to the entry of the sanctuary and lays his hand upon his head (Fig. 8.10); then he cuts its throat and collects the blood into a basin (Fig. 8.11); sprinkles the blood on the curtain of the sanctuary (Fig. 8.12); smears the blood on the corners of the altar (Fig. 8.13); pours the remaining blood onto the ground at the foot of the altar (Fig. 8.14); burns the fat, kidneys, and lobes of the liver on the altar (Fig. 8.15); and finally burns the carcass and the meat outside the camp (Fig. 8.16). Communal offerings are also made by the Elders. Although their ritual is as elaborate as the priest's, only one part of this complex ritual is represented in the Gallery of Maps: when the Elders gather in front of the altar and meet the priest standing on the opposite side of the altar. This iconography stresses the priest's mediating role, while the inscription below the fresco, "The Elders offer a calf for the sin of the entire community" ("Seniores pro omnis pop[ulo] peccato offeret vitulum d."), points to the communal nature of the sacrifice. After these scenes of commu-

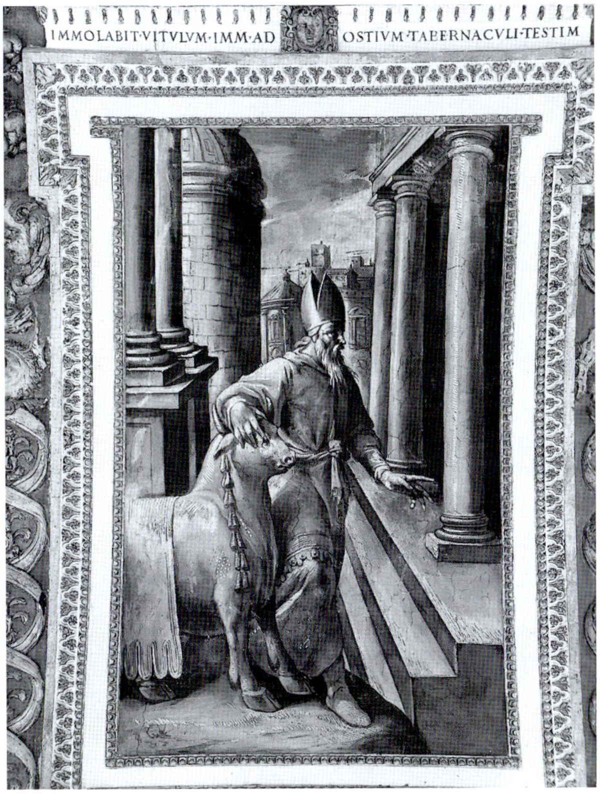

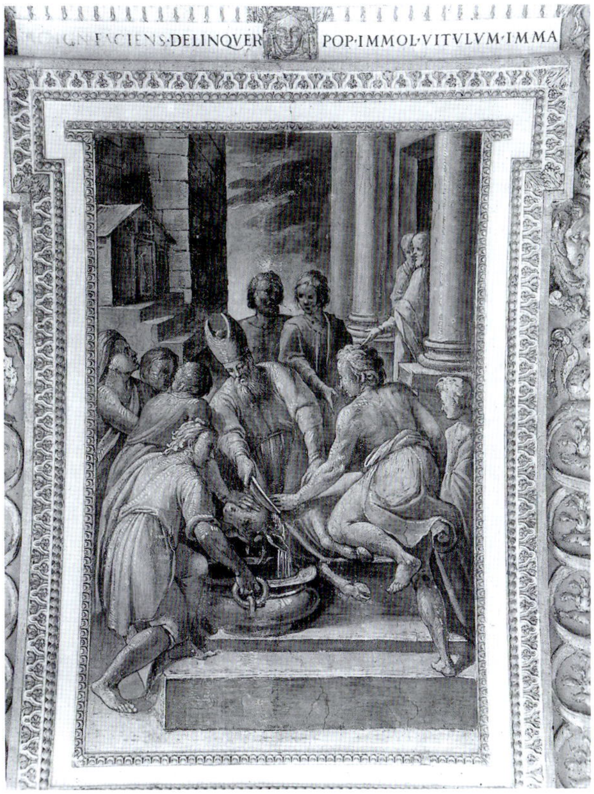

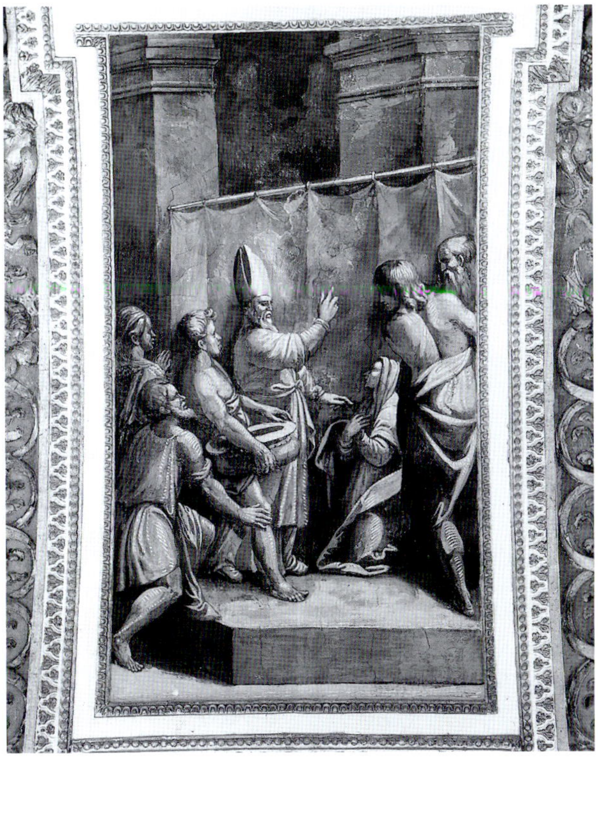

8.10 (*top left*) *The Priest Leading the Bullock to the Temple for the Communal Sacrifice*, ca. 1580, ceiling fresco. Vatican, Gallery of Maps

8.11 (*top right*) *The Priest Cutting the Throat of the Bullock and Collecting its Blood in a Basin*, ca. 1580, ceiling fresco. Vatican, Gallery of Maps

8.12 (*left*) *The Priest Sprinkling Blood on the Curtain of the Sanctuary*, ca. 1580, ceiling fresco. Vatican, Gallery of Maps

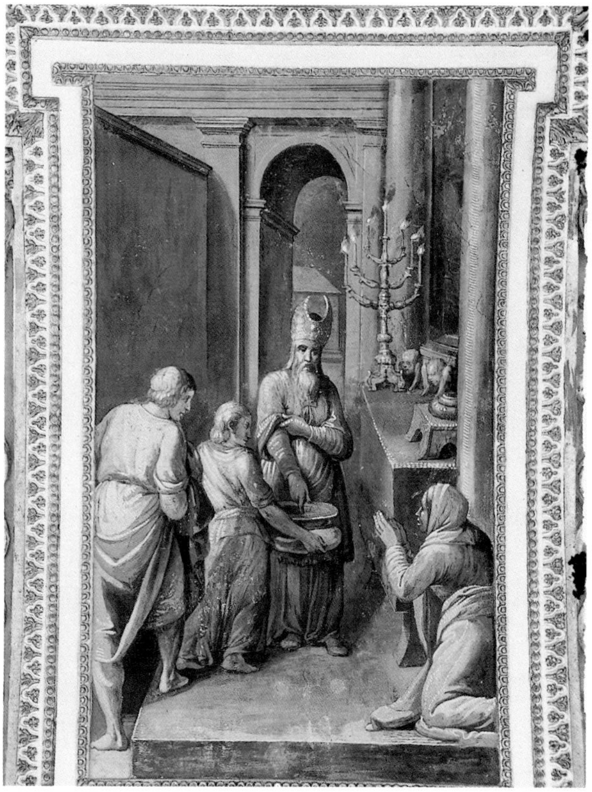

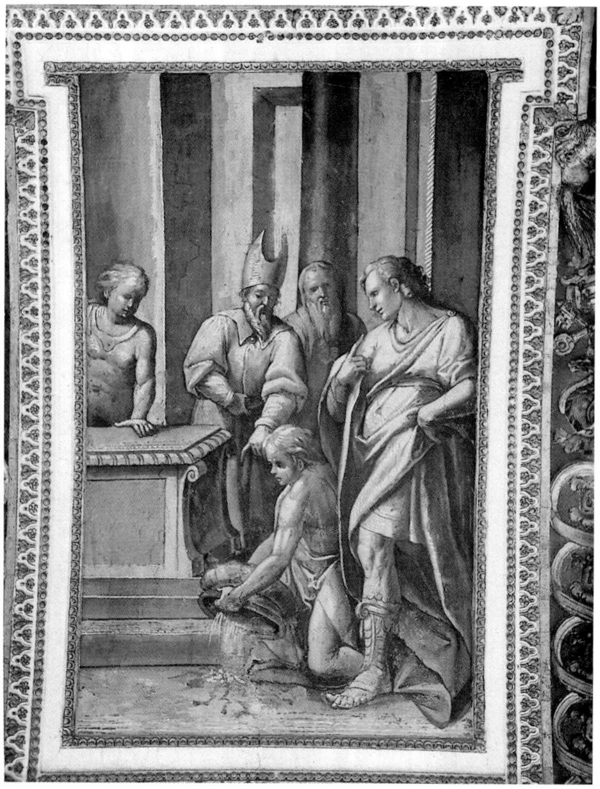

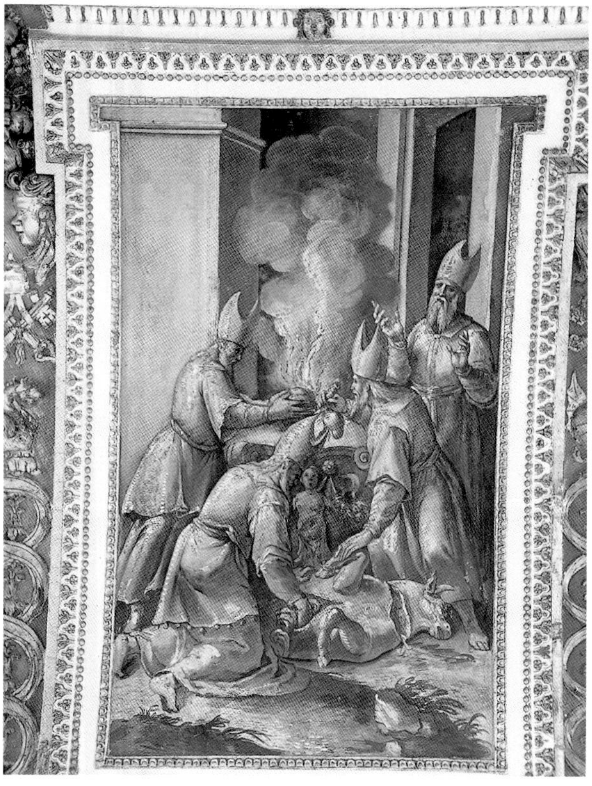

8.13 (*top left*) *The Priest Smearing the Blood on the Corners of the Altar*, ca. 1580, ceiling fresco. Vatican, Gallery of Maps

8.14 (*top right*) *The Priest Pouring the Remaining Blood at the Foot of the Altar*, ca. 1580, ceiling fresco. Vatican, Gallery of Maps

8.15 (*left*) *The Priest Burning the Fat, Kidneys, and Lobes of the Liver on the Altar*, ca. 1580, ceiling fresco. Vatican, Gallery of Maps

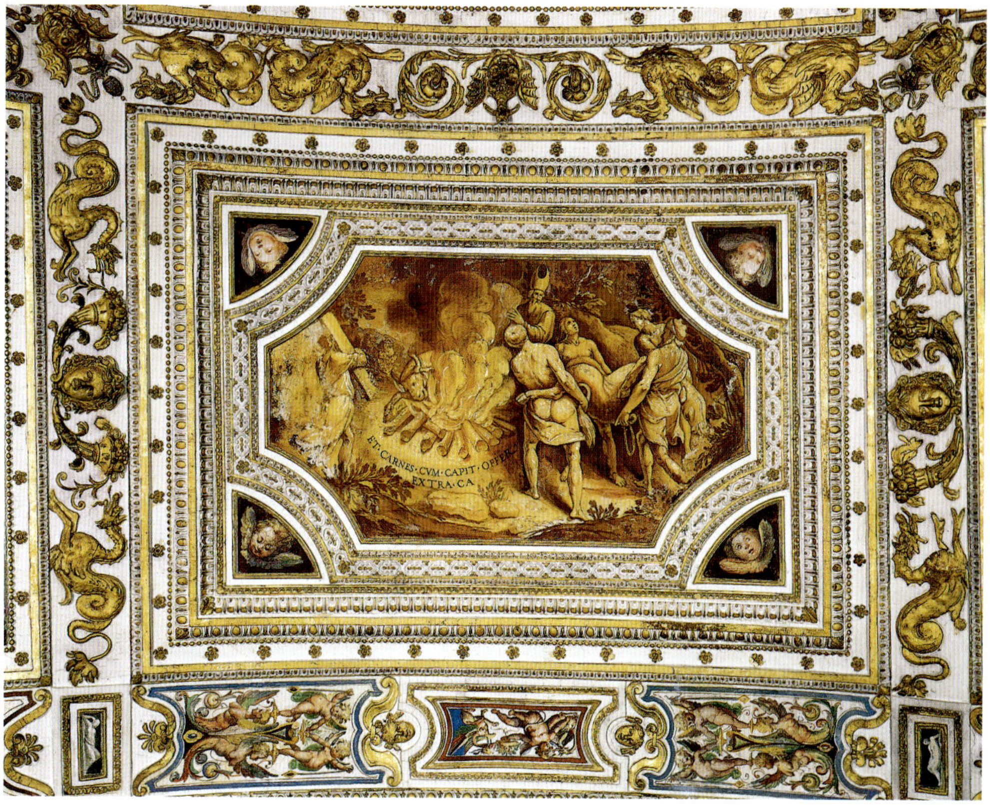

8.16 *The Priest Burning the Carcass and the Meat Outside the Camp*, ca.1580, ceiling fresco. Vatican, Gallery of Maps

nal sacrifices, there follow scenes of sacrifices made by the king and common man in atonement for their own individual sins. Iconographically, these two scenes are very similar in representing the king and the common man greeted by the priest at the sanctuary as they bring their sacrificial animal, a ram and a goat respectively. The individual offerings are followed by scenes of the rites of purification for women after childbirth and menstruation, and for men after ejaculation and leprosy, to restore their purity and thus become once again part of the community.[19]

In this detailed and unusual sequence of sacrificial offerings from Leviticus, the intermediary role of the Levites between the secular and the divine space is stressed through the representation of curtains or walls within the sanctuary. Equally impressive is the systematic distinction, within the firmly set ritual, between the act of thanksgiving and that of atonement. The act of thanksgiving, performed by the offerer, comprised the bringing of the sacrificial animal, the placing of hands on its head, and its slaughter. The act of atonement was performed by the priest and involved the pouring of blood and the destruction of whatever part had to be burned. By showing or implying the separate parts of the ritual performed by the people and by the priest, the biblical sacrifices in the Gallery of Maps emphasize the double nature of the biblical sacrifice as both an act of thanksgiving and of atonement, of praise and impetration.

In the Gallery of Maps, this series of sacrifice from Genesis, Exodus, and Leviticus documents the historical formation of the biblical sacrifice. It implies a temporal and spiritual progression from the south end of the Gallery of Maps to its opposite end. At the south end of the corridor, Adam's sacrifice of praise was the foundation of any other sacrifice, although it was a figurative offering that did not require an altar or an intermediary. The sacrifices made by Cain and Abel, Noah, Abraham, Jacob, and Moses that followed were all literal sacrifices of produce or livestock and were made on an altar appositely built (often in a place where God had spoken), although they were not performed by intermediaries. The sacrifices from Leviticus that concluded the series were the final codification of the Jewish sacrifice involving: a sacrificial gift (produce or livestock), its destruction (through slaying for livestock and pouring for libations), a fixed place for the sacrifice (an altar), a fixed building (the temple), and a qualified person to administer it (the Levite priest).

In Christian theology, since at least the Letter to the Hebrews had been written, Christ's sacrifice on the cross and its literal re-enactment in the Eucharist were regarded as the fulfillment of the offerings of the Old Testament. Like the old sacrifices, which involved both the sacrifice of praise and worship and the sacrifice for sin, the Eucharist was an offering of thanks for Christ's sacrifice as well as a literal sacrifice perpetuating Christ's sacrifice on the cross, an orthodox reading that was reconfirmed in the Roman Catechism of 1570.[20]

Biblical Sacrifices and the Eucharist

As is well known, the sacrificial nature of the Eucharist was one of the major issue of contention between Catholics and Protestants. Like Catholic theologians, Luther recognized the Eucharist as a sacrament, that is as an act of thanksgiving for God's sacrifice on the cross, but, unlike Catholics, he did not regard it as a literal re-enactment of Christ's sacrifice. Unlike the Zwinglians and the Calvinists, Lutherans accepted the idea of the real presence, but rejected the doctrine of the transubstantiation as a modern concept originating from the Fourth Lateran Council. Philipp Melanchthon had pointed out that the fathers of the church had never defined what made a true and real sacrifice, a fact that confirmed the reformer's claim that the mass was a mere sacrament, not a sacrifice. The Council of Trent had decreed in favor of transubstantiation in the Eucharist, reaffirming that it was both a sacrament and sacrifice. Catholic theologians engaged in a lively debate on the definition of the true and real sacrifice, arguing that destruction was an integral part of it. The sacrifice series in the Gallery of Maps offered an interpretation of the biblical sacrifices as consistent with the Catholic view of the Eucharist: they were communal acts of atonement and thanksgiving, performed by a priest in a sacred place. Even more pointedly, by representing graphically the destruction involved in the biblical sacrifices, the scenes of the vault visually declared that destruction was a fundamental act in offering a sacrifice, and thus challenged the Lutheran interpretation of the Eucharist as a sacrament without sacrifice.

In the Gallery of Maps the Catholic interpretation of the biblical sacrifices in relation to the Eucharist was made explicit by an anamorphosis representing a chalice with the Host, once located in the corridor and today lost. Unknown to modern scholars, the Vatican anamorphosis is recorded in the travel memoirs of Jean Tarde, a French cleric visiting Rome in 1593: "We visited . . . the Gregorian gallery, at the end of which is a mirror representing a chalice with the host, reflecting that image from a panel placed beneath it in such a way that the painting is visible only by artifice of perspective."[21] From Tarde's description we learn that the anamorphosis was located at the end of the corridor and must have been high enough to be recomposed in a mirror. This kind of anamorphoses seen from below were Danti's specialty: he described their geometrical construction in his commentary on Giacomo Vignola's *Le due regole della prospettiva pratica* (Rome, 1583), and he himself created anamorphoses, among which was a portrait of Gregory XIII.[22] It seems reasonable to suppose that Danti was the author of the anamorphosis placed in the gallery. By artifice of perspective, this anamorphosis would have revealed the Eucharistic meaning of the sacrifice series and, more generally, would have alerted viewers to the spiritual and ecclesiastical interpretation of the Gallery of Maps.

How the selection of sacrifices from Genesis, Exodus, and Leviticus came about in the vault of the Gallery of Maps is difficult to document, although it is possible that this sequence was familiar to post-Tridentine clergymen. The Letter to the Hebrews, the primary text that interpreted Christ's sacrifice and the Eucharist as the fulfillment of the biblical sacrifices, cited Abel's sacrifice and the deeds of Noah, Abraham, Jacob, and Moses, although not specifically their sacrifices, as examples of faith in divine will. Furthermore, the Letter to the Hebrews ends with a salutation that is particularly suggestive in relation to the Gallery of Maps: "Greetings to you from our friends from Italy" (Hebrews 13:24). This salutation has been variously explained as reference either to the place of origin or the destination of the Letter. In any event, it is in the Letter to the Hebrews that Italy and the biblical sacrifices were first associated. Whether the Letter to the Hebrews was ever on the mind of those who designed the Gallery of Maps is hard to say, although, admittedly, it is tempting and perhaps even obvious. It is worth remembering that the genealogy of the biblical sacrifices by Abel, Noah, Abraham, Jacob, and the Levites implied in the Letter to the Hebrews was used in one of the disputations at the Council of Trent devoted to the sacrifice of the Mass, and might have had some weight in arguing against the Lutheran interpretation of the Eucharist.[23]

But the Letter to the Hebrews also documented that an integral component of the biblical sacrifice was the custom of setting aside the tithe for the sustenance of the Levites and, even more importantly for the post-Tridentine church, that tithing was a recognition of ecclesiastical supremacy. In this respect, the biblical sacrifices in the Gallery of Maps do not simply assert the orthodoxy of the doctrine against the reformers, but they also proclaim papal ecclesiastical supremacy within the Catholic world.

Biblical Sacrifices and Ecclesiastical Supremacy

In the Letter to the Hebrews, Christ was presented as the high priest according to the order of Melchizedek, an order which was different from, and superior to, the Levite order of Aaron. The Letter related how Melchizedek greeted and blessed Abraham upon his victorious return and made an offering of bread and wine to God, which Christian theologians have invariably interpreted as a prototype for the Mass. But, the Letter continued, the superiority of Melchizedek's priestly order was demonstrated by the fact that Abraham voluntarily offered him the tithe of his enemy's spoils. Christ belonged to the order of Melchizedek, a fact commented upon in the Letter to the Hebrews and in the Roman Catechism of 1570. He therefore also had to receive the tithe in recognition of his high standing. This discussion of the tithe was of great significance to the Christian church since the obligation to pay the tithe to provide for church ministers is nowhere to be found in the New Testament. That obligation could be argued only through the exegesis of the Old Testament.

The custom of giving tithes is documented in various passages in the Bible.[24] Abraham voluntarily offered the tithe of his enemy's spoils to Melchizedek, who accepted it on God's behalf (Genesis 14:20). Jacob made a vow to give the tithe of all his possessions to God: "And of all that thou givest me, I will without fail allot a tenth part to thee" (Genesis 28:22). The Book of Deuteronomy related the complex details of the Mosaic Law, which mandated the payment of the tithe to support the priesthood and the Levites as well as to provide for a communal sacred feast that had to be consumed at the sanctuary. The Mosaic Law regulating the tithe was adapted by the Christian church. Initially, the provision for the Christian minister was provided by the spontaneous offerings of the faithful, which took place during the celebration of the Eucharist, generating, by synecdoche, the use of the world *oblatio* (offering) to indicate the entire gathering, including its ritual and liturgy.

The sacramental justification for ecclesiastical revenues to sustain the clergy was as vital as ever in the post-Tridentine church, especially in consideration of their reduction in the overall papal budget. While the papal temporal revenues grew steadily during the Renaissance and after Trent (as did the papal debt which acquired legendary proportions), the ecclesiastical revenues remained stable. Their lack of increase depended on many different factors that varied greatly from place to place. But the problem was so endemic that Gregory XIII made the increment of ecclesiastical revenues one of the decisive policies of his papacy. Economic historians have calculated that Gregory's temporal income amounted to almost three times as much as his ecclesiastical revenues, a trend that continued in the seventeenth century. These ecclesiastical revenues included common services and annuities which, however, must have been limited in the pope's time, since they were absorbed under the heading "spoils" shortly after his pontificate. More substantial were the revenues from the Dataria, which was in charge of concessions of benefices and sales of offices, but its revenues in the period between 1572 and 1576 amounted to only 68,000 gold scudi a year, having been reduced considerably as a result of the Council of Trent. Particularly significant were the resources collected thanks to the special taxation that Pius V had already imposed on the clergy and that Gregory XIII collected with diligence; although these new taxes were directed to the universal clergy, they affected the clergy in Italy to a greater extent. Even the income deriving from monastic foundations, the *commendam*, which Gregory, like other post-Tridentine popes, used extensively to remunerate members of the papal court, was collected mostly from the monasteries in Italy rather than from the foundations of Catholic Europe. The pope also revived the old institution of the spoils, the confiscation of parts of episcopal and archiepiscopal estates by the Apostolic Chamber after their death. Danti too was subjected to the Roman fiscal demands of the spoils, since he was the bishop of Alatri when he died in 1586. His manuscripts were confiscated and, since then, they have disappeared.[25]

The Boncompagni pope also vigorously sustained the battle against the alienation of ecclesiastical property, which his predecessor Pius V had already reinstated. He verified, often personally, the legal status of papal fiefs and did not hesitate to reclaim them for the Apostolic Chamber if tributes had not been paid regularly, or if the fiefs had been acquired improperly by families that did not originally retain rights. In 1580 Gregory XIII issued a papal bull urging nobles to comply with their fiscal duties by the feast of St. Peter and Paul (29 June) and entered into lengthy negotiations with the prominent families who resisted

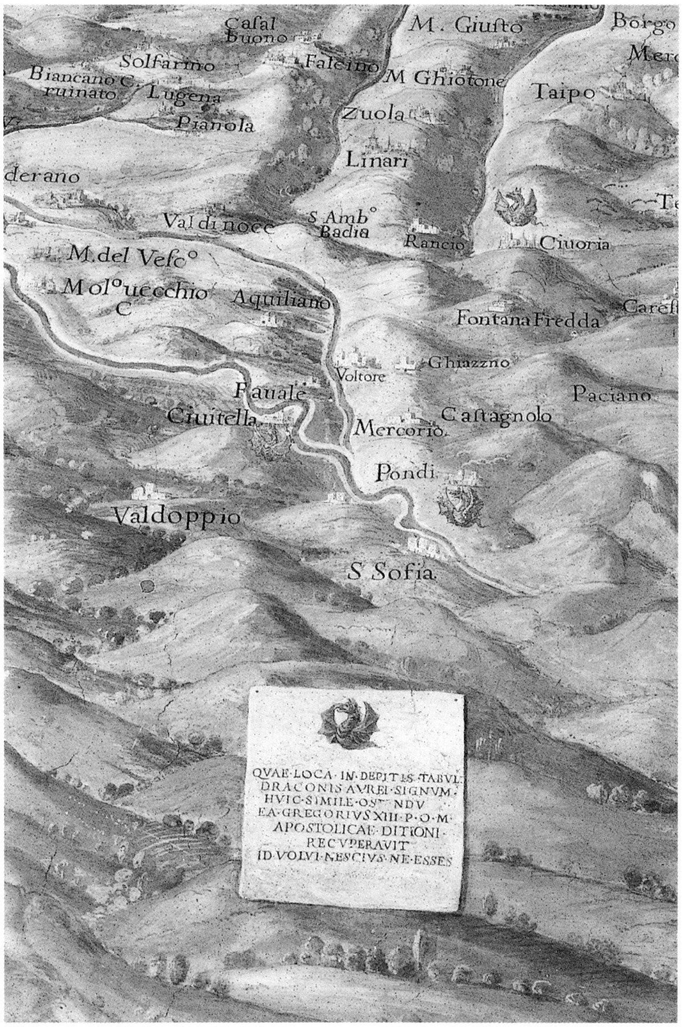

8.17 Detail of fig. 6.16: Cartouche and Gregory XIII's coat of arms

his fiscal demands. The *avvisi di Roma* voiced continuously the protests of cardinals and nobles, especially the controversy in which Orazio Savelli opposed the pope and threatened to become a bandit. Eventually he agreed to return some contested castles to the Apostolic Chamber. The changed juridical status of Savelli's castles and of many others was proudly reported on the maps of the Gallery of Maps. A small dragon, Gregory XIII's heraldic symbol, marked the places recovered for the Apostolic Chamber during his papacy (Fig. 8.17), as a cartouche on the map *Flaminia* explained: "Gregory XIII, pontifex optimus maximus, restored to the Apostolic dominion the places which show, on the painted maps, a sign of the golden dragon similar to this. In regard to this matter I wished that you not be ignorant."[26] Elsewhere, the presence of the heraldic devices of both Paul V and Urban VIII confirms that every new acquisition continued to be recorded on the maps of the Gallery of Maps in the following decades (Fig. 8.18). Also the Congregazione ai Confini, one of the most important congregations of the papal court instituted by Urban VIII in order to improve the administration of the Papal States, often copied portions of the Gregorian maps to solve territorial disputes.[27] The Gallery of Maps clearly reflected the aggressive claims for fiscal support under Gregory XIII and continued to be used as an accurate and constantly updated atlas of ecclesiastical possessions, an atlas that visualized the papal primacy in the peninsula.

In fact, the collection of ecclesiastical revenues, while indeed an ecclesiastical matter, was so intertwined with the politics and diplomacy between the pope and Catholic rulers that it became a measure of the authority the pope enjoyed within Catholic states. Indeed, most of the ecclesiastical revenues derived from the Papal States and Italy, only small amounts came from elsewhere in Catholic Europe. Traditionally, the popes shared their ecclesiastical revenues with the secular authorities, who ruled over the territories where the ecclesiastical properties were located, a practice Gregory XIII maintained.[28] But he did not hesitate to

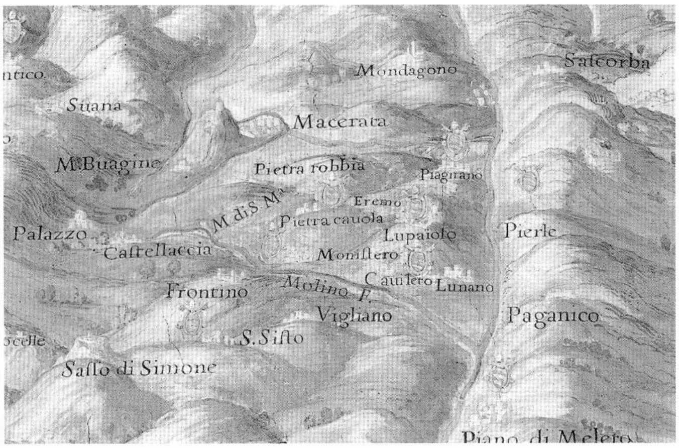

8.18 Detail of fig. 6.16. Paul V's coat of arms

revoke the privileges he had given in moments of political antagonism. When Venice stipulated a separate peace agreement with the Turks in 1573, he refused to renew the payment of the tithe from Venetian ecclesiastical institutions to the Republic, a policy he eventually revised. When Philip II retreated from his promises to attack England in 1578, the pope reminded him that the Spanish privileges on ecclesiastical revenues were meant to fight the heretics. A year later Gregory repeated his threats to stop the renewal of the ecclesiastical tithe since Spain had negotiated a separate peace with the Turks. On another occasion Philip II patiently awaited the confirmation of important ecclesiastical revenues before he appointed his chaplain as a permanent judge in Sicily, a move Gregory XIII deeply resented. Even from this brief summary, it is clear that Gregory's initiatives to increase ecclesiastical revenues often put him in opposition to Catholic rulers, nobles from the Papal States, and the clergy, particularly the Italian monasteries. The need to reassert the sacramental origin of ecclesiastical revenues was thus essential.

Cesare Baronio provided Gregory XIII with a scholarly explanation that connected etymologically the Old Testament sacrifices with the Eucharist and thus suggested indirectly the connection between biblical tithes and post-Tridentine ecclesiastical revenues. In his *Annales ecclesiastici* Baronio discussed at length the ancient names of the Eucharist, especially the names indicating the presence of the *corpus Christi*: *oblatio*, *sacrificium*, *liturgia*, and *missa*. According to Baronio, the Latin word *missa* (mass), which eventually became the most common word to denote the Eucharist, derived etymologically from the Hebrew expression *missat nidvat yadcha* (Deuteronomy 16:10), an expression that the Vulgate translated as *spontanea oblatio* (spontaneous offering). Baronio's explanation reads as follow:

> Nevertheless [it should] not pass unmentioned that given the many names of this sacrifice [of Christ], that which is the more ancient, and among the Latins the most common and more frequent, is *Missa*; this [name] indeed seems to be derived from a Hebrew or Chaldaic expression (as the Christian faith acknowledges). What in fact it reads in our Vulgate as *spontanea oblatio* (Deuteronomy 16), in Hebrew and Chaldaic is named *Missah*, which they [the Jews] offered to God as thanksgiving for the fruits of the earth, which they [the Jews] brought to the temple in memory of their earlier servitude and of their

> liberation. All these things adapt very well to the above-mentioned sacrifice instituted by Christ; even more, our forefathers used this name for some other sacrifice, especially for this reason it seems evident that the spontaneous offering made by our Lord Jesus Christ recalls the earlier ones, of which Isaiah wrote "Oblatus est quia ipse se voluit et Dominus de se ipso" (Isaiah 35); "Postestatem habeo ponendi animam meam et potestatem habeo iterum sumendi" (John 10), and that which is written in Psalm 39[Psalm 40:6–7]: "If thou hadst asked for whole-offering and sin-offering I would have said, 'Here I am.'" Therefore it is believed that in this way the name was translated by the very apostles Peter and Paul from the Hebrew sources into Latin. Jacob, Bishop of Jerusalem, himself an Apostle and brother of Christ, handed this over to us.[29]

According to Baronio, the Eucharist was the fulfillment of the spontaneous offerings of the Jews since *missa* derived from *oblatio*. Indeed the Book of Deuteronomy (16:10), to which Baronio referred in the passage quoted above, regulated the three major offerings the Jews made "in proportion to the blessing that the Lord your God has given you" at Passover, Pentecost, and Suchot. These tithes were to be brought to Jerusalem to subsidize the city, its Temple, and the Levites, and provide for the meal consumed at the temple by the family, the Levites, the poor, and the foreigners.

Baronio's etymology is today generally considered fanciful. Modern philology regards the word *missa* as the original name of the solemn early Christian rite of leave-taking still echoed in the final sentence of the liturgy, *ite missa est*. Later, by synecdoche the whole eucharistic rite was named after one of its parts. However, for the understanding of the Gallery of Maps, particularly the connection between the maps of Italy and the biblical sacrifices, it is Baronio's erroneous etymology of *missa* that is crucial. By suggesting the etymological relationship between the old sacrifices and Christ's sacrifice, Baronio was also implicitly proposing the connection between the biblical tithes and the modern ecclesiastical revenues. As the high priest Melchizedek received the tithe from the Jewish king of Israel, so the pope, who represents Christ, who himself belonged to Melchizedek's order, received the tithe from the Catholic rulers of the world.

In the vault of the Gallery of Maps, among the profusion of images and personifications surrounding the scene *Countess Matilda Donating Her Possessions to the Roman Church*, the connection between the biblical tithes and the more recent donations is made explicit (Fig. 8.19). This fresco, illustrating one of the most significant medieval donations to establish papal temporal and spiritual power, is framed by pertinent personifications: the offering of myrrh, gold, and incense to God. Also represented are the personifications related closely to the giving of tithes: the allegory of tithe itself, the high priest Melichizedek, Joshua, and the rewards that come from tithing, Gratitude and Constant Happiness.[30] The papal claim for the sacramental tithe was universal, but it was in Italy rather than in the Catholic world beyond that the pope collected it. Just like the papal claim for universal ecclesiastical supremacy was achieved more fully in Italy than elsewhere, so the universal collection of ecclesiastical revenues was mainly an Italian affair. The Gallery of Maps mapped the geographical boundaries of the lands where ecclesiastical revenues were largely collected.

* * *

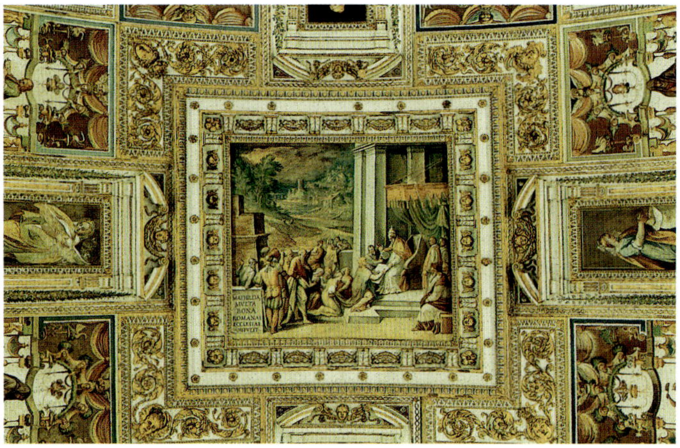

8.19 *Countess Matilda of Canossa Donating her Possessions to the Roman Church*, ca. 1580, ceiling fresco. Vatican, Gallery of Maps

Italy and the Universal Church

Gregory xiii's greater success in the collection of ecclesiastical revenues in Italy rather than in the rest of Catholic Europe was part of his concerted efforts to assert papal authority through the implementation of the decrees of the Council of Trent on the Italian peninsula. Relentlessly he tried to move towards the institutional, diplomatic, and ecclesiastical reforms to assert his authority in ecclesiastical matters within the Catholic world but obtained some successes only in Italy.[31] Indeed the Gallery of Maps not only mapped the boundaries of the largest papal ecclesiastical revenues, but it also celebrated triumphantly the primacy of Italy within the universal church.

In post-Tridentine Europe, to recognize papal authority meant first of all to accept formally the decrees of the Council of Trent and every year to publish the papal bull *In Coena Domini*.[32] Originally issued by Pius v and republished by Gregory xiii in 1580, the bull established ecclesiastical immunity and regulated the relationship between ecclesiastical and secular authorities in the Catholic world by listing the causes of papal excommunication and absolution. Although Gregory mandated that the bull be read every Maundy Thursday in the entire Catholic world, it was read only in the Papal States and in Italy and even there such an ecclesiastical matter could turn into political tension. In 1580, Michel de Montaigne was present at the solemn public reading at St. Peter's, in the presence of Gregory xiii himself; in his journal he recorded the angry reaction of the Florentine ambassador as he heard that some territories contested between the Medici and the pope had been added to the traditional list of possessions, over which the pope claimed his indirect power.

Catholic rulers had also to facilitate the implementation of the Tridentine decrees through appropriate reforms. They were expected to found seminaries for the education of the clergy, assist apostolic visitors, allow ecclesiastical tribunals to operate within their own territory, respect their decisions, and accept the papal nominations for bishops' and archbishops' seats. Ideally, Catholic Europe should have embraced the ecclesiastical reforms that the papacy requested. In fact the implementation of the Tridentine decrees was so intimately related to the politics and administration of the Catholic states that post-Tridentine popes, including Gregory XIII, had to work hard to convince even the most fervent defenders of the Catholic faith to accept the papal ecclesiastical authority within their own territories.[33]

In Catholic Spain, the bull *In Coena Domini* was never published. Philip II formally accepted the decrees of Trent, but instructed his viceroys and governors that their application could not infringe on royal authority and privilege. He also kept the right of nomination to the episcopal seats in Spain, Sicily, and Sardinia; he did not facilitate the *ad limina* report to Rome that the pope required of all bishops, and sent a royal representative to the provincial synods. He collected ecclesiastical revenues that were granted from Rome for his commitment against the Turks and the infidels, but never found sufficient funds for the creation of seminaries for the education of the clergy, another of the Tridentine decrees. The Spanish king maintained his own independent inquisition in Spain as well as in the Spanish dominions overseas, in Sicily, and Sardinia. Additionally, the Spanish crown never tolerated the minimal papal interference in the nominations in America, from which Rome always remained excluded. Obviously, Spain was deeply Catholic, but the defense of Catholicism and ecclesiastical reforms were royal prerogatives not to be delegated to others, even the pope.

In France, too, the bull *In Coena Domini* was never published and the decrees of Trent never formally accepted. Moreover, the French king did not relinquish the right to nominate bishops in his territory, since he considered it a hereditary right the pope had granted to his predecessors. Provincial synods were held in France between 1581 and 1585, but their influence was extremely limited. Apostolic visitors were truly unwelcome, and the French king even threatened to nominate a patriarch of France who would have the power to grant benefices. Although this step was never taken, papal authority was truly limited on French soil, where no inquisition, either French or Roman, ever operated. Other states of Catholic Europe also did not have an inquisition, such as the Holy Roman Empire, the lands of the Habsburgs, and Poland, where bishops were nominated by the secular authorities.

But ecclesiastical authority fared much better in Italy and the Papal States. The universal church that Gregory XIII was handed down by his predecessors was Italian at the most basic level. Generally the popes, the ambassadors, the papal administrators were Italian, as well as the majority of the members of the papal court and of the college of cardinals. The distinctively Italian orientation of the early modern papacy was determined by the papal need to resist as much as possible pressure from major powers, mainly Spain and France. Even though Gregory XIII nominated an unprecedented number of non-Italian cardinals, the predominance of Italian personnel in the papal hierarchy remained unchanged, a fact that reflected and made possible the primacy of Italy within the universal church. In the Papal States, where the pope ruled over both temporal and ecclesiastical matters, ecclesiastical reforms were introduced more smoothly than in other Italian states. There the tension that existed elsewhere between temporal and ecclesiastical authorities tended to recede because of the ecclesiastical rank of both administrators and ecclesiastical leaders. Occasionally, friction arose between bishops, who felt challenged in their ecclesiastical autonomy, and apostolic legates, who represented the centralized Roman administration. Indeed, the daily administration of power generated antagonistic feeling even with the bishops the pope most admired, Paleotti and Borromeo.[34]

In the rest of the peninsula, Gregory XIII and later popes acquired ecclesiastical privileges in exchange for favors and benefices. They convinced the rulers of Italy to accept formally the decrees of the Council of Trent and publish *In Coena Domini* every Maundy Thursday. The popes also succeeded in keeping the nomination rights for bishops' and archbishops' seats almost everywhere in Italy;[35] in having their apostolic visitors travel within the Italian states;[36] in opening seminaries for the education of the clergy in many dioceses; and in establishing ecclesiastical tribunals almost everywhere in the peninsula, although with local variations.[37] Even the king of Spain had to acknowledge the different reality of the peninsula and agreed to let the Roman Inquisition operate in his dominion in northern Italy.

If in peninsular Italy the pope could negotiate his way on ecclesiastical matters, in the islands around Italy the struggle with secular authorities was greater. The Roman church unrelentingly tried to impose its tribunals in the Spanish-dominated islands of Sicily and Sardinia, on Malta, on the Genoese island of Corsica, and Venetian Corfu. But the Spanish king regarded Sicily and Sardinia as an integral part of Spain and therefore chose their bishops and imposed his own inquisition. Gregory XIII also tried to interfere in the elections of the Order of St. John, which controlled Malta, and kept its general in Rome for many weeks, entertaining him with a visit to the Gallery of Maps. Indeed, the depiction of the Italian islands grouped together at the southern end of the Gallery of Maps suddenly acquired a more profound meaning. Contested by Spain, Venice, and Genoa, these islands were included not only as a way to conform to the Renaissance tradition of the *isolari* but also as a way to claim them under the ecclesiastical, not temporal, power of the papal court.

* * *

In the Gallery of Maps Gregory gave cartographic visualization to the primacy of Italy within the universal church. On the maps of Italy the Boncompagni pope found the minute cartographic representation of the boundaries of the Western church that had more readily submitted to his ecclesiastical authority. In the cartography of this ecclesiastical unit, the mapping of episcopal and archiepiscopal seats, feudal properties, and recent acquisitions to the Apostolic Chamber was as defining an element as it was the representation of local history and landscape. The pope did not claim political authority over the lands of Italy, but still he presided over the heroic deeds of saints and earlier popes, over the religious foundations and their revenues.

Papal primacy had distant roots in the biblical sacrifices that established the doctrinal foundation of the Eucharist and the right to collect ecclesiastical revenues. In Gregory XIII's view, the biblical recognition of papal supremacy was confirmed in late antiquity, when Constantine donated Italy to the papacy, but it had to be adapted to the new reality of the post-Tridentine universal church. In fact, although politically the post-Tridentine papacy was losing ground against the states, the universal church was acquiring an unprecedented geographical extension while the papal spiritual influence had the potential of becoming global. To assert papal supremacy on such an enlarged and varied church demanded a whole new set of actions and initiatives. While the pope oversaw his church of Italy with zeal, as he literally marched on Italy's Apennine spine, he also redesigned the papal mission towards the universal church outside of Italy and overseas.

9

The Mapping of the Universal Church

Since the Holy Spirit descended on the apostles on the day of Pentecost, the mission of the church had been to include people from around the world into the Christian community. Adapting the classical notion of universality to fit the Christian world, the universality of the Christian empire had acquired an unprecedented geographical dimension after the Portuguese and Spanish voyages, and the establishment of their colonial empires in the sixteenth century.[1] Franciscan, Dominican, and Jesuit missionaries traveled with merchants and soldiers on the trading routes to Africa, China, Japan, and the Americas, establishing their missions for the conversion of the people in the new dominions. The geographical expansion of the universal church, however, brought with it new challenges, most notably the exposure to primitive forms of religiosity and an unprecedented fragmentation of the Christian community. Protestants had spread their reformed faith and liturgy in large parts of Europe. The Turks dominated Greece and Constantinople, while the doctrine and liturgy differed profoundly between the Eastern and Western church. The missionaries in the distant lands had to adapt Catholic doctrine and liturgy to local customs and beliefs. The search for appropriate solutions to face the challenges of world evangelization was long and tortuous, but Gregory XIII, who had chosen to receive his papal crown on the day of Pentecost in 1572, made it the focus of his papacy. Going beyond the Tridentine decrees, which pertained primarily to Catholic Europe, he established a new model of operation that redesigned the papal mission *urbis et orbis.* His commitment to the world beyond Catholic Europe and to the distant lands in Asia and the Americas complemented his political goals to assert his personal power in Bologna, record and increase the financial resources of his own dominion, and celebrate the papal primacy in Italy – all policies that he characteristically expressed through major mapping projects. The universal church that Gregory envisioned was an abstract notion, an entity that was still taking shape and that was not yet under the doctrinal and liturgical authority of the Roman church. Once again, the pope resorted to cartography to visualize the boundaries of the universal church; while through astronomy, the other middle science he had depicted in his Sala Bologna, he aspired to unify the Christian empire, which was, after all, an empire without territoriality.

Gregory XIII's Universal Church

The universal church handed down to Gregory XIII by his predecessors was struggling to achieve the doctrinal and liturgical uniformity that the Council of Trent had recommended as essential for the spiritual renewal of the Catholic world. Some important steps had already been taken, especially in the revision of Catholic doctrine and liturgy that the Council of Trent had entrusted to the pope, but much more remained to be done to implement the Council's recommendations.[2]

Pius IV had worked incessantly to bring about conformity in rituals, festivals, and prayers across the Catholic world. He set up a committee for the revision of the foundational texts that explained the doctrine and regulated the relationship between the Christian community on earth and in the heavens: the very relationship that Protestant reformers had negated. The Roman Catechism, the Roman Breviary, and the Missal were published in close succession in 1566, 1568, and 1570. The Roman Catechism restated the doctrine in simple terms to educate the general public and impart lessons of good behavior and ways to correct bad ones. Written in Latin and originally meant for the education of the clergy, it was translated into the vernacular languages and became the official text for popular instruction.[3] The Revised Breviary listed the liturgical texts that had to be read during the entire year, dividing them day by day and hour by hour. It also included the list of saints, who were organized hierarchically, starting with the virgins and culminating with the apostles, and guided the faithful on the proper way to address their intermediaries in the heavens.[4] The Revised Breviary also abolished the very idea of a private office, since every Mass benefits the entire Christian community not only those who actually take part in it, a notion Protestant reformers contended but that Gregory XIII reaffirmed by having the communal biblical sacrifices depicted in his Gallery of Maps. The Missal, which included the liturgy specifically dedicated to the Mass, reaffirmed the sacrificial nature of the Mass, to which the biblical sacrifices of the Gallery of Maps also referred.

The post-Tridentine popes were able to accomplish expeditiously the revision of the doctrine and the liturgy, but they had a much harder task in convincing Catholic rulers to facilitate the implementation of the Tridentine decrees within their own states. Indeed, the primacy of Italy that the Gallery of Maps celebrated could be seen as directed at the Catholic rulers outside Italy who resisted papal reforms. Furthermore, the Tridentine reforms, which guided, at least in principle, the relationship between the pope and Catholic rulers, were insufficient in dealing with other problems faced by the papacy, primarily the missionary work to regain Protestant Europe for Catholicism and to convert the people of Africa, Asia, and America. In 1537, at the opening of the Council of Trent, Paul III wrote an official document on evangelization in America and the Dominican Marco Laureto referred to the distant lands in his address to the Council in 1546, but in fundamental ways the extra-European world, its people, new converts, and newly founded dioceses remained far from the concerns of the Council. Going beyond the Tridentine concerns and building on his predecessors' sporadic initiatives, Gregory XIII made the assertion of the papal primacy on the earth's globe a tenet of his papacy.[5]

Gregory was an attentive listener to those who suggested effective ways to propagate the faith around the globe, and in 1577 he gave audience to Jean de Vandeville, a professor of law at the University of Louvain, who presented him with his bulky manuscript on the institution of a new congregation appositely dedicated to world evangelization.[6] Contemporary reports relate how the Boncompagni pope was moved to tears as he heard Vandeville exposing his idea for a congregation divided into five geographical areas, each area characterized by a different religious problem: Scandinavia and England, central Europe and Russia, the eastern Mediterranean, Greece and Turkey, and the East and West Indies. In Vandeville's view, the new congregation had to enlist the entire Catholic community in the accomplishment of world evangelization. Merchants and sailors were expected to carry the Roman Catechism translated into various languages to remote lands and keep in contact with local Catholics. The religious orders had to commit their best members to missionary work, send them to Rome for appropriate training in the national colleges, and then dispatch them to the missions abroad. The Franciscans were expected to help in the Holy Land and Jerusalem, cardinals and bishops experienced in the Eastern rite of the Greek church had to assist in its reform to make it conform to Catholic liturgy, and papal ambassadors had to deal prevalently with ecclesiastical matters. The church hierarchy, from the pope to the parish priests, had to be fully dedicated to the salvation of souls out of pure zeal, but the people, recognizing the commitment of their pastor for their salvation, would pay the tithe voluntarily, a truly suggestive remark in relation to the Gallery of Maps. Decades later Vandeville's proposal was used as a guideline for the creation of the Congregation de Propaganda Fide established by Gregory xv in 1622 for world evangelization, but in many respects Gregory xiii had already made Vandeville's vision his own.

The reconquest of Protestant Europe was one of Gregory xiii's leading policies. His optimism was unfailing, even though disappointment was more frequent than success. His greatest achievement was in consolidating and expanding the national colleges in Rome to train missionaries in the fight against heresy in England, Germany, and Hungary.[7] He contributed papal funds to the foundation of new colleges and donated substantial revenues and a new building, still named after the Boncompagni pope, to the Jesuit Roman College, which was on its way to become the leading institution for the education of missionaries and secular elite. For a brief moment he thought he might win the Kingdom of Sweden back to Catholicism, but he had to abandon this when the king made demands of a Protestant nature. His hopes to regain Muscovy to the Catholic faith were reinforced by the success of the political mission of his emissary, the Jesuit Antonio Possevino, in negotiating the peace between Muscovy and Poland, although it soon became apparent that Ivan iv had no intention to submit to papal ecclesiastical authority. He truly believed that the Massacre of St. Bartholomew had eradicated heresy from France and triumphantly represented the events in the Sala Regia, although the French king never expelled the Calvinists from his state. He gave generous benefices to the king of Spain in exchange for his support in the fight against Protestants in England, but he was unable to stop Philip ii when he put aside the defense of Catholicism in the island and stipulated a commercial agreement with the queen of England. He sacrificed the Italianization of the college of cardinals to the eradication of heresy, electing a large number of European cardinals who had distinguished themselves in the fight against heresy. His foreign policy was dictated obsessively by the desire to reassemble the Holy League against the Turks, a desire repeatedly frustrated by the separate agreement that Venice and Spain stipulated with Constantinople in 1573 and 1582 respectively.

Relentlessly, Gregory xiii worked toward the reconciliation of the Latin and Greek church. In 1573, he founded the Greek Congregation with the aim of assisting the spirituality of Greek

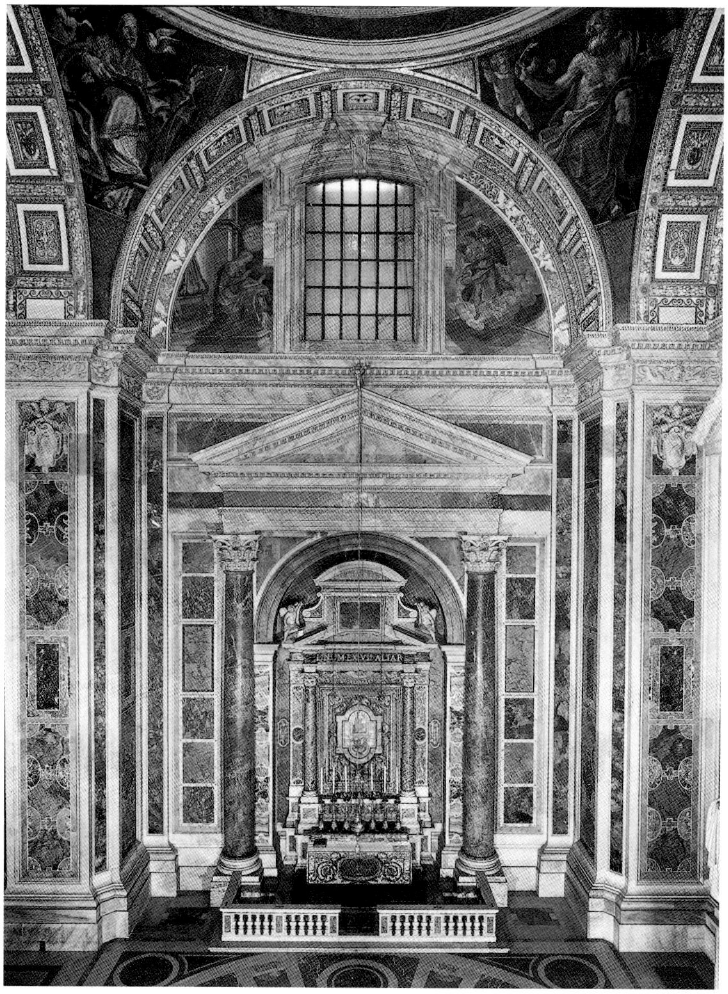

9.1 Gregorian Chapel. Vatican, St. Peter's

convents in Italy, in fact hoping to bring the practisers of the Greek rite under the Roman liturgy. Three years later he instituted the Greek College at Sant'Anastasio for the training of the clergy of the Eastern rite, to which he added later the Maronite and the Armenian colleges (1584). He venerated the Greek fathers of the church and to one of them, Gregory of Nazianzus, dedicated the new papal chapel in St. Peter's, appropriately named the Gregorian Chapel so as to conflate the reference to the Greek saint and the Boncompagni pope (Fig. 9.1). Covered from top to bottom with paintings, mosaics, polychrome marbles, antique columns, and sculptural reliefs, the Gregorian Chapel had been conceived as a monument to the common roots of the Latin and Greek liturgy, roots that Gregory XIII hoped to reunite.[8] To further stress the importance he attached to the unity of the Christian rite, the pope had Gregory of Nazianzus's body moved from the church in the convent of Santa Maria in Campo Marzio to the Gregorian Chapel in a grandiose procession attended by cardinals, aristocrats, and lay people that moved through the streets of Rome on 11 June 1580. He also enlisted eminent figures from his court to mark the event. Cesare Baronio wrote the definitive life of the saint; Francesco Panigarola, one of Gregory XIII's favorite preachers at the papal court, addressed a large crowd in Italian in the new chapel; Cardinal Santori, the protector and founder of the Greek Congregation, officiated at the first Mass in the new chapel; and Egnazio Danti had the procession painted for posterity in ten frescos in the Terza Loggia. Gregory's hopes for a closer relation with the Eastern church, or at least with the dissident groups within it, were further enhanced by the unexpected arrival of Ignatius Na'mattalah, Patriarch of Antioch, at the papal court in 1577 or 1578. Although Gregory included the patriarch in his committee for the reform of the calendar, ultimately the Eastern Church never accepted the Roman liturgy or the Gregorian calendar.

During Gregory XIII's papacy, the conversion of the people in Africa, China, Japan, and the Americas became a central concern of the papal court. Unable to establish ecclesiastical tribunals or nominate bishops in the East and West Indies, the pope enlisted the help of old and new missionary orders in the conversions of these lands. He supported the Augustinian and Franciscan missions in the Philippines and created the diocese of Manila in 1579, having instituted the diocese of Macao in charge of China and Japan three years before. He greatly favored the Jesuits in their missions in the East and West Indies and, in 1585, ignoring the protests from other missionary orders, gave them the unique right of evangelization in Japan and China as well as financial support for their seminaries in those lands. He tried to have closer ties with the Copts in Ethiopia and wrote to Malach Sagad, King of Ethiopia, asking him to convert to Catholicism and fight against the Turks (both hopes were disillusioned). Conscious that world evangelization demanded appropriate books, Gregory XIII tried to establish a papal press that would print the Bible and the Roman liturgical literature in every possible language, a project eventually accomplished by his successor Sixtus V. He did support, however, the Oriental press established by Cardinal Ferdinando de' Medici in 1584, which printed the Bible in Arabic. At the end of his pontificate his global ecumenical mission seemed to bring some rewards. A delegation of newly converted Japanese youths accompanied by their Jesuit tutors arrived in Europe and, after touring the courts of Italy, culminated their journey with a papal audience in the Sala Regia.[9]

The post-Tridentine papacy was losing political ground in world affairs, but the universal church over which it presided was extending beyond expectations. In Gregory XIII's global view, no corner of the world was beyond the concerns of the post-Tridentine papacy and, one day, the universal church could indeed cover the earth's globe. Such a global notion of the universal church, such a hope for papal supremacy, indeed such an encompassing vision of the papal mission, could find its concrete embodiment only in the realm of cartographic representation.

The Terza Loggia

Gregory XIII had his global vision of the papal mission depicted in the maps of the world in the Terza Loggia, the third storey of the Loggias of San Damaso. Pius IV had started the map cycle in the west wing of the palace in the early 1560s, setting a model in the genre of map murals, which inspired Cosimo I's Guardaroba Nuova and later Cardinal Farnese's Sala

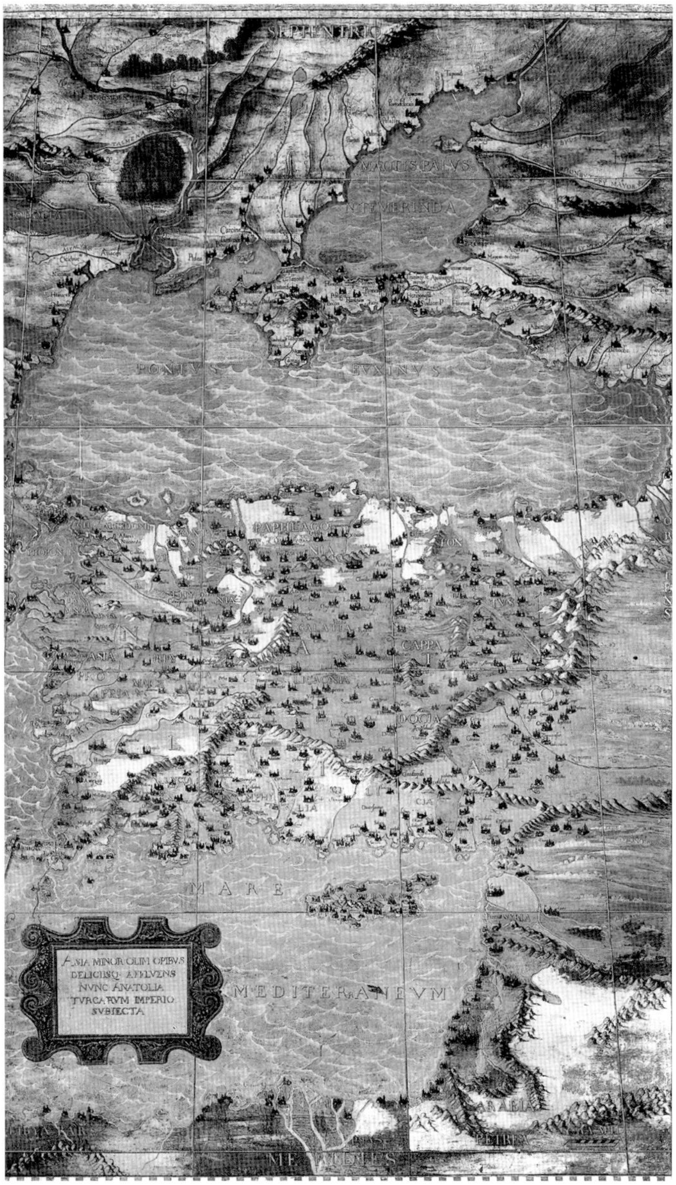

9.2 *Asia Minor* (Map of Asia Minor), 1560s, fresco. Vatican, Terza Loggia

della Cosmografia at Caprarola. The French cartographer Etienne Du Perac prepared the cartoons for thirteen maps of Europe, and the painter Giovanni Antonio Vanosino painted them in frescos (Fig. 9.2), but the map cycle was unfinished at the death of the Medici pope. Pius v did not touch it at all, but Gregory xiii, who walked daily in the Terza Loggia since his apartment and the Sala Bologna opened onto it, decided to complete it around 1580. He asked Egnazio Danti, who was finishing his work in the Gallery of Maps, to design twelve maps of Africa, Asia, and America for the north wing, and, in fulfillment of Miniato Pitti's wish of many years before, to repaint Pius iv's maps in the adjacent wing, which were already ruined (Fig. 9.3).[10] Gregory must have attached some significance to this cartographic loggia: close to his Sala Bologna and thematically connected to the Gallery of Maps, the Terza Loggia epitomized his aspirations for the universal church.

Like Cosimo i's Florentine maps, the maps of the Terza Loggia were meant to represent the modern world, which was organized according to Ptolemy's order. Du Perac used famous printed maps as his cartographic sources,[11] while Danti adopted the same sources that he had previously used for Cosimo i's Guardaroba Nuova, including Portuguese sources from the Medici collection, copies of which he must have kept after his departure from Florence.[12] As in the Gallery of Maps, each map was accompanied by views and plans of the important cities, although the city views that we see today correspond only in part to the Renaissance originals.[13] Among Danti's maps, his world map divided into two hemispheres is particularly interesting, and Danti himself regarded it as his greatest cartographic achievement (Figs. 9.4 and 9.5). He proudly described it to Ortelius as, "a world map twice as large as the one made by Your Lordship."[14] This world map is indeed an original compilation of verbal and visual sources, including Baroncelli's travel reports, which he had already used in Florence. It complements the Florentine maps Danti never finished and his monumental globe for the Guardaroba Nuova, which is illegible today.

Unlike Cosimo I's maps of the world, which celebrated the intellectual aspirations of a secular ruler, the maps of the Terza Loggia were made for the official residence of the Catholic spiritual leader and, as might be expected, would reflect the papal preoccupations for religious supremacy. Indeed, in the legend of the map of Greece it is lamented that the Turkish dominion limits papal supremacy, although no mention appears of the Christians practicing the Greek rite. Italy is praised for the quality of its doctrine and the Spanish monarchy commended for its spirituality, but no mention occurs of the religious divisions in France, Poland, and the Netherlands, of the Protestant presence in England and Germany, or of the missions in distant lands, not even the contribution of Catholic missionaries to the mapping of the modern world.

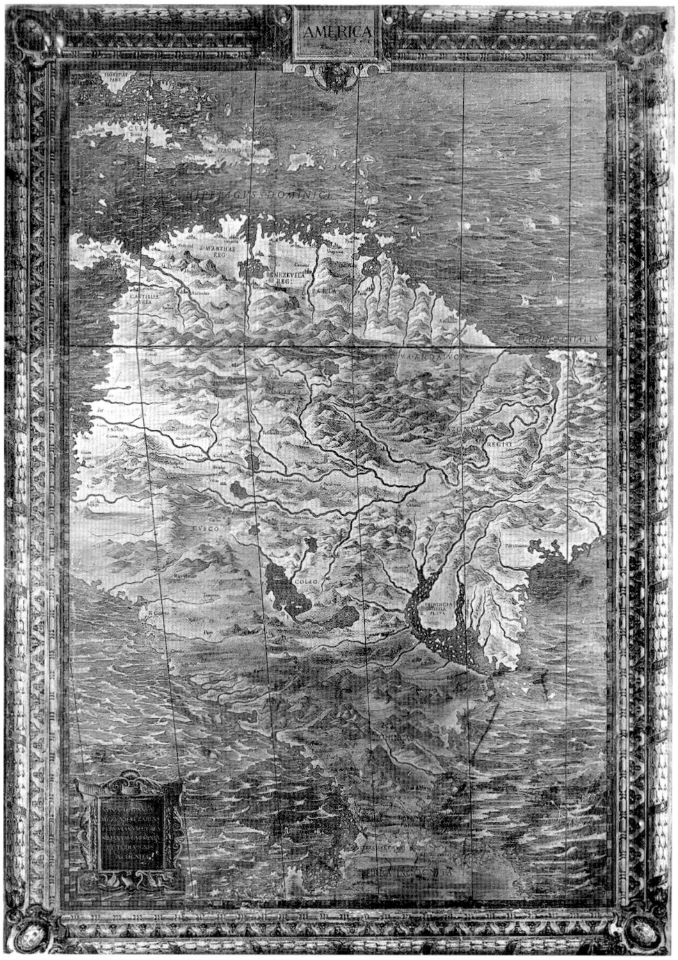

9.3 Egnazio Danti, Map of South America, 1580s, fresco. Vatican, Terza Loggia

The maps were accompanied by images painted above them, which were extremely damaged already in the early eighteenth century, when Agostino Taja wrote the first exhaustive description of the Vatican Palace. In fact, Taja could decipher only two biblical inscriptions in the wing painted under Pius IV: "Thou didst fix a boundary [to the waters] which they might not pass" (Psalm 104: 9), and: "His rule shall extend from sea to sea" (Zechariah 9: 10).[15] Pertinently, both inscriptions refer to the geographical boundaries of the Christian empire, which, it is safe to assume, was meant to apply to the regions of the world mapped below. The maps of Gregory XIII's wing, on the other hand, were completed with a frieze illustrating the procession of the translation of Gregory of Nazianzus's body to the Gregorian Chapel in St. Peter's. This frieze, divided into ten large scenes, offers a stunning view of sixteenth-century Rome and was painted by Matthijs Bril and Antonio Tempesta, two artists who had worked together in the Gallery of Maps. Danti adjusted the frieze to the space of the loggia; Bril painted the urban views; and Tempesta depicted the lively procession (Fig. 9.6).[16] In the background of Bril's magnificent views, the procession moves from the Campo Marzio through the streets of the eternal city, which had been profusely adorned with ephemeral arches, hangings, carpets, tapestries, and religious paintings. The large crowd is depicted as

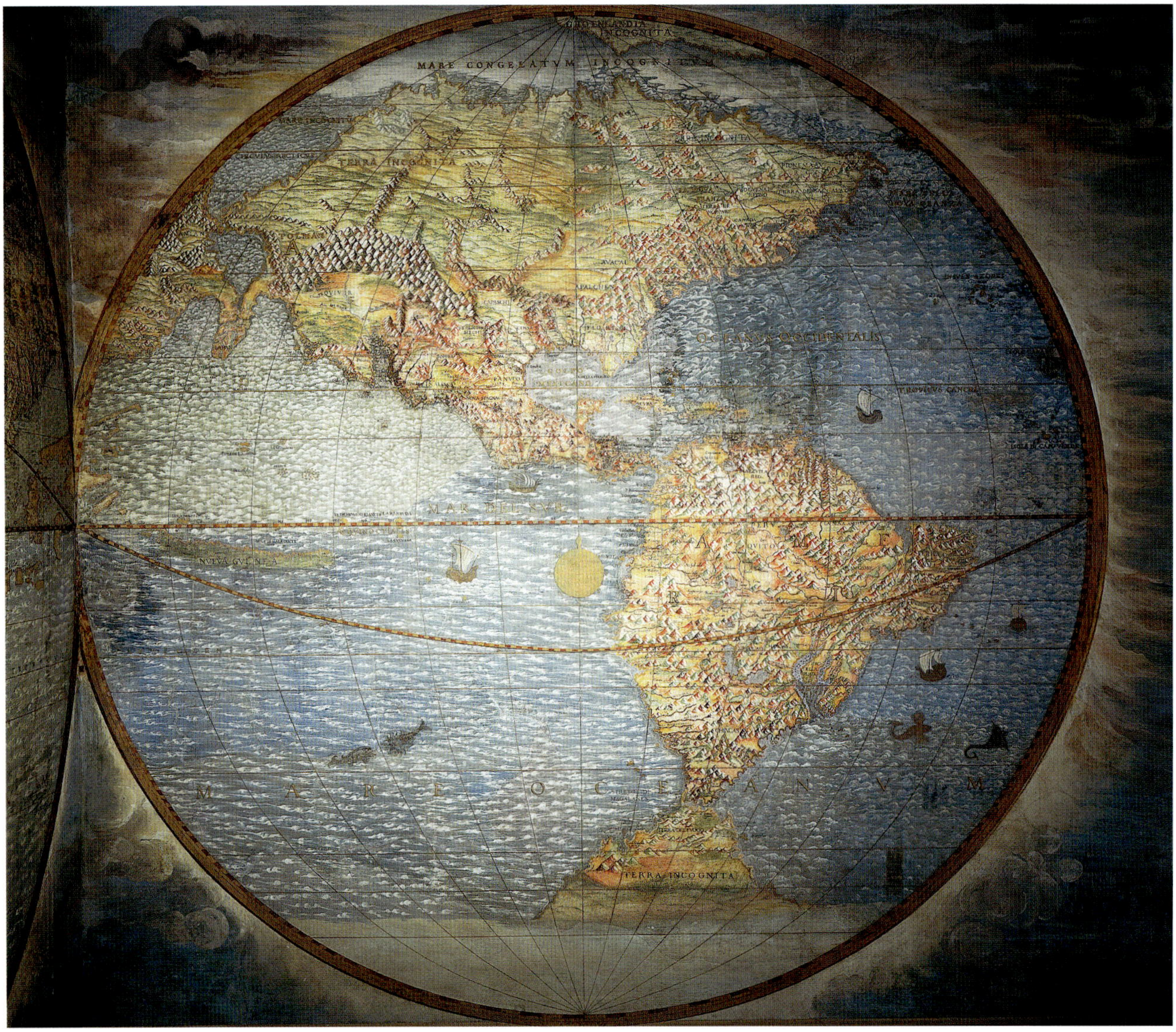

9.4 Egnazio Danti, Map of the Western Hemisphere, 1583, fresco. Vatican, Terza Loggia

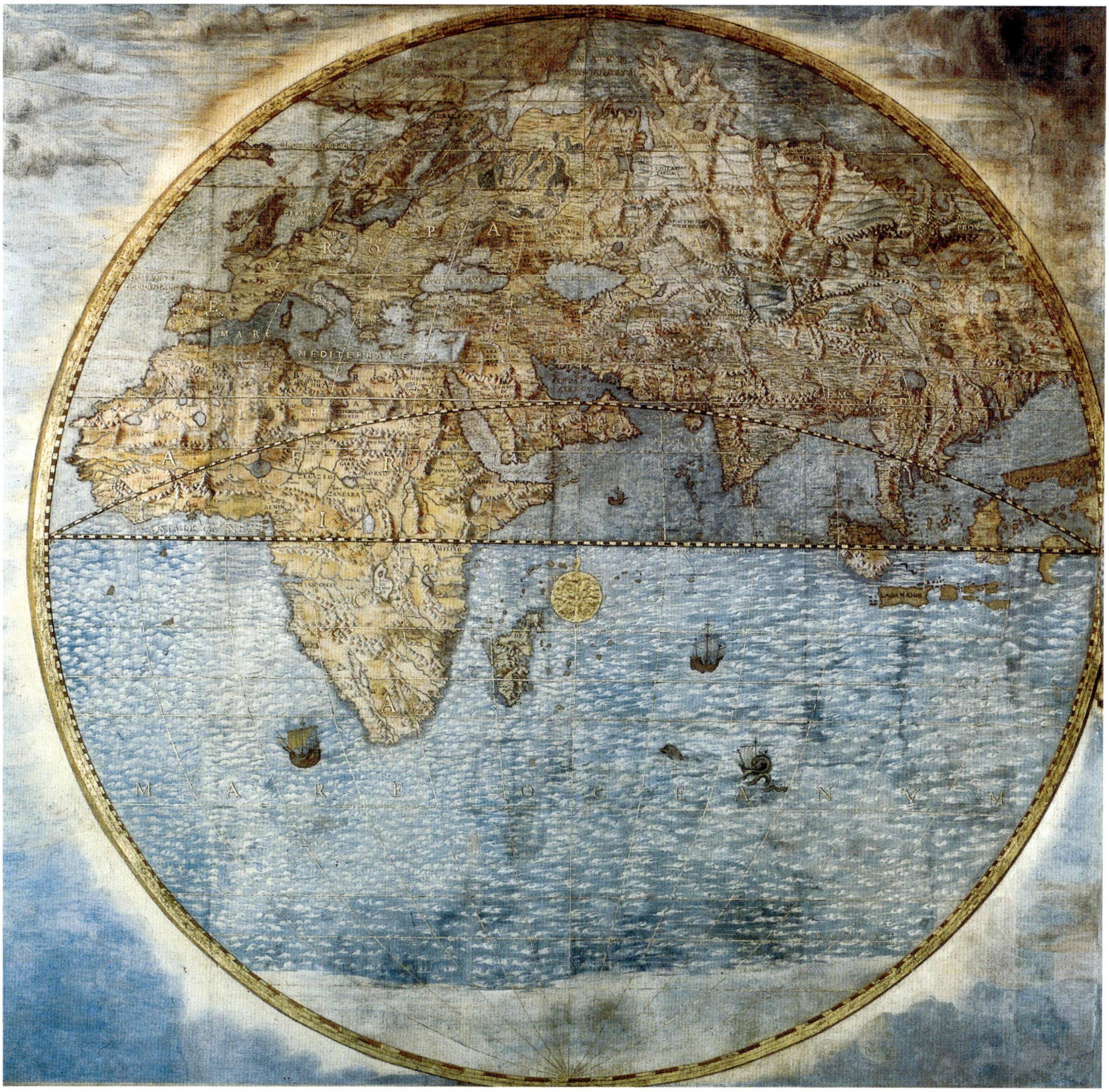

9.5 Egnazio Danti, Map of the Eastern Hemisphere, 1583, fresco. Vatican, Terza Loggia

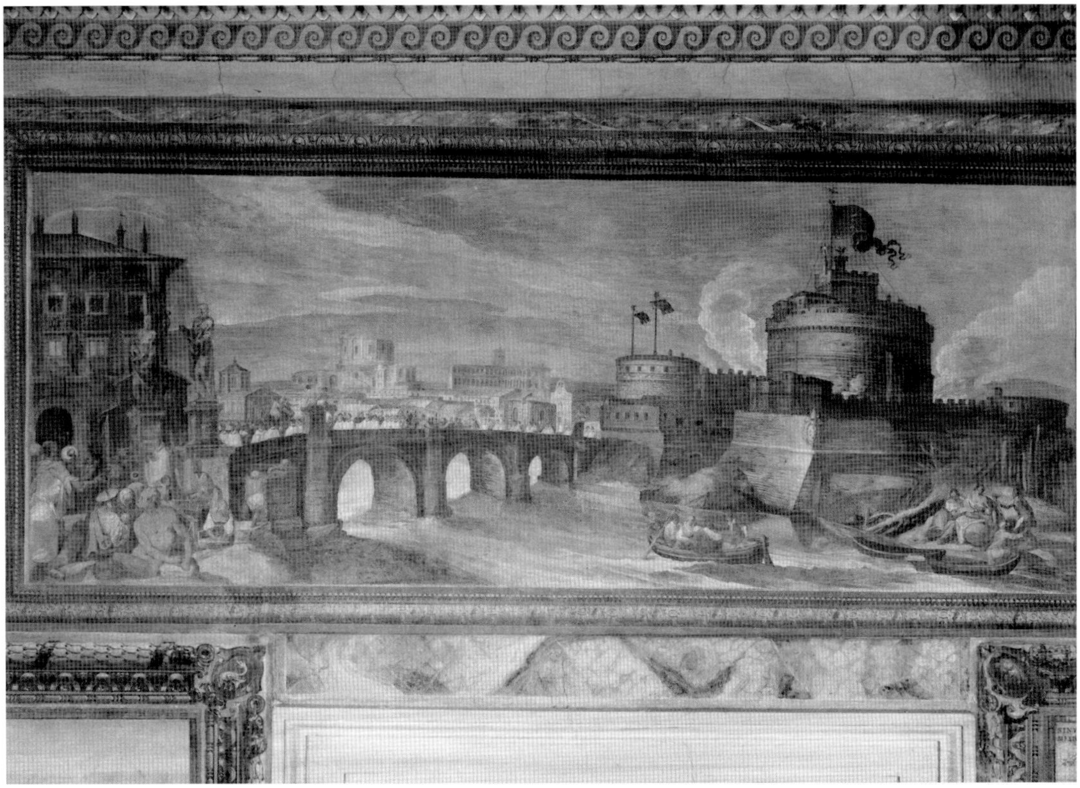

9.6 Matthijs Bril and Antonio Tempesta, *The Procession for the Translation of Gregory of Nazianzus's Body Crosses the Bridge of Castel Sant' Angelo*, 1583, fresco. Vatican, Terza Loggia

it proceeds along Via della Scrofa (views of Piazza Navona and the Pasquino statue are included in this fresco), stops in front of the churches of Santa Maria dell'Anima, Sant' Agostino, and the Oratorian church of the Chiesa Nuova, crosses the bridge of Castel Sant'Angelo, and arrives at St. Peter's, where Gregory XIII himself greets the Greek father's body in its new sanctuary. A monument to the importance that Gregory attached to this procession, to Gregory of Nazianzus, and to the common root of the Greek and Roman rites, this frieze does not at first glance relate to the cartographic theme of the Terza Loggia. Its relation to the maps of the world becomes clear once the ceiling decoration is also taken into account.

In fact, the maps of the world displayed on the walls were connected to the elaborate decoration on the ceiling (Appendix C lists the maps of the Terza Loggia and the vault scenes). Realized under both Pius IV and Gregory XIII but conceived as a unity, the ceiling decoration starts with the representation of the Trinity and continues with the personification of time, the sun, and the moon (Fig. 9.7). These images are followed by the effects that celestial motions generate on the earth's globe and the people that inhabit it. The personifications of the year and of the seasons are followed by the representation of members of the Christian community at different ages of their life: children, youths, men, women, and seniors, each group portrayed in activities exemplary of their good and bad behavior. The Christian community is reminded of the consequences of its behavior in the last scenes, which illustrate the events leading to the final judgment: the destruction of cities, the awakening of bones, the rejoining of the bones with their body, the resurrection of the dead, and the final judgement itself. The Christian community on earth is followed by its counterpart in the heavens, represented according to their hierarchy of sainthood: virgins, confessors, martyrs, doctors of the church, evangelists, apostles, angels, and the Trinity. On the ceiling, there are also inscriptions commemorating Pius IV's and Gregory XIII's deeds. Singled out among Pius IV's deeds are his restorations in Rome, his fortifications in the Papal States, his renewal of ecclesiastical discipline, his policy towards French Catholics, his promotion of science and culture, and his commitment to the conclusion of the Council of Trent, an event which was further emphasized through the depiction of its last session in a nearby fresco by Vanosino.[17] Gregory XIII instead is remembered for his Jubilee year, for the national colleges he founded, the translation of the Roman Catechism he sponsored, his Gregorian Chapel in St. Peter's, his Gallery of Maps and, above all, the new calendar he promulgated in 1582, which is further commemorated in a separate inscription.[18]

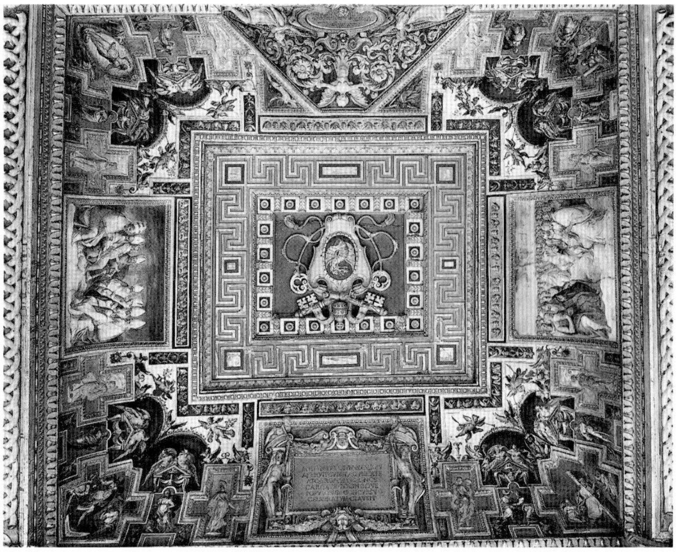

9.7 Vault, fresco. Vatican, Terza Loggia

The combination of the maps of the world, the good and bad behavior of the Christian community at different ages, the inscriptions of papal deeds, the painting of the Council of Trent, and the papal procession of 1580, were all elements that, combined together, point to to the interpretation of the Terza Loggia as the optimistic representation of the lands and people who were expected to embrace the doctrine and liturgy of the Council of Trent.

The vault decoration illustrated the intimate connection between the Christian community on earth and the community of saints in the heavens, between life and death, people and saints, the very connection regulated by the doctrine and liturgy of the Roman church which the Protestant reformers had rejected. More specifically, the vault scenes representing the Christian community on earth divided by age into examples of good and bad behavior suggestively recall the Roman Catechism, which instructed children, youths, men, women, and seniors on proper ways of behavior, connecting such instructions with the final judgment, which is also represented in the vault. The Roman Breviary is recalled in the sequence of the saints in the heavens, which is roughly depicted according to the hierarchical order of the liturgical text.[19]

By referring to the revised liturgical texts of the Roman church, the iconography of the Terza Loggia represented the Christian community as having no limits of geography or age, of space or time. It reasserted the relationship between earth and heaven, and the papal authority to dictate the proper way to regulate that relationship. In the language of cartography that Gregory XIII favored, the Terza Loggia represented the potential expansion of the universal church. It made concrete and reachable the reconquest of Protestant Europe, the reunification of the Greek and Roman liturgy, and the conversion of the peoples of Africa, Asia, and America. The pope acted in Rome, where he dictated the orthodoxy of the doctrine and the uniformity of the liturgy. But the affects of papal policy, recalled metonymically in the frieze of the procession of Gregory of Nazianzus and in the inscriptions of papal deeds, had to spread to the world mapped below. Cartography expressed the optimistic view that the geographical boundaries of the Christian empire would coincide with the earth's globe, where Catholic liturgy would dominate.

But the cartography of the universal church also made manifest the Christian empire as an empire without territoriality, a body over which the pope had no political control and minimal ecclesiastical jurisdiction. Although Gregory aimed at the global extension of the Christian community, he had full ecclesiastical power only in his Papal States and, to a minor extent, in the Italian peninsula. In the rest of the world, his ecclesiastical jurisdiction was obstructed by rulers unwilling to relinquish authority of any kind within their own states. The pope had made uniform the doctrine and the liturgy but still he had to find new imaginative ways to affect the life of people, over which he had no direct jurisdiction. Gregory XIII turned to the reform of the calendar, the old problem of Christian liturgy, as a way to assert anew his ecclesiastical supremacy worldwide.

With his characteristic insight into the potential that the *scientiae mediae* entailed for the universal church and the affirmation of papal authority, the pope embarked on a general reform of the calendar that would eventually affect both the ecclesiastical and civic calendar.

The Calendar of the Universal Church

Gregory XIII's reform of the calendar rested upon the long-standing alliance of mathematical astronomy with Christian liturgy in the definition of the correct day of Easter, the feast commemorating Christ's Resurrection, around which the entire liturgical calendar was built. Since early Christianity it was known that Easter occurred in coincidence with the first full moon after the vernal equinox, the same moon that fixed the Jewish Passover. But the bishops of the early church had established that, in order not to have the two feast days coincide, Easter had to be celebrated on the Sunday after the first full moon after the vernal equinox, and assigned to the bishop of Alexandria, who could consult the astronomers of the city's library, the task of predicting the correct date. The consultation of astronomers was essential because the celebration of Easter had been defined by mixing two incompatible calendars, the lunar calendar that regulated Jewish religious feasts, and the solar calendar promulgated by Julius Caesar, which regulated secular life.[20] Centuries later such mixing would make possible the papal regulation of both ecclesiastical and civic time, a potential that Gregory XIII seized. But until then the prediction of Easter was a vexing astronomical problem. Only expert astronomers could sort out how to match the position of the sun and the moon in the heavens with the celebration of Easter, how to compute the solar and lunar year, and how to reform the reckoning of time so that the civic and liturgical calendar would coincide.

Eminently a matter of the universal church, the computation of Easter and the reform of the calendar were addressed in church councils, starting with the Council of Nicaea convened by Constantine in 325 to fight heresy and impose liturgical uniformity within the large empire. The Council of Nicaea established that the Christian community had to celebrate Easter at the same time everywhere according to the vernal equinox of Alexandria, which was fixed to 21 March (in the Julian calendar it was fixed at 25 March). The council however did not resolve the discrepancy between the celebration of Easter and the position of the sun and the moon in the heavens since it continued to use the year's length of the Julian calendar, which was eleven minutes greater than the solar year. In the Renaissance the delay between the solar and the ecclesiastical year had accumulated considerable errors: the vernal equinox was observed in the heavens ten days earlier than predicted in the Julian calendar, and the new moon appeared five days after the date marked on the calendar according to the golden numbers, the computational system devised in the sixth century to relate the solar and the lunar year. Such a discrepancy had to be fixed with a proper reform of the calendar since there were inherent doctrinal consequences. Without a proper solution, Easter could occur, in due time, before the vernal equinox and, even worse, in coincidence with a new moon, a situation that would make the miraculous darkness that covered the universe on the day of Christ's Resurrection astronomically possible.[21]

The Council of Trent did not specifically address the reform of the calendar, although its deliberations on the uniformity of the liturgy and the education of the clergy implied it: liturgical uniformity could be obtained only through the establishment of a uniform system of time reckoning, while seminaries for the education of the clergy had to include instruction on astronomy for "the computation of ecclesiastical festivities and other useful arts" ("computi ecclesiastici aliarumque bonarum artium doctrinam").[22] Pius v included a minor reform of the calendar in his edition of the Roman Breviary (1568), which, however, pertained only to the moon calendar and thus had consequences for liturgical time but not for secular time.[23] Gregory xiii embarked on a general reform of the calendar that would eventually affect both ecclesiastical and civic time.

Around 1575, Gregory instituted a committee for the reform of the calendar and had Cardinal Sirleto, who had participated in the revisions of the liturgical texts, chair it. Sirleto must have also brought to papal attention the proposal for the reform of the calendar written by Luigi Giglio, a little-known professor of medicine who taught at the University of Perugia around 1552 and died in his hometown of Ciró in Calabria in 1576, his highest merit being possibly his provenance from Calabria, Sirleto's region. Using Giglio's work as a blueprint, the committee wrote a proposal that was printed in 1577 and circulated among European rulers who, in turn, distributed it to their official astronomers. In response to, or in anticipation of, the papal proposal, eminent European astronomers wrote pamphlets on the reform of the calendar and sent them to Rome, where they were evaluated by the two experts on astronomy that Gregory xiii had chosen for his committee, the Jesuit Christoph Clavius and the Dominican Egnazio Danti.[24]

Clavius was a teacher of mathematics at the Jesuit Roman College. Later he became a prominent figure in the Jesuit order and an important astronomer, but in the early 1570s he was only a young cleric who had published a Latin edition of Sacrobosco's *Sphere*. Nonetheless he must have been among the few competent astronomers available in Rome (the University of Rome did not have a chair in mathematics at the time), and must have joined the papal committee from the very beginning. He was instrumental in shepherding the reform through the politics of the papal court and later in defending it.[25]

Danti must have joined at a later stage, perhaps after he sent his reply to the original proposal drafted by the papal committee. As a professor of mathematics at the University of Bologna and an expert on the reform of the calendar that he had researched at the Medici court, his response must have been highly regarded. Danti's work for the papal committee has not survived, unless an anonymous report pertaining to the motion of trepidation is attributed to him. But his one-page treatise on meridian lines, *Usus et tractatio gnomonis magni* (Bologna, 1576), seems indeed a response to the papal proposal, which he might have received even before its publication in 1577. In that treatise, Danti abandoned the references to the imperial and royal attempts to reform the calendar, which he had stressed as Cosimo I's cosmographer. He echoed instead the papal view of the reform of the calendar as the long-term achievement of past church councils, from the Council of Nicaea to the Council of Trent, and of earlier popes, from Pope Anacletus, to Leo I, Ilarius I, Pius I, Sixtus IV, Innocent VIII, Julius II, Leo X, Paul III, and, obviously Gregory XIII, who was "wholly and strongly motivated to the correction and restitution of the [measurement of the] year."[26]

Both Danti and Clavius suggested the calculation of the vernal equinox at the time of Julius Caesar rather than at the time of Constantine, a sensible proposal shared by other astronomers (including Protestant astronomers) since it was the Julian calendar that had introduced the mistaken calculation of the equinox. Their proposal was disregarded in order to celebrate the Constantinian unity of the church, a leitmotiv of Gregory XIII's papacy. Also their suggestion to adopt modern measurements of the tropical year was put aside despite of the fact that both had dedicated so much effort to it. In the name of tradition, Gregory preferred to continue to use the Alfonsine tables, the planetary tables published in 1252 under King Alfonso X of Castile.

Even though Danti's and Clavius's proposals for a more comprehensive reform of astronomy were disregarded in the name of papal propaganda, their presence at Gregory's court is significant in other respects. They were both convinced supporters of the certainty of mathematics, that is of the importance of mathematics not only in the investigation of the natural world but also in the contemplation of the divine. Danti had written about mathematics and its application extensively in his Florentine treatises and had made the *scientiae mediae* the focus of his teaching at the Medici court. Later, during his stay in Bologna, he also wrote about the intermediary role of the middle sciences between physics and metaphysics. His views were not alien to the official Dominican education, which complemented the scholastic training with a selection of secular subjects (Hebrew and Greek language, mathematics, and astronomy). But departing from the Dominican curriculum, which never regarded these secular disciplines as a means in themselves, Danti shared the view of other reformed Catholics, including Cardinal Paleotti, who regarded the scientific investigation of the world as a true step towards the contemplation of the divine. Similarly, Clavius instructed future missionaries and aristocrats on how to apply mathematics in the government of the world and in the contemplation of the divine. He wrote extensively on mathematics as a link between physics and metaphysics, on the certitude of truth that mathematical demonstrations give to its praticitioners, and on their usefulness in practical life and in the study of sacred subjects.[27] Both Danti and Clavius had to defend their views within their respective orders.

The debate on the certainty of mathematics has often been interpreted as an opposition between the Dominican order, which had taught the traditional curriculum, based on Thomas Aquinas's writings, to the clergy and the aristocracy for centuries, and the new Jesuit order, which was committed to a more varied curriculum and was replacing the Dominicans in their leading educational role. Indeed, in the seventeenth century Dominicans and Jesuits found each other on opposing positions in the debates on the relationship between science and religion, astronomy and biblical exegesis. But in the second half of the sixteenth century, when the debate was still taking shape, the Dominican and Jesuit orders included both passionate advocates and stern critics of the certainty of mathematics. Earlier in the century the Dominican Giovanni Maria Tolesani had written a text against Copernicus, which Danti must have known since it was kept in the library of his own convent of San Marco in Florence. Tolesani condemned Copernicus for his disruption of the Aristotelian hierarchy of knowledge, restating the rigid Thomist curriculum, and he formulated the first accusation of heresy against Copernicus that was to find an audience over fifty years later when Tolesano's text was used by another Dominican to prepare his case against Galileo.[28] If Danti ever read Tolesano's manuscript, he must have found little inspiration in it for his teaching at the Medici court and even less for his work as Gregory's XIII's papal cosmographer. Clavius eventually succeeded in having mathematics included in the *Ratio studiorum*, the official curriculum of Jesuit colleges, but only after years of disputations with his fellow Jesuits, especially Benedict Pereira.

If advocates for the certainty of mathematics had to fight for their position within the leading religious orders dedicated to clerical and secular education, they found a favorable environment at Gregory's court. The pope understood that his reform of the calendar would have had a better chance of success if it were tied to the certainty of mathematics. He recruited these talented mathematicians, who were willing to put astronomy at the service of the universal church in order to fulfill his dream of ecclesiastical supremacy.

Clavius and Danti certainly knew each other, but the nature of their relationship is pure speculation since no trace of their interaction has come to light. Their relationship must have been built on their common scientific interests but also, it must be imagined, on considerable competition. They both opposed the Copernican structure of the universe, but embraced Copernicus's epistemological shift. They shared the internal fight within their respective orders to assert the certainty of mathematics over the traditional view expressed by other members, such as Tolesano and Pereira. They both taught the *scientiae mediae* extensively in their respective classes and edited Sacrobosco's astronomical manual – Clavius in Latin and Danti in Italian. They held different views on the motion of trepidation: Danti was in favor of it while Clavius rejected the theory. They both wished to accomplish a general reform of astronomy, although they accepted the papal view for a more modest and measured reform, and continued to serve the Boncompagni pope after the promulgation of the new calendar.[29]

In 1582, Gregory XIII finally promulgated the new calendar, asking European rulers to accept it on the ground that its astronomy was sound and its computations accurate. In order to adopt the papal calendar, European monarchs had to abolish ten days from the month of October 1582 (Thursday, 4 October was followed by Friday, 15 October), to accept the papal calculation of the equinox at the time of Constantine, and to use the cycle of the epacts, the computational system that denotes the age of the moon on 1 January which Giglio had

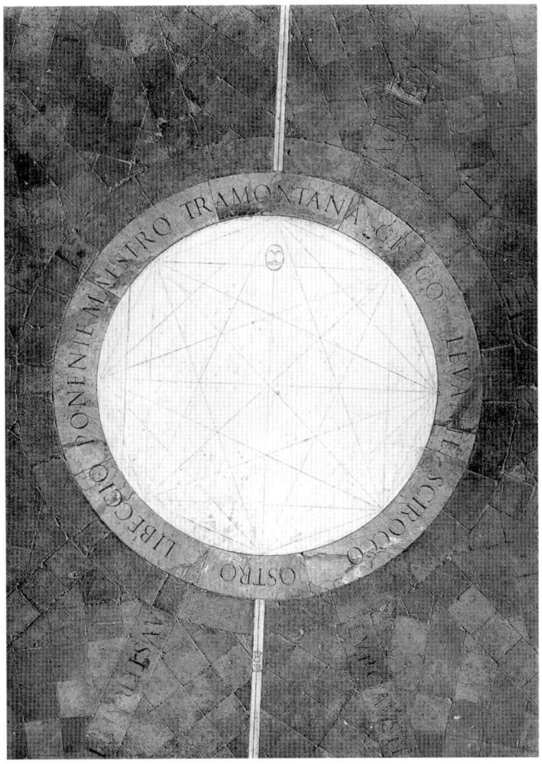

9.8 Meridian Line, 1582, stone floor. Vatican, Tower of the Winds, Sala della Meridiana

devised to relate the solar and calendar year and which astronomers across Europe had harshly criticized. European rulers also had to accept the length of the year that Gregory had fixed according to the Alfonsine tables, disregarding the most recent astronomical research, which in many instances they themselves sponsored. Monarchs in Italy, Spain, Portugal, and Poland adopted the new calendar immediately, followed, with some delays, by the rulers in France, Belgium, and the Catholic states in the Netherlands. Protestant monarchs and Greek patriarchs, however, sternly opposed the Gregorian calendar, which they rightly saw as an attempt to reestablish papal authority in religiously divided lands. Indeed, by claiming that he had fixed a miscalculation in the reckoning of time, Gregory xiii managed to establish papal ecclesiastical supremacy throughout the universal church.

Conscious of the fact that the new calendar was a fundamental step toward the achievement of liturgical uniformity, the pope ordered the celebration of a Mass every evening in the Chapel of Nicholas v in the Vatican Palace, forcing papal courtiers to attend.[30] He also continued to employ his experts Clavius and Danti, enlisting them in the defense and celebration of the papal reform. Clavius became the official defender of the reform in his writings, a task he continued under Gregory xiii's successors. Danti was asked instead to install two meridian lines to demonstrate the mistake that the Gregorian reform had eliminated. One meridian line Gregory wanted in his apartment in the Tower of Winds (Fig. 9.8). Significantly, the other was installed in the Gregorian Chapel in St. Peter's.[31] In that chapel Danti's meridian line would have expressed the papal hope that the reform of the calendar would also bring the reunification of the Greek and Roman liturgy of the early church.

The reform of the calendar put Rome at the center of the Christian empire, an utopian empire without territoriality but that nonetheless aimed at the inclusion of people from around the world. During Gregory's papacy, such a utopian universality was fully articulated through the middle sciences of cartography and astronomy. Cartography expressed the pope's optimistic view that the universal church would one day coincide with the earth's globe, while the mixed science of astronomy made possible the affirmation of papal authority on the de-territorialized Christian empire through the reform of the calendar. Ultimately, the Gregorian calendar did not unify the Christian empire but, a couple of centuries after its promulgation, it became the universal measure of secular time.

As Gregory XIII was able to give new life to the traditional alliance between astronomy and liturgy and thus regulate the ecclesiastical and civic time of Catholics worldwide, so he was able to imagine a new meaning for the traditional alliance between maps and religion, indeed a new sacred cartography.

Gregory XIII's Sacred Cartography

Sacred cartography has a long history, dating from early Christianity, when maps represented the earth's globe protectively embraced by its creator, the land divided into three continents, one for each of Noah's sons, and Jerusalem placed at its center. Filled with quotations from the Scriptures and images of Paradise, Adam and Eve, and Christ, these early maps of the Christian world served as instruments for monastic learning, as tools for the memorization of sacred texts, as emblems for meditation, as means for moral teaching and, with appropriate guidance, as steps towards divine contemplation. The close relationship between cartography and religion exemplified by these early maps was maintained throughout the centuries, and flourished during the sixteenth century, when a multitude of printed maps combined the reading of religious texts with modern geographical knowledge.

During the sixteenth century, the most common maps of sacred cartography were the historical maps of the Holy Land and the Mediterranean, which a growing map-literate public knew how to use in order to help visualize the stories of the Bible. These historical maps were produced by both Catholic and Protestant scholars and map-makers, who shared their training in humanism, antiquarianism, and biblical exegesis. Like the maps of the modern world, the historical maps of the Holy Land and the Mediterranean portrayed the land according to the most advanced geographical knowledge of the area, but unlike the maps of the modern world, they labelled the land with ancient toponyms, populated it with biblical events, and divided it according to biblical partitions. These maps were historical in the deepest sense, since they were instruments for biblical reading as well as for the antiquarian reconstruction of Jewish culture and religion.

Although both Protestants and Catholics produced sacred maps of the Holy Land, only Protestants were keen to put them in the hands of the reading public at large, and they printed them as illustrations of their vernacular Bibles. Luther specifically wished for, "a good geography and more correct map of the Land of Promise"[32] as part of the individual reading of the Bible that he advocated. In Protestant Bibles the maps worked as visual aids for the reader to understand specific texts, usually cited in a caption on the map (Fig. 9.9), and in this respect cartography was an integral part of reformed biblical exegesis. Protestants, however, did not produce maps illustrating any episode of church history later than the peregrinations of the apostles.

Catholics, on the other hand, reserved the study of their maps of the Holy Land and the Mediterranean for a selected public of educated readers, preferably clerics in training or scholars. In the rare occasions in which maps appear in Catholic Bibles, it is either in Latin Bibles of the late sixteenth century or in highly specialized volumes, such as the volumes that Benedictus Arias Montanus wrote as a commentary to the Antwerp Polyglot (Antwerp, 1572). In that monument of biblical and antiquarian scholarship, Montano's Bible maps were only one ingredient of his multifaceted reconstruction of Jewish culture, religion, and

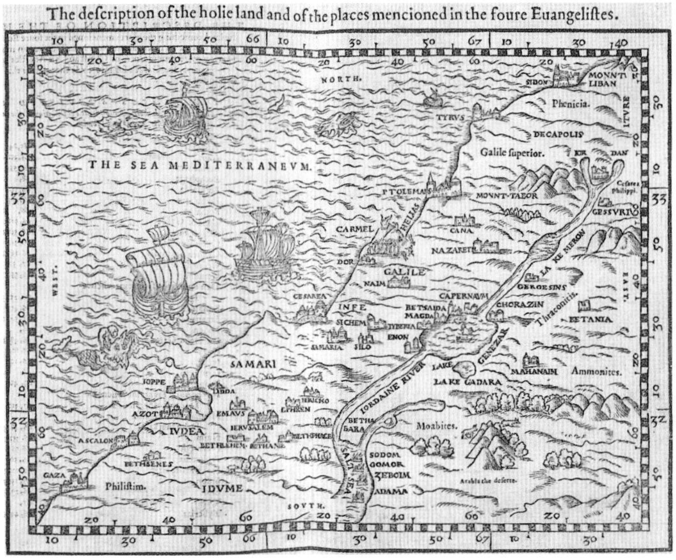

9.9 *The description of the holie land and of places mencioned in the foure Evangelists*, from The *Bible and the holy scriptures contayned in the Olde and Newe Testament*, Geneva 1560. Princeton University Library

customs achieved through a multitude of visual and verbal sources. Like other maps of sacred geography, Montano's maps of the Holy Land (Fig. 9.10) had a specific religious function in facilitating moral meditation, although Montano also imagined them as part of a philosophy entirely based on the Bible, a reading hardly shared by others. Instead many other scholars shared the view that maps could function as religious images.[33]

Ideas on the spirituality of maps circulated in sixteenth-century Europe among Protestants and Catholics alike, although Protestants were ready to give access to maps of sacred geography to the general public, while Catholics regarded the spirituality of maps as a domain attainable only by learned scholars and doctors of theology. Maps had been used as images for meditation since the Middle Ages, but only in the sixteenth century did theologians, clerics, and map-makers, perhaps inspired by the multitude of printed maps that circulated in Europe, explicitly discuss them as sacred images. As mathematics and its applications were achieving a higher status in the hierarchy of knowledge, including the knowledge of the divine, so maps were acquiring a different status and, in specific contexts, could actually function as sacred art.

Ignatius of Loyola encouraged young Jesuits to meditate on the religious stories by visualizing people, events, and places, "to see with the eye of the imagination the synagogues, towns, country, and places through which Christ our Lord preached . . . the great room and round of the world, where dwell so many and such diverse nations," or to visualize the road from Nazareth to Bethlehem as part of the meditation on the Nativity, or the way from Mount Zion to the valley of Josaphat while meditating on the Agony.[34] Suggesting his disciples imagine the world from above like in a map, Loyola was also considering maps as a stimulus for spiritual exercises. Matteo Ricci, the Jesuit who traveled to China and contributed considerably to the mapping of that part of the world, envisaged his world map not only as a memory theater but also as a tool of moral teaching to show to his Chinese listeners the site of their empire in relation to the earth's globe, and thus inspire them to convert to Christianity. Like the papal maps of the Terza Loggia, Ricci's world map did not register religious fragmentation.[35]

Ortelius himself made maps of sacred cartography, in which he explicitly invited the viewer to follow the voyages of Abraham, Moses, and Paul through an imaginary pilgrimage, which was ultimately spiritual. In fact, Ortelius imagined his maps and his *Theatrum* as emblems of moral meditation and imbued them with inscriptions that encouraged the viewer to meditate on the world as a stage of human actions or the ephemeral

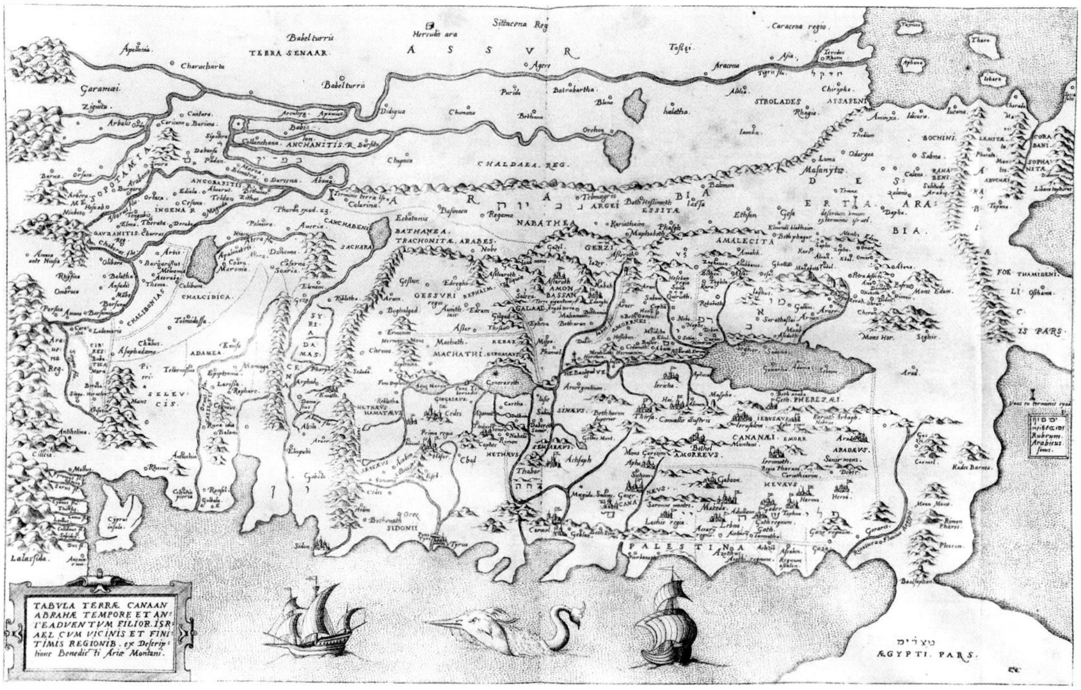

9.10 Benedictus Arias Montanus, *Tabula Terrae Canaan Abrahmea Tempore* (Map of Canaan) from Benedictus Arias Montanus, *Biblia sacra*, Antwerp 1571, vol. 8. Princeton University Library

nature of earthly things.[36] Ortelius's meditative maps were shaped by his affiliation with the Family of Love, a secret sect that stressed individual meditation over doctrine and that wished for world pacification based on religious tolerance, a utopian dream in sixteenth-century Europe as much as in the modern world, but their spirituality resonated deeply with Renaissance viewers of all denominations, who bought them across religious divides. While the acolytes of the Family of Love read Ortelius's maps circumstantially within the creed of their sect and interpreted his quotations from Cicero and Seneca surrounding Ortelius's famous heart-shaped world map as examples of irenic thought, the majority of viewers interpreted them as generic neo-stoic invitations to contemplate the world while being detached from it.

Cardinal Paleotti also envisaged a spiritual function for maps, although not one based on contempt for earthly things but rather on the Augustinian notion that the knowledge of the natural world facilitates the intuition of the perfection of its creator. As discussed in an earlier chapter, Paleotti's *Discorso* offered a theoretical framework for a more comprehensive understanding of Gregory XIII's map murals in the Vatican Palace, while Egnazio Danti's relationship with Paleotti and the papal court provided the historical circumstances that favored the application of Paleotti's views on Catholic art to the decoration of the Vatican Palace. It is in the context of Paleotti's treatise that Gregory's map murals appear not only as sacred images but also as examples of the new revised art that the cardinal envisaged. In the Sala Bologna, in the Gallery of Maps, and in the Terza Loggia, the maps were connected with images of church history, which conformed in content and style to Paleotti's view of post-Tridentine religious art. As Paleotti suggested, the historical scenes represented exemplary deeds of worthy men and saints, painted in simple and balanced compositions, with great attention paid to the details of their historical context, from the buildings to the clothes to the geographical location. Also the grotesques, which Paleotti condemned in his treatise but others, including Danti, favored in a reformed form, were simplified to adapt to the more restrained taste of the decades after Trent. But the maps that ordered the iconography of these rooms were the real protagonists of the new reformed art that Paleotti advocated.

Gregory's predilection for maps displayed on the walls of the Vatican Palace appears now in accord with the long-standing alliance between cartography and religion that was revived in the sixteenth century in the writings of Loyola, in the maps of Ortelius, Ricci and others, and above all in Paleotti's interpretation of scientific images as sacred images. But Gregory XIII also imagined a different kind of sacred cartography, one that gave concrete form to the immaterial hopes of the post-Tridentine universal church. As cartography had made visible the places of the Bible, so it could contribute to make visible the places outside the Holy Land, where the sacred history that happened after the peregrinations of the apostles took place. Indeed, for Catholics the privileged theater of sacred history shifted from Jerusalem to Rome, Italy, and the world at large, from the places of the Bible and Christ to the seat of his representative in Rome, from the Holy Land to the places across the globe where the spiritual army of saints and religious orders acted on the behalf of the Roman pontiff.

Gregory XIII's sacred cartography evolved as his pontificate progressed. In the Sala Bologna the maps were a means to affirm his personal power over the Bolognese aristocracy, the instrument of a parochial assertion of supremacy that, while essential to his own prestige, did not match his role as the spiritual leader of the Christian empire. In that dining room, however, Gregory had found a way to express his hopes to use the mixed sciences of cartography and astronomy for the universal church. The mapping of the Papal States that he commissioned from Danti for purposes of administration represented another important, albeit narrow, form of sacred cartography, one that registered the network of ecclesiastical hierarchy in his dominion as well as the resources to sustain the Papal States. But the Gallery of Maps and the Terza Loggia represent a new kind of sacred cartography, one that visualized two tenets of the post-Tridentine papacy, the primacy of Italy within the Catholic world and the extension of the universal church to every corner of the globe. Armed with the intellectual and spiritual implications that had sustained the very idea of sacred cartography throughout the centuries, Gregory XIII imagined his universal church through mapping: a Christian empire that extended to the entirety of the globe, with Italy at its center, united in celebrating the rituals dictated by the Roman church on the days mandated by his new calendar. And he left to his successors the task of bringing to fruition his cartographic utopia of the universal church.

Conclusion

Long ago Jacob Burckhardt suggested that painted map cycles, such as the map cycles at the Medici and papal courts analyzed in this book, constituted a pictorial genre of the Italian Renaissance. However, in line with the modern skepticism regarding the taxonomy of symbolic forms, I have not pursued the study of map cycles as a pictorial genre. Rather than isolating map cycles from Renaissance culture on the simplistic consideration that they all contain cartographic images, I have instead investigated the interactions between mapping and other forms of knowledge, other systems of representation, and other symbolic realms. At the Medici and papal courts, mapping was part of the ruler's broader interests in applied mathematics, scientific instruments, and astronomy, and in both courts mapping was part of the personal iconography of the Renaissance monarch. In both case studies the order of the maps was used as a system for the organization and transmission of knowledge. The metaphorical meanings of these cartographic decorations were created by the interaction of the painted maps with images of history, mythology, zoology, botany, and religion, and could be fully grasped only through the firsthand experience of the architectural spaces that contained them. But those meanings were also defined by the broader cultural context of court life, by the interactions of the patron with his court cosmographer, the network of courtiers, the study of cosmography, and obviously by the celebration of the ruler's hopes of political and religious influence. To consider map cycles exclusively from the perspective of cartography or simply as an artistic genre is to miss not only the richness of their meanings but also the deep significance of mapping in Renaissance culture.

Analysis has shown that mapping participated in the general re-evaluation of the mixed sciences in the sixteenth century, although it never achieved the level of controversy that surrounded its sister discipline, astronomy. In representing the surface of the earth, after all, it did not matter whether the terrestrial globe was placed at the center of the universe or at its periphery. Renaissance mapping was informed by the desire for a univocal representation of the world, so that any terrestrial location corresponds to one and only one place on a map. Such mathematical accuracy was made possible by the rediscovery of Ptolemy's theoretical geography, but only the invention of accurate clocks a couple of centuries later would make possible the achievement of cartographic precision. But instead of focusing on what Renaissance mapping did not achieve, I have examined the place that mapping did hold in the ongoing culture of Renaissance courts.

The detailed analysis of the language of maps has made evident the full potential of geographical description that Renaissance map-makers envisaged for their maps. Ptolemy had divided the scope of mapping into two branches, geography and chorography, but, as discussed in this study, his geographical theory was transformed into a complex, indeed fascinating affair that mixed cartography with verbal and visual chorography, symbols with icons, images with words, mythology with firsthand accounts of voyages and discoveries. The marvel of these maps came from their hybrid language, from their fusion of humanism and mathematics, their combined representation of places and sites. Like modern map-makers, Renaissance cartographers knew well that their maps resulted from the myriad selections they made in deciding what to include and what to omit. They knew that the Ptolemaic grid reduced the globe to a sequence of sites. And yet they never gave up the wish for an encompassing representation of the globe, a representation that combined sites with places, toponymy with history, geography with mythology, plants with animals, people with cities, beliefs, and customs. Cartography had the potential to achieve such an encompassing description of the world, but only in defined contexts, in which special patrons interacted with imaginative map-makers, could this potential come to full fruition. I have followed the process whereby the geographical maps of Cosimo I's Guardaroba Nuova became chorographical and, conversely, the process whereby the chorographical maps of Gregory XIII's Gallery of Maps became geographical. What has emerged is a deeper understanding that the difference between geographical and chorographical maps did not lie necessarily in the presence, or lack, of mathematics, or in the use of a particular system of representation or mode of description, but rather in the inherently relative nature of geographical description: a map was chorographical in relation to, say, the map of the world, but became geographical in relation to another map illustrating a locale contained by it.

The historical importance of painted map cycles that I have analyzed in this book is thus twofold. First, by eluding a direct political interpretation of dominion and conquest, these maps force the scholar to look for the interaction of mapping with other symbolic forms. Second, by representing the mapped lands on an unprecedented large scale and thus enhancing the descriptive possibility of Renaissance cartography, they have substantially clarified the language of Renaissance maps. But perhaps their greatest historical significance lies elsewhere. The detailed analysis of the Medici and papal map cycles shows that in the Renaissance geography was not only a favored subject for the education of men and women, but also one of the conditions for the dissemination of knowledge, both profane and sacred. In Cosimo I's Guardaroba Nuova geography provided the conceptual and visual system to organize a collection of artifacts, the structure to make sense of a world of expanding horizons. Similarly, at the papal court geography gave structure to the foundational notions of the post-Tridentine church. In the Gallery of Maps the imaginary circumnavigation of the Italian peninsula gave geographical structure to the primacy of Italy while at the same time reassuring the papal court, the cardinals, ambassadors, and the Italian rulers of their special place within the Catholic world, a place that Gregory XIII both reasserted and revised through those actions that were increasingly directed towards the universal church outside of Italy and overseas. In the Terza Loggia geography organized the new global mission of the papacy, a mission that could be expressed only cartographically.

As densely qualified and highly mathematical images, these map cycles continue to help the modern viewer to reintegrate and reconcile partial, fragmentary, and contradictory views of Renaissance culture into a coherent perceptible totality, and they do so beautifully, harmoniously, and effectively. Their hybrid and polysemic structure resonates, consciously or unconsciously, with the multimedia structures that surround us daily. But they may also be reminders of a time in which geography and place mattered in the knowledge of the world, its organization, and communication.

Appendix A

The Maps of the Guardaroba Nuova

This appendix proposes a reconstruction of the original layout of the maps of the Guardaroba Nuova. This sequence of the maps, which differs from their current display in Palazzo Vecchio, is based on Giorgio Vasari's description of the Guardaroba Nuova in his *Lives* and on the geographical order that Giacomo Gastaldi adopted in his edition of Ptolemy's *Geography* (Venice, 1548) and that Girolamo Ruscelli repeated, with minor variations, in his own edition of Ptolemy's geographical manual (Venice, 1561).

The maps, which are all painted in tempera on wood panels, are reproduced in color in Danti-Bonsignori 1995.

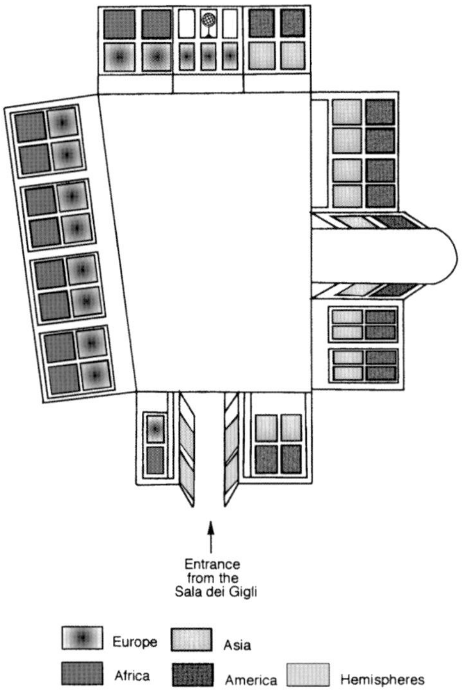

Europe: Lower Level of the Cupboards from the Main Entrance to the Right

ISOLE BRITANICHE le quali co(n)tengono il Regno di Inghileterra et di Scotia con l'Hibernia

(British Isles, which contain the Kingdom of England and Scotland, and Island), 1565, Egnazio Danti, 110 × 111 cm
Longitude 3–, Latitude 49–63

Cartographic sources: Mercator, *Map of Europe* (1554) for the mapping of Ireland. George Lily, *Map of the British Isles* (1546), most likely through one of Lily's derivative maps, perhaps the map engraved by Paolo Forlani and published by Fernando Bertelli (Venice, 1562), for the mapping of England, Scotland, and Wales, and for large sections of the legends.
Bibliography: Danti-Bonsignori 1995, 28–9.
Notes: This part of the old continent was represented in Ptolemy's first map of Europe, and in Gastaldi's *Anglia et Hibernia Nova Tabula*, the modern map added to Ptolemy's first map of Europe (Ptolemy 1548).

Thile I(nsul)a

(Thule Island) 1565, Egnazio Danti, 110 × 53 cm
Longitude 352–9, Latitude 65–9

Cartographic sources: Mercator, *Map of Europe* (1554) for the cartographic content.
Bibliography: Danti-Bonsignori 1995, 34–5.
Notes: This part of the old continent was represented in Ptolemy's first map of Europe and in Gastaldi's *Anglia et Hibernia Nova Tabula*, the modern map added to Ptolemy's first map of Europe (Ptolemy 1548).

La Spagna

1577, Stefano Bonsignori, 107 × 106 cm
Longitude 34–45, Latitude 8–14

Signature (top left): "D.S. Fl.o A.M.F. 1577."
Cartographic sources: Clusius, *Map of Spain* (1571).
Bibliography: Schilder 1986, 99–110; Danti-Bonsignori 1995, 90–1.
Notes: This part of the old continent was represented in Ptolemy's second map of Europe and in Gastaldi's *Hispania Nova Tabula*, the modern map he added to Ptolemy's second map of Europe (Ptolemy 1548).

La Francia

1576, Stefano Bonsignori, 110 × 106 cm
Longitude 41–53, Latitude 16–35

Bibliography: Danti-Bonsignori 1995, 88–9.
Notes: This part of the old continent was represented in Ptolemy's third map of Europe and in Gastaldi's *Gallia Nova Tabula*, the modern map added to Ptolemy's third map of Europe (Ptolemy 1548).

La Germania

1577, Stefano Bonsignori, 106 × 107 cm
Longitude 45–56, Latitude 27–47

Signature (top left): "D. Stephanus Flo. Mon. Monto. F.A.S. 1577."
Bibliography: Danti-Bonsignori 1995, 92–3.
Notes: This part of the old continent was represented in Ptolemy's fourth map of Europe, and in Gastaldi's *Germania Nova Tabula*, the modern map added to Ptolemy's fourth map of Europe (Ptolemy 1548).

La Schiavonia

1578, Stefano Bonsignori, 111 × 106 cm
Longitude 49–37, Latitude 40–58

Signature (top left): "D. Stephanus Florentinus monachus Montolivetanus F.A.S. 1578."
Bibliography: Danti-Bonsignori 1995, 96–7.
Notes: This part of the old continent was represented in Ptolemy's fifth map of Europe, and in Gastaldi's *Dalmacia Nova Tabula*, the modern map added to Ptolemy's fifth map of Europe (Ptolemy 1548).

L'Italia
1578, Stefano Bonsignori, 109 × 104 cm
Longitude 46–34, Latitude 29–40

Signature (top right): "D. Stephanus Florentinus F.A.S. MDLXXVIII."
Bibliography: Danti-Bonsignori 1995, 94–5.
Notes: This part of the old continent was represented in Ptolemy's sixth map of Europe, which included the Italian peninsula and Corsica, and in Ptolemy's seventh map of Europe, which included Sicilia and Sardinia; in Gastaldi's *Italia Nova Tabula*, one of the modern maps he added to Ptolemy's sixth map of Europe, and his *Sicilia Sardinia Nova Tabula*, the map he added to Ptolemy's seventh map of Europe (Ptolemy 1548).

Livonia et Littuania
not dated, Egnazio Danti, 110 × 94 cm
Longitude 40–65, Latitude 65–45

Cartographic sources: Mercator, *Map of Europe* (1554) for most of the cartographic content and for the second and third paragraph of the main legend. Anthony Jenkinson, *Nova absolutaque Russiae, Moscoviae, & Tartariae descriptio* (London, 1562), for the delta of the River Obi in the North Sea, the course of the Volga and its estuary in the Caspian Sea, the River Pechora, and the outline of the Caspian Sea.
Bibliography: Paratore 1976; Licini 1988, 119–21; Danti-Bonsignori 1995, 36–7.
Notes: This part of the old continent was represented in Ptolemy's eighth map of Europe, and in Gastaldi's *Prussia et Livonia Nova Tabula*, one of the two modern maps he added to Ptolemy's eighth map of Europe (Ptolemy 1548).

Norvegia Gotiae
around 1565, Egnazio Danti, 110 × 93 cm
Longitude 25–52, Latitude 76–55

Cartographic sources: Mercator, *Map of Europe* (1554) for the cartographic content, and Gastaldi's *Schonlandia Nova Tabula* (Ptolemy 1548) for the legend.
Bibliography: Danti-Bonsignori 1995, 54–5.
Notes: This part of the old continent was not included in Ptolemy's *Geography*, but it was mapped in Gastaldi's *Schonlandia Nova Tabula*, one of the two modern maps he added to Ptolemy's eighth map of Europe (Ptolemy 1548).

Moschovia
not dated, Egnazio Danti, 111 × 95 cm
Longitude 63–89, Latitude 71–64

Cartographic sources: Mercator, *Map of Europe* (1554), for most of the cartographic content and for the inscriptions on the map. Anthony Jenkinson, *Nova absolutaque Russiae, Moscoviae, & Tartariae descriptio* (London, 1562) for the course of the Volga and its estuary in the Caspian Sea, the River Pechora, and the outline of the Caspian Sea. Gastaldi, *Moschovia nova Tabula* (1548), and Ruscelli, *Moschovia nuova tavola* (Ptolemy 1561), for the legend.
Bibliography: Paratore 1976; Licini 1988, 119–21; Danti-Bonsignori 1995, 38–9.
Notes: This part of the old continent was represented in Ptolemy's eighth map of Europe, in Gastaldi's *Schonlandia Nova Tabula*, one of the two modern maps he added to Ptolemy's eighth map of Europe (Ptolemy 1548), and in Gastaldi's *Moschovia Nova Tabula*, the map he added to Ptolemy's second map of Asia (Ptolemy 1548).

Grecia
1585, Stefano Bonsignori, 109 × 93 cm
Longitude 45–30, Latitude 43–60

Signature (top left): "D. Stephanus Flor. F.A.S. 1585."
Bibliography: Danti-Bonsignori 1995, 122–3; Tolias 2002.
Notes: This part of the old continent was represented in Ptolemy's ninth map of Europe which included also Dacia and Macedonia, and in his tenth map of Europe, which included the Peloponnese and the Greek Islands; and in Gastaldi's *Graecia Nova Tabula*, the modern map he added to Ptolemy's ninth and tenth maps of Europe (Ptolemy 1548). A copy of this map is now in the Museum Benaki in Athens, Greece.

Africa: Upper Level of the Cupboards from the Main Entrance to the Right

Mauritania

1579, Stefano Bonsignori, 109 × 105 cm
Longitude 2–26, Latitude 38–19

Signature (top left): "D. Stephanus Floretinus monachus Monteolivetanus F.A.S. 1579."
Cartographic sources: Gastaldi, *Il disegno della geografia moderna de tutta la parte dell'Africa* (1564) for the general geographical outline. The legend is by Bonsignori.
Bibliography: Danti-Bonsignori 1995, 104–5.
Notes: This part of Africa was represented as the first map of Africa in Ptolemy's *Geography*; it also appeared in Gastaldi's *Mauritania Nova Tabula*, the modern map he added to Ptolemy's first map of Africa (Ptolemy 1548).

Affrica

(Tunisia), 1579, Stefano Bonsignori, 109 × 106 cm
Longitude 37–51, Latitude 38–18

Signature (top left): "Monachus MDLXXIX,"
Cartographic sources: Gastaldi, *Il disegno della geografia moderna de tutta la parte dell'Africa* (1564) for the general geographical outline. The legend is by Bonsignori.
Bibliography: Danti-Bonsignori 1995, 106–7.
Notes: This part of Africa was represented as the second map of Africa in Ptolemy's *Geography*; it also appeared in Gastaldi's *Africa Minor Nova Tabula*, the modern map he added to Ptolemy's second map of Africa (Ptolemy 1548).

L'Egitto

1578, Stefano Bonsignori, 108 × 107 cm
Longitude 51–66, Latitude 22–33

Signature (top left): "D. Stephanus Floren(tinu)s Monachus Montoliveta(n)us F.A.S. MDLXXVIII."
Cartographic sources: Gastaldi, *Il disegno della geografia moderna de tutta la parte dell'Africa* (1564). The legend is by Bonsignori.
Bibliography: Danti-Bonsignori 1995, 98–9.
Notes: This part of Africa was represented as the third map of Africa in Ptolemy's *Geography*; it also appeared in Gastaldi's *Marmarica Nova Tabula* and *Aegyptus Nova Tabula*, the two modern maps he added to Ptolemy's third map of Africa (Ptolemy 1548).

Libia interiore

1580, Stefano Bonisgnori, 107 × 106 cm
Longitude 9–27, Latitude 4–19

Signature (top left): "D. Stephanus Florentinus Monachus Montolivetanus F.A.S. MDLXXX."
Cartographic sources: Gastaldi, *Il disegno della geografia moderna de tutta la parte dell'Africa* (1564). The legend is by Bonsignori.
Bibliography: Danti-Bonsignori 1995, 108–9.
Notes: This part of Africa was represented as the fourth map of Africa in Ptolemy's *Geography*; it also appeared in Gastaldi's *Africa Nova Tabula*, the modern map he added to Ptolemy's third map of Africa (Ptolemy 1548).

Parte dell'Agisamba

1580, Stefano Bonsignori, 107 × 102 cm
Longitude 27–45, Latitude 20–1

Signature (top left): "D. Stephanus Monachus Montolivetanus F.A.S. MDLXXX."
Cartographic sources: Gastaldi, *Il disegno della geografia moderna de tutta la parte dell'Africa* (1564). The legend is by Bonsignori.
Bibliography: Danti-Bonsignori 1995, 110–11.
Notes: This part of Africa was represented as the fourth map of Africa in Ptolemy's *Geography*; it also appeared in Gastaldi's *Africa Nova Tabula*, the modern map he added to Ptolemy's third map of Africa (Ptolemy 1548).

Nubia

1579, Stefano Bonsignori, 111 × 109 cm
Longitude 48–58, Latitude 21–1

Signature (top left): "D. Stephanus Florens. monachus Montolivetanus F.A.S. MDLXXIX."

Cartographic sources: Gastaldi, *Il disegno della geografia moderna de tutta la parte dell'Africa* (1564). The legend is by Bonsignori.
Bibliography: Danti-Bonsignori 1995, 102–3.
Notes: This part of Africa was represented as the fourth map of Africa in Ptolemy's *Geography*; it also appeared in Gastaldi's *Africa Nova Tabula*, the modern map he added to Ptolemy's third map of Africa (Ptolemy 1548).

Trogloditica

1579, Stefano Bonsignori, 110 × 108 cm
Longitude 69–88, Latitude 21–1

Signature (top left): D. Stephanus Mon(nacus) Montolivetanus Florentinus F.A.S. MDLXXIX."
Cartographic sources: Gastaldi, *Il disegno della geografia moderna de tutta la parte dell'Africa* (1564). The legend is by Bonsignori.
Bibliography: Danti-Bonsignori 1995, 100–1.
Notes: This part of Africa was represented as the fourth map of Africa in Ptolemy's *Geography*; it also appeared in Gastaldi's *Africa Nova Tabula*, the modern map he added to Ptolemy's third map of Africa (Ptolemy 1548).

Par(te) d'Affrica nuova, et Regno di Manicongo

1580, Stefano Bonsignori, 108 × 105 cm
Longitude 42–58, Latitude 1–18

Signature (top left): "D. Stephanus Bonsignoris Monachus Montoliv. F.A.S. 1580."
Cartographic sources: Gastaldi, *Il disegno della geografia moderna de tutta la parte dell'Africa* (1564). The legend is by Bonsignori.
Bibliography: Danti-Bonsignori 1995, 112–13.
Notes: This part of Africa was represented as the fourth map of Africa in Ptolemy's *Geography*; it also appeared in Gastaldi's *Africa Nova Tabula*, the modern map he added to Ptolemy's third map of Africa (Ptolemy 1548).

Parte d'Affrica

(Ethiopia), 1581, Stefano Bonsignori, 110 × 105 cm
Longitude 57–76, Latitude 0–19

Signature (bottom left): "D. Stephanus Bonsig(no)ris Flor(entinus) Monac(us) Monteolivetanus faciebat A. S. 1581."
Cartographic sources: Gastaldi, *Il disegno della geografia moderna de tutta la parte dell'Africa* (1564). The legend is by Bonsignori.
Bibliography: Danti-Bonsignori 1995, 114–15.
Notes: This part of Africa was represented as the fourth map of Africa in Ptolemy's *Geography*; it also appeared in Gastaldi's *Africa Nova Tabula*, the modern map he added to Ptolemy's third map of Africa (Ptolemy 1548).

Parte di Buona Speranza

not dated, Egnazio Danti, 109 × 95 cm
Longitude 58–83, Latitude 12–36

Cartographic sources: Gastaldi's *Cosmographia universalis* (1561) for the general geographical outline of the African continent and the coast of Madagascar. Ortelius, *Map of the World* (1564) for the legend. Danti pasted together different inscriptions that Ortelius had placed in three separate areas of his map. Paolo Forlani, *Africa* (Venice, 1562). Danti might have used this map too, which was in turn heavily based on Gastaldi's *Map of the World* for the geographical outline.
Bibliography: Danti-Bonsignori 1995, 52–3.
Notes: This part of Africa was represented as the fourth map of Africa in Ptolemy's *Geography*; it also appeared in Gastaldi's *Africa Nova Tabula*, the modern map he added to Ptolemy's third map of Africa (Ptolemy 1548).

Parte d'Affrica

(Cape of Good Hope), 1581, Stefano Bonsignori, 110 × 106 cm
Longitude 37–59, Latitude 18–38

Signature (bottom left): "D.o Stephanus Bonsig(nor)is Monach(u)s Monteolivetanus Faciebat A. V. P. MDLXXXI."
Cartographic sources: Gastaldi, *Il disegno della geografia*

moderna de tutta la parte dell'Africa (1564). The legend is by Bonsignori.
Bibliography: Danti-Bonsignori 1995, 116–17.
Notes: This part of Africa was represented as the fourth map of Africa in Ptolemy's *Geography*; it also appeared in Gastaldi's *Africa Nova Tabula*, the modern map he added to Ptolemy's third map of Africa (Ptolemy 1548).

I(sola) di S(an) Lorenzo

1565, Egnazio Danti, 110 × 53 cm
Longitude 77–87, Latitude 11–30

Cartographic sources: The general geographical outline of Madagascar does not follow any printed map but is very close to the geographical representation of the island found in Portuguese nautical charts such as Bartolomeo Velho, *Four nautical charts of the world*, Museo di Storia della Scienza, Florence (previously in the Accademia del Disegno, Florence). The legend and inscriptions are derived from Ortelius, *Map of the World* (1564).
Bibliography: Danti-Bonsignori 1995, 24–5.
Notes: This part of Africa was represented as the fourth map of Africa in Ptolemy's *Geography*; it also appeared in Gastaldi's *Africa Nova Tabula*, the modern map he added to Ptolemy's third map of Africa (Ptolemy 1548).

Asia: Lower Level of the Cupboards from the Main Entrance to the Left

Natolia

before 1565, Egnazio Danti, 110 × 95 cm
Longitude 54–70, Latitude (unreadable)

Cartographic sources: Gastaldi, *Il disegno della prima parte dell'Asia* (1559) for the outline of the southern coast of the peninsula. The northern coast of the peninsula and the island of Cyprus are different from Gastaldi's map. Also the west coast, Rhodi, and other smaller islands, which were not included in Gastaldi's map, seem to be derived from nautical charts.
Bibliography: Danti-Bonsignori 1995, 30–1.
Notes: This part of Asia was represented as the first map of Asia in Ptolemy's *Geography*; it also appears in Gastaldi's *Natolia Nova Tabula*, the modern map he added to Ptolemy's first map of Asia (Ptolemy 1548). This map, cut into two halves and extremely damaged, must have been superseded by the other map *Natolia* that Danti made in 1565.

Natolia

1565, Egnazio Danti, 112 × 95 cm
Longitude 55–71, Latitude 47–31

Cartographic sources: Gastaldi, *Il disegno della geografia moderna della provincia di Natolia* (Venice, 1564). Danti's map is different from Gastaldi's in the representation of the sourthern coast of Cyprus, the outline of the islands of Samos and Rhodi, as well as in the southern coast of the peninsula.
Bibliography: Danti-Bonsignori 1995, 32–3.
Notes: This part of Asia was represented as the first and fourth map of Asia in Ptolemy's *Geography*; it also appeared in Gastaldi's *Natolia Nova Tabula*, the modern map he added to Ptolemy's first map of Asia (Ptolemy 1548), and in Gastaldi's *Soria et Terra Sancta Nova Tabula*, the modern map he added to Ptolemy's fourth map of Asia (Ptolemy 1548).

Armenia

not dated, Egnazio Danti, 112 × 75 cm
Longitude 70–86, Latitude 48–29

Cartographic sources: Gastaldi, *Il disegno della prima parte dell'Asia* (1559) for the general geographical outline. The legend is Danti's.
Bibliography: Danti-Bonsignori 1995, 40–1.
Notes: This part of Asia was represented as the third map of Asia in Ptolemy's *Geography*. The legend was added later since it paints over the frame with the degrees of latitude and longitude as well as some other cartographic details.

Persia
not dated, Egnazio Danti, 112 × 97 cm
Longitude 85–100, Latitude 45–23

Cartographic sources: Gastaldi, *Disegno della seconda parte dell'Asia* (1561) for the general geographical outline. The legend is adapted from Ruscelli's *Della Persia ventesima Tavola Nuova* (Ptolemy 1561).
Bibliography: Danti-Bonsignori 1995, 42–3.
Notes: This part of Asia was represented as the fifth map of Asia in Ptolemy's *Geography*; it also appeared in Gastaldi's *Persia Nova Tabula*, the modern map he added to Ptolemy's fifth map of Asia (Ptolemy 1548). The legend was added later since it paints over the frame with the degrees of latitude and longitude as well as some other cartographic details.

Arabia
1575, Egnazio Danti, 112 × 75 cm
Longitude 72–99, Latitude 59–12

Cartographic sources: Gastaldi, *Disegno della seconda parte dell'Asia* (1561) for the cartographic content. The legend is by Danti.
Bibliography: Danti-Bonsignori 1995, 78–9.
Notes: This part of Asia was represented as the sixth map of Asia in Ptolemy's *Geography*.

Sogd(i)ana
not dated, Egnazio Danti, 112 × 98 cm
Longitude 102–17, Latitude 45–21

Cartographic sources: Gastaldi, *Disegno della seconda parte dell'Asia* (1561) for the general geographical outline. The legend is by Danti.
Bibliography: Danti-Bonsignori 1995, 44–5.
Notes: This part of Asia was represented as the seventh map of Asia in Ptolemy's *Geography*.

No title: Central Asia (Arachosia)
not dated, Egnazio Danti, 112 × 36 cm
Longitude 114–22, Latitude 47–25

Cartographic sources: Gastaldi, *La terza parte dell'Asia* (1561) for the general geographical outline. The legend is by Danti, except for the reference to the "balassci bellissimi" which is taken from one of Gastaldi's inscriptions on the map.
Bibliography: Danti-Bonsignori 1995, 46–7.
Notes: This part of Asia was represented as the seventh map of Asia in Ptolemy's *Geography*.

No title: Central Asia (Deserto di Camul)
not dated, Egnazio Danti, 114 × 36 cm
Longitude unreadable, Latitude 48–23

Cartographic sources: Gastaldi, *La terza parte dell'Asia* (1561) for the general geographical outline. The legend is by Danti.
Bibliography: Danti-Bonsignori 1995, 48–9.
Notes: This part of Asia was represented as the seventh map of Asia in Ptolemy's *Geography*.

Indosta(n) fuori al Gange
not dated, Egnazio Danti, 112 × 55 cm
Longitude 134–56, Latitude 60–21

Cartographic sources: Gastaldi, *La terza parte dell'Asia* (1561) for the general geographical outline and the legend. Danti combined in a single legend the many inscriptions Gastaldi had included in his map.
Bibliography: Danti-Bonsignori 1995, 50–1.
Notes: This part of Asia was represented as the eighth map of Asia in Ptolemy's *Geography*.

Parte del India Dentro al Gange Ho(ggi) detta Indostan
1575, Egnazio Danti, 112 × 66 cm
Longitude 115–35, Latitude 38–0

Cartographic sources: Gastaldi, *La terza parte dell'Asia* (1561) for the general geographical outline and the inscription referring to the body of St. Thomas along the coast. Ortelius, *Asiae orbis partium maxime nova descriptio* (1567) for the legend pertaining to the island of Ceylon.

Bibliography: Danti-Bonsignori 1995, 72–3.
Notes: This part of Asia was represented as the ninth and tenth map of Asia in Ptolemy's *Geography*; it also appeared in Gastaldi's *Calecut Nova Tabula*, the modern map he added to Ptolemy's tenth map of Asia (Ptolemy 1548).

No title: Indochina and Indonesia

1573, Egnazio Danti, 110 × 62 cm
Longitude and latitude unreadable

Cartographic sources: Gastaldi, *La terza parte dell'Asia* (1561) for the general geographical outline, except for the regions below the equator, which were taken from Gastaldi, *Cosmographia universalis* (1561). Ortelius, *Asiae orbis partium maxime nova descriptio* (1567) was used for the first paragraph of the legend.
Bibliography: Danti-Bonsignori 1995, 70–1; Suarez 1999, 130–57.
Notes: This part of Asia was represented as the eleventh and twelfth map of Asia in Ptolemy's *Geography*; it also appeared in Gastaldi's *India tercera Nova Tabula*, the modern map he added to Ptolemy's eleventh and twelfth maps of Asia (Ptolemy 1548).

China

1575, Egnazio Danti, 112 × 62 cm
Longitude 149–71, Latitude 53–16

Cartographic sources: Gastaldi, *La terza parte dell'Asia* (1561) for the general geographical outline and the inscription along the coast pertaining to the fishing of pearls.
Bibliography: Danti-Bonsignori 1995, 74–5.
Notes: This part of Asia was unknown to Ptolemy and thus not represented in his *Geography*. It appeared, however, in Gastaldi's map *India tercera Nova Tabula*, the modern map he added to Ptolemy's eleventh and twelfth maps of Asia (Ptolemy 1548). The cartouche with the legend was added after the map was completed since it was painted over some islands and the frame, covering the degrees of longitude and latitude.

No title: China and Japan

not dated, Egnazio Danti, 110 × 63 cm
Longitude 173–97, Latitude 62–26

Cartographic sources: Gastaldi, *Cosmographia universalis* (1561) for the coast of China and Japan; Ortelius, *Asiae orbis partium maxime nova descriptio* (1567) for the legend (Danti disregarded Ortelius's new representation of Japan in this map and followed instead Gastaldi's). The first paragraph of Danti's legend is his original composition, while the second paragraph is taken from Ortelius's map. It combines Ortelius's legend pertaining to the city of Quintai with his inscription on porcelain in the province of Quinci.
Bibliography: Kish 1951; Danti-Bonsignori 1995, 76–7.
Notes: This part of Asia was unknown to Ptolemy and thus not represented in his *Geography*. It appears in Gastaldi's map *India tercera Nova Tabula*, the modern map he added to Ptolemy's eleventh and twelfth maps of Asia (Ptolemy 1548).

L'isole Moluche con l'altre circu(m)vicine che producano le gioie et le spetierie, mdlxiii

1563, Egnazio Danti, 110 × 63 cm
Longitude 158–80, Latitude 26–0, 0–12

Cartographic sources: Gastaldi, *Il disegno della terza parte dell'Asia* (1561) for the regions north of the equator; and Gastaldi's *Cosmographia universalis* (1561) for the regions below the equator.
Bibliography: Danti-Bonsignori 1995, 76–7; Suarez 1999, 130–57.
Notes: This part of Asia was unknown to Ptolemy and thus not represented in his *Geography*. It appears, however in Gastaldi's map *India tercera Nova Tabula*, the modern map he added to Ptolemy's eleventh and twelfth maps of Asia (Ptolemy 1548)

Parte di Scitia

1582, Stefano Bonsignori, 110 × 107 cm
Longitude 60–100, Latitude 74–54

Signature (top right): "D.s Stephanus Florentinus Monachus Montolivetanus F.A.S. 1582."
Cartographic sources: Mercator, *Map of the World* (1569) for the general geographical outline. The legend is by Bonsignori.

Bibliography: Danti-Bonsignori 1995, 118–19.
Notes: This part of Asia was unknown to Ptolemy and thus not represented in his *Geography*, nor did it appear in Gastaldi's maps of Asia (1559–61).

No title: Siberia
1586, Stefano Bonsignori, 110 × 109 cm
Longitude 110–210, Latitude 76–46

Signature (top right): "D. Stephanus Monachus Montolivetanus F.A.S. MDLXXXVI."
Cartographic sources: Mercator, *Map of the World* (1569) for the general geographical outline. The legend is by Bonsignori.
Bibliography: Danti-Bonsignori 1995, 124–5.
Notes: This part of Asia was unknown to Ptolemy and thus not represented in his *Geography*, nor did it appear in Gastaldi's maps of Asia (1559–61).

America: Upper Level from the Main Entrance to the Left

Ultime parti note nel Indie Occide(nta)li
May 1564, Egnazio Danti, 109 × 95 cm
Longitude 210–40, Latitude 51–19

Cartographic sources: Gastaldi, *Cosmographia universalis* (1561), and Ortelius, *Map of the World* (1564) for the general geographical outline and Japan.
Bibliography: Danti-Bonsignori 1995, 22–3

Nuova Spagna
1565, Egnazio Danti, 110 × 95 cm
Longitude 40–63, Latitude 33–8

Cartographic sources: Gastaldi, *Cosmographia universalis* (1561), and Ortelius, *Map of the World* (1564) for the general geographical outline; Ruscelli's *Nuova Hispania, tremptesimaprima Tavola Nuova* (Ptolemy 1561) for the legend.
Bibliography: Danti-Bonsignori 1995, 26–7.

No title: Gulf of Mexico
not dated, Egnazio Danti, 112 × 76 cm
Longitude 263–90, Latitude 43–4

Cartographic sources: Gastaldi, *Cosmographia universalis* (1561), and Ortelius, *Map of the World* (1564) for the general geographical outline.
Bibliography: Danti-Bonsignori 1995, 58–9.
Notes: The cartouche was added after the compilation of the map since it paints over some geographical details. Some retouchings are evident around the islands although it is unclear whether they are by Danti himself or by a later restorer. The map was not completed and has no legend in the cartouche.

No title: Central America
dated after 1569, Egnazio Danti, 112 × 76 cm
Longitude 281–14, Latitude 39–1

Cartographic sources: Gastaldi, *Cosmographia universalis* (1561), and Ortelius, *Map of the World* (1564) for the general geographical outline.
Bibliography: Danti-Bonsignori 1995, 60–1.

No title: South America, Peru
not dated, Egnazio Danti, 114 × 69 cm
Longitude 280–305, Latitude 5–30

Cartographic sources: Gastaldi, *Cosmographia universalis* (1561) and Ortelius, *Map of the World* (1564) for the general geographical outline.
Bibliography: Danti-Bonsignori 1995, 62–3.

No title: Peru
not dated, Egnazio Danti, 111 × 68 cm
Longitude 229–323, Latitude 5–30

Cartographic sources: Gastaldi, *Cosmographia universalis* (1561) and Ortelius, *Map of the World* (1564) for the general geographical outline.
Bibliography: Danti-Bonsignori 1995, 64–5.
Notes: The map was not completed and the cartouche, which was added later, has no inscription.

No title: North Brazil
Egnazio Danti, 113 × 30 cm
Longitude 324–36, Latitude 1–35

Cartographic sources: Gastaldi, *Cosmographia universalis* (1561) and Ortelius, *Map of the World* (1564) for the general geographical outline.
Bibliography: Danti-Bonsignori 1995, 66–7.
Notes: The cartouche has no inscription.

No title: Strait of Magellan
not dated, Egnazio Danti, 113 × 57 cm
Longitude 196–226, Latitude 25–55

Cartographic sources: Gastaldi, *Cosmographia universalis* (1561),and Ortelius, *Map of the World* (1564) for the general geographical outline. Portuguese nautical charts, possibly Bartolomeo Velho's four nautical charts (1561) Museo di Storia della Scienza, Florence, or Mercator, *Map of the World* (1569) for the geographical outline of the Tierra del Fuego and of the Strait of Magellan.
Bibliography: Danti-Bonsignori 1995, 68–9.
Notes: This map must have been superseded by the other map of the Strait of Magellan that Bonsignori made in the early 1580s.

No title: Strait of Magellan
(?)1581, Stefano Bonsignori, 112 × 57 cm
Longitude 296–313, Latitude 28–53

Signature (bottom left): "Stephanus (?)1581."
Bibliography: Danti-Bonsignori 1995, 120–1.

Greenland and the North Pole

Gronla(n)dia
not dated, Egnazio Danti, 110 × 90 cm
Longitude 295–360, Latitude 76–53

Cartographic sources: Nicoló and Antonio Zeno, *Carta da navegar* (Zeno Map) (Venice, 1558), perhaps through one of its derivative maps, possibly the map published in Ptolemy 1561. Like the map in Ptolemy 1561 but unlike the original Zeno Map, Danti's map does not show any connection between Greenland and Europe.
Bibliography: Danti-Bonsignori 1995, 56–7

No title: North Pole
after 1585, attributed to Antonio Santucci, 110 × 48 cm
Longitude 140–230, Latitude 90–74

Cartographic sources: Mercator, *Map of the World* (1569) for the general geographical outline and for the legends and inscriptions.
Bibliography: Danti-Bonsignori 1995, 128–9.

No title: North Pole
after 1585, attributed to Antonio Santucci, 120 × 47 cm
Longitude 50–140, Latitude 90–73

Cartographic sources: Mercator, *Map of the World* (1569) for the general geographical outline and for the legends and inscriptions.
Bibliography: Danti-Bonsignori 1995, 130–1.

No title: North Pole
after 1585, attributed to Antonio Santucci, 110 × 48 cm
Longitude 310–50, Latitude 90–73

Cartographic sources: Mercator, *Map of the World* (1569) for the general geographical outline and for the legends and inscriptions.
Bibliography: Danti-Bonsignori 1995, 126–7.).

No title: North Pole
after 1585, attributed to Antonio Santucci, 110 × 47 cm
Longitude 230–320, Latitude 90–73

Cartographic sources: Mercator, *Map of the World* (1569) for the general geographical outline and for the legends and inscriptions.
Bibliography: Danti-Bonsignori 1995, 132–3.

Appendix B

The Gallery of Maps

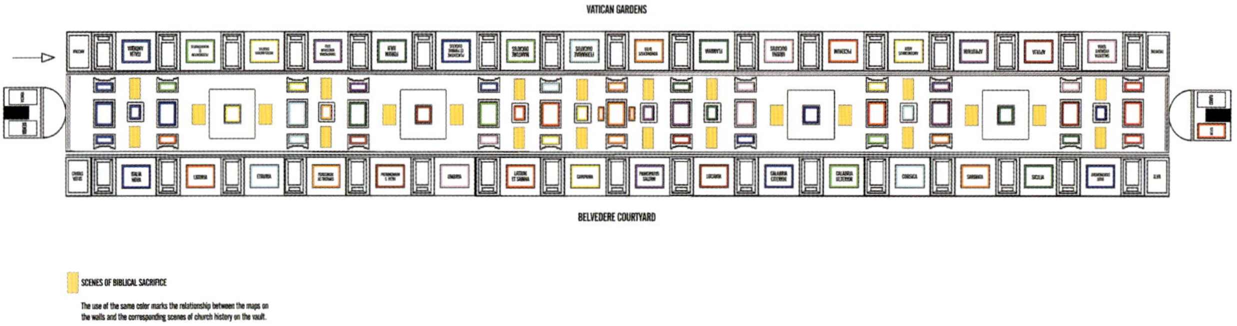

Italy and Church History

This appendix shows the close relations between the maps of Italy in the Gallery of Maps and Leandro Alberti's *Descrittione di tutta l'Italia* (Bologna, 1577).

First, the regional maps of the Vatican corridor are organised according to the narrative sequence of Alberti's chorography book, that is, according to the perspective of an imaginary traveler who is circumnavigating the peninsula. Like Alberti's book, in the Gallery of Maps the representation of the peninsula starts from the northern regions, on the side of the Tyrrehnian Sea, which is depicted at the end of the corridor – toward the old Vatican Palace. It continues with the depiction of southern Italy, represented along the east wall of the Vatican corridor, and of the regions around the boot of the peninsula, painted at the end of the gallery towards the Belvedere Hill. The mapping of Italy continues with the regions on the side of the Adriatic Sea, which are represented as proceeding from south to north on the north wall of the Vatican corridor. It concludes with the depiction of Venice at the end of the gallery.

Second, the appendix reinforces the connection between the maps and the vault historical scenes by listing the maps with their corresponding historical scenes. Pertinent references to Leandro Alberti, *Descrittione di tutta l'Italia* (Bologna, 1577), to Cesare Baronio, *Annales ecclesiastici* (Cologne, 1624), and Gambi–Pinelli 1994 are provided for both maps and historical scenes.

GENUA

Historical vignettes on the map: None
Vault scenes: None
Bibliography: Gambi–Pinelli 1994, no. 405.

CIVITAS VETUS
Historical vignettes on the map: None
Vault scenes: None
Bibliography: Gambi–Pinelli 1994, nos. 399, 400, 403, 404.

ITALIA NOVA
Historical vignettes on the map: None
Vault scenes:
Emperor Constantine Building the Basilica of St. Paul. Inscription: *Constantinus Imp. S. Paulo Apostolo Basilicam Erigit.*
Pope Sylvester Baptizing Emperor Constantine. Inscription: *S. Silvester P.P. Constantinu. (i)mp. Baptizat.*
Emperor Constantine Building the Basilica of St. Peter. Inscription: *Constantinus Imp. S. Petro Apostolo Basilicam Erigit.*
Emperor Constantine Holding the Reins of Pope Sylvester's Horse. Inscription: *Constantinus Imp. S. Silvestri Equi Frenum Tenet.*
Constantine's Vision of the Cross Before the Battle at the Milvian Bridge. Inscription: *In Hoc Vince.*
Bibliography: Gambi–Pinelli 1994, nos. 3–30, 554–6, 558, 566.

LIGURIA
Historical vignettes on the map: None
Vault scenes:
The Translation of St. John's Ashes to Genoa. Inscription: *S. Io. Baptistae Cineres Genuam Deferuntur* (Alberti 1577, 15v).
Bibliography: Gambi–Pinelli 1994, nos. 27–41, 567.

ETRURIA
Historical vignettes on the map: None
Vault scenes:
St. Romuald Founding the Hermitage at Camaldoli. Inscription: *S. Romualdus Camaldulensen Eremum Instituit* (Alberti 1577, 50v).
St. Francis Receiving the Stigmata at Mount Verna. Inscription: *S. Franciscus in Avernia Stigmatibus Decoratur* (Alberti 1577, 51).
Bibliography: Gambi–Pinelli 1994, nos. 48–59, 600–1.

PERUSINUS AC TIFERNAS
Historical vignettes on the map:
Hannibal Winning over the Romans at Lake Trasimeno. Inscription: *Hic Hannibal consule interfecto Romanos profligavit A. 537* (Alberti 1577, 65v).
Vault scenes:
Bishop St. Constantius Healing the Weak with the Sign of the Cross. Inscription: *S. Constantius Episc. Signo Crucis Infirmos Sanat* (Baronio, 1624, year 175).
Bibliography: Gambi–Pinelli 1994, nos. 62–5, 604.

PATRIMONIUS S. PETRI
Historical vignettes on the map: None
Vault scenes:
Countess Matilda Donating Her Possessions to the Roman Church. Inscription: *Mathilda Multa Bona Romanae Ecclesiae Obtulit* (Alberti 1577, 35).
Pope Urban IV Celebrating in Orvieto the Miracle of the Corpus Christi that Took Place at Bolsena. Inscription: *Christi Corporis Miraculum Vulsini Actum in Urbe Vetu ab Urbano IIII celebratur* (Alberti 1577, 62 on the celebrations at Orvieto, 70 on the miracle at Bolsena).
Bibliography: Gambi–Pinelli 1994, nos. 71–5, 613, 622.

UMBRIA
Historical vignettes on the map:
Hannibal Unsuccessfully Besieging Spoleto (Alberti 1577, 92v).
Winigis Rescuing Pope Leo III and Bringing Him to Spoleto.
Vault scenes:
St. Clare Freeing Assisi from Siege by the Saracens. Inscription: *S. Clara Assisium a Saracenis Obsessum Liberat* (Alberti 1577, 91, mentioned St. Clare in Assisi but not specifically this episode).
Bibliography: Gambi–Pinelli 1994, nos. 84–92, 647.

LATIUM ET SABINA
Historical vignettes on the map: None
Vault scenes:
The Fall of Simon Magus. Inscription: *Simon Magus S. Petro Debitas Poenas Persolvit* (Voragine 1993, I, 340–50.

Baronio 1624, year 68).
Christ Appearing to St. Peter at the Via Appia. Inscription: *Domine Quo Vadis?* (Voragine 1993, I, 340–50).
Bibliography: Gambi–Pinelli 1994, nos. 95–100, 650, 658.

CAMPANIA
Historical vignettes on the map:
Pope John X Defeating the Saracens near the River Garigliano (Alberti 1577, 185, 193v).
Vault scenes:
The Miracle of St. Januarius's Blood in Naples. Inscription: *S. Ianuarii Sanguis Neapoli Ebullit* (Alberti 1577, 184v. Baronio 1624, year 305; and Baronio 1588, 423).
St. Benedict Unmasking the Disguised King Totila. Inscription: *S. Benedictus Totilam Regem Non Agnitum Detegit* (Alberti 1577, 274, on the meeting between St. Benedict and Totila, but not on this specific episode. Baronio, 1624, year 543).
Bibliography: Gambi–Pinelli 1994, nos. 104–7, 659, 662.

PRINCIPATUS SALERNI
Historical vignettes on the map: None
Vault scenes:
The Manna Coming from St. Andrew's Body. Inscription: *Manna e Corpore S. Andreae Fluit* (Alberti 1577, 195v. Baronio, 1624, year 1208).
Rainulph Winning His Enemies Thanks to St. Bernard's prayers. Inscription: *Rainulphus S. Bernardi Precibus Hostes Vincit* (Baronio, 1624, year 1137).
Bibliography: Gambi–Pinelli 1994, nos. 113–15, 676, 684.

LUCANIA
Historical vignettes on the map:
Marcellus Killed by Hannibal's Cavalry near Venosa (Alberti 1577, 245v, mentioned Venosa in relation to Hannibal's victory at Cannae, although not the details mentioned in the inscription).
Vault scenes:
St. Laverius Liberating a Man from Demons. Inscription: *S. Laberii Precibus Fugantur Daemones.*
Bibliography: Gambi–Pinelli 1994, nos. 125–31, 685.

CALABRIA CITERIOR
Historical vignettes on the map:
Alexander of Epirus Killed in the Battle at Pandosia Against the Bruttians (Alberti 1577, 201–201v.).
Vault scenes:
St. Francis of Paola Unharmed from Fire. Inscription: *S. Franc. De Paula ab Igne Non Laeditur.*
St. Francis of Paola Using His Cloak as a Boat. Inscription: *S. Franciscus de Paula Pallio Utitur Pro Navigio* (Alberti 1577, 203v, mentioned the saint in relation to the town of Paola, but did not specifically refer to any miracle).
Bibliography: Gambi–Pinelli 1994, nos. 135–42, 697, 706.

CALABRIA ULTERIOR
Historical vignettes on the map: None
Vault scenes:
Emperor Henry II Reinforcing His Soldiers with the Corpus Christi. Inscription: *Henricus Imp. Christi Corpore Milites Suos Corroborat* (Baronio, 1624, year 1022).
Bibliography: Gambi–Pinelli 1994, nos. 145–6, 731.

CORSICA
Historical vignettes on the map: None
Vault scenes:
The People of Corsica Acknowledging Gregory VII's Sovereignty through His Legate. Inscription: *S. Gregorium VII Dominum Suum Per Legatu[m] Agnoscunt* (Baronio 1624, year 1077).
Bibliography: Gambi–Pinelli 1994, nos. 146–64, 734.

SARDINIA
Historical vignettes on the map: None
Vault scenes:
Pope Symmachus Sending Aid to Exiled Bishops. Inscription: *Symmachus Papa Epis. Relegatis Necessaria Subministrat* (Baronio, 1624, year 504).
Bibliography: Gambi–Pinelli 1994, nos. 168–96, 743.

SICILIA
Historical vignettes on the map: None

Vault scenes:
King Theodoric Falling into Hell. Inscription: *Theodoricus Rex in Infernum Delicitur* (Baronio 1624, year 526).
The Stopping of the Lava on Aetna Thanks to St. Agatha's Veil. Inscription: *S. Agathae Tenui Velo Aetnae Ignis Propulsatur* (Alberti 1577, 34; Baronio 1624, year 254; and Baronio, 1588, 71).
Bibliography: Gambi–Pinelli 1994, nos. 178–89, 752, 777.

AVENIONEN(SIS) DITIO ET VENAISINUS COMITATUS
Historical vignettes on the map:
Pope Gregory XI Returning the Papacy from Avignon to Rome.
Vault scenes:
Emperor Valentinian Compelled by Divine Intervention to Submit to St. Martin Bishop. Inscription: *S. Martino Ep[. . .]o Adventari Valentinianus Imp. Reverenter Assurgere Divinitus Compellit* (Baronio 1624, year 375).
Bibliography: Gambi–Pinelli 1994, nos. 192–5, 780.

ILVA
Historical vignettes on the map: None
Vault scenes: None
Below the map: Reconstruction of the ancient port at Ostia
Bibliography: Gambi–Pinelli 1994, nos. 439–41.

MELITA
Historical vignettes on the map: None
Vault scenes:
St. Paul Unharmed by the Bite of a Viper. Inscription: *S. Paulum Viperam Illaesus Excutit.*
St. Paul Healing Publius's Father in Malta. Inscription: *Paulus Publi Parentem Melitae Sanat.*
St. Paul Embarking for Rome. Inscription: *S. Paulus Romam Venturus Navem Consce[n]dit.*
Below the map: The Siege of La Valletta
Bibliography: Gambi–Pinelli 1994, nos. 430–8, 788, 789, 790.

CORFU
Historical vignettes on the map: None
Vault scenes: None
Below the map: The Battle of Lepanto
Bibliography: Gambi–Pinelli 1994, nos. 430–8.

TREMITAE
Historical vignettes on the map: None
Vault scenes: None
Below the map: View of the port at Ostia
Bibliography: Gambi–Pinelli 1994, nos. 428–9.

SALLENTINA HYDRUNTI TERRA
Historical vignettes on the map: None
Vault scenes:
St. Anselm Arguing with the Heretics at the Council of Bari. Inscription: *S. Anselmus in Concilio Haereticos Arguit* (Baronio 1624, year 1097).
Bibliography: Gambi–Pinelli 1994, nos. 384–93, 778.

APULIA
Historical vignettes on the map:
Roman Disaster at Cannae (Alberti 1577, 246).
Vault scenes:
St. Michael the Archangel Appearing on Monte Gargano. Inscription: *S. Michael in Monte Gargano Apparet* (Alberti 1577, 249–250v; Baronio 1624, year 493).
Bibliography: Gambi–Pinelli 1994, nos. 372–80, 776.

APRUTIUM
Historical vignettes on the map: None
Vault scenes: None
St. Bernardine of Siena Ordering the Bonfire of the Vanities. Inscription: *S. Bernardinus Vitiorum Illecebras Conburit* (Alberti 1577, 263v, on the saint's tomb in L'Aquila, but not specifically on this episode; Baronio 1624, year 1424).
The Hermit St. Peter of Morrone Being Elected Pope. Inscription: *S. Petrus de Murano Anacoreta in Pontificem Eligitur* (Alberti 1577, 259; Baronio 1624, year 1294).
Bibliography: Gambi–Pinelli 1994, nos. 361–9, 742, 744.

ANCONITANUS AGER

Historical vignettes on the map: None
Vault scenes:
St. Marcellinus Saving the City of Ancona from Fire. Inscription: *S. Marcellinus Anconam Urbem Incendio Eximit.*
Bibliography: Gambi–Pinelli 1994, nos. 330–3.

The Holy Spirit Designating St. Severus Archbishop of Ravenna. Inscription: *S. Severus a Spiritu[m] Ravennae Archiep. Designatur* (Alberti 1577, 308–308v, mentioned St. Severus together with the other archbishops of Ravenna designated through the intervention of the Holy Spirit; Baronio 1624, year 337).
Bibliography: Gambi–Pinelli 1994, nos. 305–13, 686–688.

PICENUM

Historical vignettes on the map:
Picenum Sending 15,000 Soldiers to Protect Rome and the Pope.
Vault scenes:
The Translation of the Holy House of Loreto. Inscription: *Angeli Mariae Virginis Domum Circuferunt* (Alberti 1577, 282v–283; Baronio 1624, year 1295).
Bibliography: Gambi–Pinelli 1994, nos. 334–7, 730.

URBINI DUCATUS

Historical vignettes on the map:
Consul Claudius Nerus Killing Hasdrubal and 56 000 Carthaginians at Metauro (Alberti 1577, 288v).
Vault scenes:
St. Ubaldus Defeating His Enemies with the Sign of the Cross. Inscription: *S. Ubaldus Signo Crucis Hostes Fugat* (Alberti 1577, 89–89v, mentioned the cult of St. Ubaldus in Gubbio, his church, the yearly celebrations and Dante's reference to the saint, Par. XI. Baronio 1624, year 1154).
Cardinal St. Peter Damian Writing the Rule for His Hermit Community. Inscription: *B. Petrus Damianus Card. Reculam Eremiticam Conscribit.*
Bibliography: Gambi–Pinelli 1994, nos. 320–7, 696, 698.

FLAMINIA

Historical vignettes on the map:
The Battle of the French Army and the Holy League at Ravenna (Alberti 1577, 310, 360v).
Julius Caesar Crossing the River Rubicon (Alberti 1577, 301).
Vault scenes:
St. Antony of Padua Preaching to the Fish. Inscription: *S. Antonio praedicante pisces ex aqua emergunt.*

BONONIENSIS DITIO

Historical vignettes on the map:
The Bolognese Defeating the Visconti at San Ruffillo (Alberti 1577, 330).
The Formation of the Second Triumvirate (Alberti 1577, 339–339v).
Vault scenes:
Angels Distributing Bread to St. Dominic in Bologna. Inscription: *S. Domenico Panem Bononiae Angeli Ministrant* (according to the saint's Life this episode took place in Rome rather than in Bologna).
St. Petronius Reviving a Dead Man in Bologna. Inscription: *S. Petronius Bononiae Mortuum Suscitat* (Alberti 1577, 328, recalled the construction of the church of St. Petronius as an endless project. Baronio 1624, year 430).
Pasce Oves Meas (Bible 1970b, John, XXI, 15. Baronio 1624, year 34).
The Monastery of San Michele in Bosco near Bologna. Inscription: *Celeberrimum S. Michaelis Monasterium in Nemore Propre Bononiam.*
The Monastery of the Virgin Mary at the Monte della Guardia near Bologna. Inscription: *Effigies B.M. Virginis in Monte Guardiae S. Lucae Opus.*
Bibliography: Gambi–Pinelli 1994, nos. 289–97, 670–4.

FERRARIAE DUCATUS

Historical vignettes on the map:
Pope Julius II Entering Mirandola (Alberti 1577, 360v).
Vault scenes:
Modena Being Freed Thanks to St. Geminianus's prayers. Inscription: *Mutina S. Geminiani Precibus Servatur.*
Bibliography: Gambi–Pinelli 1994, nos. 273–81, 660.

MANTUAE DUCATUS
Historical vignettes on the map:
Leo I Stopping Attila the Hun (Alberti 1577, 389).
Vault scenes:
Leo I Stopping Attila near Mantua. Inscription: *S. Leo Pont. Max. Attila Furentem Reprimit* (Alberti 1577, 389; Baronio 1624, year 452).
Bibliography: Gambi–Pinelli 1994, nos. 262–6, 646.

PIACENTIAE ET PARMAE DUCATUS
Historical vignettes on the map:
The Carthaginians Defeating the Romans at Trebbia (Alberti 1577, 372, 374v).
Vault scenes:
Parma Being Freed From Frederick II's Siege Thanks to Pope Innocent IV's intervention. Inscription: *Parma Urbs Federici II Innocentii IIII Obsidione Eximitur Pont. Max. Opera* (Alberti 1577, 369, mentioned the siege of Parma during Innocent IV's papacy but not the direct papal intervention; Baronio 1624, year 1247).
Bibliography: Gambi–Pinelli 1994, nos. 250–9, 648.

FORUM IULII
Historical vignettes on the map: None
Vault scenes:
Emperor Frederick II Pledging Obedience to Pope Alexander III in Venice. Inscription: *Federicus Imp. Alexandro III Obedientia[m] Praestat* (Alberti 1576, 86; Baronio 1624, year 1177).
Bibliography: Gambi–Pinelli 1994, nos. 237–8, 612.

TRANSPADANA VENETORUM DITIO
Historical vignettes on the map: None
Vault scenes:
St. Francis Appearing to St. Antony of Padua at Arles. Inscription: *S. Antonio Praedicanti S. Franciscus Apparet.*
Bibliography: Gambi–Pinelli 1994, nos. 227–34, 614.

MEDIOLANENSIS DUCATUS
Historical vignettes on the map:
Cornelius Scipio Defeated by Hannibal at Pavia (Alberti 1577, 422).
Count de St. Paul Overcoming Pavia (Alberti 1577, 423v).
Marcus Marcellus Defeating the Insubrians at Clastidium (Alberti 1577, 414).
Charlemagne Defeating the Lombards and Capturing Their King Desiderius at Pavia (Alberti 1577, 422).
Vault scenes:
St. Ambrose Repelling Emperor Theodosius from Entering Church. Inscription: *S. Ambrosius Theodosium Imp. ab Ingressu Ecclesiae Repellit* (Baronio 1624, year 390).
St. Ambrose Expelling the Enemies from Milan. Inscription: *S. Ambrosius e Mediolano Ecclesiae Hostes Expellit.*
Bibliography: Gambi–Pinelli 1994, nos. 215–24, 576, 602.

PEDEMONTIUM ET MONSFERATUS
Historical vignettes on the map:
The Battle between the French and the Spanish at Ceresole (Alberti 1577, 384v, reporting on the many battles between the French and the Spanish in this area, but not specifically on the battle at Ceresole).
Vault scenes:
The Exposition of the Holy Shroud in Turin. Inscription: *Christi Imago Ostenditur Sancta Sindone Expressa* (contemporary event attended by numerous clergymen, including Cardinal Charles Borromeo).
Bibliography: Gambi–Pinelli 1994, nos. 115–24.

ITALIA ANTIQUA
Historical vignettes on the map: None
Vault scenes:
Emperor Constantine Building the Basilica of St. Paul. Inscription: *Constantinus Imp. S. Paulo Apostolo Basilicam Erigit.*
Pope Sylvester Baptizing Emperor Constantine. Inscription: *S. Silvester P.P. Constantinu. (i)mp. Baptizat.*
Emperor Constantine Building the Basilica of St. Peter. Inscription: *Constantinus Imp. S. Petro Apostolo Basilicam Erigit.*
Emperor Constantine Holding the Reins of Pope Sylvester's Horse. Inscription: *Constantinus Imp. S. Silvestri Equi Frenum Tenet.*
Constantine's Vision of the Cross Before the Battle at the

Milvian Bridge. Inscription: *In Hoc Vince.*
Bibliography: Gambi–Pinelli 1994, nos. 3–30, 554–6, 558, 566.

ANCONA
Historical vignettes on the map: None
Vault scenes: None
Bibliography: Gambi–Pinelli 1994, no. 414.

VENETIA
Historical vignettes on the map: None
Vault scenes: None
Bibliography: Gambi–Pinelli 1994, no. 413.

Appendix C

The Terza Loggia

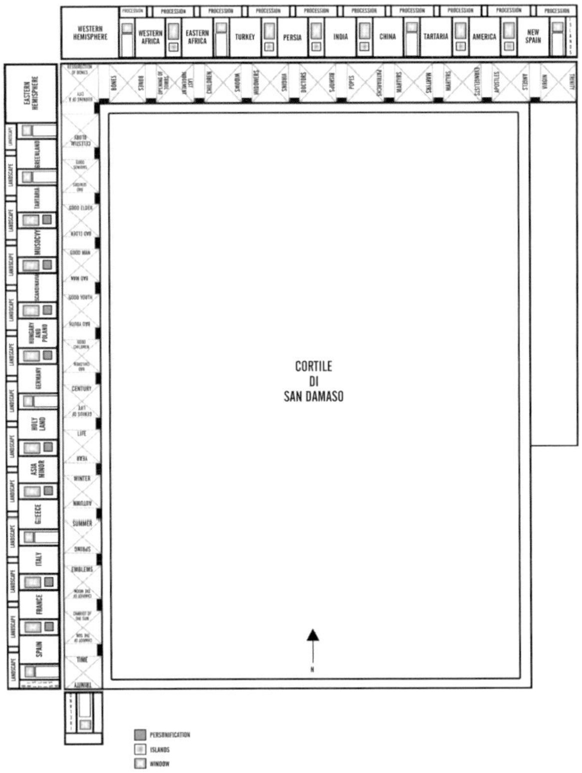

Maps and Religion

This appendix shows the relations between the maps painted on the wall of the Terza Loggia and the religious scenes depicted on the ceiling. It lists the decorations vault by vault, starting from the western wing that Pius IV had painted in the early 1560s and proceeding to the northern wing that Gregory XIII completed in the early 1580s. The decorations of the vault are listed before those of the walls. It is worth mentioning that the maps are not depicted directly underneath the vaults, but rather in the areas between two consecutive vaults. The titles given in brackets are the ones documented in Agostino Taja, *Descritione del Palazzo vaticano* (Rome, 1752). Most of the city views that once complemented the general maps are impossible to identify today and many were unidentifiable already in the early eighteenth century, when Taja wrote his guidebook to the Vatican Palace.

First vault
Scenes: *Trinity* and *Time*
Inscription: *Pius IIII Medices Mediolanensis Pontifex Maximus*
Coat of arms: Pius IV

On the wall
Map of the British Isles

Map of Ireland

Second vault
Scenes: *Chariot of the Sun* (repeated in two separate scenes).
Inscription: *Pacis auctor et custos parta securitate rerum omnium copiam et vilitatem introduxit.*
Coat of arms: Pius IV

On the wall
Map of Spain
View of the Monastery of San Justo in Castile, where Emperor Charles V spent his last days (with inscription).

Third vault
Scenes: *Chariot of the Moon* and emblems relating to planetary motions.
Inscription: *Iustitiam colit causarum explicatione revocata iudiciis ex aequo et bono constitutis.*
Coat of arms: Pius IV

On the wall
Map of France
Landscape with the following inscription: *Terminus posuisti, quem non transgredientur.*

Fourth vault
Scenes: *Spring* (*Ver*) and *Summer* (*Aetas*)
Inscription: *Egentes homines molesta et fallaci mendicitate sublata certis locis distributos alit.*
Coat of arms: Pius IV

On the wall
Map of Italy
Landscape with river and the following inscription: *Potestas eius a mare usque ad mare.*

Fifth vault
Scenes: *Autumn* (*Autumnus*) and *Winter* (*Hyems*)
Inscription: *Portam et viam Pias publicae commoditati aperuit et munivit.*
Coat of arms: Pius IV

On the wall
Map of Greece
Landscape

Sixth vault
Scenes: *Year* (*Annus*), and *Life* (*Vita*)
Inscription: *Thermas diocletianas Mariae Virgini de' dicatas magnificentissimo templo exornari mandavit.*

On the wall
Map of Asia Minor
Landscape with city view

Seventh vault
Scenes: *Genius of Life* (*Genius Vitae*) and *Century* (*Saeculum*)
Inscription: *Portam et viam Flaminias reficiebat Adriani molem nova munitione cingebat.*
Coat of arms: Pius IV

On the wall
Map of the Holy Land
Landscape with city in the background

Eighth vault
Scenes: *Bad Children* (*Pueritia mala*) and *Good Children* (*Pueritia bona*)
Inscription: *Anconam Hostiam Centumcellas moenibus portubus arcibus muniebat.*
Coat of arms: Pius IV

On the wall
Map of Germany
Landscape

Ninth vault
Scenes: *Bad Youth* (*Iuventus mala*) and *Good Youth* (*Iuventus bona*)
Inscription: *Viam Aureliam pene in viam restituibat*

aquam Saloniam deducebat.

On the wall
Map of Hungary and Poland
Landscape with Polish soldiers

Tenth vault
Scenes: *Bad Men* (*Virilitas mala*) and *Good Men* (*Virilitas bona*)
Inscription: *Virtutem et liberalia studia honesta bat. Imprimendi artem in urbem inducebat.*
Coat of arms: Pius IV

On the wall
Map of the Scandinavian Peninsula
Landscape

Eleventh vault
Scenes: *Bad Elders* (*Senectus mala*) and *Good Elders* (*Senectus bona*)
Inscription: *Concilio Tridenti indicto principes ad rem adiuvandam cohortatus omnibus evocatis.*
Coat of arms: Pius IV

On the wall
Map of Muscovy
Landscape

Twelfth vault
Scenes: *Bad Seniors* (*Decrepitas mala*) and *Good Seniors* (*Decrepitas bona*)
Inscription: *Christianae religionis disciplinam multis locis perturbatam omni ope et contentione corrigendam restituendamque curabat.*
Coat of arms: Pius IV

On the wall
Map of Tartaria
Painting by Giovanni Antonio Vanosino representing the Council of Trent convened in its last session (with inscription) now removed.

Thirteenth vault
Scenes: *Death* (*Mors*) and *Celestial Glory* (not documented in Taja)
Inscription: *Sac. Concilii Tridentini legibus firmatis pro earum observantia stenue laboravit* (not documented in Taja).
Coat of arms: Urban VIII

On the wall
Map of Greenland
Landscape

Fourteenth vault (at the corner between the western and northern wing)
Scenes: *The Burning of a City* (*Finis per ignem*) and *The Resurrection of Bones* (*Ossa vivent*).
Inscription: *Galliam dissidio religionis exorto a domesticis hostibus vexatam opibus copiis pecuniis adiuvit ann. IV.*
Inscription: *Avenionem et regionem universam praesenti periculo et calamitate liberavit.*
Coat of arms: Pius IV

On the wall
Map of the Western Hemisphere
Map of the Eastern Hemisphere

Fifteenth vault
Scenes: *The Restoration of Bones With Bones* (*Ossa ad ossa*) and *The Restoration of Bones With Flesh* (*Carnes succrescunt*).
Inscription: *Iustitiam restituebat mores emendabat sacerdotia ad veteris formam instituti revocabat.*
Coat of arms: Pius IV

On the wall
Map of Western Africa (*Africae pars, regna diversa continens*)
Three city views (*Senega lignea civitas portatilis, Fessanova civitas, Fessavetus Mauritaniae urbs Regia*).
Inscription on the Translation of the Relics of St.

Gregory of Nazianzus: *Gregorius* XIII *pont. max. Beati Gregorii Nazianzeni corpus ex sacrum Virginum templo Dei Genitrici Mariae ad Campum Martium dicato in basilicae vaticanae Sacellum a se ornatum celeberrima quam vides pompa transtulit* III *Idus Junii* MDLXXX.
Translation of the Relics of Gregory of Nazianzus (first scene).

Sixteenth vault
Scenes: *The Opening of the Tombs* (*Aperientur tumuli*) and *The Last Judgment* (*Dies Domini*).
Inscription: *Ravennam urbem civium numero et virtute instructam fortissimis etiam propugnaculis muniebat.*
Inscription: *Generale Concilium ante* XXVI *annos inchoatum summo omnium consensu feliciter absolvit comprobavitq.*
Coat of arms: Pius IV

On the wall
Map of Eastern Africa (*Africae altera pars, Abyssinorum Regna diversa continens*)
Map of Cyprus
View of a mountain in Ethiopia (*Amara mons Aethiopiae*)
Three city views (*Cassum Reginae Candacis sedes*)
Translation of the Relics of Gregory of Nazianzus (second scene)

Seventeenth vault
Scenes: *Children in Heaven* and *Widows in Heaven*
Inscription: *Galliae regem contra seditiosos haereticos pecunia iuvit dissentientes cenvensium partes per antiquiorem card legatum a se missum composuit tumultus metu ab Italia sublato.*
Inscription: *Gregorius* XIII *Pont. Max. Pii Papae* V *sanctionem de S.R.E. locis non alienandis iure iurando firmavit.*
Coat of arms: Gregory XIII

On the wall
Map of Egypt (*Aegyptus*)
Map of Arabia (*Arabia felix*)
Map of Turkey (*Turcicum extra Europam Dominium*)
Three City Views (*Alexandria Aegypti* and *Damascus*)
Translation of the Relics of Gregory of Nazianzus (third scene)

Eighteenth vault
Scenes: *Widowers in Heaven* and *Virgins in Heavens*
Inscription: *Res et loca quae S.R.E. dominii sunt iniuste occupata evixe recuperabat eius ditionem augebat.*
Inscription: *Pontem Senatorium diu interruptum iactis molib. restituit alterum in periculosis Palliae vadis via Cassia fecit.*
Coat of arms: Gregory XIII

On the wall
Map of Persia (*Persiae Regnum*)
City Views (*Babilon magna*)
Translation of the Relics of Gregory of Nazianzus (fourth scene)

Nineteenth vault
Scenes: *Doctors of the Church in Heaven* and *Saint Bishops in Heaven*
Inscription: *Annum Iubileum* MDLXXV *aperuit summaq. pietate atq. erga peregrinos caritate magnoque populorum concursu celebratum clausit.*
Inscription: *Viam Flaminiam et quae area perpetum diversis . . . fecit Anconae arcem dilatavit amplissimum adveniis pestilentia suspectis mercibusque expurgandis aedificium ad Portum struxit Centumcellarum munitionem perfecit.*
Coat of arms: Gregory XIII

On the wall
Map of India (*India intra Gangem et pars extra*)
Map of Ceylon (*Zeilam*)
Three city views
Translation of the Relics of Gregory of Nazianzus (fifth scene)

Twentieth vault
Scenes: *Patriarchs in Heaven* and *Popes in Heaven.*

Twenty-First vault
Scenes: *Martyrs in Heaven* (repeated in two separate scenes).
Inscription: *Aquam virginem in celeberrima urbis loca deduxit lacubusq. e marmore summo artificio constructis fontes publicae utilitati et iucunditati aperuit.*
Inscription: *Imprimendi artem haebraicis graecis chaldaeis arabicis armenii illyricis caracterib ab se praecipuae instructam ita promovet ut iam Orientis et oppressae Turcis et Schismaticis Europae populi missis catechismi propria ad pietatem lingua instituantur.*
Coat of arms: Gregory XIII

On the wall
Map of Tartaria (*Tartarorum Regiones*)
Map of a Battle
Translation of the Relics of Gregory of Nazianzus (seventh scene)

Twenty-Second vault
Scenes: *Martyrs in Heaven* and *Four Evangelists in Heaven*
Inscription: *Immensam basilicae S. Petri aedificationem maxime urgebat magnam partem a fundamentis excitatam altius tollebat.*
Inscription: *Sacellum in Basilica S. Petri admirabili opere exornavit proventu addictis sacerdotib. comparato Dei genitrici et D. Gregorio Nazianzeno dicavit.*
Coat of arms: Gregory XIII

On the wall
Map of South America (*America, sive Peru*)
Three city views
Translation of the Relics of Gregory of Nazianzus (eighth scene)

Twenty-Third vault
Scenes: *Virgin and Apostles in Heaven* and *Choir of Angels*
Inscription: *Ambulationem in Vaticano pictis in ea totius Italiae regionibus urbibus oppidis Rom. Ecclesiae et principatum ditionibus magnifice extruxit.*
Inscription: [unreadable] *sanctissimi Corporis Christi sacellum in Laterano aedificavit Constantini baptisterium ibidem et pro aede Sanctae Mariae Maioris porticum alia atque alia pia loco instauravit.*
Coat of arms: Gregory XIII

On the wall
Map of New Spain (*Hispania Nova*)
Map of Japan (*Japan*)
Map of the Antilles Islands (*Cuba*)
Three city views
Translation of the Relics of Gregory of Nazianzus (ninth scene)

Twenty-Fourth vault
Scenes: *Virgin in Glory* and *Trinity*
Inscription: *Litteras et praeclara in quovis doctrinae genere ingenia fovebat academiam romanam sumptuosa et nobilium disciplinarum studiosis commoda aedificatione augebat.*
Inscription: *Gregorius XIII Pont. Max. exornavit anno domini MDLXXXII Pontificalis sui XI.*
Coat of arms: Gregory XIII

On the wall
Map of Islands (*Insulae circa primum meridianum*)
Inscription on Gregory XIII's reform of the calendar: *Gregorius XIII Pont. Max. ut sanctum Pascha suo in perpetuum tempore celebratur rationem anni diuperpeturbatam restituit modumque adhibuit quo futuris saeculis in preaescriptae definitionis statu sine confusione perseveret anno MDLXXXII Pont. sui XI.*
Inscription on Gregory XIII's renovations in the Vatican Palace: *Palatii hanc partem tum a Pio Papa IV tum a se fondatum ad fastigium totam perduxit et picturis conspicuam fecit consistoriale aulas instauravit Constantinianam altius sublatam testudine iunxit regiam vario marmore stravit s. Pauli sacellum pictis et plastico opere fictis imaginibus et altari constructo decoravit.*
Translation of the Relics of Gregory of Nazianzus (tenth scene)

Inscription: *Publica in Diocletiani Thermis horrea ad levandam rei frumentariae inopiam construxit Portam Caelimontanam a solo in ampliorem formam aedificavit.*
Inscription: *Germanos Britannos Pannonios Graecos Neophytos in constitutis a se urbanis collegiis doctrina excolendos curabat collegia alia alibi instituebat.*
Coat of arms: Gregory XIII

On the wall
Map of Malaysia (*Malaca*)
Map of Western Asia and China (*India extra Gangem pars altera, sive Cina*)
Three city views (*Quinzai Civitas*)
Translation of the Relics of Gregory of Nazianzus (sixth scene).

Notes

ABBREVIATIONS

AB	Art Bulletin
ASB	Archivio di Stato, Bologna
ASF	Archivio di Stato, Florence
ASP	Archivio di Stato, Perugia
ASM	Archivio di Stato, Mantua
ASV	Architio Segreto Vaticano
BAB	Biblioteca dell' Archiginnasio, Bologna
BAV	Biblioteca Apostolica Vaticana
BM	Burlington Magazine
BNF	Biblioteca Nazionale Centrale, Florence
BUB	Biblioteca Universitaria, Bologna
DBI	Dizionario Biografico degli Italiani
JWCI	Journal of the Warburg and Courtauld Institutes

CHAPTER 1

1 For example Anthony Grafton's many studies, particularly Grafton 1991; Grafton 1992; and Grafton 1997. Ingrid Rowland's approach is exemplified in Rowland 1998.

2 Edward Casey investigated the notion of place in western philosophy (Casey 1998); he also applied the notions of place and site to the interpretation of landscape painting and maps in general (Casey 2002).

3 Casey 2002, 353.

4 Among Barbara Stafford's numerous studies on strategies of visualization, see: Stafford 1991; Stafford 1994; and Stafford 1999. Horst Bredekamp's approach to visual evidence is exemplified in Bredekamp 1995. Martin Kemp has dedicated numerous studies to the epistemology of visualization in the Renaissance, of which see especially: Kemp 1990; Kemp 1992; and Kemp 2000.

5 Denis Cosgrove has studied the Renaissance mapping process in Cosgrove 1992; and Cosgrove 1993. For his assessment on mapping in general, see: Cosgrove 1999; and Cosgrove 2001.

6 The reference here is specifically to Anthony Grafton, *Defenders of the Text: The Traditions of Scholarship in an Age of Science, 1450–1800* (Cambridge, Mass., 1991).

7 On the language and audiences of Renaissance maps, see: Harley 2001; Woodward 1996; Delano Smith 1985; and Woodward 2005.

8 On scientific illustrations in the Renaissance and the debate of their validity in the study of the natural world, see: Prinz-Beyer 1987; Olmi 1992, 21–163; Baigrie 1992, especially the essays by Bert S. Hall, Martin Kemp, Brian Baigrie, and David Topper; Mazzolini 1993; Kusukawa 1997; Jones-Galison 1998; Johns 1998; Frasca Spada-Jardine 2000; and Freedberg 2002.

9 For a semiotic analysis of maps, see Jacob 1992.

10 This traditional view, according to which the history of Renaissance maps is the evolution toward accuracy and the reduction of art, has been radically revised in the multi-volume *History of Cartography*, edited by David Woodward and Brian Harley (the fourth volume on Renaissance cartography is in preparation at the time of writing). This view also informed the art-historical studies by Jurgen Schulz and Svetlana Alpers. Although methodologically distant, the studies of Schulz and Alpers both viewed maps in terms of dichotomies – description and rhetoric, art and cartography – that obstruct rather than facilitate the understanding of the polysemic structure of maps. Schulz interpreted Renaissance maps as moralized cartography, a notion that implies, as a counterpart, a kind of cartography that has no metaphorical meaning, a cartography that certainly did not

exist in the Renaissance and, one wonders, whether it could ever exist. Alpers instead interpreted Dutch art as different from Italian art because northern artists were gifted with a mapping impulse that, inspired by contemporary Dutch cartography, prompted them to produce a descriptive mode of representation in opposition to the rhetorical mode of Italian art. The mapping impulse is a stimulating concept that, however, refers to a general and basic human need to describe the surrounding space, to measure it, and therefore know and master it. It is questionable whether the chronological and geographical boundaries of the mapping impulse could be limited to the Dutch republic of the seventeenth century, a point remarked by reviewers of Alpers's influential book.

11 Foucault 1970. Interesting in relation to geography and mapping are Foucault's brief statements on the place of geography in the archeology of knowledge, a point he did not develop further in his writings (Foucault 1994).

12 Casey 1998, 300–20; and Casey 2002, 227–30. For a post-formalist, art-historical analysis of cartography, see Summers 2003, chapters 2 and 5.

13 Conceptually, the scale of representation of a map is unrelated to its size, but size is germane to the definition of scale in three-dimensional map cycles. On the importance of scale, see Cosgrove 1999, 9–10.

14 Among the studies that I found particularly important for interpreting the intellectual nexus of Renaissance mapping at the Medici and papal courts, even though they do not deal specifically with maps, are: Olmi 1992 and Findlen 1994, for their interpretation of scientific culture in Renaissance Italy; Greenblatt 1980 for his delineation of strategies of self-fashioning; Summers 1987 for the discussion of the role of the mixed sciences in Renaissance thought; Biagioli 1989 and Biagioli 1993 for the role of astronomers and astronomy at court; Westman 1980 and Westman 1986 for the relations between science and religion; Warnke 1993 for court patronage; Kaufman 1993 for the intersection between art, science, and patronage at the imperial court; Headly 2000 for the relations between geography and the Christian empire; Gillies 1994 for the relations between geography and the poetic imagination; and Freedberg 2002 for a recent assessment on the role of images in the investigation of the natural world.

15 Boccaccio's manuscript that Danti owned is in the Biblioteca Nazionale, Florence. On Danti's copy of Giacomo Vignola's perspective treatise, see Fiorani 2003. On Danti's copy of Leonardo da Vinci's *Treatise on Painting*, see Pedretti 1977, 18, 29, 31. The drawings by Francesco di Giorgio Martini have been tentatively identified by Gustina Scaglia (Scaglia 1992, 270–3). Danti mentioned these drawings in 1583: "... i suoi stupendi disegni [of Francesco di Giorgio Martini], de' quali me ne sono stati donati alcuni da M(esser) Oreste Vannocci da Siena, hoggi Architetto del Serenissimo Duca di Mantova" [his (of Francesco di Giorgio Martini) stupendous drawings, some of which Messer Oreste Vannocci of Siena, Architect of the Most Serene Duke of Mantua, gave to me as a gift] (Vignola–Danti 1583, 72). The two drawings by Bartolomeo Passerotti owned by Danti, about which Raffaele Borghini wrote in 1584 (Borghini 1584, 566), are published in Ghirardi 1990, 217, who also published Danti's portrait by Passerotti (216, and Tavola XII; Fig. 6.10 in this book).

16 According to the Dominican Serafino Razzi, who was Egnazio's contemporary and the author of a detailed history of the Order of Preachers, Danti intended to donate his books to the Dominican convent in Perugia, but his sudden death prevented him from drafting his will (Razzi, *Cronica della provincia romana dell'ordine dei frati predicatori*, Biblioteca Mediceo Laurenziana, Florence, MS San Marco 873, fol. 57v, hereafter quoted as *Cronica*). Danti's books were instead inherited by his nephew, Giulio Danti, son of his brother Girolamo, who later wrote to the mathematician Bernardino Baldi offering the books for sale. Baldi, in turn, recommended the acquisition of Danti's books to the Duke of Guastalla, but, as far as we can tell, the sale did not take place. The correspondence on Danti's books is in Affó 1780, 39, and Ronchini 1873, 26–8.

17 The foundational study on Renaissance map murals remains Schulz 1990. An overview on painted map cycles in the Renaissance is in Fiorani forthcoming (with bibliography).

18 Burckhardt 1992, 109–13.

19 The passage from Livy is quoted in Schulz 1990, 55. The reference to Aelian comes from Gabriele Paleotti in his *Discorso sopra le immagini sacre e profane* (Bologna, 1582), published in Barocchi 1961, vol. 3, 356.

20 Pliny 1949, XXXV, 116–17. On the importance of Pliny in the Renaissance, see: Findlen 1994.

21 Pliny 1949, CI, 77. On the *Forma urbis Romae* (approx. 18 × 13 m), see: Dilke 1985, 104–05, who also refers to the drawings of the *Forma urbis Romae* contained in a Renaissance manuscript; Nicolet 1988, 173.

22 Comments on map murals by Renaissance authors are in: Alberti 1966, Books 8–9, 767; Cortesi 1980, 45–119; Armenini [1587] 1972, 192, 206; and Paleotti in Barocchi 1961, vol. 3, 356.

23 Dee 1975, AIIII.

CHAPTER 2

1 Adriani 1583, 73: "che era principe assoluto, et arbitro del Governo, e torre l'animo a coloro che presumessero che fusse diviso il governo della Città da quello della famiglia de' Medici." On Giovambattista Adriani see: Giovanni Miccoli in *DBI*. For the historical account of Cosimo I's election, see: Spini 1980, especially 22–30; Cochrane 1973, 13–92, especially 17–21; Hale 1977, 118–43. On Cosimo I's life, politics, and

culture, see: Carcerieri 1926–9; Diaz 1976; Spini 1980; and Elena Guarini Fasano in *DBI*.

2 On the Medici Palace, see: Kent 2000, 217–38 (with bibliography).

3 Benedetto Varchi reported this anecdote in his *Storia fiorentina* (Varchi 1963, Book XV, Chapter XVI, 572): "Desinó quella mattina nel palazzo, quasi pigliandone la possessione, benché la sera, nella quale si fecero i fuochi, e sonarono le campane a festa e gloria, se ne tornó a cenar colla madre in casa sua, e quivi, parendoli dover aver rispetto a madama Margherita, si stette." [That morning he had lunch in the palace, almost as a way to take it into his possession. However, in the evening, when the fireworks were set off and the church bells rang in celebration, he returned to his house to dine with his mother, thinking that he ought to show respect for Madama Margarita (Margarita of Austria).]

4 For the history of the Medici collections from the fifteenth century until Cosimo I, see: Allegri–Cecchi 1980, 287–302, with bibliography; *Palazzo Vecchio* 1980; Massinelli–Tuena 1992; Acidini Luchinat 1997; *Magnificenza* 1997; and Chapter Four.

5 Built between 1299 and 1315 as the seat of the Republic of Florence and known as the Priori Palace, this building became the Medici ducal palace from 1540 until the 1570s, when the Medici court moved to the larger Pitti Palace. Since then this building has been known as the Palazzo Vecchio, that is the old ducal palace. The literature on the Palazzo Vecchio (as the Medici ducal palace is known today) is so extensive that only the most significant studies can be listed: Muccini 1991; Muccini 1992, and Rubinstein 1995 for the history of the palace as the seat of the republic; Conti 1893; Allegri–Cecchi 1980; Cox Rearick 1993; Klieman 1993, 68–76; and Cox Rearick 2002 on the history of the building as the Medici ducal palace.

6 Vasari in Frey–Frey, 1923–30, vol. 1, 722. The English translation is taken from Partridge and Starn 1993, 188. Built by Girolamo Savonarola in the last decade of the fifteenth century as the meeting hall of the Cinquecento, and thus also known as the Salone dei Cinquecento, this large hall became Cosimo I's audience hall. The ceiling was painted between 1563 and 1565, while the wall canvases were made at a later date. Vasari himself explained the iconography of the Sala Grande in his *Ragionamenti* (Vasari–Milanesi 1906, vol. 8, 199–223). On the Sala Grande, see: Rubinstein 1968; Forster 1971; Cox Rearick 1984, 229–92; Muccini 1991; Muccini 1992; Partridge and Starn 1993, 149–210; Morel 1993; Kliemann 1993, 74–7; Van Veen 1998; Williams 1998; and Scorza 1998.

7 On Cosimo I's policy in the 1560s, see: Cochrane 1973, 88–92. For a discussion on the place of the Sala Grande in relation to other Renaissance halls of power see: Partridge and Starn 1993, 149–210; and Klieman 1993, 69–78.

8 Vasari–Milanesi, vol. 8, 220, the central tondo is "... la chiave e la conclusione delle storie." The initial plans for the ceiling of the Sala Grande are admirably discussed in Rubinstein 1968.

9 On the central tondo, see: Forster 1971; Cox Rearick 1984, 281–2, who interpreted the replacement of Florence with Cosimo I as an unmitigated celebration of the duke and suggested to entitle the tondo *The Apotheosis of Cosimo I*; and Van Veen 1998, who discussed the tondo in relation with the inscription surrounding it, which reads: SPQF OPTIMO PRINCIPE CONSTITUTA CIVITATE AUCTO IMPERIO PACATA ETRURIA.

10 Partridge and Starn 1993, 178.

11 Morel 1993 interpreted the ceiling of the Sala Grande as a diagram of the administration of the state and as a map of its territory. He also showed that the four quarters of Florence (San Giovanni, Santa Maria Novella, Santo Spirito, and Santa Croce) and the cities of the dominion were grouped together according to the administration of justice under Cosimo I or, to use Vasari's words, "according to the order of the judges of the Ruota" (*secondo l'ordine dei giudici di ruota*) (Vasari–Milanesi, 1906, vol. 7, 200). The Ruota was a tribunal in charge of the Duchy and was composed of judges coming from the four quarters of Florence, two judges for each quarter. As Morel pointed out, the quarters of Florence and the gonfalons, which were to be prominently represented in the early plans for the ceiling, were eventually moved to the margins in the final version. Despite this move, however, the role of the Florentine quarters was enhanced in the final ceiling, since they were represented by their administrative functions within both the city of Florence and the Duchy of Tuscany.

12 On the Guardaroba Nuova, see: Vasari–Milanesi 1906, vol. 7, 633–6; Vasari–de Vere 1996, vol. 2, 891–3; Allegri-Cecchi 1980, 303–16, who published the documents pertaining to the room's construction and decoration; Almagià 1952, 13–16, and Schulz 1990, 97–122, who analyzed the room in relation to other Renaissance map cycles; Danti-Bonsignori 1995 published a complete set of handsome illustrations of the maps; Schaefer 1976, 149–53, and Morel 1982, 185–205, studied the room in connection with the Studiolo of Francesco I de' Medici and in relation to the Medici collections; Liebenwein 1988, 118–30, considered it within the context of European Kunstkammer and Renaissance studioli; and Fiorani 2003 explored it in relation to Cosimo I's imagery.

13 Vasari described the Guardaroba Nuova as, "a new hall of some size" that the duke walled (*ha murato*) "expressly as an addition to the guardaroba," suggesting that he simply walled up a pre-existing area (Vasari–de Vere 1996, vol. 2, 891). Most likely the area that became the Guardaroba Nuova was originally the landing area for the stairway that Battista del Tasso had built in the Cortile della Dogana around 1550, traces of which are still visible. When Vasari

demolished this old stairway, he did not dismantle the obsolete landing loggia, but rather seized the opportunity to add a new room to the palace. He closed two of the three grand arches looking into the Cortile della Dogana, the three pointed arches opening into Machiavelli's Chancellery, and the mullioned window, one of the original external windows of the medieval palace, which opened into the Sala dei Gigli. Vasari retained the three doors of the landing loggia: the main door leading to the Sala dei Gigli, another one entering into Machiavelli's Chancellery, and the last one descending to Cosimo I's private apartment. He transformed, however, the last two doors into secret passages and hid them inside the cupboards. On the building history of the stairways in Palazzo Vecchio, see: Lensi 1929, 124 (specifically on Battista del Tasso's stairway); Allegri–Cecchi 1980, 9; Pillsbury 1974; Rubinstein 1995, 19, 30, 37.

14 On the ducal guardaroba, see Allegri–Cecchi 1980, 287–302; *The Medici and Michelangelo* 2002, and Chapter Four.

15 Vasari–de Vere 1996, vol. 2, 891. Vasari's full description of the Guardaroba Nuova reads:

Vincenzio [Danti] has a brother in the Order of the Preaching Friars, called Fra Egnazio Danti, who is most excellent in matters of cosmography and of rare genius, insomuch that as Duke Cosimo de' Medici is having him execute a work which none greater or more perfect has ever been done at any time in that profession, which is as follows: His Excellency, under the direction of Vasari, has walled a new hall of some size expressly as an addition to the guardaroba, on the second floor of the apartments in the Ducal Palace; and this he has furnished all around with cupboards seven braccia high, with rich carvings of walnut-wood, in order to deposit in them the most important, precious, and beautiful things that he possesses. Over the doors of those cupboards, within their ornaments, Fra Egnazio has distributed fifty-seven pictures about two braccia high and wide in proportion, in which are painted in oils on the wood with the greatest diligence, after the manner of miniatures, the Tables of Ptolemy, all measured with perfect accuracy and corrected after the most recent authorities, with exact charts of navigation and their scales for measuring and degrees, done with supreme diligence; and with these are all the names, both ancient and modern. His distribution of these pictures is as follows: At the principal entrance of the hall, on the transverse surfaces of the thickness of the cupboards, in four pictures, are four half-spheres in perspective; in the two below is the Universe of the Earth, and in the two above is the Universe of the Heavens, with its signs and celestial figures. Then as one enters, on the right hand, there is all Europe in fourteen tables and pictures, one after another, as far as the center of the wall that is at the head, opposite to the principal door; in this center is placed the clock with the wheels and with the spheres of the planets that every day go through their motions, which is the so famous and renowned clock made by the Florentine Lorenzo della Volpaia. Above these tables is Africa in eleven tables, as far as the said clock; and then, beyond that clock, Asia in the lower range, which continues likewise in fourteen tables as far as the principal door. Above these tables of Asia, in fourteen other tables, there follow the West Indies, beginning like the others from the clock, and continuing as far as the same principal door; and thus there are in all fifty-seven tables. In the base at the foot, in an equal number of pictures running right round, which will be exactly in line with those tables, are to be all the plants and all the animals copied from nature, according to the kinds that those countries produce. Over the cornice of the cupboards, which is the crown of the whole, there are to be some projections separating the pictures, and upon these are to be placed such of the antique heads in marble as are in existence of the Emperors and Princes who have possessed those lands; and on the plain walls up to the cornice of the ceiling, which is all of carved wood and painted in twelve great pictures, each with four celestial signs, making in all forty-eight, and little less than life-size, with their stars – there are beneath, as I have said, on those walls, three hundred portraits from life of distinguished persons for the last five hundred years or more, painted in pictures in oils (and a note will be made of them in the table of portraits, in order not to make too long a story here with their names), all of one size, and with one and the same ornament of carved walnut-wood – a very rare effect. In the two compartments in the center of the ceiling, each four braccia wide, where there are the celestial signs, which open with ease without revealing the secret of the hiding-place, in part after the manner of a heaven, will be accommodated two large globes, each three braccia and a half in height. In one of them will be the whole earth, marked distinctly, and this will be let down by a windlass that will not be seen, down to the floor, and will rest on a balanced pedestal, so that, when fixed, there will be seen reflected all the tables that are right round in the pictures of the cupboards, and they will have a countermark in the globe wherewith to find them with ease. In the other globe will be the forty-eight celestial signs arranged in such a manner, that it will be possible with it to perform all the operations of the Astrolabe to perfection. This fanciful invention came from Duke Cosimo, who wished to put together once and for all these things both of heaven and earth, absolutely exact and without errors, so that it might be possible to see and measure them separately and all together, according to the pleasure of those who delight in this most beautiful profession and study it; of which, as a thing worthy to be recorded, it has seemed to me my duty to make mention in this place on account of the art of Fra Egnazio and the greatness of the Prince, who holds us worthy to enjoy such honorable labors, and also to the end that it may be known throughout the whole world. (Vasari–de Vere 1996, vol. 2, 891–3).

16 On the compilation of Vasari's *Lives*, see Rubin 1995. On its fictional aspects, see: Barolsky 1991; Hope 1985; and Hope 1995.

17 The maps are discussed in detail in Chapter Five, while Appendix A lists them according to their original sequence.

18 Letter of Egnazio Danti to Polidoro Castelli, 23 September 1571 (in Tiraboschi 1833, vol. 3, 497): "una palla . . . tutta armata dentro di ferri, perché sì gran globo non si reggeria da per sé . . . è fatto con invention nuova talmente che con un sol dito sí gran macchina si muove per tutti i versi, et si fa alzare et abassare i poli con facilità gran(dissi)ma." [The interior of the globe is entirely armed in iron, because such a large globe could not sustain itself . . . (the globe) is made with such a novel invention that this grand machine can be moved in any direction by the touch of a single finger, and it is possible to raise and lower the poles with the greatest ease.] The metal structure of the globe, made by Antonio Lupattino, was covered with many layers of different materials: sheets of thin metal, a spiral of ropes, layers of papier mâché, and finally a layer of parchment, on which the map was painted in tempera. Antonio Lupattino made also the pedestal for the globe (Frey 1940, 180). In 1595, Danti's globe was restored by Antonio Santucci, a mathematician and astronomer at the ducal court and later professor of mathematics at the University of Pisa, but, even after the restoration, it continued to be displayed outdoors, first in the Loggia della Cosmografia in the Uffizi Palace and later in the Museo Galileo. On Danti's globe, see: Fiorini 1889, 172–81, who analyzed its cartographic content; Allegri–Cecchi 1980, 308–09, who published the pertinent archival documents and a recent restoration report; Collaretta 1980, 118, who published a document showing that in 1570 the gold lettering of the globe was still under way; Heikamp 1970, 3–25, who documented its display in the Uffizi Palace from 1589 onward; and Righini Bonelli 1978, 59–67, who examined Santucci's restoration. The interior of Dante's globe is currently being scanned.

19 Document published in Frey 1940, 179–80.

20 Danti's leather for the celestial globe is recorded in the inventory of the ducal palace dated 1574, ASF G. 87, fol. 34v: "Pelli di Danti intere n.o tredicij; Pezi di pelli di Danti di piu grandeze n.o ventiquattro; dissono che stan.o loro in conto p. pella dua 1/2; Piu ritagli di detta pelle di danti piccolissimi, che no. buonj a niente." [Thirteen whole pieces from Danti's leather; Twenty-four pieces of different size from Danti's leather, which correspond to two whole pieces and a half; Many small-size pieces from Danti's leather, which are good for nothing.] This "pieces from Danti's leather" can refer only to the celestial globe, since the terrestrial one had already been completed in 1574. Danti's celestial globe is also mentioned in Spini 1570, "Dedication letter to cardinal Ferdinando de' Medici," not paginated. On Antonio Santucci's armillary sphere, now kept in the Museo di Storia della Scienza in Florence, see: Meucci 1876; Righini 1951; Heikamp 1970; and Miniati 1991, 104–5. On Santucci in general, see: Righini Bonelli 1978; and Settle 1984. On the Sala della Cosmografia in the Uffizi Palace, which was decorated with maps of the Duchy of Florence and Siena made by Stefano Bonsignori and Ludovico Buti, see: Heikamp 1970; Morel 1993, 127–8; Rombai 1993, 98–101; and Fiorani (forthcoming).

21 Lorenzo della Volpaia's armillary sphere can be identified with the following entry in the inventory of the palace dated 1570, ASF, G 73, c. 59: "uno apamondo dotone dorato con un suo piede meso in una vesta di cuoio rosso" [a gilded brass globe with its own pedestal in a case of red leather.] See also: Vasari–Milanesi 1906, vol. 2, 593–4; and Allegri–Cecchi 1980, 312, for payments to Dionigi di Matteo Nigetti.

22 These portraits are recorded in the inventory of the ducal palace dated 1570, ASF, G. 73, c. 59: "Dugentoventinove quadri di pitura tutti a una misura e piu personaggi . . . Quatro tende de tela verde che cuoprono detti ritratti" [237 paintings of people, all of the same size . . . Four green curtains covering the said portraits.] The same arrangement is recorded in the inventory of 1574, ASF, G 87, c. 57: "Nella stanza dell'horiuolo sopra gli armarj Quadri di pittura sopra gli armarj. Quadri di pittura tutti a una misura di piu personaggi con ornamento di nocie n.o Duecentotrentotto. Tende di tela verde et loro funi che quocrono intorno tutti li detti quadri et sono quattro pezzi." [In the Room of the Clock, above the cupboards, 238 paintings of people, all of the same size with walnut frames . . . Four green curtains and their ropes, covering the said portraits all around.] In the 1574 inventory the *stanza dell'horiuolo* (Room of the Clock) should be identified with the Guardaroba Nuova rather than the Sala dei Gigli. In earlier inventories, the *stanza dell'horiuolo* indicated the Sala dei Gigli, where indeed Lorenzo della Volpaia's armillary sphere was displayed, but, as we know from other sources, the instrument had been moved to the Guardaroba Nuova. These portraits copied from Giovio's collection are also listed by Vasari in an appendix at the beginning of the third part of the 1568 edition of his *Lives* (republished, with few omissions, in Allegri–Cecchi 1980, 310–12). On these portraits, see: Prinz 1988, 603; and Mosco 1980. Occasionally some of the portraits were removed from the Guardaroba Nuova to furnish the rooms prepared to host special guests. In 1577, 131 portraits were distributed in 21 rooms in the Pitti Palace as part of the special display made for the Venetian ambassador who came to congratulate Francesco I for the imperial recognition of his grand ducal title (see Mosco 1980, who, however, erroneously suggested that Cosimo I displayed the portraits made after Giovio's collection in the Pitti Palace. The inventories of 1570 and 1574 show that, at the time of Cosimo I's rule, they were still kept in the Guardaroba Nuova). It is worth mentioning that Danti himself marked the location of Giovio's museum near Como on the map *Mediolanensis Ducatus* that he painted in the Gallery of Maps in the Vatican Palace around 1580, on which see: Gambi-Pinelli 1994, vol. 1, 286, no. 215.

23 On Cosimo I's interests in the world's flora and fauna, see: *Corte, mare, mercanti* 1980, 195–216. On the botanical gardens in Pisa and Florence, see the numerous studies by Tongiorgi Tomasi listed in *Flowering of Florence* 2002; see

also *Giardino dei Semplici* 1991; and Tongiorgi Tomasi–Garbari 1995. On Piero di Cosimo's *libro d'animali*, later inherited by Francesco de' Medici, see: Borghini 1584, 307; and Berti 1967, 236–7. Numerous books on animals are quoted in the inventories of the Medici guardaroba: ASF, G 87 (inventory of 1574), fol. 42: "Sei libri di piu sorte d'animali dipinti in vari modi" [six books of animals from different species painted in different manners]; ASF, G 126 (inventory of 1587), fol. 68: "un album di disegni di vari animali." [An album of drawings of various animals.] On Giovanni Neri, see: Ulisse Aldrovandi, who himself reported the presence of Neri's drawings at the Medici court before the arrival of the painter Giacomo Ligozzi in 1577: "attorno . . . erano attaccati al muro più di cento legni di uccelli dipinti dal mio pittore il sig. Gian Triulxi . . . avanti che 'l Gran Duca avesse condotto il sig. Jacomo Ligozzi." [All around hung over hundred panels of painted birds made by my painter Sig. Gian Triulxi before the Grand Duke brought Sig. Giacomo Ligozzi (to Florence)] (Aldrovandi 1907, 25). On the identification of Gian Triulxi as Giovanni Neri, see: Olmi, 1992, 65. On the Tribuna's socle with animals, now lost, see: Heikamp 1963, 193–268, 198–200, who was the first to note that Johann Zoffany recorded this detail in his painting *The Tribuna of the Uffizi* (1772–8).

24 On Cosimo I's astrological chart, see: Cox Rearick 1984, 204–12.

25 Vasari–de Vere 1996, 891. On Egnazio Danti, see: Razzi 1596, 123–7, 346–7; Razzi, *Cronica*; Marchese 1878–9, vol. 2, 351–77; Del Badia 1881; Palmesi 1899, 81–125; Righini 1977; Righini Bonelli–Settle 1979; Frangenberg 1988; Frangenberg 1994; Settle 1990; Gambi-Pinelli 1994; Dubourg Glatigny 2002; Fiorani 2003, and Fiore in *DBI*. On Danti's activity as iconographic adviser, see: Brink 1983; Pillsbury 1974; Mancinelli-Casanovas 1980; Courtright 1995; and Fiorani 1996. On Danti's portrait by Bartolomeo Passerotti, reproduced here as (Fig. 6.10), see: Ghirardi 1990, 216–19.

26 The literature on the relationship between Cosimo I and Vasari in connection with the restoration and decoration of the ducal palace is extensive, but see in particular: Hall 1979; Allegri–Cecchi 1980; Barocchi 1980, 121–8; Collaretta 1980; Davis 1985b; Rubin 1995; and Le Mollé 1995. Their correspondence is in Frey–Frey 1923–30 and Frey 1940.

27 The friendship between Vasari and Miniato Pitti dated from at least 1530. Not only was Miniato Pitti instrumental in getting commissions to Vasari in the early stages of the artist's career, but he also initiated the marriage arrangements with the family of Vasari's future wife. Their relationship is documented by numerous letters published by Frey–Frey 1923–30 and Frey 1940. After Miniato Pitti died in 1566, Egnazio Danti bought some of his scientific instruments (Rosen 2003).

28 Sforza Almeni had also been instrumental to secure ducal commissions for Egnazio's brother, the sculptor Vincenzo Danti, on which see: Summers 1969, 65–6. On Sforza Almeni, see: Davis 1985a. Settle 1990, 28, pointed out that Danti must have spent his training years in Florence since the Dominican Studium for the Roman Province had been moved from Perugia to Florence in 1555. The Studium was moved back to Perugia in 1563, but Danti must have remained in Florence since, from September 1563, his name appears in Vasari's letters and in the records of the ducal court.

29 Razzi, *Cronica*, fol. 56: "Dopo un certo tempo datesi allo studio della Cosmografia, riusci tale, aiutato da i principij da lui appresi intorno al dissegnare, sotto la disciplina, quando era nel secolo, di maestro Giulio Danti, suo padre." [After spending sometime studying cosmography, he became so talented, helped by the principles of drawing that he had learned from his father Giulio Danti, when he was still in the secular world.] Danti himself commented on his early training on two later occasions: in his dedication letter to Diomede della Cornia, Marquis of Castiglione, published in Danti 1571a, not paginated; and in his dedication letter to Giacomo Boncompagni, 14 November 1583, published in Vignola-Danti 1583, not paginated.

30 The maps of the Terza Loggia were completed only twenty years later and, due to a strange turn of events, by Egnazio Danti himself. On the Terza Loggia, which is discussed in detail in Chapter Nine, see: Hess 1967a; Banfi 1951; Almagià 1955; Redig de Campos 1967, 161–2, 169–174; Schulz 1990, 97–103; and Bragaglia Venuti 2002.

31 Letter of Antonio Compagnoni of Macerata in Rome to Ulisse Aldrovandi in Bologna, 17 April 1563 (published in Olmi 1992, 134): "Oggi si è ragionato nel palazo del papa de huomini che si dilettano di minerali piante et d'ogni sorte de animali et quasi tutti dicevano il Mathiolo è raro, essendo io a cerchio con quelli letterati ragionai di V.S. a talch'uno ch'ha cura di fare una loggia dove serrà tutta la cosmographia volendo adornare alcuni quadri sopra le tavole di Cosmographia, io glie dissi ch'V.S. poteva darli et piante et ucelli ritratti dal naturale et me estesi in narrarli il vostro studio, mi ha pregato strettamente ch'io voglia haver da V.S. una pianta et uno ocello con el nome del uno et delaltro e che sia ritratto dal naturale acció si possi monstrare al Cardinale Amulio che piacendoli ne parlerà con Sua Stà perché l'Amulio ne ha cura di adornare questa loggia che serrà la piu bella che sia al mondo et di qui ne potrà nascere occasione che V.S. verrà in cognitione di S. Stà et se non si facesse altro si potria havere qualche favore nel studio vostro di augumentarvi la condutta et perder non vi si po, perhó io prego V.S. la si degni mandarmi el ritratto d'una pianta et uno occello di belli ch'abbiate acció io portandola al car.le Amulio se ne compiaci et includetele in un plico et indrizatele a me in sto Pietro nel arcipresbiteriato et venendomi occasione per quella non mancheró far quanto sia possibile." Olmi identified the cardinal mentioned in the letter as Cardinal Marcantonio da Mula, then superintendent to the Vatican Palace.

32 Letter of Miniato Pitti to Cosimo, 1564 (Lorenzoni 1912, 175):

the maps of the Terza Loggia, "sono fatte et si fanno alla giornata dove sono di grandi sciarpelloni ma solo un bene vi è che essendo all'aria et dipinte a tempra, fra pochi dì bisognerebbe rifarle et si correggeranno."

33 The documents pertaining to the wood model of the Guardaroba Nuova are published in Frey–Frey, 1923–30, 2, 69, 70–1, 73–5; and Allegri–Cecchi 1980, 309. The best source to define Dionigi di Matteo Nigetti's tasks as Vasari's *sottoarchitettore* is Vasari's letter to Cosimo I, 19 December 1561 (Frey 1940, 64–5). On his activity as a woodworker, see Trionfi Onorati 1985, 371–82.

34 Letter from Vasari, in Florence, to Cosimo I, at Poggio a Caiano, 1 September 1563 (Frey–Frey, 1923–30, vol. 2, 2): "Intanto Don Miniato andra spartendo le tavole di Tolomeo per la guardaroba, che tutto andra bene, et si dara ordine al far fare le palle di tiglio." [In the meantime Don Miniato will divide the tables of Ptolemy for the guardaroba, and everything will be well, and I will order the palle di tiglio.]

35 Letter of Vasari to Cosimo I, 18 April 1564 (Frey, 1940, 75): "Il frate con le tavole dj Tolomeo lavora gagliardo, et fra Miniato Pittj l'à aiutato e insegnato molte cose; et sarà cosa rara." [The friar works vigorously on the maps of Ptolemy, and Fra Miniato Pitti has helped him and taught him many things.]

36 Letter from Cosimo I's secretary in Pisa to Vasari in Florence, 23 November 1563 (Frey, 1940, 57): "quanto alle tavole di Tolomeo, havendo Fra Miniato mandato l'ordine che si ha tenere, dicie S.E.I. che si attenda à tirare avanti, che non accade mandare altro disegno." [As for Ptolemy's maps, Fra Miniato sent the order that has to be followed; his Illustrious Excellency said to go ahead, and that there is no need to send additional drawings.]

37 Letter from Vasari to Cosimo I, 4 October 1563 (Frey–Frey, 1923–30, vol. 2, 7): "Il frate per le tavole dj Tolomeo dal priore di San Marco se ottenuto, et lan dato volentierj, perche non vogliono perder Santa Maria Novella piu tosto il frate." [The prior of San Marco gave the friar for the maps of Ptolemy, and they (the Dominicans of San Marco) gave him willingly because they would rather lose the friar than Santa Maria Novella (the main Dominican church in Florence)]. Danti's move from the convent of San Marco to that of Sta Maria Novella is also reported in a letter that Giovan Bernardino Rastelli, one of Danti's friends from Perugia, wrote to Prospero Podiani, 18 July 1564 (ASP, miscellanea 103, c. 23c).

38 Letter from Vasari, in Florence, to Cosimo I, in Pisa, 20 Janaury 1565 (Frey–Frey, 1923–30, vol. 2, 145): "Perche i frati gli anno finalmente date certe stanzuccje, vj o fatto spendere par.cj scudi, perche stia comodo et sano a lavorarvj; ne sigli mancha provederlo di quanto a bisogno." [Since the friars have finally assigned him (Egnazio Danti) certain small rooms, I spent numerous scudi so that he can work there comfortably and healthily, and I provide him of whatever he needs.]

39 Serafino Razzi, *Cronica*, fol. 56, reported that Danti, "disegnò le tavole tutte di Tolomeo secondo le vere misure, e descrizioni e le fe egregiamente dipignere da suoi servati." [He drew all the maps of Ptolemy according to their true measures and descriptions, and had them painted by his apprentices.] On the other hand, numerous payments to carry the panels from Sta Maria Novella to the ducal palace suggest that Danti painted the maps in his cell at the convent (the payments are published in Allegri–Cecchi 1980, 309).

40 Letter from Miniato Pitti to Vasari and Cosimo I, 10 April 1564 (Lorenzoni 1912, 175): "Il frate di S. Marco che per lei lavora la Cosmografia, si porta egregiamente et è stata buona elettione, et io non manco di aiutarlo et comunicare seco quanto gli bisogni in servitio di V.S. Ill.ma et tengo certo che ne sara contenta et ben servito."

41 Letter from Vasari to Cosimo I, 20 January 1565 (Frey–Frey, 1923–30, vol. 2, 145): "Et gia a per una facciata tuttj e quadrj fattj, et anche e finito gli ornamenti di nocie, et va seguitando il far la palla grande: Che tutto sara cosa che inporta." [He has already completed the panels for one wall, and also the walnut ornaments will be finished; he is still working on the large globe, and everything will be something of importance.]

42 A note from Vincenzo Borghini, Cosimo I's iconographic adviser for the wedding pageantry, to Vasari is vague (Frey–Frey, 1923–30, vol. 2, 218): "Io mando hora al frate degli epitafij, et stasera gli hara tutti." [I send now some epitaphs to the friar, and tonight he will have them all.] On the other hand, Borghini listed Danti among the many people awaiting for compensation on the work for the triumphal entry. Danti presented a bill for 25 florins but obtained a payment of only 10 florins (Frey, 1940, 180). On the 1565 pageantry, see: Ginori Conti 1936; and Scorza 1981. On Vincenzo Borghini, see the many studies by Richard Scorza, all listed in Belloni-Drusi 2002, which should also be consulted for earlier bibliography.

CHAPTER 3

1 On Cosimo I's political imagery, see Forster 1971; Richelson 1974; Cox Rearick 1984, especially 229–92 for Cosimo I's cosmic imagery; Crum 1989; and Cox Rearick 2002, 36–37.

2 Cosimo I used the Capricorn as his impresa for the first time in 1539 on the occasion of his marriage to Eleanor of Toledo. Vasari considered the Capricorn as, "the sign that astrologers regarded as appropriate to the greatness of Illustrious Princes and as their ascendant; as it was Augustus' [ascendant sign], so it is our Duke Cosimo's" [segno appropriato dalli astrologi alla grandezza de' Principi Illustri e ascendente loro; come fu Augusto; cosi è ancora del Duca Cosimo nostro] (Vasari–Milanesi 1906, vol. 8, 66). The historian Paolo Giovio explained the political meaning of Cosimo I's device exclusively in reference to Charles V, "under whose

protection flourishes the state of the above mentioned Signor Duca" [sotto la protettione [di Carlo v] fiorisce il principato del prefato Signor Duca] (Giovio 1556, 38). For a detailed discussion of Cosimo i (born on 11 June 1519 under the sign of Gemini), his Capricorn in ascendancy, and its relation to both Augustus (born on 23 September 63 B.C. under the sign of Virgo) and Charles v (born on 24 February 1500 under the sign of Pisces), see Richelson 1974, 25–78; and Cox Rearick 1984, 257–72.

3 Vasari wrote his *Ragionamenti* in close consultation with Cosimo Bartoli, who may indeed have written extensive parts or, at the very least, suggested to Vasari appropriate historical and mythological readings. Cosimo i personally revised it and Vasari surely meant to publish it, although the book appeared only posthumously in 1588. It is republished in Vasari–Milanesi 1906, vol. 8, 9–225. On Vasari's *Ragionamenti*, see McGrath 1985, who brought Vasari's contrived explanations to the attention of scholars who study the genesis of Medicean iconography, and the studies of Paola Tinagli, which explained the meaning of Vasari's expression *secondo il senso nostro* as well as the scope of the treatise (Tinagli Baxter 1985; Tinagli 2000; and Tinagli 2001).

4 On the Sala degli Elementi, see Vasari–Milanesi 1906, vol. 8, 19–35; Allegri–Cecchi 1980, 63–73, for a detailed description of each painting; and Davis 1985b, for the genesis of the iconography.

5 Vasari–Milanesi 1906, vol. 8, 22.

6 Vasari–Milanesi 1906, vol. 8, 32.

7 On the symbolism of the globe, see Schramm 1958; Lippincott 2002 (with bibliography); Brotton 1997, 17–23, 155–161; Brotton 1999; and Cosgrove 2001.

8 The myth of Hercules was well known in Renaissance Florence through the writing of Coluccio Salutati. On the use of Hercules in the iconography of the Florentine republic and of the early Medici, see Utz 1971. On Cosimo i's use of Hercuelan imagery, see Richelson 1974, 79–106; and Fiorani 2001.

9 On the Sala di Ercole, which is part of the Apartment of the Elements, see Vasari–Milanesi 1906, vol. 8, 76–84. Allegri–Cecchi 1980, 97–101; *Palazzo Vecchio* 1980, 68 (on the *arazzi*); Richelson 1974, 90–1; and Tinagli 2001, 70–1 (in relation to Vasari's *Ragionamenti*).

10 Vasari–Milanesi 1906, vol. 8, 83: ". . . come Ercole, la palla del mondo posta in sulle spalle a Atlante, il quale non é altro che l'aiuto de' principi nel governo loro, fatti simili a Dio nella pietà, nella clemenza, nella giustizia, e nelle altre virtù, le quali membra fortissime sostengono la palla del mondo: che sará ora in Vostra Eccellenza lo aiuto che dovrete dare al duca nostro nel governo di questo stato."

11 The pageantry *La genealogia degli dei* also included a float entitled *Il carro del cielo*, which represented Saturn, the father of the Titans, standing on a gigantic celestial globe and surrounded by his sons including Atlas. A surviving sketch is the BNF, CB III. 53, fols. 6–15, on which see Nagler 1964, 29; Gaeta Bertelà-Petrioli Tofani 1969, 21–2; and *Palazzo Vecchio* 1980, 395–6. On this pageantry, see Nagler 1964, 24–35.

12 Vincenzo de Rossi's statue of *Hercules Holding the Globe* was originally designed as the crowning element of a fountain dedicated to Hercules's labors for a Medici villa. In the following decades, the statue continued to be associated with grand dukes named Cosimo. In 1592, on the occasion of the baptism of Cosimo, first son of Grand Duke Ferdinando, it was moved to the Sala Grande in the Palazzo Vecchio. After 1620, during the reign of Cosimo III, it was placed at the entrance to the Medici villa at Poggio Imperiale, where it is still located today. On Vincenzo de' Rossi's statue, see Utz 1971; *Palazzo Vecchio* 1980, 344–5; Allegri–Cecchi 1980, 381–5. The theme of Hercules Holding the Globe appeared with a very different meaning in Raphael's tapestries for Leo x, on which see Cox Rearick 1984, 186–7.

13 Biagioli 1993, 144–7.

14 Although no comprehensive study of Charles v's patronage of cartography exists, Eisler 1983 and recently *Karl v* 2000 provide a useful overview and bibliography. See also: *Jahrbuch der Kunstsammlungen der Aller. Kaiserhauses in Wien*, 9, 1890, II, 1, on Charles v's rights for Matteo da Verona's planisphere. On Charles v's political iconography, see Checa Cremades 1999; Burke 1999; *Karl v*, 2000 and *Carlos v* 2000. On Charles v's Palace in Granada, see Marías 2000 (with bibliography).

15 On Charles v's emblems, see Rosenthal 1971; Rosenthal 1973; and Rosenthal 1974. On the festivities organized for Charles v during his trips to Italy, see *Il potere e lo spazio* 1980, 34–8; on Charles v's entrance into Genoa in 1530, see Gonzaga 1892; Varchi 1963, vol. 2, 23–5; Fagiolo 1980; and Mitchell 1979. On Charles v's entrance into Milan in 1541, see Eisler 1983; and Checa Cremades 1999. The triumphal entry into Antwerp of the future king Philip II in 1549 is illustrated in Strong 1973, fig. 63.

16 Schramm 1958, 114. It is relevant to mention Sebastiano del Piombo's drawing illustrating the emperor and the pope in Bologna around a multitude of globes and scientific instruments (*Italian drawings* 1983, 167).

17 Vasari-De Vere 1996, vol. 1, 322; and Vasari–Milanesi 1906, vol. 5, 229: "un fanciullo in forma d'un Ercole piccolino . . . porgeva [a Carlo v] il mondo, quasi dandogliene il dominio." On this portrait by Parmigianino, see Chiusa 2001, 95–6 (with bibliography).

18 Cirni 1560, 4v: "uno dei felici e santi successori di san Pietro." On Cosimo i's entrance into Siena in 1560, see *Il potere e lo spazio* 1980, 37, 41; and Cox Rearick 1984, 279–80.

19 Cirni 1560, 4v: "un mondo e di dietro sotto ad esso queste lettere greche alludendo con bella invenzione al nome del Duca *Kosmos kosmou kosmos* che secondo mi hanno riferito che dicono d'intenderle, significano che il Duca Cosimo honora il mondo, e 'l mondo lui, o vero, che'l mondo é di Cosimo et egli di lui." On the Cosimo–cosmos theme, see Cox Rearick 1984, 279–83, who perceptively suggested that the Cosimo–cosmos theme also offered a poignant connec-

tion between Cosimo I and his ancestor Cosimo the Elder; and Crum 1989.

20 Document quoted in Settle 1990, who provides an excellent discussion of Danti's teaching at the Medici court. Although it was not until 1571 that Danti received a formal teaching position, a fixed salary of 3 scudi, and the official title of "Cosmographer of the Grand Duke of Tuscany," which he proudly used on the frontispiece of his treatises, circumstantial evidence suggests that his teaching appointment may date to an earlier period, possibly as early as 1567. As for the institutional affiliation of Danti's lectureship, it still remains unclear whether it was somehow connected to the Florentine Academy or was instead fully administered within the court. Cosimo I transferred the old Florentine studium to the University of Pisa in 1544, but maintained two lectureships in Florence – one in Italian literature, and another in theology. In 1571, the latter was suppressed and Danti's lectureship established instead. On the Medici patronage of cosmography and astronomy, see Galluzzi 1980, and Biagioli 1993.

21 Danti 1573a, 51: "il terrazzo . . . sopra il tetto del Casino di Belvedere ne Pitti, il quale sta al medesimo piano [del campanile fiorentino]."

22 Danti 1569, Dedication to Cardinal Ferdinando de' Medici, not paginated: "egli [Cosimo] no(n) pure mi habbia mantenuto, & mantenga in stato, che io possa seguitare questo mio studio . . . ma si è etiandio degnato di aiutarmi con ottimi avvertimenti, & avvisi particulari à cotale essercitio appartenenti."

23 These documents are published in Del Badia 1881, 6; and Lorenzoni 1912, 177–81.

24 On the smuggling of this famous statue, still today in the Museo Archeologico in Florence, see Cristofani 1979.

25 On the University of Pisa in the sixteenth century, see the numerous studies by Charles B. Schmitt, gathered in Schmitt 1981 and Schmitt 1983; and Grendler 2002, 70–7. Filippo Fantoni taught mathematics at Pisa from 1560 to 1567, and again from 1582 to 1589; Giuseppe Nozzolino held the same chair from 1567 to 1581. Astrology was also a crucial part of mathematics at the University of Pisa, in a way that it was not in Danti's classes. On Renaissance mathematics and mathematicians, see Rose 1973; Rose 1975; Boas Hall 1982; Westman 1980; and Biagioli 1989. On the teaching of mathematics in Renaissance universities, see Grendler 2002, 408–29.

26 Danti's discussion of the *scientiae mediae* is in Danti 1583, Proemio, not paginated.

27 Danti acquired a manuscript copy of Vignola's *Le due regole della prospettiva pratica* from Niccoló Gaddi, annotated it with his own commentary and, many years later, published it. On the Vignola–Danti book, see Fiorani 2003.

28 A manuscript convincingly attributed to Danti is preserved in the Biblioteca Riccardiana, Florence, MS 2834 (Baccini 1888). It includes a series of fragmentary notes on military architecture, which Danti partially published later, reporting that he took the notes "from the treatise of Messer Antonio Alberti" [dalli discorsi di M. Antonio Alberti] (Danti 1577, 55, 57). On Antonio Alberti (1495–1555), a member of the distinguished Florentine family and a late heir of the more famous Leon Battista, see Passerini 1869, vol. 1, 173–6.

29 The reason for the missed edition of Ptolemy's *Geography* may have been because Prince Francesco did not grant Danti the ban of foreign editions in Tuscany, an act that would have compromised the editorial success of Danti's project (Del Badia 1881, 10–11).

30 On Danti's *Le scienze matematiche ridotte in tavole* (Bologna, 1577), see Daly Davis 1982; Settle 1990; Frangenberg 1988; and Frangenberg 1994.

31 On the debate on the certainty of mathematics in the late sixteenth century, see De Pace 1993; Feldhay 1995, 162–70, and Feldhay 1998.

32 On the early reception of Copernicus's *De revolutionibus*, see Westman 1986. On the diffusion of Copernicanism in Italy, see Bucciantini–Torrini 1997 and Pepe 1996.

33 Danti 1573b, "Delle lodi della prospettiva," not paginated: ". . . conoscere perfettamente il moto, la quiete, il sito, la grandezza, et qualità delle cose naturali intorno alle quali consiste tutta la sua speculatione." In Danti's edition, Euclid's text was extensively annotated with Danti's own commentary that summarized sixteenth-century Aristotelian commentaries, but almost completely disregarded the medieval optics, that is the Arab translations of original Greek works. On Danti's optics, see Frangenberg 1988.

34 Danti 1573b, "Delle lodi della prospettiva," not paginated: ". . . ridurre in piano, in forma o ovata ó circulare, & in diverse altre maniere proportionevolmente il sito di tutta la terra insieme, & delle provincie particulari & ci fa havere perfetta cognitione delle distanze de' luoghi facendoci conoscere chiaramente non solo la proportione, & convenienza di sito, che ha una regione con l'altra, ma con il cielo."

35 On Galileo's use of optics to study the surface of the moon, see Ronchi 1958 and Winkler-Van Helden 1992.

36 Serafino Razzi, *Cronica*, fol. 56v: ". . . i più nobilissimi giovani, et anco alcune illustri signore."

37 Danti dedicated his *Trattato dell'uso et della fabbrica dell'astrolabio* (Florence, 1569; second edition: Florence, 1578) to Cardinal Ferdinando de' Medici. On the artistic and scientific patronage of Cardinal Ferdinando de' Medici, see Butters 2002 (with bibliography).

38 Danti 1571b, Dedication to Isabella Medici Orsini, 18 November 1571, not paginated: "tanto di questa nobilissima, & piacevole scienza si diletta." Danti also dedicated *La sfera di Proclo Liceo* (Florence, 1573) to Isabella Medici Orsini, in which he called her, "most knowledgeable of this discipline" (*intendentissima di questa faculta*), and thanked her for the "infinite benefices, which I receive continuously from her, in whatever major occasion" (infiniti benefizij, che del continuo da quella ricevo, in qual si voglia maggiore occasione) (Danti 1573a, Dedication to Isabella Medici Orsini, not pag-

inated). It seems that because Cosimo I was very ill and Cardinal Ferdinando was in Rome, Isabella had become Danti's major supporter in the close Medici family.

39 Danti 1571a, Dedication to Cesare della Penna, 12 March 1570, not paginated. Danti published a second edition of this book in 1574 and dedicated it to Diomede della Cornia, Marquis of Castiglione and brother of Cesare della Penna, who meanwhile had died fighting against the Turks. Danti 1574, Dedication to Diomede della Cornia, 24 October 1573, not paginated: ". . . while he [Cesare della Penna] was here at the service of my Most Serene Lords, I [showed] him the fifteenth book of Euclid's Geometry, and many things about Astronomy . . ." (. . . Perche mentre egli [Cesare della Penna] stette qua al servizio de' miei Sereniss(imi) padroni havendoli io mostrato i quindici libri della Geometria di Euclide, & molte cose di Astronomia . . .). Diomede della Cornia also had good relations with Cosimo I, who wrote to console him on his father's death (ASF, FM 238, c. 37v). On aristocratic warriors in need of mathematical training, see Biagioli 1989, 44–5.

40 Danti 1569, Giunti's Dedication to Cosimo I and Ferdinando de' Medici, not paginated: "siano cose molto necessarie a chi pratica il mare." According to the Giunti, "many more gentlemen and servants of His Excellency" urged the publication of Danti's treatise in Italian in order to gain a sound knowledge in astronomy.

41 Spini 1570, Dedication to Cardinal Ferdinando, 10 November 1569, not paginated: "Having this past summer . . . attended some lectures by the above mentioned cosmographer [Egnazio Danti] on the treatise of the astrolabe that he wrote and dedicated to you, together with a short summary on the sphere, I collected and annotated [here] some useful comments on mathematics" (essendomi questa state passata . . . posto a udire dal sopradetto Cosmografo (Egnazio Danti) alcune lezioni sopra l'trattato che egli compose, e à lei dedicò dell'Astrolabio, con un breve compendio della sfera, ho raccolto, & annotato alcune utili esposizioni attenenti alle Mathematiche). On Spini's treatise on architecture, see Wazbinski 1987, 215–34. The treatise, which remained in manuscript, is now available in Acidini Luchinat 1980A.

42 On Danti's possible involvement with the Accademia del Disegno, see Wazbinski 1987; Settle 1990; Barzman 1989; and Barzman 2000, 151–7. Danti's brother, Vincenzo, was a member of the academy, and wrote his *Trattato delle Perfette Proporzioni* (Florence, 1567) as a consequence of his teaching anatomy to young artists, and it is likely that Egnazio was involved with a similar task of teaching perspective. According to Raffaele Borghini (Borghini 1584, 522), Egnazio planned to publish his brother's treatise, on which see Summers 1969; Rossi 1980, 123–45; and Daly Davis 1982. On the Accademia del Disegno, in general see Wazbinski 1987; Goldstein 1996; and Barzman 2000.

43 On Cosimo I's cultural politics, see Eisenblicker 2001. On Cosimo I's promotion of the vernacular and the Florentine Academy, see Plaisance 1973; Plaisance 1974; Plaisance 1992 on vernacular books dedicated to Cosimo I; and Ricci 2001.

44 On Clavius's Latin edition of Sacrobosco, see Lattis 1994, 30–45.

45 Danti, *La Prospettiva di Eliodoro Larisseo* in Danti 1573b, Dedication to Cardinal of Lorraine, 11 June 1573, not paginated: "When Your Illustrious Lord came to Florence recently and was kind enough to favor these rooms with his presence and see my works, he seemed to have desired Heliodorus' treatise on optics. Therefore, I sent you [the treatise] as I received it from the Vatican Library together with the translation that I made to accompany Euclid's Optics, which I also translated. Although the gift is small, I hope you will not despise it. You will know well its beauty and excellency since you are knowledgable not only in the most noble science of Theology but also in the other parts of Philosophy, including in Mathematics." (Quando V(ostra) S(ignoria) Illustriss(ima) passò ultimamente per Firenze & con la presenza sua si degnò favorire queste stanze, & veder l'opere mie: mostrò di desiderare il trattato della prospettiva di Eliodoro Larisseo. Perciò lo mando hora à lei tale quale io l'hebbi dalla Libreria Vaticana, et la tradutione appresso, che io ne ho fatta per accompagnare la Prospettiva di Euclide tradotta similmente da me. Et se bene il dono é piccolo spero che non sará sprezzato da lei, si come quella, che conoscerá molto bene la bellezza, & eccellenza di esso, poiche non solo ella ottimamente possiede la nobilissima scienza della Theologia, ma fra l'altre parti della Filosofia, questa delle Matematiche.)

46 On the humanist editions of classical texts, see Kenney 1974. On the humanist editions of ancient scientific texts, see Rose 1975; Grafton 1991, especially 1–46; and Grafton 1997. The books quoted in the texts are: Ptolemy, *Geographia*, translated and commented by Sebastian Münster (Basel, 1540); Federico Commandino, *Planisphaerium* (Venice, 1558); F. Commandino, *Claudi Ptolomei Liber de Analemmata instauratus* (Rome, 1562). On Commandino, see Rose 1975, 185–221.

47 On the social background of Renaissance mathematicians and astronomers, see Rose 1975; Boas Hall 1983; and Biagioli 1989, 56–67. On the self-fashioning of Renaissance writers, see Greenblatt 1980.

48 Danti 1569, 193–4. For a penetrating discussion on the uses of Renaissance scientific instruments, see Field 1988, who distinguished between instruments used as demonstration pieces and actual astronomical tools; Kemp 1991, who introduced the definition "intellectual ornaments;" and Kemp 1995. On the Medici collection of scientific instruments, see Heikamp 1970; Miniati 1991; and Bedini 1995.

49 Danti's surviving correspondence documents that, apart from the monumental globe for the Guardaroba Nuova, he produced other small terrestrial globes, none of which has

survived or is identifiable: one he made for Francesco Bolognetti, another for Francesco de' Medici in 1571. Danti also made numerous astrolabes, some of which survived. The large astrolabe, kept in the Museo di Storia della Scienza, Florence (inv. no. 3361) and dated 1571–5 was used for astronomical observation (Righini Bonelli–Settle 1979). Other astrolabes have been attributed to Danti, but some of them might more likely be recent copies (see Settle 1990, 32, n. 29). The astrolabes positively attributed to Danti are kept in the following museums: Museum of American History, Washington D. C. and Musée de la Vie Wallonne, Liège. Three more are considered nineteenth-century copies and are now kept in the following museums: Museo della Navigazione, Genova-Pegli; Science Museum, Oxford; Old Royal Observatory, Greenwich. The astrolabe that was once in the Staatliche Mathematisch-Physische Salon, Dresden, has been missing since 1945. Often Danti had other Florentine craftsmen build the instruments that he had designed (Turner 1995).

50 Danti's letter to Count Polidoro Castelli, 23 September 1571 (Tiraboschi 1833, vol. 3, 497).

51 Danti 1578a, 322: "un Tritone di metallo, che nella destra mano teneva una verga con la quale mentre era girato dal Vento fermandosi sopra una di quelle teste mostrava quale di essi spirava."

52 Danti wrote a treatise on the anemoscope, entitled *Anemographia* (Bologna, 1578), dedicated to Cardinal Gabriele Paleotti, Bishop of Bologna (Fig. 6.11). He published an Italian translation of the same text, dedicated to Pietro Ghislieri, in the 1578 edition of his *Trattato dell'uso et della fabbrica dell'astrolabio* (Danti 1578a, 250–81). Danti's text also survives in a Latin manuscript, a beautiful presentation copy Danti made for Pope Gregory XIII Boncompagni (BAV, MS Vat. Lat. 5647, now published in Courtright 2003, 219–42). For Gregory XIII, Danti built an anemoscope in the Tower of Winds in the Vatican Palace, on which see Chapter Nine.

53 Danti 1569, Quinta parte, 38: ". . . far tali strumenti di quella maggior grandezza, che possibile sia, perché quanto saranno maggiori, tanto piu minutamente saranno divisi, & tanto piu agevolmente, & bene si opererà con essi. Come espresso si vede in uno Astrolabio di metallo, che attualmente fo condurre al fine di piedi quattro di Diametro, per l'Eccellenza del S. Duca mio Signore." Archival sources show that the actual making of the instrument was delegated to other craftsmen. Danti's monumental astrolabe is mounted on an octagonal walnut table and has a hinge that can block its position at various angles to the horizon, making it possible for one person to operate it. On this astrolabe, see Righini 1977 and Righini Bonelli–Settle 1979, 54–6.

54 Danti 1569, 106: "la Geometria di Orontio, il cui libro dal Mag(nifico) M(esser) Cosimo Bartoli è stato felicemente in questa nostra lingua tradotto." The book Danti quoted is Cosimo Bartoli, *Opere di Orontio Fineo del Delfinato* (Venice, 1587), on which see Bryce, 1980; and Bryce 1983, 176–83.

55 Danti 1569, Quinta Parte, 3: "Tolomeo, & Giordano, i Planisferi de' quali sono non mediocreme(n)te illustrati da i Commentarii del Commandino, dal Barbaro succintame(n)te ridotti nella sesta parte della sua prospettiva." The books mentioned by Danti are Federico Commandino, *Planisphaerium* (Venice, 1558) and Daniele Barbaro, *La prospettiva pratica* (Venice, 1568).

56 Danti 1569, not paginated: "un libro di strumenti quale Alfonso Re di Spagna d'Arabo, & Caldeo tradusse in Castigliana lingua, & hor di Castigliana lingua, mutato è già gra(n) te(m)po in volgar fiore(n)tino, che appresso di me si tiene."

57 Danti may have known Rojas's projection directly from Juan de Rojas Sarmiento, *Commentariorum in astrolabium quod planisphaerium vocant libri sex* (Paris, 1550) or from Cosimo Bartoli, who had included a detailed discussion in his *Del modo di misurare* (Venice, 1564).

58 Spini 1570, 27.

59 On the early history of the calendar, see Pederson 1983; North 1983; Heilbron 1999, 24–46; and Howell 2002, 13–38.

60 On the history of the reform of the calendar, see Coyne–Hoskin–Pederson 1983, especially the essays by North 1983, 86–97 on the early history; and Moyer 1983, 171–88 on the Gregorian reform. See also: Maiello 1989; Heilbron 1999, 47–80; and Courtright 2003, 33–40.

61 Danti 1578A, 323: "Et perche il Gran Duca Cosimo hebbe sempre l'animo Eroico inclinato a imprese grandissime, essendo sempre stato emulatore de' fatti egregij degli antichi Eroi, & conoscendo (oltra l'utile universale) quanto di gloria habbia apportato a Giulio Cesare la riforma, che egli fece dell'anno se bene non fu perfetta, si era proposto nell'animo di adoperarsi con ogni suo potere attorno questa honoratissima impresa, la quale (have[n]do proposto a Sua Beatitudine) l'havria senza alcun dubio condotta a perfettione, se pria della importuna, & longa infermità, & poi dalla morte non fosse stato prevenuto." On Danti's research in connection with the reform of the calendar, see Righini Bonelli–Settle 1979 and Heilbron 1999, 47–80.

62 Danti's quadrant is a large marble slab placed perpendicular to the church's façade and sustained by a large bracket with two inscriptions: one celebrating Cosimo I's patronage of cosmography (*Cosm. Med. Magn. Etr. Dux Nobilium Artium Studiosus Astronomiae Studiosis Dedit. Anno* D. MDLXXII), the other reporting some astronomical measurements pertaining to the inclination of the ecliptic and the distance from the tropic, which, however, were not taken with this instrument (*Dilige[n]ti Observatione Perspecta Tropicop Distantia G.* XLVI.LVII.XXXIX.L. *Et Angulo Sectionis G.* XXIII.XXVIII.XXXXVIIII.LV.) The two sides of the quadrant are both marked with lines referring to five solar clocks for

as many different systems to count time: the *horae ab occasu* (*H. Ab Occ*) used in Italy which counted the hours starting from sunset; the *horae ab ortu* (*H. Ab Ort.*) used in Bohemia which counted the hours starting from sunrise; the *horae astronomicarum*, (*H. Astr.*) which divided the day into 24 hours and started from noon; the *oriolo comune degli Oltramontani* which divided the day into two sequences of twelve hours each and started from noon (this last solar clock is on the quadrant's bracket); and the system of the planetary hours (*H. Plan*) marked on the church's façade in Roman numbers. On Danti's quadrant, see Danti 1578a, 285–8; and the exemplary study, on which my discussion is based, in Righini Bonelli–Settle 1979. Righini Bonelli and Settle discussed the relationship between the instrument and the preparatory drawing, possibly by Danti, now in the Uffizi (Gabinetto dei Disegni e delle Stampe degli Uffizi, no. inv. 3946A). They also showed that Danti's technique in drawing the transversals, preceding that invented by Tycho Brahe, is similar to the graduation of the large astrolabe in the Museo di Storia della Scienza, Florence (inv. no. 3361) (Fig. 3.10).

63 Spini 1570, 31.

64 Danti 1578a, 285: "l'osservazione degl'Equinozij, per poter con essi ritrovare la vera quantità dell'anno."

65 Danti 1578a, 284: "acciocchè quando si fa l'osservazione dell'Armilla si possa in un subito vedere nel Quadrante l'hora giusta che corre."

66 Danti 1578a, 323: "Callippo, & poi Hipparco, Sosigene, & Tolomeo . . . [fecero] le piu importanti osservazioni, attorno alla grandezza, & misura dell'anno." On 20 January 1574 Cosimo I gave Danti and his students the sum of 35 scudi they had requested to build the armillary sphere; two months later, on 19 March, the grand duke authorized an additional payment of 24 scudi for the marble slab to be placed on the façade. The instrument, placed above the east portal of the church, was used for astronomical observation pertaining to the measure of the tropical year. The instrument's inscription, commemorating Cosimo I's patronage and the first astronomical observation, reads: *Cosmus Medices Magnus Etru(riae) Dux Post Antiquos Aegiptiorum Reges Primus Astronomiae Studiosis Posuit.* MDLXXIIII. *V Idus Martii Hora* XXII M. XXIIII *P.M. Ingrediente Sole Primum Arietis Punctum.* The same measurement was repeated the following year in front of a group of learned men. Danti regarded his armillary sphere as incomplete. On 4 June 1577 he sent from Bologna to Jacopo Dani, ducal secretary, in Florence, "a drawing of the armillary [sphere], which I had begun for the facade of the church of Santa Maria Novella" [uno schizzo dell'armilla cominciata sulla facciata della chiesa di S. Maria Novella] (Del Badia 1881, 49). On Danti's armillary sphere, see Danti 1578, 318–25; and Righini Bonelli–Settle 1979.

67 Danti mentioned the meridian lines of Sta. Maria Novella and San Petronio in his treatises: "And with a similar invention I made a large meridian line in the church of San Petronio in Bologna, similar to the [meridian line] I began in the oculus and vault in Sta. Maria Novella, which remained unfinished because of the death of Grand Duke Cosimo." [Et con simile intenzione ho fabbricato nella chiesa di S. Petronio nella medesima città di Bologna il Gnomone grande Astronomico, simile a quello che in S. Maria Novella cominciai, nella volta, & nell'occhio di essa Chiesa, & per la morte del G(ran) D(uca) Cosimo restò imperfetto.] (Danti 1578, 325). A few years later Danti also recalled his meridian line in St. Peter's (Danti 1586, 105). Danti's meridian line in San Petronio was later substituted with another one by Giovan Domenico Cassini, while the one in St. Peter's has not been located yet. On Danti's meridian lines, see Danti 1577; Danti 1578, 298–317; Righini Bonelli–Settle 1979, 11; Field, 1988, 4–8; Thomas Settle, "Egnazio Danti as a Builder of Gnomons. An Introduction," paper delivered at the Villa I Tatti, Florence, 1990 (I thank the author for sending me a copy of his unpublished paper); Heilbron 1999, 40–80; and Courtright 2003, 212–17. On Danti's meridian lines in Bologna as well as his treatise of 1576 or 1577, see Chapter Six.

68 Danti 1569, Dedication to Cardinal Ferdinando de' Medici, not paginated: "Cosimo dal quale tutto quello veramente procede, che io ho apparato di cosi gentile, & piacevole professione com'é la Cosmographia."

69 Danti 1569, introductory pages, not paginated.

Quel ch'era prima in mille, e mille carte
Confuso e sparso, hor qui vostra merced,
Breve, & chiaro si scorge, & con tal fede
Che de gl'honor nel Nilo è l'Arno a parte.
Tal suo figlio da vulgo hoggi diparte,
E'l ciel tutto misura, intende, & vede,
Che senza mai d'oblio ritrarre il piede,
Fora oscuro ov'ei s'erge alto erdisparte.
Felice voi, che del GRAN COSMO havete
L'età sortita: ond'ogni oltraggio e danno
Ben potrete scivar di Lethe avaro.
Cosi fermando al vostro nome andrete
(Tuo malgradorio tempo) alto riparo
Che sei dell'altrui gloria empio tiranno.

70 Cox Rearick 1984, 279, discussed Danti's impresa as part of Cosimo I's cosmic imagery, although she did not connect it to the Guardaroba Nuova. She was, however, the first scholar to point out that Cosimo's imperial imagery became increasingly stylized after 1569. On this impresa, see also Crum 1989 and Fiorani 2001.

71 On the events that led to Cosimo I's title as Grand Duke of Tuscany, see Carcerieri 1926–9, vol. 2, 169–247, which also includes a detailed description of his coronation in the Sala Regia in the Vatican Palace; and vol. 3, for the reception of the Medici's new title by other rulers, including the imperial recognition, which did not occur until 12 February 1575, well

after Cosimo I's death. The papal bull, dated 27 August 1569 and published on 13 December 1569, described the crown as a "radiate like that of the Eastern kings" (Forster 1971, 69). The crown is illustrated in the painting by Alessandro Fei del Barbera, *The Goldsmith Workshop*, Palazzo Vecchio, Studiolo of Francesco I, 1570, which includes both the drawing enclosed with the papal edict and the coronet being made after the papal design. On the crown see Forster 1971 and Sframeli 1997.

72 Danti 1569, introductory pages, not paginated. On Giovan Battista Strozzi the Elder (1504–1571) see Minor–Mitchell 1968, 27–30; Bianconi–Vassalli 1983, vol. 2, 439–55. A modern edition of his madrigals is in Strozzi 1975, which also includes biographical information. I thank Konrad Eisenbichler for kindly translating Strozzi's poem.

73 Vasari–Milanesi 1906, vol. 8, 22.

CHAPTER 4

1 The literature on collecting in early modern Europe is extensive, but the most significant studies are the following: Balsiger 1970; Impey–MacGregor 1985; Lugli 1990; Pomian 1990; Kaufman 1993, 174–94; Findlen 1994; and Bredekamp 1995.

2 The Medici collections have been studied extensively, especially as individual groups of statues, drawings, paintings, decorative objects, scientific instruments, exotica, etc. Listed here are the most important studies. On the collections of the early Medici, see Müntz 1888; Spallanzani 1996; and Merisalo 1999, for fifteenth-century inventories of the Medici collections; Dacos–Giuliano–Pannuti 1972; and Heikamp 1974, for Lorenzo the Magnificent's gems and vases. On the ducal collections, see Conti 1893, Müntz 1895, Gaeta Bertelá 1997, and Barocchi–Gaeta Bertelá 2002, for the inventories of the Medici ducal palaces; *Palazzo Vecchio* 1980; *Corte mare mercanti* 1980; Massinelli–Tuena 1992; *Magnificenza* 1997; Acidini Luchinat 1997; and *Medici and Michelangelo* 2002, for the collections. Important also are the following specialized catalogues: Cora–Fanzani 1986; and Spallanzani 1994, on Medici ceramics and porcelain, and Massinelli 1991 on Cosimo I's statuettes.

3 An inventory of Cosimo I's ducal palace, taken in 1553 and published in the nineteenth century (Conti 1893), recorded the arrangement of rooms in the palace during Cosimo I's rule and the display of his collections. Specifically on his apartment, which consisted of six rooms and was located on the first floor, see Conti 1893, 33–7. Eleanor's apartment was above Cosimo's, on the second floor, while the children of the ducal couple lived on the third floor, their rooms being connected to their mother's apartment through stairs. On the private rooms of the ducal family, see Allegri–Cecchi 1980, 184.

4 On Cosimo I's guardaroba, see Conti 1893; Müntz 1895, 98–105; and Allegri–Cecchi 1980, 287–302. According to the 1553 inventory of the ducal palace, the ducal guardaroba was divided into three sets of rooms, a partition maintained also in the following years. The first set of rooms was actually a guest apartment known as the "new apartment of the guardaroba where visitors lodge" (*appartamento nuovo della Guardaroba dove alloggiano i forestieri*), on which see Conti 1895, 79–100. The second set of rooms was the general guardaroba, which spread throughout the palace; most of these rooms cannot be identified today because of the palace's later transformation, including Vasari's elevation of the ceiling of the Sala Grande (Conti 1895, 101–36). The third part consisted of one single room, the secret guardaroba (*guardaroba segreta*), on which see note 5 below.

5 In the secret guardaroba, ancient and modern statues in marble and bronze, paintings, and reliefs were displayed above the fourteen cupboards that run around the room. In the cupboards the objects were stored typologically: gold and silver fabrics and purple velvets were stored in the first and second cupboards; oriental carpets and tapestries were rolled in the fourth; silverware was in the fifth; wall hangings were folded in the sixth; tapestries were rolled in the seventh; more wall hangings were in the eighth; ceremonial pillows, elaborate tapestries, and comfortable mattresses were in the ninth; ceremonial dresses for the duke were in the tenth; porcelain ware was stored in the eleventh; miniature objects, ancient and modern sculptures and reliefs, mother-of-pearl, corals, semi-precious stones, fossils, and even a dry chameleon were kept in the twelfth (many objects from this cupboard were moved to the Studiolo di Calliope in 1559–60); crystal and silverware were in the thirteenth; ceremonial dress in metal for both people and horses were stored in the fourteenth together with other precious small objects. On the secret guardaroba, see Conti 1895, 137–94, and Allegri–Cecchi 1980, 287–302.

6 Cellini 2002, 343: "In those days the new apartments towards the Via dei Leoni were built, so that when His Excellency wanted to retire to more private quarters, he had a little chamber [*un certo stanzino*] arranged in these recently built apartments, and he ordered me to go through his guardaroba, where I was to pass secretly above the ceiling of the great hall [the Sala Grande] and then go through certain cubby-holes to this most secret chamber." Cellini's account is dated between 1553, when the Etruscan statuettes were discovered, and April 1554, when his bronze statue *Perseus* was publicly displayed. The artist recounted this episode in his self-serving autobiography not to document his work on the statuettes, which he considered prosaically as a huge waste of time which kept him from the bronze casting of his *Perseus*, but rather to report publicly the high esteem that the duke reserved for him allowing him to enter into his private apartment. On Cosimo I's own cleaning work on the statuettes, see: Frey–Frey 1923–40, vol. 1, 394; and Cellini 2002, 342–3.

7 The Studiolo di Calliope (2.9 × 2. 35 m), decorated and furnished by Vasari between 1555 and 1558, was named after the Muse Calliope depicted on the ceiling panel. A typical collector's study, it was small in size and isolated from the other rooms of the apartment. It featured a secret access to assure privacy, a stairway leading to the lower floor hidden behind the wall paneling, but it could also be reached publicly from the Sala di Cerere. It was illuminated by suffused light which was filtered by a stained-glass window, the perfect illumination for reading and observing (Fig. 4.1). The room was appropriately furnished with shelves and drawers to display small-size objects. Crimson velvet covered the walls, three rows of shelves ran around the walls, cedar wood drawers were put below the shelves (perhaps attached to them), and large cupboards and drawers were placed on the floor. On the Studiolo di Calliope, see Vasari–Milanesi 1906, vol. 8, 58–62, for the description of the room; Liebenwein 1988, 120–2, for the analysis of the ceiling panel; Allegri–Cecchi, 1980, 80–2, for the pertinent archival documents; and Massinelli 1991, 9–22, for a discussion of its furnishing and collections. The objects Cosimo I selected to be moved from the twelfth cupboard of the secret guardaroba to the Studiolo di Calliope are listed in an inventory of the ducal palace dated 1559–60 (ASF, G 37, c. 13 and following pages; the inventory is also published in Müntz 1895, 132–40). For a general assessment of Renaissance studioli, see Liebenwein 1988; and Thornton 1997, especially 99–126.

8 At the opposite end of the Apartment of the Elements, the Studiolo di Minerva was decorated between 1557 and 1562 and named after the ceiling panel by Vasari illustrating the *Birth of Minerva from Jupiter's Head.* Although there is no record of the objects that were displayed in this study, a note from Vasari to the duke suggests that it was conceived as a pendant to the Studiolo di Calliope: "I still have to report that I completed the study on the terrace, and that I do not know whether Your Illustrious Excellency wants there a desk, or drawers or shelves. Let me know. I already ordered the boxes with shelves to display the small statues and the reliefs in marble." (Restami a dirvi che lo scrittoio d'in sul terrazzo l'ó finito, et che non so se V(ostra) E(ccellenza) I(llustrissima) ci vuol banco o cassette o scafali, che melo avisi: perché i palchetti per metter le statue piccole di marmo già gli ó ordinati.) (Vasari's letter to Cosimo I, 8 March 1562, Frey–Frey 1923–40, vol. 1, 665). On the Studiolo di Minerva, see also Allegri–Cecchi 1980, 110.

9 On the visual experience provided by the Kunstkammer, see Bredekamp 1995.

10 Some of these artifacts that Cosimo I selected for his Studiolo di Calliope have been identified with objects still kept in Florentine museums, on which see Massinelli 1991 (with bibliography and illustrations). Nicola Pisano's head is now lost, but may be the same head that Lorenzo the Magnificent kept in his own study (Schaefer 1976, 132).

11 On the importance of the Etruscan myth during Cosimo I's reign see: Cipriani, 1975; Cristofani 1979; Cipriani 1980a; and Cipriani 1980b.

12 On the miniature portraits of the Medici family by Bronzino's workshop, each 12 × 15 cm and now in the Uffizi, see Vasari–Milanesi 1906, vol. 7, 603; Langendijk 1981–7, vol. 1, 100–03; *Palazzo Vecchio* 1980, 197; and Acidini Luchinat 1997, 60, with illustrations.

13 The two spherical incense burners, recorded in the 1553 inventory and now at the Museo Nazionale del Bargello, Florence, were stored in the twelfth cupboard of Cosimo I's secret guardaroba until 1559. Lorenzo the Magnificent had similar objects, possibly the same objects, in his study in the Medici Palace. On these artifacts, see Conti 1893, 193; *Palazzo Vecchio* 1980, 191–2; Spallanzani 1980b; Curatola-Spallanzani 1981, 3–4, 7–12; and Massinelli 1991, 20–1. The miniature Aztec animal heads, an amethyst dog head and an onyx dog head, are now kept in the Museo di Mineralogia, Florence (inv. nos. 13221/206 and 13364/209), together with two Aztec statuettes that also belonged to Cosimo I. The animal heads, which originally were trinkets, dated 1430–1520, are documented in Cosimo's guardaroba since 1553. They were stored in the twelfth cupboard. On these Aztec statuettes, see *Palazzo Vecchio* 1980, 168, nos. 310 and 311; Heikamp 1972, 13–14, 44; Acidini Luchinat 1997, 71 with illustrations; and Massinelli 1991, 16–17. In the Studiolo di Calliope were also displayed larger objects such as cameos, sardonyx, cornelian, and oriental crystals in the cupboards on the floor, while modern miniatures mainly by Giulio Clovio, the most renowned miniaturist of the time, were to be displayed on a wall panel (Vasari–Milanesi 1906, vol. 7, 567; and *Palazzo Vecchio* 1980, 194).

14 The *Chimaera* was too large a statue to be displayed in the Studiolo di Calliope and indeed continued to be displayed in the Sala di Leone X, next to the Sala Grande. Not only did its larger size demand a larger room, but its interpretation as "a bronze lion with the head of a capricorn" (*uno lione di bronzo con una testa di chapricorno*) dictated its location in the room celebrating Leo X (*Leone* in Italian). On the political meaning of the *Chimaera*, see Vasari 1568, Proemio, not paginated; and Collaretta 1980, 116, who recorded Vasari's project to move the statue into the Studiolo di Calliope.

15 The Tesoretto (2.88 m on each side) was furnished all around with eleven cupboards in marble marked by imposing doors, which were used for the storage of Cosimo I's books, papers, and classified documents. On the ceiling, around a large globe, the symbols of the evangelists intermingled with allegorical personifications of the arts (architecture, painting, sculpture, and music), and scenes illustrating the activities of the sciences of philosophy, astronomy, poetry, and geometry. On the *Tesoretto*, see Allegri–Cecchi 1980, 191–4, for the pertinent archival documents; and Liebenwein 1988, 122–3, for the interpretation of the ceiling's iconography.

16 Vasari–Milanesi 1906, vol. 7, 631. On this study, see Allegri–Cecchi 1980, 189–90, who, however, erroneously identified the room as part of what is today the Studiolo of Francesco I; Davis 1985b; and Muccini 1992, 59, who both correctly identified Cosimo I's study with the room below the *Tesoretto*. On Vincenzo Danti's bronze relief, see: Summers 1969; Davis 1985b; and *Medici and Michelangelo* 2002, 199.

17 Vasari 1568, not paginated. On the hall of ancient statues in the Pitti Palace, see Daly Davis 1985. On Cosimo I's collection of ancient statues, see: Cristofani 1985. On Renaissance collections of ancient statues and their display, see: Haskell–Penny 1981.

18 ASF, G 73 (for the 1570 inventory); and G 87 (for the 1574 inventory). Later inventories of the palace are of little use since Cosimo I's sons, Francesco I and Ferdinando, substantially revised the organization and display of the ducal collection, reserving for them the larger rooms and galleries available on the third floor of the Uffizi Palace.

19 Ortelius, *Map of the World* (Antwerp, 1564), cartouche entitled "Catalogue of some foreign goods, and of the places from which they are brought to us" (*Mercium aliquot peregrinarum, et locorum unde ad nos adferuntur, catalogus*). On Ortelius's map, see Schilder 1987.

20 Danti's legend to the map *Hindustan Outside the Ganges*: "Questa parte dell'India fuor del Gange si stende in longhezza sino al Catai et contiene molte p(ro)vinzie nelle quali si trovano molto notabili cose. Et in Ergimut di detta provinzia si trova il perfettissimo Muschio. Nei monti di Ava si trovano bellissimi rubini et nei monti di Satgaym oltre ai rubini si trovano anche dei diamanti. In Caindú et nelle città vicine si spendano 15 coralli in cambio dei danari, et nel lago Salzo si pescano bellissime perle. Nel territorio di Carazan si trovano serpi di stupenda grandezza, et nei monti di lachi vi sonno Lioni ferocissimi. In Agrigaia pro(vincia) vi si trovano Buoi della grandezza degli Elefanti li quali hanno la lana finissima come seta. Si trovano anche Rinoceronti nei monti di sotto a Carazan. Dicono che nel deserto di Camul si trovano molti spiriti che ingannano li viandanti mostrandoli il falso camino per farli perdere nel deserto."

21 The recipe for the technique of the making of porcelain was a much disputed issue in Renaissance Europe involving distinguished scholars such as Girolamo Cardano and Joseph Scaliger. Danti's recipe repeated that of Duarte Barbosa, whose book had been published in Italian by Giovanni Battista Ramusio in the first volume of his series on voyages. In the late 1570s the pottery workshops sponsored by the Medici found a way to imitate Chinese porcelain. On these Renaissance disputes over the making of porcelain, see Lightbown 1969, 230–1.

22 On the Studiolo of Francesco I, see Berti 1967, for its relation to Francesco's personality; Schaefer 1976, for a detailed analysis of its iconography; Allegri–Cecchi 1980, 323–50, for pertinent archival documents; Morel 1982, for the symbolic interpretation of its iconography and the relation to the Guardaroba Nuova; Bolzoni 1985; Bolzoni 1995, 245–69, for its interpretation as a memory theater; and Feinberg 2002, for a recent reassessment.

23 Borghini's letter to Vasari, not dated: "L'inventione, mi pare che si dimandi conforme alla materia, et alla qualità delle cose, che vi si hanno a riporre, talché la renda la stanza vaga, et non sia interamente fuor di questo proposito, anzi serva in parte, come un segno, et quasi inventario da ritrovar le cose, accennando in un certo modo le figure et le pitture, che saranno sopra, et intorno, et negl'armadij quel che e serbono dentro da loro" (Vasari–Del Vita 1938, 47–8, but see also 52–60).

24 Borghini's letter to Vasari, not dated: "intendendo sempre di cose singulari, et che portin seco una certa maraviglia che le comuni non s'hanno a riporre in simil luoghi" (Vasari–Del Vita 1938, 51).

25 Borghini's letter to Vasari, 3 October 1570: ". . . bisognerebbe che io sapessi . . . sopra tutto che si fussi fermo le cose che hanno a ire negli armarij (notate bene tutta questa parte che io dico hora) per poter accomodare le picture di fuori alle cose di drento; i quali armarij, se fussin corrispondenti alle cose di sopra, tornerebbono anche molto meglio, cioè: le cose del fuoco, come stillamenti et lavori fatti a fuoco di ferro, acciaio, vetri et simili fussino sotto la statua di Vulcano et l'elemento del fuoco; le medaglie, i lavori d'oro, rubinj e miniere di simil cose sotto quella di Pluto et d'Ope et sotto lo elemento della terra; et cosi le gioie sotto Giunone, i cristalli sotto Zephiro, le perle sotto Venere, et cosí il resto; et ferma questa distintione, mi bisognerebbe saper,' quante storie bisogna per armario et quel che v'ha a ire, come dir': Ece 1.o armadio, che ha dua quadri di fuori da dipigner,' ove staranno medicine pretiose, come balsami, corno d'unicorno, rimedij di veleni et simil cose; in un'altro v'hara verbi gratia un quadro solo et drento staranno medaglie, et cosi del resto" (Frey–Frey, 1923–40, vol. 2, 530).

26 Borghini's letter to Vasari, 3 October 1570: "certe donne come provincie: che verbi gratia nella Magna [Germany] si lavora d'horioli et d'ingegni di ferri eccellentemente, in Francia di cose d'oro etc.; in quegli armadij dove fussino di simili cose provi in uno quadretto una Germania con quegli ingegni in mano, come se fussin rovesci di medaglie, et cosi negli altri" (Frey–Frey, 1923–40, vol. 2, 532–3).

27 For a general discussion on the exotica in European collections, see Lightbown 1969; Impey–MacGregor 1985 (especially the essays by Christian Feest, Ezio Bassani and Malcolm McLeod, Julian Raby, John Ayers, Oliver Impey, and Robert Skelton); Feest 1990; and Shelton 1994.

28 On the Medici collection of exotica, see Heikamp 1972; Bassani 1975; Spallanzani 1978, especially 55–68, 176–93; Bassani 1980; Spallanzani 1980, 73–94; *Palazzo Vecchio*, 1980,

161–80, 191–2; Spallanzani 1994; Scalini 1996; Scalini 1997; Scalini 1998; and Heikamp 1998.

29 On the celadon plate and vase now in the Museo degli Argenti, Pitti Palace, Florence see Lightbown 1969, 229; *Palazzo Vecchio* 1980, 178; Ayers 1985, 261–2; and Scalini 1997, 145. On the Chinese porcelain in the Medici collection, see: Spallanzani 1978, esp. 55–68, 176–93; Spallanzani 1980; *Palazzo Vecchio* 1980, 173–80; Cora–Fanfani 1986; and Spallanzani 1994, especially 29–58. It is worth noting that Medici inventories never recorded the geographical provenance of porcelain. But the association between China and porcelain was so secure in the sixteenth century that, it is safe to assume, the label "green porcelain" (*porcellana verde*) or "white porcelain with flowers" (*porcellana bianca a fiori*) carried also the indication of the piece's geographical provenance. Although far less expensive than other luxurious goods such as drapery, silverware, carpets, and jewels, Chinese porcelain was still highly regarded and used for special banquets at the Medici court.

30 The label "Turkish" (*turco*) could apply to myriad artifacts, while the labels "damascene" and "after the damascene manner" (*domaschino* and *alla domaschina*) were used unsystematically and interchangeably for objects coming from either Syria or Persia, as well as for European artifacts made after the oriental prototypes. On Cosimo I's oriental artifacts, see Conti 1895, 114–15, 193; and Scalini 1997, 154. For a discussion of *domaschino* and *alla domaschina*, see Spallanzani 1980, 96. On the falcon hood, see Scalini 1997, 147.

31 The best and most comprehensive study on Cosimo I's exotica from the West Indies remains Heikamp 1972, who also listed the American exotica that Cosimo possessed in 1539 (Heikamp 1972, 34). The surviving mosaic mask, now at the Museo Etnografico Nazionale Luigi Pigorini, Rome (inv. no. 4213) and recorded in Cosimo I's collections since 1553, has been identified as an artifact of the Mixteca Puebla culture, on which see Conti 1895, 174; Heikamp 1972, 12; *Palazzo Vecchio* 1980, 170; and Heikamp 1998, 399. It is worth noting that Medici inventories recorded the artifacts from the West Indies with the generic label Moorish (*moreschi*) or as coming "from the Indies" (*dalle Indie*), with no further specification if they came from the East or West Indies. Two surviving cloaks of feathers are now kept in the Museo di Antropologia e Etnografia, Florence (inv. nos. 284b and 281), on which see Heikamp 1972, 10; and *Palazzo Vecchio* 1980, 316–17.

32 The oliphant bearing the Medici-Toledo coat of arms is at the Museo di Antropologia e Etnografia, Florence, (inv. no. 22181); the other two are in the Museo degli Argenti, Pitti Palace, Florence (inv. nos. consegna dal Bargello 1925, nos. 2 and 3). On these oliphants, documented in Cosimo I's collection since 1553, see Bassani 1975; and *Palazzo Vecchio* 1980, 161–3. The Portuguese-African spoons used by Eleanor are currently divided between the Museo di Antropologia e Etnografia, Florence, which keeps three of them, and the Museo Etmographico Nazionale Luigi Pigorini, Rome, which keeps the other two. Similar items also featured in the collections at Ambras, Dresden, and Braunschweig. On these African spoons, recorded in the Medici inventories from 1560, see Bassani 1975, 9–19; and *Palazzo Vecchio* 1980, 162–5.

33 On Ptolemy's *Geography*, see Dilke 1985. After decades of wide circulation in manuscript versions, some of which also included modern maps, the Latin translation of Ptolemy's *Geography* was printed in Vicenza in 1475, although without maps. Two years later it was republished in Bologna and Rome and was illustrated with the twenty-seven Ptolemaic maps as designed by Nicolaus Germanus. On the significance of Ptolemy's *Geography* in the Renaissance, see Crone 1978, 35–41; Brown 1979; Broc 1989, 9–15 (with bibliography). On the Renaissance editions of Ptolemy's text, see Codazzi 1945; Campbell 1987, 122–40; and Babicz 1994. Ptolemy's translation into English is in Ptolemy 1932, while a new, more accurate, translation of selected parts appeared recently in Ptolemy 2000. On the role and importance of Pliny in Renaissance Italy, see Findlen 1994. On Vitruvius in the Renaissance, see Pagliara 1986; Rowland 1998, 139–40, 223–6; and Vitruvius 1999, for a modern English translation. On the relationships between Vitruvius and Ptolemy in the Renaissance, see Cosgrove 2003a.

34 The manuscript tradition of Ptolemy's *Geography* handed down two different sets of maps, one that included sixty-four maps and another that included only twenty-six. But the smaller set corresponded better to the text of the manual, a fact that Renaissance editors did not fail to notice and indeed, early on, they selected the sequence of twenty-six regional maps, plus the general map of the known world, as the standard set of Ptolemy's maps.

35 On the conception of the world as made up of continents, see Lewis–Wigen 1997, 21–46. On the addition of the New World into Renaissance world maps see Mignolo 1995, 259–314.

36 Campbell 1987 showed that at least half of the maps printed before 1500 were related to Ptolemy's text.

37 On the relevance of Ptolemy's *tabulae modernae* in Renaissance cartography, see Akerman 1995 and Akerman 1998.

38 Gastaldi's edition of Ptolemy (Ptolemy 1548) was notable for many reasons: it was the first octavo edition, the first translation into Italian, and the first edition illustrated with engraved maps. Moreover, for the first time the modern maps made by Gastaldi were placed in conjunction with the corresponding Ptolemaic map, rather than being assembled at the end of the volume as in previous editions. The ordering of maps in Gastaldi's edition is further discussed below. On Gastaldi's edition of Ptolemy, see Fahy 1993.

39 Ruscelli's commentary on Ptolemy's text (Ptolemy 1561) accounts for more than half of this lengthy publication, while his maps are reprinted mainly from Gastaldi's edition of 1548. He added, however, two other modern maps, which

indeed appeared also in the Guardaroba Nuova: the Zeno Map illustrating Greenland and the map of the world divided into two hemispheres, on which see Shirley 1983, 99. On Ruscelli, see Eamon–Paheau 1984.

40 On Münster's *Cosmographia universalis*, see Strauss 1959, especially on Münster's treatment of Germany; Hodgen 1964, 111–206, for its ethnographic and anthropological content; Burmeister 1964, for Münster's biography; Broc 1986, 51–6, for its significance in Renaissance descriptive geography; Grafton 1992, 97–111 for an effective overview; Levesque 1994, 6–10, for a brief consideration of its geographical order; and Cormack 1997, 129–62, for the role of descriptive geography in English universities of the early modern period. Montaigne recalled in his diary that he wished he had taken along a copy of Münster's book on his trip to Italy in order to be alerted "of the remarkable and rare things of each place" (*des choses rares et remarcables de chaque lieu*) (Montaigne 1955, 33–4).

41 Münster 1550, preface, not paginated.

42 In the Renaissance, the word "theater" (*theatrum*) in a book's title denoted the comprehensive treatment of the subject: the theater of anatomy, the theater of instruments, or, in Ortelius's case, the theater of the world (Ong 1958, 114, listed some 30 books published in the sixteenth century with the word theater in the title, while Blair 1997, 153–79, provided a recent discussion on the theme of the "theater of the world"). The bibliography of Ortelius's *Theatrum orbis terrarum* is extensive, but special attention should be given to the following studies: Koeman 1964; Akerman 1995; Akerman 1998; Van der Broecke–Van der Krogt–Meurer 1998; Mangani 1998a; Mangani 1998b; and Melion 1999.

43 Ortelius 1609, preface, not paginated.

44 These composite atlases included: first, the map of the world, and continued with maps of Europe, Asia, Africa, and the New World, then maps of regions, cities, islands, and battlefields. Over 30 examples of composite atlases are still kept today in European and American libraries, on which see Tooley 1939; and Woodward 1997.

45 Bolzoni 1995, 80.

46 On Jean Bodin, see Couzinet 1996 and Blair 1997.

47 In Appendix A the Florentine maps are listed according to the order established by Gastaldi's and Ruscelli's editions of Ptolemy's *Geography*, rather than according to their current display in the Guardaroba Nuova.

48 The author of the *Rhetorica ad Herennium*, the most important classical text on artificial memory once attributed to Cicero and widely read in the Renaissance, recommended as the *loci* of memory a series of architectural backgrounds, sufficiently distinct from each other, against which were fixed the images of words and things to remember. Aristotle in his *De memoria* suggested a memory scheme based on the alphabet, a practice used extensively in the Middle Ages and extended to different alphabetical sets, Greek, Latin, Coptic, Hebrew, Scythian, and even invented alphabets. Peter of Ravenna in his *Fenix*, an early-Renaissance influential manual on artificial memory, substituted the letters of the alphabet with human figures. Hugh St. Victor proposed the use of a mnemonic system based on numbers and numerical series to break down texts and things, a system elementary and widespread that later also informed the page layout of medieval manuscripts as well as the printed pages of Renaissance bibles with their numbered chapters and verses. Thomas Bradwardine and John of Garland had students memorize the Zodiac, the Bestiary, and the *voces animantium*, a list of the characteristics of animals derived from the Bestiary, as potential heuristic mnemonic schemes with which knowledge could be located. On the art of memory, see Yates 1966; Rossi 1983; Carruthers 1992 (specifically on the use of images in heuristic systems); Bolzoni 1995; and Carruthers 1998.

49 Citolini 1561, 550–1: "Ne la terra videro tutte le parti sue, tutti i paesi del Mondo e in generale e in particolare, e perfino nel centro de la Terra penetrarono . . . Parve poi loro di entrare nel piu compiuto giardino, che mai veduto havessero. Imperoche non era né herba, ne arbore, che quivi piantato non fosse . . . in un'altra parte videro quanti animali solevano, l'aria, e quali l'acqua; e quanti calcano la terra . . . E in cotal forma vedendoli, non i nudi nomi soli, ma la vera essenzia, e figura loro a comprender venivano, e non pur di loro ma di tutte l'altre cose ancora, che forma patiscono." This passage is also quoted in Bolzoni, 1995, 252. The English translation is from Bolzoni 2001, 243. Citolini imagined his *Tipocosmia* as a six-day reunion in the house of an aristocrat, who instructed a group of guests on the order of universal knowledge. On Citolini, see Yates 1966, 144, 239; Bolzoni 1995, 257–9; Bolzoni 2001, 240–6.

50 Citolini 1561, 548: "una grandissima palla, ne la quale entrar vi si potea. E quivi entrati si videro dintorno il cielo, e nel mezzo la terra, e videro quivi le cose ordinate in modo assai più grato a l'occhio de'l corpo che a quello de l'intelletto. E dopo molti discorsi, e ragioni udite da l'Conte [the host], tutti finalmente conclusero, queste esser cose più tosto da fanciulli, che da desiosi di sapere." This passage is quoted in Bolzoni 1995, 252. The English translation is from Bolzoni 2001, 243.

51 The description of Camillo's theater is in Camillo 1550. On Giulio Camillo and his theater, see: Yates 1966, chapters 6 and 7; Camillo 1991; Bolzoni 1995; Margolin 1991, 459–82; and *DBI*.

52 On Samuel Quicchelberg, see Quicchelberg 1565; Hajos 1963, 207, on the relations between Camillo and Quicchelberg; Lugli 1990, 83–4; Hauger 1991; Falguières 1992; Bolzoni 1995, 245–6, specifically on the relation between the art of memory and collecting; and Woodward 1996, 92–3, for the presence of geographical prints in Quicchelberg's museum.

CHAPTER 5

1 Vasari – de Vere 1996, vol. 2, 891.

2 Nicolaus Germanus used the trapezoid projection in the manuscript copy of Ptolemy's *Geography* that he made for his patron Borso d'Este, Duke of Ferrara, on which see Crone 1978, 34–6; and Akerman 1995, 9–10. Although Germanus's projection correctly interpreted Ptolemy's intention to maintain, as much as possible, the correspondence between the sphericity of the earth and the flat paper, it did not correspond to any of Ptolemy's projections since the parallels and meridians were represented as straight lines. Nonetheless, Germanus's projection was widely used in the fifteenth and sixteenth centuries. On Nicolaus Germanus, see Karrow 1993, 255–65. On Ptolemy's projections, see Jones 2000. On the cartographic grid, see Summers 2003, Chapter 5.

3 Ptolemy distinguished geography from chorography: geography represents the known world fixing the position and the outline of its most general features, while chorography portrays, "particular regions and describes each separately, representing practically everything of the lands in question, even the smallest details such as harbors, villages, districts, streams branching from the principal rivers and the like." (English translation from Cohen–Drabkin, 1958, 162–4).

4 Ptolemy 1561, 354–5: "Da poi che dunque, noi habbiamo fatta una competente ó convenevole descrittione di tutta la terra abitabile, in una tavola, da potersi comprendere dalla vista nostra, si deve ragionevolmente esporre il modo di far le sommarie descrittioni, dividendo essa terra, ó Tavola generale in più Tavole, per potervi mettere tutti i luoghi conosciuti, con chiarezza, & con giusta misura dall'uno all'altro. Percioche conviene nell'universale descrittione, per serbar le proportioni delle parti della terra fra loro, che alcune per la moltitudine de' luoghi, che vi hanno à mettere, si allarghino, & altre per non esservi luoghi da mettere, si ristringano, ó si lascin vote. . . Percioche non è in niun modo necessario, che le Tavole sien cosí grandi l'una come l'altra fra loro, ma basta, che solamente in ciascuna d'esse si servi la proportione, che hanno i luoghi l'un all'altro, sí come quando noi dipingiam solamente un capo, serviamo solamente la ragione & la proportione di quelle cose, che son dal capo, ó in una sola mano quelle solamente, che son della mano. Et in quelle del capo non miriamo quelle delle mani, se non quando noi facciamo sotto una figura tut'un'uomo intero. Ma sí come in un tutto, niuna cosa ci vieta, che non possiamo farlo maggiore & minore, ó crescerlo & mancarlo, cosí parimente nelle parti non si vieta, che quando stanno per se sole, non si possano crrescere & diminuire secondo la capacità delle Tavole, in che s'hanno à mettere."

5 Chorographical maps appeared only in Matthew Ringman and Martin Wandesmueller's edition of Ptolemy's *Geography* (Strasbourg, 1513), in which were added four maps representing, at a larger scale, regions that had appeared in more general maps within the same book. These chorographical maps, entitled *Quattuor deinde Chorographiae particulariores magis extensea praedictis tabulae*, confirm the relative nature of chorography to geography. On chorography, see Chapter Seven.

6 Ptolemy 2000, Book 2, Chapter 1, 95.

7 On medieval world maps, see Woodward 1987; Gautier Dalche 1991; and Edson 1997. On the legacy of medieval moralized cartography in Renaissance maps, see Schulz 1990, 23–39.

8 On the Renaissance interpretations of Ptolemy's geographical theory, see Alpers 1983; Lestrignant 1991; Cosgrove 1992; and Woodward 1996.

9 Ptolemy 1475, 2v.

10 Guarino Veronese's letter to an anonymous correspondent, 20 May 1412, is quoted in Fischer 1932, vol. 1, 205.

11 On Francesco Berlinghieri's edition of Ptolemy's *Geography*, see Almagià 1961, 497–526; Skelton in Ptolemy 1482; and Brotton 1997, 87–98.

12 Apian 1533, 2.

13 Vasari–de Vere 1996, vol. 2, 891.

14 Apian 1533, 3.

15 For the discussion of Gastaldi's definitions of cosmography, geography, and chorography, see Woodward 1996, 5–6. For Münster's understanding of geography, see Hodgen 1954; Strauss 1959; Strauss 1961; Hodgen 1964; and Chapter Four.

16 Ptolemy 1561, 4: ". . . interamente comprendere la vera significazione di questa voce Geografia, & non prendervi errore, quando troveranno per aventura per questo volume usata la detta parola terra in significationi tra lor diverse."

17 Ptolemy 1561, 4: "Cosmografia sia quella, che senza curarsi della particolar quantità ó misura delle lontananze de' luoghi, attenda a descrivere & narrar le nature & proprietà de' paesi, & delle cose, che in esse sono, i costumi, i popoli, le cose notabili accadute in tempo in tempo." A similar concept of geography belonged to Münster, on which see note 15 above.

18 The Latin translation of Strabo's *Geography*, made by Guarino Veronese in 1469, was printed in 1472 (Strabo 1472). The classic English translation is by H. L. Jones (Loeb Classic Library, Harvard and London, 8 vols.). On Strabo's geography, see also Aujac 1966 and Clarke 1999.

19 Ptolemy 1561, 5: "all'opera di Strabone, ove già son posti, & ancor molte volte replicati i nomi delle provincie, & de' luoghi [aggiungere] solamente i numeri delle graduationi, che sono in questo volume di Tolomeo & le XXVI Tavole, che egli ha fatto."

20 Danti 1577, 44: La descrittione della terra ó veramente è trattata dalla GEOGRAFIA, che la descrive senza havere riguardo nessuno al cielo come fece Strabone, & Pomponio Mela, che la descrissero solo con le distanzie itinerarie, nel qual modo

è stata descritta anco da Domenico Mario Negro, & da molti altri. Et specialmente da Tolomeo, che se bene ad alcuni pare, che si debba chiamare più tosto Cosmographia per descriverla con li gradi della longitudine, & latitudine, nondimeno chi considererà, che egli piglia detti gradi non in cielo, ma nell'Equinoziale & nei meridiani della terra, vedrà che al libro suo conviene il nome della Geografia per descrivere la semplice terra, & non della Cosmografia, che tratta anco del Cielo.

IDROGRAFIA, che descrive il mare universalmente, togliendo dalla Geografia la descrittione delli liti marittimi per vedere i confini, che il mare ha con la terra tanto nel contine(n)te, quanto, che nell'Isole, & Scogli come si vede nelle carte marine.

COROGRAFIA, che suona in lingua nostra descrittione di luogo particulare, come si vede nella corografia della città di Perugia con il paese circo[n]vicino fatta da Giulio Danti mio Padre ove è levata la pianta del tutto & con ogni dilige[n]za sono disegnati i luoghi dal naturale, con ogni strada, casa alberi, & altre cose, che all'intorno vi sono segnalate. Onde chiaro si scorge, che la corografia ancorche possa essere esercitata dal Geometra, & sia scienza subalternata alla Geometria, può anco essere come arte mecanica esercitata dal semplice Pittore, che la Geografia non può se non dal Geometra essere appresa, non havendo bisogno alcuno della Pittura, poiche essa per una città descrive un punto, & per un fiume ò una costa di mare, una linea retta, ò curva."

TOPOGRAFIA, che anco essa descrive i luoghi particulari, ma facendo cotale descrittione co(n) parole non ha bisogno ne della Geometria, ne del disegno. In questa maniera Polibio, ava(n)ti, che narri la rotta, che Annibale diede à Romani appresso Perugia descrive maravigliosamente il lago Trasimeno, simile descrittione è anco quella che T. Livio fa di Canne in Puglia. (English translation of part of this quotation is from Frangenberg 1994, 55–56). On Renaissance chorography and Danti's definition of chorography, see Chapter Seven.

21 Mercator 1961, 46. On Mercator's *Map of the World*, see Karrow 1993, 389–90 (with bibliography).

22 Mercator 1961, 56–7.

23 The systematic analysis of the maps of the Guardaroba Nuova has never been attempted before, although some individual maps have received close scrutiny: Kish 1951 analyzed the map *China*; Paratore 1976 and Licini 1988, 119–22, studied the maps of Muscovy; Tolias 2002 discussed the map of Greece; and Danti–Bonsignori 1995 published a complete set of color illustrations. Appendix A provides a detailed description of the individual maps.

24 While we have no information at all concerning Bonsignori's personal maps, Danti himself reported later in a letter to Abraham Ortelius that his personal maps "got ruined in moving them from Bologna to Rome" (*sono andati male nel portarli da Bologna a Roma*) (Egnazio Danti to Abraham Ortelius, 24 December 1581, published in Ortelius 1887, letter no. 100). As for the maps themselves, some of them show traces of decay in the degrees of latitude and longitude marked on their borders, which at times disappeared or were erroneously restored in the following centuries, but otherwise they are largely well preserved.

25 Cornelius Koeman estimated that about fifteen hundred wall maps, each at least about two meters by three, existed by the mid-sixteenth century, and suggested that their widespread use in the sixteenth century rather than their marginality was the main cause of their survival in such small numbers (Koeman 1964, 24). The display of wall maps is recorded in numerous sources, including the following: Ortelius 1887, 232, on the maps that Charles V hung in his private rooms; Koeman 1964, 24, on the 52 maps that the lords of Brederote displayed in their residence at Vianen in south Holland, and on the maps that Vigliuss of Aytta, president of the Council of State in the Netherlands, hung in his library and study; Ambrosini 1981, on the world maps displayed in the houses of Venetian merchants and patricians; Alpers 1983, 119–68, on the Dutch interiors painted by Jan Vermeer documenting the display of wall maps among the luxury goods of wealthy merchants; Conti 1893, 36, 134, 141, 188; ASF, G 73, ff. 5, 30, 71, 75v, 78; and ASF, G 87, ff. 31v, 41v, 42, 58v, 66, 75v, and 76v, on Cosimo I's maps, many of which were displayed in public rooms; Conti 1893, 73, for the city views of Genoa and Pisa that Eleanor had painted in the small terrace of her apartment. For a general assessment on wall maps, see Schilder 1986, and Woodward 1996, 79–87.

26 Giacomo Gastaldi's *Cosmographia universalis* (Venice, 1561) is known only through a seventeenth-century reprint, now at the British Library, London. At the center is the hemisphere of Europe and Africa, while the other hemisphere, representing Asia and America, is divided into two halves and attached to each side of the first hemisphere. Gastaldi's world map accompanied a booklet, which Danti consulted, entitled *Universale descrittione di tutta la terra conosciuta fin qui* (Venice, 1561), which survives in numerous copies. On Giacomo Gastaldi and his maps, see Cosgrove 1992; Cosgrove 1993; Karrow 1993, 216–49 (with bibliography); and Minella 1993. Specifically on Gastaldi's *Cosmographia universalis*, see Shirley 1983, 122–3; Minella 1993, 79–82; and Karrow 1993, 240–1.

27 The ducal holdings included the famous Salviati planisphere, a nautical chart donated by Charles V to Cardinal Giovanni Salviati earlier in the century, which later entered the Medici collections, and beautiful maritime atlases, such as those made by Battista Agnese and Francesco Ghisolfo respectively, which were both dedicated to Cosimo I. Battista Agnese's nautical atlas is today in the Biblioteca Mediceo Laurenziana (MS Palatino 245), while Francesco Ghisolfo's is kept in the Biblioteca Riccardiana (MS 3636). On the Medici collection of geographical texts and maps, see: *Corte*

mare mercanti 1980, 221. The bibliography on manuscript nautical atlases is extensive, but for a general, perceptive assessment, see Wallis 1997.

28 On these maps by Bartolomeo Velho, now kept in the Museo di Storia della Scienza, Florence, see Cortesao–Teixeira da Mota 1960–2, vol. 3; and Miniati 1991, 104–07, with illustrations. It is worth mentioning that Cosimo I's collections of nautical charts also included a map by Lopo Homen, currently in the Museo di Storia della Scienza, Florence, although it seems that Danti did not use it (see Cortesao–Teixeira da Mota 1960–2, vol. 2; and Miniati 1991, 104–07, with illustrations).

29 Egnazio Danti to Abraham Ortelius, 24 December 1581 (published in Ortelius 1887, letter no. 100).

30 Bernardo Baroncelli's portolan is in BNF, MS Magliabechiano XIII, 6, on which see Crinó 1931, who also transcribed Baroncelli's text; Rostagno 1923; and *Mostra Vespucciana* 1954.

31 On Danti's world map in the Vatican Palace, see Chapter Nine.

32 Danti's cartouche for the map *Norvegia Gotiae*. Danti must have compiled his maps of northern Europe as a group possibly in, or around, 1565 because of the cartographic similarities between *Thile Insula*, dated 1565, and *Norvegia Gotiae* and *Groenlandia*, which instead are both undated.

33 Danti's maps of Europe are: *Isole Britanniche* (1565), *Thile Insula* (1565), *Norvegia Gotiae*, *Grecia*, and *Livonia et Lituania*. Later, from 1576 to 1585, Bonsignori completed the series: *La Spagna* (1577), *La Francia* (1576), *La Germania* (1577), *La Schiavonia* (1578), *L'Italia* (1578), *Grecia* (1585). Bonsignori's and Danti's maps of Europe are still kept in the Guardaroba Nuova. A sixteenth-century copy of Bonsignori's map of Greece is today kept in the Museum Benaki, Athens, Greece, on which see Tolias 2002.

34 On Mercator's *Map of Europe* (Duisburg, 1554), see Heyer 1889, 379–89, 474–87, 507–28, with full transcriptions of the legends and inscriptions; Mercator 1891 (this is a nineteenth-century facsimile of the only known copy, once in the University Library in Breslau, and now destroyed); Karrow 1993, 386–7; and Van der Gucht 1994, 145 on its display in the study of Viglius ab Aytta. In 1572 Mercator published a revised edition of his map of Europe introducing a number of changes pertaining mostly to the cartography of northern Europe. On the revised edition of Mercator's *Map of Europe* (1572), see Karrow 1993, 396; and Dürst 1998.

35 The title of Danti's map of the British Isles, *Isole Britanniche le quali co(n)tengono il Regno di Inghilterra et di Scotia con l'Hiberia*, translated almost word for word the title of Lily's map: as printed in Venice in 1562: *Britanniae Insulae quae duo regna continet Angliae et Scotiae cum Hibernia adiacente*. Also Danti's legends translated almost verbatim Lily's legends. Danti's map of the British Isles has the north at the top, whereas the legend pertaining to Ireland is placed above the island. On the other hand, it seems that Danti did not have access to Mercator's eight-sheet of the British Isles (Duisburg, 1564), which superseded all previous maps of the region. On Lily's map, see Shirley 1991, 20–2 and Karrow 1993, 270, especially for its relation to Paolo Giovio's *Descriptio Britanniae*. On the Bertelli–Forlani map of the British Isles, see Woodward 1990; and Shirley 1991, 34–5, plate 38. On Mercator's depiction of the British Isles in his maps of Europe (1554 and 1572) and in his map [*Angliae Scotiae et Hiberniae Nova Descriptio*] (Duisburg, 1564), see Shirley 1991, 36–9, plate 40; Karrow 1993, 387–9; and Barber, 1998. I thank Rodney W. Shirley for redirecting my thoughts on Danti's map of the British Isles in a fruitful conversation in Lisbon in 1997.

36 The ambivalent position of Russia as a land between Europe and Asia is shown in numerous sixteenth-century maps. In his edition of Ptolemy's *Geography*, Gastaldi represented Russia in two modern maps of Europe, *Prussia et Livonia Nova Tabula* and *Schonlandia*, which he added to Ptolemy's eighth map of Europe, but he also represented Russia in another modern map, *Moschovia Nova Tabula*, which instead he added to Ptolemy's second map of Asia (Ptolemy 1548, table XXXVIII). In 1554 Mercator included the entire land of Russia in his map of Europe. On the mapping of Russia in sixteenth-century cartography, see Paratore 1976, 21; Licini 1988; and Broc 1989, 72–5. On Jenkinson's *Map of Muscovy* and its later editions, see Karrow 1993, 317–20. On Danti's maps of Russia, see Paratore 1976, 21; and Licini 1988, 119–22, who showed that Danti translated into Italian every single legend and inscription written by Mercator in Latin on his *Map of Europe* (1554). On the representation of the Mountains of Hyperborea in Renaissance cartography, see Licini 1988, 17–38.

37 On the comparison between Carolus Clusius's *Map of Spain* (1571) and Bonsignori's panel map in the Guardaroba Nuova, see Schilder 1986, 99–100 (with illustrations).

38 Of these maps of Africa Danti made only two, *Isola di San Lorenzo* (1565), and *Parte di Buona Speranza* (datable after 1565). Bonsignori was responsible for the remaining ten maps, which he made between 1578 and 1581: *L'Egitto* (1578), *Affrica* (Libia) (1579), *Mauritania* (1579), *Trog-loditica* (1579), *Nubia* (1579), *Libia interiore* (1580), *Parte dell'Agisamba* (1580), *Parte dell'Affrica Nuova, et Regno di Manicongo* (1580), *Parte d'Affrica* (Cape of Good Hope) (1581), *Parte d'Affrica* (Ethiopia) (1585). The maps of Africa that were eventually made correspond very closely to Vasari's original plan, especially if we consider that the representation of the Cape of Good Hope, today duplicated in two different maps made by Danti and Bonsignori respectively, would have appeared only in one map since Danti's map of this area must have been considered outdated only a few years after its execution.

39 The literature on the Portuguese explorations and their

maps is extensive, but for a brief and lively assessment of the topic see Brotton 1997, especially 46–86, 119–49. On the mapping of Africa, see McIlwaine 1996; and Relano 2001.

40 Danti's legend of the map *Parte di Buona Speranza.*

41 The four nautical charts made by Bartolomeu Velho (died in 1568), now kept in the Museo di Storia della Scienza, Florence, are dated 1561 and represent the Pacific Ocean, the New World, Europe, and Asia. They were recorded in the collection of the Accademia del Disegno, Florence since the early days of its foundation in 1562, although why and how they arrived there has never been fully explained. It is suggestive to imagine that Danti might have been instrumental in bringing them to the academy. He consulted these maps for the Guardaroba Nuova and might have decided to discuss them in the lectures on mathematics he very likely taught at the academy. In this case, it is also possible that Velho's nautical charts were actually part of the Medici collection, as the high quality of their design might suggest. On Velho's nautical charts, see Miniati 1991, 104 (with illustrations). Additionally, it should be mentioned that Danti's *Isola di San Lorenzo* corresponded quite closely also to the representation of the island in the navigational world charts made by the Portuguese Diogo Homen in the 1550s and 1560s. Diogo Homen's world atlases featured in the libraries of many rulers and aristocrats throughout Europe and possibly also in Italy, since many of them were actually made in Venice. But, so far, there is evidence that only Diogo Homen's nautical chart of the Mediterranean was available in Florence in the 1560s, not the chart of the world. The Biblioteca Nazionale Centrale, Florence, has a copy of Diogo Homen's nautical charts of Europe and the Mediterranean dated 1563 (Port. 26), although its Medici provenance remains uncertain. On Diogo Homen, see Cortesao–Teixeira da Mota 1960–2, vol. 2.

42 Gastaldi was an expert on the mapping of Africa. Previously, he had made small maps of Africa to illustrate the 1548 edition of Ptolemy's *Geography.* In 1549, he made a large cartoon representing the continent for the map mural in the Ducal Palace in Venice. In 1554 he made another map of Africa for Ramusio's book *Delle Navigazioni et viaggi.* In reduced size, Gastaldi's Africa was included in the Roman and Venetian composite atlases and, from 1578 to 1593, in De Jode's atlas *Speculum orbis terrarum*, one of the most popular map books up to the end of the sixteenth century. On Gastaldi's map of Africa and its many editions, see Biasutti 1920; Karrow 1993, 243–5; and Relano 1995.

43 Today the Guardaroba Nuova includes sixteen maps of Asia, although according to Vasari's original plan only fourteen had to be painted. Danti made almost all of them. His maps are: *Natolia* (this map is painted in two different versions, one dated 1565, the other before 1565), *Arabia* (1575), *Armenia, Persia, Moschovia, Sogdiana, Indostan drento il Gange* (this map is divided in two different panels), *Indostan fuori al Gange*; Indonesia (1573), India (1575), China (1575), Japan, and the Moluccas Islands (1563). Later Bonsignori added two maps representing the northern regions of Asia, which Danti had omitted; Bonsignori's additions are entirely based on Mercator's *Map of the World* (1569).

44 Gastaldi's maps of Asia were divided as follows: the first map, *Il Disegno Della Prima Parte Delasia* (Venice, 1559) represented in two sheets Anatolia, and was accompanied by a list of ancient and modern toponyms of the area. The second part, *Il Disegno della Seconda Parte dell'Asia* (Venice, 1561) showed in two sheets the area from Egypt to the mouth of the Indo River at the same scale of the first map of 1559. The third part, *Il Disegno della Terza Parte Dell'Asia* (Venice, 1561), showed in two sheets India and southeastern Asia, but not the islands south of the equator (Fig. 4.16). It differed somewhat from the previous two maps. It was a much larger copper engraving; it adopted a scale of representation larger than that of the previous two maps; it included many inscriptions on the topography of the map, inscriptions that were almost absent in the other two parts, and a list of toponyms of the area since they were considerably less than for the previous two parts of Asia. It seems, however, that the two sheets representing the Pacific Islands were added somewhat later, possibly around 1565. Gastaldi's large maps of Asia were a major editorial success. Each map went through several new editions in the following years. In reduced sizes, they were included in the Roman and Venetian composite atlases and, from 1573 to 1598, they appeared in De Jode's *Speculum orbis terrarum.* On Gastaldi's maps of Asia, see Karrow 1993, 231–2, 238–40; and Suarez 1999, 130–57. On the relationship between Gastaldi and Ramusio, see Schilder 1986, and 64–5 Karrow 1993, 227–8.

45 Ortelius's eight-sheet *Map of Asia* (Antwerp, 1567) was almost entirely based on Gastaldi's, but, unlike Gastaldi's, its sheets can be neatly joined together to form a wall map. On Ortelius's wall map of Asia, see Schilder 1986, 59–84, who also provided a facsimile edition and a detailed comparison with Gastaldi's maps of Asia; and Karrow 1993, 3–4.

46 Ortelius's inscription, placed between Ceylon and Sumatra, reads: "The island called Zeilan by the Arabs and Persians, is called by the Indians Tenerisim, which means the land of delights, for it has a very temperate and healthy climate, making its inhabitants live for more than hundred years almost entirely without diseases. The best cinnamon grows there. It has elephants and is very rich in almost all kind of jewels. It seems to correspond to Ptolemy's Taprobana, and the Island Sumatra to his Golden Chersonese, since it is very rich in gold. For those who think that Malacca is Ptolemy's Golden Chersonese are mistaken, since there is no gold there apart from that which is imported from somewhere else. It is true that Sumatra is not a peninsula now, but it is very likely that it was separated from the continent by the force of the ocean after Ptolemy's time. Moreover, if you imagine

Sumatra joined to Malacca with an isthmus, it will correspond very well with the shape of the Golden Chersonese as described by Ptolemy" (English translation from Schilder 1986, 73).

47 Danti's legend to the map illustrating China and Japan: "Gran diversitá é tra i geografi nel situare questa costa della China, e l'isola del Giapan, le quali nella presente tavola si sono poste secondo l'opinione de più moderni, se bene par che si dovesse far correre questa costa dal capo Litampo fino a Paugiu per Scirocco et Maestro et non per Greco et Libeccio come sta qui. Et l'isola del Giapan dovria correre per Ostro et Tramontana et no per Levante et Ponente. Ma perchè nel gran globo si sono posti (co[n] la determinatione del Seren.mo Cosimo Gran D[uca] di Toscana) nel modo che li descrive Marco Polo et nel modo che la pone il S. Giovanni do Baros il quale spressamente dice che il più orientale capo dell'Indie scoperte li é capo Lampo il quale é vicino alla città di Nimbo dal quale é chiamato anche capo Ninsu. Senza che si veda quella costa della China posta per scirocco e maestro in uno antico Mappamondo fatto a simiglianza d'uno che giá fece in Venetia Marco Polo, il quale scrisse molte cose di quella provincia le quali se bene alora non se li credevano si é poi trovato non di meno tutto esser vero."

48 Inscription on Danti's map illustrating China and Japan. In this inscription Danti translated and shortened one of Ortelius's legends in his *Map of the World* (Antwerp, 1564): "The northern shores of America have not yet been explored and are utterly unknown. It is not yet agreed whether it is sea or land here" (English translation from Schilder 1986, 77).

49 Danti's letter to Bernardo Neri, 28 October 1569: ". . . di havere dal S. Giova[nni] de Baros la copia della cosmografia della cina ch'esso S[ignore] Giovanni dice di haver fatta tradure di lingua cinese in lingua portoghese, et se detto S[ignore] avesse carta alcuna di detto paese et provincia della Cina, del Mangi, et del Catai, et particularme[n]te notitia alcuna della gran citàß del qunisai. V[ostra] S[ignoria] veda di avere copia perche so per certo che no[n] puo portare al Ecc[ellenza] del Duca cosa piu grata di questa, no[n] ci essendo delle sop[ra] dette cose notitia alcuna, che buon sia. Io no[n] parlo delle carte da navigare ordinarie, perche il S[ignor] D[uca] ne ha pure assai ma di carte che vi fosse descritto il paese fra terra delle sop[ra] dette p[ro]vincie del India" (Houghton Library, Harvard University, MS Port 17.1).

50 This legend, dated September 1575, is the last Danti wrote before his departure from Florence: "Nel descrivere li liti marittimi di questa tavola non ho osservato le misure delle carte marine dei Portoghesi per esser false et da essi falsate in prova per voler includere nel territorio loro l'Isole Moluche essendo fuori." On the disputes between Portugal and Spain on the demarcation line, which in 1494 was placed at 370 *leghe* west of the Island of Capo Verde, see Julien 1948; Broc 1989, 30, 129; and Brotton 1997, 125–6.

51 According to Vasari's original plan, the continent of America had to be represented in fourteen maps, but only nine were ever made. Danti made the majority of them: *Ultime Parti note nel Indie Occidentali* (1564), *Nuova Spagna* (1565), Mexico, Antille (after 1569), Coast of Peru, Peru, North Brazil, and Strait of Magellan. In 1581 Bonsignori mapped again the Strait of Magellan, although this duplication most likely would not have appeared in the final project. On the mapping of America, see Burden 1996.

52 Danti's legend of the map *Nuova Spagna*: "La presente tavola si è cavata quanto ai contorni dalle carte Marine fatte dai Castigliani. Et il resto tra terra si è tratto dalle relationi del Cortese et altri che vi sono stati et da alcune carte di tal provincia."

53 Ortelius's *Map of the World* (Antwerp, 1564) must have been available to Danti right after its publication in 1564 since his first map of the West Indies, *L'ultime parti note nel Indie Occidentali* is dated 1564. The date might extend to 25 March 1565 according to the Florentine calendar. On Ortelius's *Map of the World* (Antwerp, 1564), see Schilder 1986, 33–58, who also reproduced it in facsimile; Karrow 1993, 2; and Mangani 1998b.

54 On Mercator's *Map of the World* (1569), see Mercator 1561 (with facsimile edition); Karrow 1993, 389–91; and Pastoureau 1998.

55 The Zeno Map, which was based on the equally fantastic map of northern Europe made by Olaus Magnus (Venice, 1539), illustrated the book *De i commentari del viaggio di M. Catarino Zeno il K. & delle guerre fatte nell'imperio Persiano dal tempo di Ussancassano in quà. Libri due. Et dello scoprimento dell'Isole Frislanda, Eslanda, Estotilanda, & Icaria fatto sotto il Polo Artico, da due fratelli Zeni, M. Nicolló il K e M. Antonio. Libro uno. Con un disegno particolare di tutte le dette parti di Tramontana da lor scoperte* (Venice, 1558). The letters reporting on the purported travel of the Zeno brothers were reprinted in Ramusio's collection of travel reports (Ramusio 1978–83, vol. 4, 139–202), while the map was also published and sold separately. On the Zeno Map, see Karrow 1993, 600–02; and Burden 1996, 31–2.

56 Mercator 1961, 61. For these maps of the North Pole Mercator's *Map of the World* (Duisburg, 1569) must have been consulted, rather than the map *Septentrionalium terrarum descriptio* made by Mercator's son Rumbold in 1595 and inserted in Mercator's *Atlas*. In fact, this last map omitted the continents surrounding the Arctic and also Mercator's legends, which are instead reported on the maps of the Guardaroba Nuova.

57 Vasari–de Vere 1996, vol. 2, 893.

58 Danti mentioned Niccoló Gaddi in 1583 (Danti, Dedication letter to Giacomo Boncompagni, in Vignola–Danti 1583, not paginated, on which see Fiorani 2003). Danti's relationship with Giuliano de' Ricci is documented in the letter that Danti wrote from Bologna to Ricci, 2 November 1575 (BNF, Palatino 1187, fol. 251).

59 Numerous letters document the relationship between Danti and Prospero Podiani: Letter of Danti, from Florence, to Prospero Podiani in Perugia, 24 March 1563 (ASP, Miscellanea 103a, fol. 17); letter from Danti in Prato, to Prospero Podiani in Perugia, 24 September 1563 (fol. 17a); letter from Danti in the convent of San Marco in Florence, to Prospero Podiani in Perugia, 20 December 1563 (17b); letter from Danti in Florence, to Prospero Podiani in Perugia, 15 July 1565 (17c); other letters from Giovan Bernardino Rastelli in Perugia, to Prospero Podiani in Poggio di Prepo contain additional information pertaining to Danti's activity as a book dealer for Podiani (ASP, Miscellanea 103a, nos. 23, 23a, 23b, 23c 23d, 23e, 23f, 29, 31). On these letters see Cecchini 1978, 350. No record remains of the meeting between Varchi and Danti, although it is possible that Danti might have attended the public lectures of the Accademia Fiorentina that met at the Sala dei Papi in the convent of Sta Maria Novella in the early 1560s. On Danti's private collection of drawings and books, see Chapter One, nos. 15 and 16.

60 On Vincenzo Borghini's library, see Testaverde Matteini 1983, 611–43. On Borghini's relations with Danti, see the documents published in Lorenzoni 1912, 180. On the Guardaroba Nuova and the 1568 edition of Vasari's *Lives*, see: Frey–Frey 1923–40, 150.

61 Borghini 1584, 522, published the sonnet. Piero di Gherardo Capponi (1504–1568) was nominated Cosimo I's private counselor on 22 May 1566 (E. Luzzati in *DBI*).

62 Danti himself reported the fact that he attended the performance of the play *La Vedova* in the ducal palace in 1569. He also reported that the scene changed twice. The Florentine urban view at the bridge of Santa Trinita changed into the Tuscan countryside surrounding the Medici villa at Arcetri, and then again to the view of the ducal palace seen from the "canto agl'Alberti." While the scene was changing, the audience was distracted by "bellissimi intermedij" written by Strozzi and accompanied by the music of Alessandro Striggio, by the actors acting and dancing, and by the clouds coming from above and from the side to cover the scene (Vignola–Danti 1583, 92). Baldassarre Lanci worked for Cosimo I in many different capacities. He ordered the carriage for the masquerade *La genealogia degli dei* in 1565, designed the set for the comedy *I Fabii* by Lotto del Mazzo, performed in the ducal palace in 1569 to celebrate the baptism of Eleanor de' Medici, first daughter of Francesco and Johanna, and built scientific instruments, one of which is still kept in the Museo di Storia della Scienza, Florence (inv. 152). Later Danti wrote on Lanci's perspective instrument in Vignola–Danti 1583, revising the erroneous account provided by Daniele Barbaro in *La prospettiva pratica* (1567). On the performance of *La Vedova* by Giovan Battista Cini, see: Gaeta Bertelà–Petroli Tofani 1969, 25–9, fig. 1; and Nagler 1964, 36–48, who discusses more generally Lanci's involvement in Medici festivals. On Lanci and his instrument, see Barbaro 1568; Maltese 1980; Kemp 1992, 175–6; Miniati 1991, 46–7; and Fiorani 2003.

63 In 1569 Danti had to apologize formally to the Dominican father of his province, as he himself reported in a letter from Prato to Cosimo I in Florence, 30 September 1569 (ASF, Carteggio universale, F 545, fol. 133). In 1571, Cosimo I had to write to the General Father of the Dominican Order, 21 July 1571, as well as to Pope Pius V, 25 July 1571 (both letters are published in Lorenzoni 1912, 182–3). Two days later Cosimo I wrote also to his son, Cardinal Ferdinando, in Rome to intercede with the General of the Dominican for Danti (Del Badia 1881, 14). Danti's exceptional position within the Dominican order as Cosimo I's protege was already evident years before, when he obtained the permission not to attend the Order's general assembly, as he had been commanded (Lorenzoni 1912, 179–80).

64 Francesco I to Serafino Cavalli, 23 September 1575 (ASF, Mediceo del principato, 667, c. 183), and Cavalli's answer to Francesco I, 27 September 1575 (Lorenzoni 1912, 182–3).

65 Serafino Cavalli to Cardinal Ferdinando in Rome, 23 September 1575: ". . . desiderando da lui informazioni che lui solo mi puó dare" (Lorenzoni 1912, 183).

CHAPTER 6

1 On Gregory XIII and the Holy Year 1575, see Pastor 1936–55, vol. 19, 197–214; *Roma 1300–1875* 1985, 178–96; *Roma Sancta* 1985, 266–74; Jacks 1989; Wisch 1990; Freiberg 1991; Freiberg 1995. On Gregory XIII's papacy in general, see Pastor 1936–55, vols. 19 and 20; Borromeo 2000; and Borromeo 2002.

2 On the works in the Vatican Palace during Gregory XIII's papacy, see Redig de Campos 1968, 168–84; Pietrangeli 1992, 151–68; Kuntz 1998 (on the ceiling of the Cappella Paolina); and Eitel-Porter 1995 (on the Sale dei Foconi).

3 Partridge–Starn 1990, 31, who quoted a sixteenth-century document explaining the criteria for selecting the scenes for the Sala Regia. In the Sala Regia, Vasari and his workshop painted two scenes representing the deeds of earlier popes named Gregory, *Gregory IX Excommunicating Frederick Barbarossa*, and *Gregory XI Returning from Avignon*, as well as three scenes pertaining to the Massacre of St. Bartolomew. On the Sala Regia, see Roettgen 1975; Fehl 1974; Herz 1986; Partridge–Starn 1990; Boeck 1997; and Jong 1998.

4 For the little vaults on the first floor of the Palace of San Damaso Lorenzo Sabatini invented fictional pergolas of fruits and flowers populated by birds and animals that visually recall Giovanni da Udine's adjacent wing. For the second loggia, which was actually completed after 1575, he designed grotesques on the walls and stories of the life of Christ on the vaults (four episodes for each vault) that related visually and thematically to Raphael's loggia. The loggia on the third floor known still today as the Terza Loggia, also completed after 1575 under the supervision of Egnazio Danti, was embellished with maps of the world and, thus, complemented the iconography of its adjacent wing that illustrated the maps of Europe commissioned by Pius IV. The con-

struction of the east wing of the Palace of San Damaso had to be postponed until after the jubilee and was actually completed by Sixtus v, Gregory xiii's successor. On Raphael's loggias and the organization of his workshop, see Dacos 1986; on the loggias during Gregory xiii's papacy, see Hess 1967a; Redig de Campos 1968, 169–74; and Pietrangeli 1992, 22–223.

5 On the Sala Bologna, see Comelli 1914, especially 4–16; Redig de Campos 1968, 173–4; Pietrangeli 1992, 154–5; and Ghizzoni 2003 (specifically on the perspective plan of Bologna). On the sky map, see Warner 1971; Hess 1967b; and Partridge 1983.

6 As a cardinal, Gregory xiii had worked at a new edition of the *Corpus iuris canonici*; he continued to work on it as a pope, announced its new publication in 1580 but actually printed it only in 1582. On Gregory xiii's edition of the Decretals, see Pastor, 1936–55; vol. 19, 279–80. A few years after the completion of the Sala Bologna, in 1577 or 1578, Gregory xiii decorated the Sale dei Foconi in the Vatican Palace with scenes from the life of Gregory the Great (Eitel-Porter 1995).

7 The astronomers on the ceiling of the Sala Bologna were painted by Sabatini, although their accurate foreshortening *dal sotto in su* and their close relationship with the painted loggia strongly suggest Mascherino's intervention in the overall design. Also the sky map was painted mostly by Sabatini, but Mascherino may have contributed a few figures such as the putti at the corners of the sky map and the group representing *The Fall of Phaeton*, a drawing of which, now in Boston and traditionally attributed to Pellegrino Tibaldi, is instead related to the Sala Bologna and stylistically close to the work of Mascherino (McTavish 1980). The cartographic source of the sky map has been convincingly identified in the very popular star map made by Franciscus Demongenet after 1552. On Mascherino's loggia in perspective, see the important comments by Egnazio Danti in Vignola–Danti 1583, 88–90; and Kemp 1990, 71–2. On the sky map of the Sala Bologna, see Hess 1967b; Warner 1971; and Partridge 1983.

8 On the Sala della Cosmografia in the Farnese Palace at Caprarola, see Lippincott 1990; Partridge 1995; and Quinlan-McGrath 1997. On the artistic patronage of Cardinal Alessandro Farnese, see Robertson 1992. Gregory xiii himself was entertained in the Sala della Cosmografia when he visited Cardinal Farnese during his summer holiday in 1578, on which see Ehrlich 2002.

9 Through the Bolognese ambassador in Rome, the pope requested that Scipione Dattili, a civic engineer who worked extensively for the Bolognese Senate, update an older map of the city and/or of the county of Bologna, then kept in the Bolognese Cathedral of St. Peter. This old map is now lost but Gregory xiii surely knew it well. Both the old map from St. Peter and Dattili's updated version were sent to Rome through the artist Pellegrino Tibaldi, although it remains unclear whether he simply carried the map or also intervened on its content. The map was then sent back to Bologna to be revised according to the pope's wishes and the requirements of the dining hall specified by Sabatini. On the cartographic sources of the maps in the Sala Bologna, see Comelli 1914, 4–16; and Ghizzoni 2003 for its correspondence to the reality of the urban fabric of sixteenth-century Bologna.

10 The reference here is to Erwin Panofsky's seminal essay of 1923, *Die Perspektive als symbolische Form* (English trans.: *Perspective as Symbolic Form*, Cambridge, Mass., 1991).

11 For a concise but penetrating overview of the different currents of post-Tridentine spirituality, see: O'Malley 2000 (with bibliography).

12 The best study on Gabriele Paleotti remains Prodi 1959–69. On the cultural and scientific milieu of sixteenth-century Bologna, see Olmi 1992, 21–163; Findlen 1994; and Findlen 1999. For an overview of the spiritual life of the city, see: Mazzone 1997.

13 On the ideal post-Tridentine bishop, see Jedin–Alberigo 1985; and Bergin 1999 (with bibliography). On Charles Borromeo, see Prodi 1957 (specifically on the relationship between Borromeo and Gregory xiii); *San Carlo e il suo tempo* 1986; Headley–Tomaro 1988; Buzzi–Zardin 1997; and Rossi 1999. On the clashes between Paleotti and Gregory xiii, see Prodi 1959–69, 215–68; and Wright 2000, 231–69, who revised Prodi's interpretation.

14 Prodi 1959–69, vol. 1, 154: "quattro gran volumi dove ha più di mille e cinquecento piante tutte aglottinate, colte a Trento, Roma, Bologna e altrove."

15 On the map of Bologna by Agostino Carracci, see De Grazia 1984.

16 On the importance of visitations, see Peter Burke, "The Bishop's Questions and the People's Religion," in Burke 1987, 40–7. On the maps that Borromeo had made in connection with his visitations, see Brivio 1985.

17 Paleotti started to work on his treatise on sacred and profane images in January 1578 and circulated drafts to his friends until its publication in 1582. Generally, the historiography on Paleotti's treatise focused on the identification of paintings and iconography that reflect Paleotti's view on sacred images and their abuses. Federico Zeri identified the *arte senza tempo* of Scipione Pulzone as an example of post-Tridentine painting along Paleotti's lines (Zeri [1957] 1997). Paolo Prodi, Charles Dempsey, and Carl Goldberg stressed the connection between Paleotti's ecclesiastical reform and the artistic reform of the Carracci (Prodi 1962; Dempsey 1977; and Goldstein 1988). Stephan Ostrow saw the influence of Paleotti's spirituality in the Sistine Chapel that Sixtus v had built and decorated in S. Maria Maggiore (Ostrow 1996, 225–6). Only Paolo Prodi and Pamela Jones wrote about the bishop's interest in scientific images and his positive approach towards the knowledge of nature achieved through the study of scientific images (Prodi 1962 and Jones 1995). For the dis-

cussion of Paleotti's theory on art, see: Prodi 1962; Scavizzi 1989; and Jones 1995. Paleotti's treatise is published in Barocchi 1961, vol. 2, 117–509; and in facsimile in Paleotti 1990. The literature on the debate of sacred images in the sixteenth century is vast, but particularly significant are the following studies: Mâle 1958; Prodi 1962; Zeri [1957] 1997; Freedberg 1982a; Freedberg 1982b; Scavizzi 1989; Scavizzi 1992; Jones 1995; and Hecht 1997.

18 Paleotti in Barocchi 1961, vol. 2, 356: "tavole di geografia, navigazioni delle Indie, descrizzioni del cielo e delle stelle, disegni di città e paesi, espugnazioni di fortezze, vittorie massime contro infedeli, effigie d'animali, di piante, di pietre, cose di architettura, di edificii, disegni di fortificazioni, di essicazioni di valli e fiumi; et in somma tutte quelle cose che sono permesse che nelle academie publiche si leggano et insegnino agli altri e che s(ant') Agostino (*De Doctrina Christiana*, II, 3r) vuole cha ad un dottore ecclesiastico siano non solo non proibite, ma anco necessarie." The title of the entire chapter is: "Delle pitture profane che rappresentano varie cose come guerre, paesi, edifici, animali, arbori, piante e simili."

19 Paleotti, in Barocchi 1961, vol. 2, 177.

20 Paleotti, in Barocchi 1961, vol. 2, 356.

21 On scientific illustrations in the Renaissance and the debate of their validity in the study of the natural world, see: Prinz–Beyer 1987; Olmi 1992, 21–163; Baigrie 1992, especially the essays by Bert S. Hall, Martin Kemp, Brian Baigrie, and David Topper; Mazzolini 1993; Kusukawa 1997; Jones–Galison 1998; Johns 1998; Frasca-Spada–Jardine 2000; and Freedberg 2002. Now considered methodologically unsatisfactory is Shirley–Hoeniger 1985.

22 Paleotti, in Barocchi 1961, 356: "oltre il proprio loro uso, possono ancora servire alle volte ad altri effetti et utilitá della vita, sí come, a proposito di geografia, si legge (Eliano, *Varr. hist.*, III, 28) che Socrate, volendo reprimere l'altezza di Alcibiade gonfio per le richezze e copia di possessioni ch'ei godeva, gli disse, mirando una tavola dove era tutto il circuito del mondo, che li mostrasse dove erano posti i suoi campi; i quali non ritrovando egli in alcuno luogo dissegnati, soggionse Socrate: 'Dunque senza ragione t'innalzi di quello che non é parte alcuna, nè pure apparisce nella terra'."

23 Paleotti in Barocchi 1961, vol. 2, 363: "Altro [abuso] será del luoco, come chi ponesse Roma in Asia, o Betlehem fuori della Giudea . . . nell'Alpi d'Italia dipingesse gli elefanti e cammelli, e nelle valli di Chioggia gli abeti e i faggi." The same *abuso di luogo* was mentioned by Giovanni Andrea Gilio in his *Due Dialoghi* (Camerino, 1564), another influential post-Tridentine treatise on art (Barocchi 1961, vol. 2, 19).

24 Paleotti, in Barocchi 1961, 301: "le pitture profane, potendo portare a noi varii giovamenti, non devono essere subito ributtate, ma misurate bene con sano giudicio, acchiocché, a guisa di api che da vari fiori colgono il mele, sappiamo valerci di queste pitture ad uso di pietá."

25 On Danti's work in the Dominican convent in Bologna, see: Righini Bonelli–Settle 1979, 11; and Heilbron 1999, 68–76 (on the meridian line). An eighteenth-century source documents Danti's inscription on the meridian line: "Egnatius Dantis Perusinus Bonon. anno jubil. 1575. Qui observat ventos non seminat, et qui considerat nubes nunquam metit. Ecclesiast." (Mariotti 1797). In addition, Danti was involved in the restoration of the reliquary containing the head of St. Dominic, the most important relic of the Dominican order that was then kept in the church of the Bolognese convent.

26 Danti's meridian line in San Petronio was later substituted with another one by Giovan Domenico Cassini, but Danti described it in detail in a short treatise specifically dedicated to it (Danti 1577). Thomas Settle found the only surviving copy of Danti's treatise on the use of meridian lines, *Usus et tractatio gnomonis magni quem in Bonomia ipse, in Divi Petronii templo confecit* (Bologna, 1577) in the University Library in Lucca (*Corte mare mercanti* 1980, 187). Danti translated into Italian his brief treatise on meridian lines and published it in Danti 1578a, 298–317, dedicating it to Cardinal Gabriele Paleotti.

27 On Danti's appointment to the University of Bologna as professor in *disciplinas Mathematicas et dependentes*, see Zaccagnini 1930, 259; and Simeoni 1947, 59–60.

28 Danti sought other courtly appointments but to no avail. He tried to resume his relationship with Francesco de' Medici and even when it was abundantly clear that it was irredeemably compromised he continued to inform the Medici grand duke of his activities. He sent Francesco his report on the famous comet that appeared in June 1576 and inspired a flood of pamphlets and essays. Danti's essay is lost, but the critical response to his analysis written by Antonio Lupicini on Francesco's behalf survives (Antonio Lupicini, *Osservazione prima fatta nel* MDLXXVI *alli 17 di giugno per trovare il vero luogo della Cometa apparsa alli 14 del medesimo mese*, BNF, Classe XIX, n. 83, which also contains Danti's response). Danti received "most loving and favorable offers" (*amorevolissime et benignissime offerte*) from Francesco Maria della Rovere, Duke of Urbino, which the Dominican authority forced him, however, to reject. There is no way to know what exactly those offers entailed, but the surviving correspondence between Danti and the Duke of Urbino shows a certain familiarity and common interests in mathematics and applied sciences. Danti's surviving letters to the Duke of Urbino are as follows: 25 September 1576 (Arrighi 1963, 473); 27 October 1576 (Del Badia 1881, 30); and 8 December 1576 (Sapori 1981, 12). On mathematics at the court of Urbino, see Rose 1973 and Rose 1975.

29 On the teaching of astrology in Renaissance universities, see Westman 1980, 118–19. On Danti's classes, see Zaccagnini 1930, 260. Danti was able to maintain his chair at the uni-

versity until 1581 and obtain salary raises, even though he did not teach in the last years because of his commissions at the papal court.

30 For the Senate of Bologna, Danti made his *Disegno della Palata*, signed and dated 1579, illustrating the controversial border area between Ferrara and Bologna (ASB, Reggimento, Ass. Acque e Confini, Mappe, vol. 6) (Roversi 1974, 260–2; and Trento 1981, 23–5, with illustration).

31 Danti's signature is in the guest register of Aldrovandi's museum (BUB, MS Aldrovandi 41, not paginated). Antonio Gigante's sonnet in praise of Danti is in Danti 1578a, not paginated.

32 Danti 1578a, 324: "mi posi in animo di persuadere all'Illustriss(imo) Card(inale) Paleoto, che facesse un cotale strumento [the armillary sphere], il che facile mi fu, essendo quel Signore perito d'ogni sorte di scienza, & havendo l'animo molto inclinato a tutte le virtuose attioni."

33 On Danti's anemoscope for Cardinal Paleotti, see letter from Pietro Antonio Cataldi to Lorenzo Costa, 1 October 1577 (Danti 1578a, 252); and Danti's own description in Danti 1578a, 275. Danti's anemoscope for Paleotti is mentioned also by Ulisse Aldrovandi (*Historia de Onobrychi*, Bologna, Biblioteca arcivescovile, MS D 21, quoted in Fantuzzi [1781–94] 1965, vol. 1, 395). A seventeenth-century source mentions another anemoscope that Danti made in Palazzo Bianchetti at Ozzano, of which, however, there are no traces (Masini 1666, 455). Danti published the Latin edition of his treatise on the anemoscope in 1578 (Danti 1578b). Pietro Antonio Cataldi, who replaced Danti at the University of Bologna in 1584, translated the text into Italian and dedicated it to Lorenzo Costa, the first patron who commissioned an anemoscope from Danti in Florence. On Pietro Antonio Cataldi, see De Ferrari in *DBI* (with bibliography).

34 Cataldi's letter to Lorenzo Costa, 1 October 1577, in Danti 1578a, 252: "si affaticava molto . . . di tirare alla fine altre cose del Cardinale che avea alle mani."

35 In 1579 Danti gave these relics to the Compagnia del SS. Sacramento and the Compagnia del Rosario della Terra both in Perugia (pertinent documents published in Bertolotti 1965, 50–1).

36 These letters pertaining to the comments on Paleotti's treatise by Sigonio and Aldrovandi, now in the Archivio Isolani, Bologna, are discussed in detail in Prodi, 1959–69, vol. 2, 531–5. Aldrovandi's notes are also published in Barocchi 1961, vol. 2, 511–20. The letters by Danti, which Prodi attributed to Pirro Ligorio, are in Archivio Isolani, Bologna, F. 30/16. For Aldrovandi's comments on the grotesque, see Farago 1995, Introduction.

37 Danti 1577, not paginated: "innalzano l'intelletto, & assottigliano l'ingegno, alla contemplatione delle cose divine."

38 Danti in Bologna to Vincenzo Borghini in Florence, 23 November 1577, BNF, MS Magliabechiano, classe xxv, n. 551, fol. 104. In 1577 Danti mentioned his work for a chorography of Tuscany, but it is unclear whether he ever started it. The chorography of Tuscany was published in 1583 by Stefano Bonsignori, Danti's successor in the Guardaroba Nuova. For regional mapping projects similar to Danti's, see Cristoforo Sorte's chorography of the Venetian inland (Schulz 1990, 79–95); Marco Antonio de'Pasi's survey of the Duchy of Ferrara (Federzoni 2002); and Vincenzo Civitali's survey of fortifications in Lucca (Camerota 1999).

39 Danti's mural map of the Perugino, once located in the Governor Palace in Perugia, in the room adjacent to the old Decemviri Chapel, was destroyed during an eighteenth-century restoration. An earlier source recorded its inscription: "9 fortresses. 234 walled casltes. A very great number of large villas. 20 lively springs below the city. 6 wonderful stone bridges above the Tiber. And it has a perimeter of 138 miles." (Rocche viiii. Castela murate ccxxxiv. Ville grosse grandissimo numero. Fontane vive alle radici della città xx. Ponti di pietra bellissimi sopra il Tevere vi. E gira all'intorno miglia cxxxviii). This inscription is quoted in Bonazzi 1959, vol. 2, 208. On Danti's map of the territory of Perugia in the Gallery of Maps, see Gambi–Pinelli 1994, vol. 1, nos. 62–5. On the engraved version of Danti's map *Descrittione del Territorio di Perugia* (Rome, 1580) see Almagià 1948, table xxxiii. In addition, Danti may have built an anemoscope for the Governor's Palace in Perugia, pieces of which have recently been discovered (Levi-Donati 2003).

40 Danti's letter to Vincenzo Borghini, 15 February 1578: "Thursday I received the license from Cardinal S. Sisto [Filippo Boncompagni] which Jacopo Boncompagni had sent me, in which he orders the Governor of Bologna and the President of Romagna to provide me with everything I and three other helpers will need during the trip. And, with the blessing of God, I will start the first day after Easter in this county. The enterprise will be demanding, and I hope that, after having done the work in Perugia, it will be less difficult. The people in Rome have seen the chorographic drawings of Perugia and they were satisfied, and I believe they will be even more satisfied with this other chorography, since I want to execute it with much more diligence." (Io hebbi giovedì le patenti del Cardinale S[an] Sisto [Filippo Boncompagni] che mi ha mandate il Sig. Jacopo Bonc[ompagn]i nelle quali comanda al Governatore di Bologna et al Presidente di Romagna che mi provedino di quanto mi fara bisogno nel viaggio, per me et 3. servitori. Et mi cominciero con l'aiuto di Dio il p[rim]o dì dopo pasqua in questo contado, l'impresa sara grande, e spero da quello che ho fatto à Perugia che mi riesca meno difficile. A Roma hanno visto i disegni di quella di Perugia, et li hanno sodisfatti, et credo che di quest'altra li sodisferanno anco piu perche le voglio usare molto piu diligentia . . .) (BNF, MS Magliabechiano, classe xxv, 551, fol. 111).

41 The letters between Danti and Francesco are as follows: Danti to Francesco, 8 March 1578; Francesco to Danti, 25 March 1578; and Danti to Francesco, 6 April 1578, (Barocchi–Gaeta Bertalà 1993, 145–6).

42 The chronology of Danti's survey of the Bolognese was established by Mario Fanti in his study of Danti's sketchbook (BAB, MS Gozzadini 171) (Fanti 1967).

43 On Danti's maps *Bononiensis Ditio* (Fig. 6.15) and *Flaminia* (Fig. 6.16), see Gambi–Pinelli 1994, vol. 1, nos. 289–97 (Bolognese), and 305–13 (Flaminia).

44 BAV, MS Urb. Lat. 831, fol. 85v: "di tutto la S(anti)ta V(ost)ra havrà facilm(en)te perfetta notitia, se con questa scrittura accompagnerà il dissegno, ch'io feci fare al Pa(dr)e Ignatio Danti di questa Provintia, il quale ho mandato ultimam(en)te a V[ost]ra Beatitud[in]e." Ghislieri's report is not dated, but circumstantial evidence dates it in the second half of 1578, or at the beginning of 1579, certainly before 12 February 1579, when the dispute on the borders between Ferrara and Bologna was resolved (the dispute is still unresolved in Ghislieri's account).

45 On Philippe van Winghe's letter to Ortelius, see Ortelius 1887, letter no. 217; and Schuddenboom 1996, 39. On Mario Cartaro's engraving of Danti's map *Urbis Veteris Antiquae Ditionis Descriptio*, see Almagià 1948, table XXIV. On Lukas Holste's maps, *Patrimonium Sti. Petri* and *Latium*, see: Gambi–Pinelli 1994, vol. 1, nos. 71–5 (*Patrimonium Sti. Petri*) and nos. 95–100 (*Latium*).

46 According to this method, an area had to be divided into a series of triangles based on a line of known length, so that accurate measurements of distances and directions may be calculated by the application of trigonometry. The writings of Peter Apian, Gemma Frisius, and Oronce Finè were available in Italian translation in Bartoli 1564, especially Book IV. Danti frequently quoted Bartoli's text, as in Danti 1578a, 105–6: "the geometry of Oronce, whose book has been successfully translated by Messer Cosimo Bartoli." (. . . la Geometria di Orontio, il cui libro dal Mag. M. Cosimo Bartoli é stato felicemente in questa nostra lingua tradotto).

47 On Danti's scientific instruments, see Chapter Three.

48 Danti 1586, 107–8: "per comandamento della santa & gloriosa memoria di Papa Gregorio XIII. l'anno 1577. & 78. & 79. & 80. hò levato con questo medesimo strumento agevolissimamente la pianta di più provincie come fu tutta la Romagna, una parte del Umbria & del Latio, e di Sabina, & tutto il Territorio di Perugia, & di Bologna, & grandissima commodità mi arecò il poterlo piegare & spiegare senza smontare dà cavallo, & anco il poterlo portare all'arcione, senza impaccio alcuno." The *radius latinus* is a quadrilateral made of four adjustable rulers. Two rulers of the same size called *bracci* are hinged at the end of the main arm. The other two rulers of the same size, called *gambe*, are fixed at the extremities of the *bracci* and, hinged together, move along the main arm. The *gambe* carry the altimeter and shadow scales and a sight at the end. The main arm carries the degrees of a circumference, while the back of the arm is divided into equal parts. A compass could be attached to the instrument and might or might not be leveled with the instrument. To level the compass, a plumb line had to be attached to the instrument while taking measurements. Danti recommends a plumb line heavy enough to be unaffected by the wind. On the *radius latinus*, see Gambi 1994; and Fiorani 1996.

49 Danti 1586, 107.

50 Danti 1586, Dedication to the Readers, not paginated: with the *radius latinus* "si opererà in quel sito, & positione che piu ci piace, anzi in quello stesso modo, che ci verrà preso in mano, ó à livello, ó à piombo, ó inclinato . . . [other instruments] oltre che in fabbricarli ci é molta briga, riescono le più volte poco giusti, & vi bisogna usare gran diligenza, & consumarvi del tempo, in collocarli nel sito giustamente dove si ha da operare con essi."

51 Latino Orsini, natural son of Camillo Orsini, a member of the side branch of the Orsini family, was a condottiere who fought for the pope and the Venetian Republic. In 1571 he became Governor of the Island of Candia where he died. Orsini might have written his treatise in relation to the military academy he envisioned but never realized. Orsini's manuscript must have circulated in selected military groups at the papal court. Danti had access to it during his years in Bologna, possibly through the mediation of Giacomo Boncompagni, who may have given him a copy in connection with the mapping of the Papal States. It is certain that Danti and Orsini, given their shared interests, had met and a note by Danti confirms such contact: "the *radio latino* was shown to me by its inventor." (. . . si come [il radio latino] a me dall'autore fu mostrato) (Danti 1586, 59). On Latino Orsini, see Colonna 1955, 278–9.

52 Danti 1586, 110: "al medesimo tempo [as surveying] si trovava la pianta nel foglio bella e segnata."

53 Danti 1586, 110: "This operation can be done very well with our *radius latinus*, positioning it closed on the plane table, or rather after having opened its *braccia* and *gambe* the stylus is positioned on the plane table, and without making a hole in it, the instrument is always used at the same degree; attaching the cord as mentioned, one proceeds as with the rule, and one can eventually use the compass, which will be a very convenient thing to do." (Questa operatione si potrà far benissimo con il nostro radio, ponendolo sopra la tavolozza serrato, ò vero spiccandone le braccia & le gambe si ficchi lo stile sopra la tavolozza, & senza bucare lo strumento vi si accosti sempre ad un medesimo grado, attacca[n]dovi il filo, come si è detto, si opererà nel medesimo modo che co[n] la riga, & ci potremo servire della bussola, che sara ta[n]to più commoda cosa.) Mary Pedley pointed out to me that with the compass, the *radius latinus* becomes a sort of theodolite.

54 BAB, MS Gozzadini 171. The manuscript was published in facsimile in 1967 (Fanti 1967) and later attributed to Danti (Roversi 1974, 260). See also: Trento 1981, 23–4; and Fiorani 1996, in relation to Danti's mapping of the Bolognese.

55 There is very scanty evidence that Danti did astronomical measurements during his survey. However, he had measured astronomically the latitude and longitude of Florence, which he discussed in his correspondence with Vincenzo Borghini in 1578 (BNF, MS Magliabechiano, 551, fol. 12, 132–3).

56 Danti in Bologna to Vincenzo Borghini in Florence, 23 November 1577, BNF, MS Magliabechiano, classe xxv, n. 551, fol. 104.

57 Danti 1577, 50: "Quello che cerca le regioni forestiere per terra, et per mare, et per fiumi . . . se vole riportar frutto delli viaggi suoi fa mestieri che osservi [the list of 26 subjects follows]."

58 Danti 1577, 50: *Osservationi de' viaggi.* For a discussion on, and an English translation of Danti's *Osservationi de' viaggi*, see Frangenberg 1994, 52, 56–8.

CHAPTER 7

1 *Avvisi di Roma*, 22 July 1578 (BAV, MS Urb. Lat. 1046, fol. 288v): "per due hore [ha] dato aud(ien)za caminando sempre indefessamente." On the Belvedere Courtyard, see the classic study by Ackerman 1954. Specifically on the construction of the Gallery of Maps, see Ackerman 1954, 103–09; Redig de Campos 1968, 174–8; Wasserman 1966, 19–24; and Pinelli 1994a. On the development of the gallery as an architectural space in Renaissance palaces, see Settis 1983 and Cieri Via 1988.

2 In October 1578 the Mantuan ambassador in Rome, Giovanni Antonio Odescalchi, reported the beginning of the new corridor to his Duke Guglielmo Gonzaga (Smith 1977, 27).

3 Vignola–Danti 1583, 144: ". . . riuscita molto bella."

4 Four drawings by Mascherino pertaining to the Gallery of Maps survive in the Mascherino archive now kept in the Accademia di San Luca, Rome. The oldest drawing, datable to 1578, documents the early phase of the project: the *galaria grande* is marked at the bottom but the size of the Tower of Winds is still undecided (inv. no. 2470). Another drawing pertains to the stairways of the Gallery of Maps (inv. no. 2475). A third drawing is a study for the pavement of the Sala della Meridiana in the Tower of Winds (inv. no. 2471). The last drawing, datable around 1582–5, documents the last phase of construction, when the corridor was extended to the third terrace of the Belvedere Courtyard, *al piano della Vignola* (inv. no. 2476). Three additional drawings by Mascherino, also in the Accademia di San Luca, pertain to the central Serlian window (inv. nos. 2472, 2473, and 2474); like Mathijs Bril's view, these drawings document the intention of further enhancing the middle of the corridor by repeating the Serlian window with balcony in the lower levels. On these drawings, see Ackerman 1954, 106, 231–3, Wasserman 1966, 147–8; Marconi–Cipriani–Valeriani 1974; and Pinelli 1994a, 46–52.

5 In June 1579 Gregory xiii "spent half an hour walking alone with his chaplain in the Belvedere, [discussing] the new construction" (*Avvisi di Roma*, 27 June 1579, BAV, MS Urb. Lat. 1047, fol. 248v). He visited the site again two years later when the corridor was almost completed: "His Holiness went for a walk in his gallery on Monday" (*Avvisi di Roma*, 24 August 1581, BAV, MS Urb. Lat. 1049, fol. 349). In December 1580 he postponed other projects: ". . . since the expense [for the new aqueduct] will be more than sixty thousand scudi, His Holiness postponed [the construction] until the Spring, in consideration of the excessive expense for the corridor in Belvedere, which is truly a stupendous thing" (*Avvisi di Roma*, 24 December 1580, BAV, MS Urb. Lat. 1048, fol. 452v). The corridor's chief inscription, quoted in the text, dates the completion of the decoration to 1581. It is indeed documented that, after that date, most of the artists were busy working on other commissions in Rome and elsewhere. It is, however, possible that some details were added *a secco* later, especially the coordinates of latitude and longitude in the maps, which were largely repainted in the following centuries.

6 The poem *Ambulatio Gregoriana* has been recently published by Ferri 1994, who, however, did not know its author. A list of Gregory xiii's books, made at his death, documents that Giovan Battista Stella was the author of this poem (ASV, Fondo Boncompagni, Indice vii, fol. 30v). Descriptions of the Gallery of Maps also appeared in two other encomiastic poems: *De Gregorij xiii Beneficentia Panegiricus* (BAV, MS Boncompagni D 46, fols. 97–115, especially fols. 103r–v), and *Beatissimus Pater* (BAV, MS Boncompagni D5, fols. 312–13).

7 The inscription on the northern portal reads: "Italia Regio totius orbis nobilissima ut natura ab Apennino secta est hoc itidem in duas partes alteram hinc Alpibus et supero alteram hinc infero mari terminatas dividitur a Varoque flumine ad extremos usque Brutios ac Salentinos regniis provinciis ditionibus insulis intra suos ut nunc sunt fines dispositis tota in tabulis longo utrimque tractu explicatur. Fornix pia sanctorum virorum facta locis in quibus gesta sunt ex advorsum respondentia ostendit. Haec ne iucunditati deesset ex rerum et locorum cognitione utilitas Gregorius xiii Pont. Max. non suae magis quam Romanorum Pontificum commoditati hoc artificio et splendore a se incoata perfici voluit anno mdlxxxi." On the Gallery of Maps, see Almagià 1952, a seminal study on the cartographic significance of the Gallery of Maps, which, however, entirely disregarded the scenes of church history on the vault, although they are clearly related to the maps. See Schulz 1990, a pioneering study stressing the moralizing message of the Gallery of Maps. Cheney 1989 was the first study to investigate the symbolic meaning of the ceiling narratives and of the biblical sacrifice, although it did not explain the connection between the maps of Italy and the biblical sacrifices. Schütte 1993 pro-

vides a detailed analysis of the ceiling decoration and its literary sources; Gambi-Pinelli 1994, provided color illustrations of maps and church narratives but treated cartography apart from history. (See also the reviews of the Gambi-Pinelli volume in Scafi 1995 and Courtright 1997). Fiorani 1996, for an analysis of the cartographic language of the maps; Goffart 1998, for the interpretation of the Gallery of Maps as a historical atlas; and Battisti 1999, for the sources of the biblical sacrifices.

8 It should be noted, however, that the orientation of the peninsula within the corridor does not correspond to its real orientation, that is northern Italy is represented at the south end of the corridor, rather than at the north. Nonetheless, within the corridor itself, the illusion of walking along the peninsula from north to south is achieved.

9 For the list and discussion of the ceiling's scenes, see: Schütte 1993, 59–148; Gambi-Pinelli 1994, vol. 1, 407–504 (vol. 2, for handsome reproductions). Today the only vault scene that does not conform to the geographical order of the entire gallery is the fresco *Feed My Sheep* (*Pasce oves meas*) at the very center of the corridor, in correspondence with the Serlian window and the maps of Gregory XIII's homeland. This fresco was repainted by Giovan Battista Romanelli in the seventeenth century and may not reflect the original subject (Gambi-Pinelli 1994, vol. 1, no. 670).

10 BAV, MS Boncompagni D 5, fol. 240v: "Della Galeria maggiore ne fù architetto Ottaviano Mascharino. La pittura delle tavole di cosmografia fù fatta con ordine e disegno del R[everendissi]mo Padre ignatio Danti Vescovo d'Alatri. Le pitture della volta furono ordinate da Gieronimo Mutiano, mà disegnate da Cesare Nebbia da Orvieto" (Pastor 1936–55, vol. 20, 651, and Pinelli 1994a, 70). This note from the Boncompagni archive concurs with the writings of such early art critics as Raffaele Borghini, Giulio Mancini, and Giovanni Baglione (Borghini 1584, 577; Mancini 1956, vol. 1, 280; and Baglione 1642, 51, 116).

11 On Muziano, to whom Gregory XIII had previously commissioned the ceiling painting *Pentecost* for the Consistory, see *Pittura in Italia. Il Cinquecento* 1988, vol. 2, 779 (with bibliography); and Valone 1993 (specifically on his painting *Pentecost*). On Nebbia, see *Pittura in Italia. Il Cinquecento* 1988, vol. 2, 781. On the collaboration between Muziano and Nebbia, see Mack 1974; Strinati 1980, 16–17; and Pinelli 1994a, 136–9, who also published Nebbia's surviving drawings for the Gallery of Maps. One more drawing by Nebbia is published here as Fig. 7.12.

12 Baglione listed the following seventeen artists: the brothers Matthijs and Paul Bril, Niccolò Circignani, Baldassare Croce, Antonio Danti, Marco da Faenza, Donato da Formello, Matteo da Siena, Giovanni Battista della Marca, Ottaviano Mascherino, Paris Nogari, Giacomo Palma, Cristoforo Roncalli, Lorenzo Sabatini, Jacopo Semenza (or Sementa), Giacomo Stella, Antonio Tempesta. Some names can certainly be excluded from Baglione's list. Lorenzo Sabatini died in 1576. Antonio Danti was most likely confused with the painter Girolamo Danti, brother of Egnazio, who died in Perugia in 1580 (Borghini 1584, 524; Baglione 1642, 57; Sapori 1981; F. P. Fiore in *DBI*; and Pinelli 1994a, 64, who suggested that this Antonio mentioned by Baglioni was not Egnazio's brother, but rather the painter Giovanni Antonio da Varese, known as Vanosino, who had no family relation to Danti, although he was an expert in map murals). Giacomo Palma had left Rome in 1576 and was painting in the ducal palace in Venice (Baglione 1642, 183; Hess, 1967a; and Mason Rinaldi 1984, 10). Circignani worked in the Tower of Winds and later in the Terza Loggia under Danti's supervision (Van Mander 1604, 310; Baglione 1642, 41; *Pittura in Italia. Il Cinquecento* 1988, vol. 2, 680; and Courtright 2003, 69–103). Roncalli, who worked under Circignani's supervision, was involved more likely only in the Tower of Winds (Baglione 1642, 288; Turcic 1981; and Courtright 2003, 269). Mascherino did not paint in the Vatican Gallery but in the Loggias of San Damaso, although he was the architect of the corridor, a point Baglione omitted (Baglione 1642, 99). To complete Baglione's long list, Taja 1750, 288, added four more artists: Pasquale Cati, Giovanni Guerra, Girolamo Massei, and Raffaellino da Reggio. But among these artists, only Guerra and Cati could have participated in the decoration of the Gallery of Maps, although Guerra might have worked only on its restoration in the 1590s (*Pittura in Italia. Il Cinquecento* 1988, vol. 2, 674). Raffaellino da Reggio was dead, while Massei is still such an undefined painter that it is impossible to have an informed judgment on his possible involvement in the Gallery of Maps. Van Mander 1604 and Mancini 1956 also dealt with the decoration of the Gallery of Maps, but they are less reliable than Baglione 1642. Van Mander, who had left Italy in 1576, dated the beginning of the decoration to the mid-1580s, while documentary evidence shows that it was completed at that date. Mancini instead confused the decoration of the Gallery of Maps with that of the Terza Loggia, and, from his text alone, it is impossible to establish what decoration he was actually describing.

13 Baglione 1642, 22. Marco Marchetti da Faenza (1528–1588) had worked in the Loggias of San Damaso with Sabatini and after Sabatini's death was the only painter responsible for the grotesques in the Vatican Palace. In 1580, he returned to his native town of Faenza. He might have been responsible for the grotesques in the Gallery of Maps, although, it should be noted, these are much simpler and less imaginative than those he had previously painted in Florence and Rome. On Marco da Faenza, see Cecchi 1977; and *Pittura in Italia. Il Cinquecento* 1988, vol. 2, 762.

14 On Matteo da Siena (1533–1588), who was also an expert in monochromes, see Baglione 1642, 44, 47; *Pittura in Italia. Il Cinquecento* 1988, vol. 2, 766. On Giovanni Battista Lom-

bardelli (died in 1592), see Sapori 1981 and *Pittura in Italia. Il Cinquecento* 1988, vol. 2, 750.

15 It has long been debated whether both Mathijs and Paul Bril participated in the decoration of the Gallery of Maps, as Baglione reported (Baglione 1642, 296), a view shared by Pinelli 1994a, 64–5. A different, more convincing opinion is in Courtright 1990 and Courtright 1995. On the influence of Mathjis Bril on his brother Paul, see Dacos 1964; Courtright 1995; Wood Ruby 1999, 11, 15–19; and Courtright 2003, 105–08.

16 Taja 1750, 293, 336–7, identified Antonio Tempesta as the author of most of the ceiling narratives and of the ochre monochromes. On Tempesta (?1555–1630), see *Pittura in Italia. Il Cinquecento* 1988, vol. 2, 849.

17 On Donato da Formello (born, c.1530), see Baglione 1642, 16; and *Italian Drawings* 1983, 69. On Paris Nogari (1536–c.1601), who had worked in the Loggias of San Damaso in 1575–7 and was busy with individual commissions in Rome from 1580, see Baglione 1642, 87; Strinati 1980; *Pittura in Italia. Il Cinquecento* 1988, vol. 2, 784; and Litteri 2000.

18 On Baldassarre Croce (1558–1628), see Baglione 1642, 297; Strinati 1980; *Pittura in Italia. Il Cinquecento* 1988, vol. 2, 690. On Giacomo Stella (1545–c.1630), see Baglione 1642, 336; and *Pittura in Italia. Il Cinquecento* 1988, vol. 2, 847. On Jacopo Semenza, see Baglione 1642, 17; and Hess 1967a.

19 Some of the processions with accurate depictions of urban views and of vivacious crowds not only are the most accomplished of the vault, but they also recall in style and composition the later frieze illustrating *The Translation of the Body of St. Gregory of Nazianzus* in the Terza Loggia, where Tempesta painted the procession and Bril the architecture (Fig. 6.3). Such accomplished scenes are, for instance, *Emperor Frederick II Pledging Obedience to Pope Alexander III in Venice*; *The Translation of St. John the Baptist's Ashes to Genoa*; *The Exposition of the Holy Shroud in Turin*; *Modena Freed Thanks to St. Geminianus's Prayers*. On Bril and Tempesta's frieze in the Terza Loggia, see Chapter Nine.

20 Baglione 1642, 315: "per le sue opere di cavalcate, di caccie, e di battaglie."

21 Inscription of the map *Aveniensis Ditio* (Gambi-Pinelli 1994, vol. 1, no. 193). The papacy had acquired the County of Avignon from the French royal house in the early fourteenth century and made it its official residence for most of the fourteenth century.

22 Cheney 1989, 24–5; and Markus 1994. On the iconography of the Gallery of Maps, see also: Pinelli 1994b; Milanesi 1994; and Pinelli 1996.

23 On the scenes of biblical sacrifices in the Gallery of Maps, see: Pinelli 1994b, 140–3.

24 It is worth recalling that also for the decoration of the Sala Regia and of the Sale dei Foconi Gregory XIII had selected the deeds of earlier popes named Gregory. On Giovan Battista Romanelli's *Feed my Sheep*, see Cheney 1989 and Gambi-Pinelli 1994, vol. 1, no. 670.

25 Baglione 1642, 57: "tra gli artefici, a cui egli [Danti] il lavoro divisava."

26 Palmesi 1899, 44: "per papa gregorio f[elice] m[emoria] ho fatto la Gallaria, la Loggia de Venti, l'altre camore vicine la sala di Svizzeri, et quella de Palafrenieri con l'altra Loggia grande sopra l'offitio della camora avanti la Bologna." Danti wrote this note when he was Bishop of Alatri, a title he held from 1583 until his death in 1586.

27 The artists who worked in the above mentioned cycles were chosen among the team of the Gallery of Maps: Tempesta, Nogari, and Stella worked in the Sala Vecchia degli Svizzeri and in the Terza Loggia (where M. Bril was also involved). Bril and Tempesta, in particular, painted the famous frieze together, *The Translation of the Body of St. Gregory of Nazianzus*. Danti was also responsible for the iconography of the frescos for the apse of the church of Santo Spirito in Sassia in Rome, in which Nogari and Stella worked. On Danti's activities as an iconographic adviser, see: Brink 1983; Pillsbury 1974; Mancinelli–Casanovas 1980; Courtright 1990; Pinelli 1994b; Courtright 1995; Battisti 1999; and Courtright 2003, 31–2.

28 Danti, *Anemographia . . . ad Sanctiss(imus) ac Beatiss(imus) Patrem D(ominum) N(ostrum) Gregorium XIII P(ontificem) O(ptimum) M(aximum)* (Rome, 1581) (BAV, MS. Vat. Lat. 5647). This manuscript (now available in print and in English translation in Courtright 2003, 219–42) was a presentation copy to Gregory XIII, who kept it in his private bedroom. Most of the text is a copy of the treatise on the anemoscope that Danti had already published in 1578 (Danti 1578a, Sesta Parte, 250–81), but it included the discussions of the anemoscope Danti made for Gregory XIII in the Tower of Winds and of the pictorial decoration surrounding it. On the Sala della Meridiana, see Courtright 1995 and Courtright 2003, 69–103, 212–18.

29 Danti's letter to Ortelius, 24 December 1580 (Ortelius 1887, letter no. 100): "Havendo divisa l'Italia p[er] il mezzo del Mo[n]te Apennino o posta da una banda della Galleria quella parte che è bagnata dal Mare Ligustico, et Tirreno, et dall'altra, quella che è cinta dall'Adreatico, e dall'Alpi, diuidendola poi seco[n]do gli stati et le prefetture de gouernj in quara[n]ta parti, seco[n]do che la Galleria è diuisa in .40. quadrj di tanta grandezza che sono andati .64. fogli di carta reale per quadro nel fare i cartonj. Hora uo riducendo ogni cosa in un libro, oue le parti d'Italia sara[n]no fino a .48. et ui saranno apresso da .80. storie di figure, che sono dipinte nella uolta della Galleria sop[ra] ciascuno quadro rapprese[n]tando qualche segnalato miracolo occorso in quella prouincia. Ne ho uoluto dare auiso a V[ostra] signoria, accio se lej hauesse uolunta a farmj intagliare questo libro che me lo faccia sapere, che io opererej co[n] il papa che ci porgesse qualche aiuto, sara apresso ciascuna tauola qualche particulare, et noua annotatione, co[n] le inscrittioni che in essa Galleria io ho fatte, et sono state tutte riuiste et racco[n]cie dal S[ignor]. Moreto, dal

S[ignor] Bargeo, et dal Frizzolio . . . resta ancora che le dicha, che ho messo insieme fra città, et altre terre grosse d'Italia ritratte in pianta dal naturale fino a cento pezzi, che sar[n]no co[n] il medesimo libro."

30 A number of maps that arrived in Rome too late to be included in the map murals would possibly have found their place in the book. On these late maps, see: Almagià 1952, 7.

31 Of the many maps and episodes of church history that make up the Gallery of Maps, only three printed versions are known, though their connection to the book remains unclear: Danti's maps of the territory of Perugia (Rome, 1580) and of Orvieto (Rome, 1583), and the portrait of Andrea Doria from the map of Liguria (*Celebrazione di Gian Andrea Doria*, 1583, an engraving attributed to Adamo Scultori, on which see Bellini 1991, 137).

32 On the philosophical concept of *chora* and place, see Casey 1998. For a penetrating analysis of chorography, see Casey 2002. On Renaissance chorography, see Broc 1989, chapters 8 and 15; Frangenberg 1994; Nuti 1988; Nuti 1994; Nuti 1999; and Woodward 1996, 7–12. The connection between time and place must have been clear to some map-makers. For instance, Gerard Mercator wrote a book titled *Chronologia* (Basel, 1577) discussing topics such as the origin of the world or the location of paradise, a book that Mercator regarded as complementary to his cartographic work.

33 Renaissance definitions of chorography are in Apian 1533, fols. 1 and 2; Gastaldi 1561, as quoted in Woodward 1996, 6; and Ruscelli in Ptolemy 1561, 9. Four chorographical maps are in the 1513 edition of Ptolemy's *Geography*. As the title itself explained, these maps represented, on an enlarged scale, areas that were mapped on a smaller scale in another map of the same book (*Quattuor deinde Chorographiae particulariores magis extensea praedictis tabulae*). The titles of the chorographic maps were: *Chorographia eremi Elveticorum*, *Chorographia provinciae Rheni*, *Chorographia Cretae*, and *Chorographia Lotharingiae*. On the relationship between geography and chorography see also Chapter Five.

34 Sources contemporary with the Gallery of Maps identified this map cycle with different names, such as "cosmography of Italy," or "geography of Italy," but Danti did not hesitate to describe it as the chorography of Italy in the cartouche of the map *Sallentina Hydrunti Terra*, which is quoted below.

35 Danti 1577, 44. The full quotation of Danti's definitions is in Chapter Five. A similar definition by Danti is also in BAV, MS Boncompagni N 1, fol. 30. In the Boncompagni manuscript, the exemplary chorography is not the map of Perugia, Danti's hometown, but rather the map of Sora, a city then ruled by Giacomo Boncompagni, Gregory XIII's son. Girolamo Ruscelli refuted the interpretation of topography as a verbal description (Ruscelli in Ptolemy 1561, 6).

36 On Flavio Biondo, see: Gambi 1977; Fubini in *DBI*; and Fubini 1999. On Leandro Alberti, see Redigonda in *DBI* (with bibliography). On Raffaele Maffei, see D'Amico 1983.

37 Biondo 1453, 65.

38 Alberti clearly stated the scope of his book in the introductory pages (Alberti 1577, 7v).

39 Alberti 1577, 8. Also Raffaele Maffei had defined a city a place with a bishopric in his *Commentariorum urbanorum* (Rome, 1506).

40 Jacopo Strada to Guglielmo Gonzaga, Duke of Mantua, 4 October 1577: "le cose che io volevo che lui [Leandro Alberti] havesse messo erano queste: Prima erano le cose che apartengono alla cosmographia, cioè che ad hogni regione vi fosse posta la sua mappa avanti alla sua descrizione, et poi in quella regione vi fossero anche le figure di tutte le città pertinente ad essa regione, ben ritratte dal naturale; ancora a caduna città vi fossero le figure di tutte le cose più principali, come sarebe a dire gli edifitij antichi et ancor li moderni, con le loro piante et profili ben misurati con le sue misure scritte alli suoi lochi. Poi le inscriptioni antiche che nei marmi vi si trovano ritratte a punto come stanno nei detti marmi con le figure et altri ornamenti che anno d'intorno. Apresso la memoria di tutti gli huomini dotti con li loro nomi con tutte le opere che loro hanno scritte, stampate et da stampare. Ultimo tutte le arme dico di cadauno principe per cadauna città, cominciando alla prima della sua casa, et di mano in mano insino a quella del principe che hoggi vive. Apresso a queste quelle di tutti li nobili di cadauna città et così havesse finito. Hora questo ordine sudetto ch'l frate à lassato fuori io lo fo, et di già sono tanto avanti che di essa prima Italia sudetta è accresciuta più delli doi terzi et saranno tre gran volumi." (*Jahrbuch der Kunsthistorischen Sammlungen des allerhochsten Kaiserhauses*, 16: 2 (1895), document no. 14001). See also Strada's letter to Dodoens and Plantin, Vienna, 13 August 1578, and Plantin's reply, 10 October 1578, all published in Denucé 1916, 2–3, and 32–3. On Strada's antiquarian activities, which were well known in Florence and Rome, see Jansen 1982; Jansen 1987; and Marchis–Dolza 2002 (I thank the late Ruth Rubinstein for bringing to my attention Strada's project on Leandro Alberti).

41 Ruscelli 1562, 1v: "Et perche in effetto il più proprio, il più uago, il più uero, & il più util modo di far tal descrittione è quello delle Tauole ò figure, distese poscia con le parole nella quantità & qualità loro, uengo procurando, che in ciascuno di detti quattro uolumi grandi habbia da essere quanto maggior numero di Tauole ò di figure sarà possibile, cominciando sempre dall'universali, & uenendo alle particolari, & particolarissime delle prouincie, & di parte in parte, in modo, che ogni uolume sia per contener'almen cento Tauole, là oue in Tolomeo con XXVI sole si uede più accennata con dito in uniuersale, che descritta nel particolare tutta la terra." On Ruscelli, see Chapters Four and Five.

42 Alberti's text was amply used in the Gallery of Maps, especially for the content of the maps' inscriptions, some of which were copied word for word (i.e. the original inscription of the map *Patrimonium S. Petri*, the inscription of the maps *Principatus Salerni* and *Sicilia*, or the cartouche iden-

tifying the Battle of San Ruffillo in the map *Bononiensis Ditio* (Gambi–Pinelli 1994, vol. 1, nos. 72, 114, 178).

43 Roberto Almagià identified Danti's cartographic sources and Gambi and Pinelli have recently updated his research (Almagià 1952, 17–75; and Gambi–Pinelli 1994, vol. 1, 117–384). Specifically, on the maps cited in this paragraph, see: Gambi–Pinelli 1994, no. 48 (*Etruria*); no. 178 (*Sicilia*); no. 227 (*Transpadana Venetorum Ditio*); and no. 237 (*Forum Iulii*).

44 ASV, Nunziatura di Venezia, 21, fol. 582: "un ruolo, che é la descrittione dell'Isola di Termiti desiderata dal P[ad]re Ignazio Danti per l'opra della Galleria."

45 On the maps cited in this paragraph, see: Gambi–Pinelli 1994, vol. 1, no. 36 (*Liguria*); no. 84 (*Umbria*); no. 104 (*Campania*); no. 155 (*Corsica*); no. 168 (*Sardinia*); no. 192 (*Avenionensis Ditio*); no. 215 (*Mediolanensis Ducatus*); no. 227 (*Transpadana Venetorum Ditio*); no. 250 (*Placentiae et Parmae Ducatus*); no. 313 (*Mantuae Ducatus*); no. 320 (*Urbini Ducatus*); no. 349 (*Anconitanus Ager*); and no. 428 (*Tremitae*).

46 The Latin inscription reads: "Cum in conficienda hac Italiae Chorographia iis authoribus qui plurima Italiae loca terrestria maritimaque (certis longitudinum latitudinumque differentiis observatis) descripserunt ac variis valdeque dubiis eorum traditionibus qui particularia loca peragrarunt standum esset, mirum nemini videri debet si minus nota oppidula hic adamussim posita non reperiantur. Curabamus autem ut longitudinum latitudinumque gradus et minuta insignoribus locis (quo ad Chorographia ferre poterat) exacte responderent. Atque id F[rates] Egnatius Dantes Perusinus Ord[inis] Praed[icatorum] admonitum esse volebat." (See also Gambi–Pinelli 1994, vol. 1, no. 392).

47 Almagià offered the fascinating suggestion that Danti had provided a unified system of coordinates for the entire peninsula. But the loss of the one supporting document, Danti's general map of Italy, makes this hypothesis impossible to corroborate. The general maps of Italy in the Gallery of Maps are not Danti's original maps; they were drawn by Lukas Holste in the seventeenth century. On Holste's maps of Italy, see: Almagià 1942, 102–20; Almagià 1952, 77–9; and Gambi–Pinelli 1994, vol. 1, nos. 3 and 6.

48 The astronomical measurements of the Papal States were still an issue in the mid-eighteenth century, on which see Pedley 1993.

49 "Hanc Mediolani latitudinem a peritissimo viro observatam, cum aliis quoque probari viderim hic ponendam duxi, licet ad universali descriptione aliquantulum discrepare videatur." (Gambi–Pinelli 1994, vol. 1, no. 224).

50 Danti based the scale of his maps on the Roman mile, which consisted of a thousand paces. Almagià calculated that Danti's Roman mile corresponded to about 1,480 m and to a latitudinal minute (Almagià 1952, 10).

51 For a complete list of the historical vignettes on the maps, see Goffart 1998 and Appendix B.

52 The order of the Punic Wars as it appears in the Gallery of Maps starts with Hannibal's battle at Lake Trasimeno, it continues with the siege at Spoleto, the battle near Venosa, the disaster at Cannae, the battle at Trebbia, and it concludes with the battle at Pavia. Chronologically, the sequence should be: the battles of Pavia, Trebbia, Lake Trasimeno, Spoleto, Cannae, and Venosa.

53 Goffart 1998 dealt with the historical vignettes in detail. He divided them into three periods (Roman, medieval, and early modern), considered their significance in relation to a general history of western civilization, and concluded that they are an unsuccessful, odd, and unrepresentative sequence of events. He found the rationale of the historical vignettes by connecting them with the vault scenes, and interpreted the Gallery of Maps as informed by the view of history of the Hispano-Roman Orosius, one of the earliest Christian historians who narrated the "disasters of worthless men" alongside the "miracles of saints." According to Goffart, in the spirit of Orosius, the vault illustrated a "perfect history of saints and miracles," while the walls illustrated a "grimmer view of human affairs and defeats."

54 Alberti 1577, 339r–v (on the location of the second triumvirate); 201, 201v, and especially 209v, 214v (on the etymology of the Brutii); 185v (on Pope John X fighting against the Saracens); 330 (on the Battle of San Ruffino, which Alberti named and dated wrongly; these mistakes are repeated in the legend of the corresponding Vatican map, *Bononiensis Ditio*).

55 This modern inability to connect maps and vault scenes has also influenced the historiography of the Gregorian corridor, which generally kept the maps separate from the wall. Almagià limited his study to the maps, excluding the vault scenes because they were beyond his competence as a historian of cartography: of a different language from the maps, he regarded the vault scenes as extraneous to Danti's chorography of Italy. Gambi and Pinelli too followed Almagià's approach in their recent book. Although they wrote on both the maps and the figurative scenes, they treated the two parts of the iconography separately. Goffart connected the vault scenes only with the historical vignettes of the maps, that is with only one feature of the maps' cartographic language.

56 Goffart 1998, 796, has rightly noted that these contemporary definitions, miracles and pious deeds, are both inaccurate, loose descriptions of the events depicted on the ceiling. Some of the vault's scenes can unequivocally be described as miracles since they involved divine intervention in human affairs, while many others were historical events that, although exceptional, did not require divine intervention.

57 Pinelli 1994b has meticulously traced the vault scenes to Baronio's monumental history of the church, but he did not fully articulate the notion that the *storie di miracoli* were actually part of the Catholic view of church history. On Baronio, his work as a church historian, and his influence

on art and architecture, see Pullapilly 1975; Orella Y Unze 1976; Prodi 1977; Cochrane 1981, 445–78; Jedin 1982; *Baronio storico* 1982; *Baronio e l'arte* 1984; Scavizzi 1989; Scavizzi 1992; Zen 1994; and Fubini 1999.

58 On Carlo Sigonio, see Prodi 1977; McCuaig 1989 for an excellent intellectual biography; and McCuaig 1994.

59 The legends of the maps *Italia antiqua* and *Italia nova*, partially quoted in the text, are published in full in Gambi–Pinelli 1994, vol. 1, nos. 7 and 8. Another inscription in the map *Italia nova* mentions the great artists and scholars who were born in Italy, and is crowned by Pope Gregory the Great, the namesake of the Boncompagni pope.

60 Goffart 1998 regarded the Gallery of Maps as a historical atlas, albeit an unsuccessful one. On Renaissance historical atlas in general, see Goffart 1995 and Goffart 2003, 1–50.

61 The word *parergon* is documented also in Pliny, who reported that ancient painters used the word to describe the subordinate elements of their paintings (Melion 1999, 50).

62 Shirley 1998, 169; On Ortelius's *Parergon* and his other historical maps, see Koeman 1964; Black, 1997, 9–10; Goffart 1995; Meurer 1998; Shirley 1998; Van der Broecke–Günzburger 1998; Melion 1999; and Goffart 2003, 1–50.

63 Only three maps in the entire Vatican gallery, the two maps illustrating the Claudian port at Ostia and the map *Italia antiqua*, were meant and used as historical maps (Gambi–Pinelli 1994, nos. 3, 428 and 429; and see also below).

64 Vignola–Danti 1583, 81: "Volse in tanto, che io levassi la pianta di tutte le rovine che hoggi vi sono rimaste, & disegnatone l'alzato per l'appunto lo dipingessi (come feci) nella Galleria, che à sua Beatitudine ho fatta nel suo palazzo in Vaticano, per vederselo tuttavia avanti gl'occhi, & andar divisando, come potesse ridurre al pristino uso si degna, & mirabile opera" (English translation from Frangenberg 1994, 60, n. 23).

65 In 1587 the painters Cesare Nebbia and Giovanni Guerra supervised the retouching of the gallery's ornamentation, of one of the scenes on the ceiling, and of the maps *Italia nova*, *Etruria*, and *Forum Iulij*. From 1592 to 1596 the painter Pietro Oldrado retouched the cartouches, inscriptions, and toponyms of the maps *Perusinus ac Tifernas*, *Piacentiae et Parmae Ducatus*, *Mantuae Ducatus*, *Mediolanensis Ducatus*, and *Etruria*. Between 1599 and 1605 the gallery's roof was largely restored to prevent rainwater damages. In the 1590s in the map *Etruria*, the city view of Livorno was updated to include the urban renovations initiated by Giovanni de' Medici. In the seventeenth century the view of Genoa, both in the large fresco dedicated entirely to the city and in the map *Liguria*, was retouched to represent additions to its fortifications, while the map *Calabria ulterior* was substantially retouched by Giovan Battista Magni. The map *Etruria* was updated after Stefano Bonsignori published his new chorography of Tuscany in 1584. The plan of Milan in the map *Mediolanensis Ducatus et Monsferatus* was entirely redrawn in the seventeenth century, although the original might have been based on the map of Milan that Mascherino kept among his papers and now in the Accademia di San Luca, Rome (Kahn–Rossi–Franciolli 1999, 76–7). An inscription on the southern portal of the Gallery of Maps mentions Urban VIII's restorations: "Urban VIII, pontifex maximus, protected the foundation of the Gregorian gallery from the harm of water slipping below, [and] the walls and the vaults from the injury of rains and time; he restored paintings, wearing away virtually every day; he set right and added the geography in many places; he restored the entire work and the roof repaired to its original beauty in the year of our Lord 1631, the eight [year] of his pontificate." Lukas Holste made entirely new frescos only for the maps of *Italia antiqua*, *Italia nova*, *Latium et Sabina*, *Patrimonium S. Petri*, and for the city views of Civitavecchia and Ancona. His revisions of the other maps were limited to inscriptions and to minor geographical details, mainly of an antiquarian nature. Holste added the views of Loreto and Humana Antica in the map *Anconitanus Ager*, the view of the ancient ruins of Helvia Ricina in the map *Picenum*, and the Fortezza Urbana, built by Urban VIII, to the map *Mantuae Ducatus*. On the restorations of the Gallery of Maps, see: Almagià 1952; and Franzoni 1994. For a detailed account of Holste's restorations, see Almagià 1942, 102–22. On Romanelli's fresco, see Gambi–Pinelli 1994, vol. 1, no. 670.

66 On 23 December 1583, the *Avvisi di Roma* reported that the Grand Master of the Knights of Malta was admitted in the corridor: "Last week the Grand Master of Malta, coming from the visit to the gallery of the pope . . . [Tornando il gran m(aest)ro di Malta la sett(ima)na passata da veder la galeria del Papa . . .] (BAV, MS Urb. Lat. 1049, fol. 472).

67 This event is reported in Boccazzi 1986, 50.

68 Although the Vatican Gallery was not mentioned in Renaissance and Baroque guidebooks to Rome, yet it was visited by numerous, famous travelers, such as: Montaigne 1955; Furttenbach 1971, 113–14; Montesquieu 1971, 200; Brosses 1885, vol. 2, 170; Goethe 1932, 225–6; Bolkmann 1777, vol. 2, 135; Lalaonde 1786, vol. 3, 227–8. For a different evaluation of the Vatican Gallery's fame, see Sereno 1994. On the visits by later cartographers, see Almagià 1952, 8–10.

69 Filippo Pigafetta's comments on Danti's map of Latium are in his letter to Ortelius, 31 July 1591 (Ortelius 1887, letter no. 200). Philippe van Winghe wrote to Ortelius on 24 December 1589; at the bottom of this letter Ortelius listed things to ask van Winghe in Rome, including "Egnatio Danti" (Ortelius 1887, letter no. 170). Six months later van Winghe reported to Ortelius that he had found Danti's map of the territory of Perugia (Ortelius 1887, letter no. 185).

70 On Giovanni Antonio Magini's comment on the Gallery of Maps, see Almagià 1922, 155–6. Magini's atlas *Italia* was printed in Bologna in 1620.

CHAPTER 8

1 The role of Italy within the post-Tridentine universal church is addressed more extensively in Wright 2000, especially 102–45, on which my discussion is based. Wright also mentioned the Gallery of Maps in relation to papal primacy in the peninsula.

2 The first inscription, "the most noble region of the whole world," is on the northern portal; the second, "the principal regions of the world," and the third, "the nurturer of all the lands," appeared instead in the original cartouche of *Italia antiqua*; later the second inscription was changed in *Regionum orbis princeps* (chief among the regions of the world), while the third was completely abolished (they are both documented in BAV, MS Barb. Lat. 1803, fols. 1–2). On the maps *Italia antiqua* and *Italia nova*, see Gambi–Pinelli 1994, vol. 1, nos. 1–27.

3 Pliny 1949, vol. 2, 30–3. This inscription is taken, with minor changes, from Pliny, *Historia naturalis* (Book III, 5, 6): "Italia omnium terrarum alumna eadem et parens numine dei electa quae coelum ipsum clarius faceret sparsa congregaret imperia ritusque molliret et una cunctarum gentium in toto orbe patria fieret."

4 Unlike in the post-Tridentine years, earlier in the sixteenth century it was legitimate to maintain the plural "deorum" in the discussion of the Christian God (Rowland 1998, 130–1).

5 On Sebastian Münster and his questioning of Italy's role in world history, see Münster 1550; Strauss 1959; Strauss 1965; Burmeister 1964; and *Sebastian Münster* 1988.

6 On the scenes representing episodes of Constantine's life, see Gambi–Pinelli 1994, vol. 1, nos. 554, 555, 556, 558, and 566.

7 The classic study on Constantine is Coleman 1914, but see also *Costantino il Grande* 1992 for a recent assessment on the historiography on the emperor, especially the essays by Vincenzo Aiello, Riccardo Fubini, and Mario Mazza. Lorenzo Valla's treatise on the Donation of Constantine is in Coleman 1993. On the Renaissance revival of Constantine, see Aiello 1988; McCuaig 1989, 251–90; and Freiberg 1995, who was the first scholar to suggest the connection between Laureti's ceiling for the Sala di Costantino and the Gallery of Maps.

8 "Li venne in animo [to Tommaso Laureti] di far attioni del medesimo imperatore et in particulare quelle che fece in honore e beneficio di s[anta] chiesa, et havendo esso Tommaso vista in una delli parieti della medesima sala la donatione d'Italia fatta da Costantino a S. Silvestro e suoi successori rappresentata per una figuretta non molto intellegibile, pensó di fare l'istessa Italia distinta in 8 provincie secondo l'ordine de Strabone per più intelligenza di tal donatione." (BAV, MS Boncompagni D 5, fol. 241v; Pastor 1891–1953, vol. 20, 653; and Pinelli 1994a, 71). Originally, Laureti had planned the representation of Constantine ordering the destruction of the pagan idols, a story that was similar in meaning to the one that was actually painted, but much more dramatic because of the many figures it required. When Sixtus V, who succeeded Gregory XIII in 1585, refused to pay the exorbitant fee Laureti demanded, the project was scaled down and no figures were depicted in the final fresco. On the ceiling of the Sala di Costantino, see Partridge 1983; Wohl 1992; *Raffaello* 1993, 167–201; and Freiberg 1995.

9 Coleman 1993, 17. Coleman also suggested that the "western regions" should be identified with northern Italy. On the debates pertaining to the Donation of Constantine during the Boncompagni papacy, see McCuaig 1989, 251–90, on which my discussion is based.

10 On Eusebius's *Life of Constantine*, see Eusebius 1999.

11 On the Magdeburg *Centuriae*, see Headley 1963; Orella Y Unze 1976, 335–40; and Scavizzi 1989, 29–46.

12 ". . . ut a piis hominibus creditur" (McCuaig 1989, 254). On Sigonio's historical work on medieval Italy, see McCuaig 1989, Chapter Two.

13 For instance, in his book *De antiquo iure Italiae* (Venice, 1560) Sigonio disregarded the chronological order of Roman law and favored instead the organization of the subject according to geographical areas. Although Sigonio's discussion of Roman law according to geography corresponds to the juridical organization of the Roman provinces and thus would seem obvious, it was in fact a novelty at that time. It was also a novelty that Sigonio decided to include a map of Italy in his tract.

14 Antonio Pinelli proposed a strict political reading of the Gallery of Maps as expressing the papal wish to unify Italy under his leadership (Pinelli 1994b and Pinelli 1996).

15 On the series of biblical sacrifices in the Gallery of Maps, see Iris Cheney, "Sacrifice and Sacrament: Old Testament Sacrifice in the Galleria delle Carte Geografiche at the Vatican," paper delivered at *The Twenty-seventh International Congress on Medieval Studies*, Western Michigan University, Kalamazoo, Mich., 1992 (I thank the late Iris Cheney for sending me a copy of her paper); Pinelli 1994b, especially 140–3, and Gambi–Pinelli 1994, vol. 1, nos. 563, 564, 570, 594, 609, 610, 616, 640, 655, 667, 668, 681, 682, 693, 694, 700, 724, 739, 740, 746, 770, 785, 786. On the earlier series of biblical sacrifices in Raphael's Loggias in the Vatican Palace, see Dacos 1986 and Davidson 1985. On biblical sacrifices, see *Encyclopedia of Early Christianity*, especially the following entries (with bibliography): Altar, Sacrifice, Offerings, Sacraments. For an analysis of the liturgy of the Mass, see the classic Bossy 1983.

16 This scene has erroneously been interpreted as illustrating Adam's daily offering of spices, frankincense, and galbanum, an offering related in the pseudepigraphical *Book of Jubilees* (III, 26–7) (Cheney "Sacrifice and Sacrament"; and Pinelli 1994b, 414–15). But the inscription below the fresco says instead that Adam is offering his praise rather than the

produce of his labors. Part of the inscription is a direct quotation from Psalm 50 (verse 14), rather than Psalm 49 as often repeated. Thomas Aquinas's comments on the sacrifice of praise are in Aquinas 1922, II–II:85:3. All biblical quotations are from Bible 1970a (Old Testament) and Bible 1970b (New Testament).

17 It is worth mentioning that one of the rare instances in which Abel's and Cain's respective sacrifices are represented in conjunction with Cain's killing Abel occurs in Raphael's Loggias (Dacos 1956, 295; and Davidson 1985, fig. 26).

18 On Raphael's Loggias, see Dacos 1986 and Davidson 1985.

19 On the scenes of sacrifice pertaining to female purification, it should be noted that one scene conflates two separate cases of female impurity: in the background a woman offers a bird, probably after menstruation, while another in the foreground is having the ritual ablutions after intercourse (Leviticus, 15). In the second scene a woman, impure after childbirth, offers a lamb and a bird to be used for the holocaust and for sin respectively (Leviticus, 12). The last two sacrifice scenes illustrate two moments of the complex ritual lepers had to go through to be readmitted into the community, most likely the initial presentation of two birds (Leviticus 14: 4–6) and the subsequent ablution (Leviticus 14: 8). Cheney 1989 and Pinelli 1994b noted that the narrative sequence of Leviticus, which corresponds to the scenes of sacrifice in the vault, has been altered in the final scenes, so that the purification of lepers appears after the purification of women. They also suggested that this variation was introduced in order to stress the correspondence between the two ends of the Gallery of Maps, particularly the connection between Constantine, who allegedly converted to Christianity after he had been cured from leprosy, and the purification of lepers.

20 On sixteenth-century debates on the relationship between biblical sacrifices and the Eucharist, see McHugh 1991 and Cavenaugh 2001.

21 Tarde 1984, 40: "nous avons vu . . . la galérie grégorienne, au bout de laquelle est un miroir qui représente un calice avec l'hostie, prenant telle image d'un tableau qui est au-dessous, sans que telle peinture y soit, mais c'est par un artifice de perspective." Tarde stayed in Rome for four months and although he said he visited "tous les théâtres, amphitéâtres, temples . . . , termes, montagnes, aqueducts, ponts, sépulcres et autres antiquités romaines qui étaient digne d'être vues" (Tarde 1984, 42), he actually only described in detail the Gallery of Maps and a few other scientific and curious objects in Roman private collections. He was in Italy again in 1615, when he visited Galileo in Florence.

22 Danti wrote on anamorphoses in Vignola-Danti 1583, 94–5. Given the overall decoration of the Gallery of Maps, the anamorphosis and the mirror could only have been located above the southern portal, toward the old Vatican Palace, where today there is the inscription celebrating Urban VIII's restoration of the Gallery of Maps. The northern portal was occupied by the original inscription mentioning Gregory XIII's decoration. The mirror had to be placed below the painted image.

23 See Battisti 1999, 74 on the connection between the Tridentine disputation on the sacrifice of the Mass and the Gallery of Maps. Battisti noted that many episodes of church history depicted on the vault reported the miracles that the Blessed Host performed throughout Italy: St. Laverius of Agrumentum exorcised demons with the Host; St. Clare scared the Saracens away by displaying the Host from the walls of Assisi under siege; a bleeding wafer restored a priest's faith in Bolsena; the troops of Emperor Henry II gained strength from the Holy Host they received just before fighting the enemies of the Christian faith in southern Italy.

24 On the institution of the tithe, see *Encyclopedia of Early Christianity* (Tithing); and *Catholic Encyclopedia* (Tithes).

25 On papal revenues in the Renaissance, see Delumeau 1957, especially 183–203; Partner 1980; and Wright 2000. Specifically on Gregory XIII's policy pertaining to ecclesiastical revenues, see Pastor 1891–1955, vol. 19, 340; and Borromeo 2000, 191. On the Spanish contribution to papal finances, see Daudelet 2001.

26 Inscription on the map *Flaminia*: "Quae loca indepictis tabulis Draconis aurei signum huic simile stendunt ea Gregorius XIII P.O.M. recuperavit. Id volui nescius ne esset." The function of the Boncompagni dragon is registered also in Ciappi 1596, 7; in Baglione 1642, 57; and in the *avvisi di Roma*: *Avvisi di Roma*, 13 May 1581 (BAV, MS Urb. Lat. 1049, fol. 177): "It is said that the Apostolic Chamber has recently acquired three castles from Signor Honorio Savello. If this is true, they [the castles] will have to be painted as others were, in the new and beautiful Gallery that his Holiness is having done in the Belvedere." A few days later the same *Avvisi di Roma* reported more specifically how these new papal acquisitions had to be singled out in the Gallery of Maps, *Avvisi di Roma*, 17 May 1581 (BAV, MS Urb. Lat. 1049, fol. 183): "in order to distinguish all the places that His Holiness acquired for the Church from all others, he [His Holiness] is having a dragon depicted above each of them, so that in the future it will be possible to recognize the benefice that the Apostolic Chamber received from His Holiness. Already the Chamber has gained possession of three castles taken from Signor Honorio Savello, and moreover proceeds against the Farnese for its claim over the town of Castro."

27 ASV, Archivio Borghese, 8606, no. 69; and 8607, no. 69; Congregazione dei Confini, I, 27, fol. 675, a manuscript copy of Danti's printed map of the territory around Perugia.

28 For instance, Gregory XIII conceded the full amount of clerical levies as subsidies to the Republic of Venice and to the Duchy of Piedmont, but kept part of the same taxes in the Grand Duchy of Tuscany and the Kingdom of Naples. He gave to the Spanish king a papal subsidy, the tithe from the

Spanish clergy, and income from another tax on parochial wealth as compensation for the Spanish commitment to fight the Turks and the heretics (Borromeo 2000 and Borromeo 2002).

29 Baronio 1624, vol. 1, col. 179. On the origin of the Mass, see *Encyclopedia of Early Christianity*, especially the following entries (with bibliography): Eucharist, Mass, Sacraments, Sacramentary, Altar, and Sacrifice; *Catholic Encyclopedia*, entries for Eucharist, Sacrifice of the Mass, Offerings, Sacrifice. On Baronio's historiography, see Chapter Seven.

30 Gambi–Pinelli 1994, vol. 1, nos. 622–32. One of the personifications (no. 628) is identified with Generosity, although this identification is uncertain.

31 The papacy of Gregory XIII is usually passed over as a stalling period between the much more assertive papacies of Pius IV, Pius V, and later Sixtus V. Even A. Wright, whose studies have been vital in moving from the monographic study of single popes to the study of the early modern papacy, omits Gregory XIII from any list of significant post-Tridentine popes. His argument on the ecclesiastical primacy of the post-Tridentine papacy in Italy, however, is applicable to the Boncompagni papacy, and I follow Wright's argument in the following pages.

32 On the papal bull *In Coena Domini*, see *Dictionnaire de Droit Canonique*, vol. 12, 1132–6.

33 At times the situation between the pope and Catholic rulers was so tense that Gregory XIII inaugurated the new policy, followed also by later popes, of having the papal ambassadors, who enjoyed diplomatic immunity, handle ecclesiastical matters. He also increased the diplomatic body, adding new permanent nunciatures in Lucerne (1579), Graz (1580), Cologne (1584), and the Netherlands (1577) to the already-existing nunciatures in Venice, Turin, Florence, Madrid, Lisbon, Paris, and Prague (Borromeo 2000 and Borromeo 2002).

34 On the administration of the Papal States, see Prodi 1982; and Wright 2000, 231–69.

35 In the Duchy of Mantua the pope had the full right to nominate his bishops, although in exchange he had to grant to the Gonzaga the right of a private chapel in their ducal palace, a right that went against the Tridentine decrees. In Venice, the republic nominated the patriarch of Venice while Rome nominated the other bishops (Roman nominations tended to conform with Venetian expectations). Similarly, in Lombardy, the Spanish governor nominated one bishop while Rome chose the others, although they had to receive approval from the secular authorities, a fact that encouraged the Roman Curia to select them with an eye to Spanish interests. In Naples the Spanish viceroy nominated a handful of bishops, including in Sicily and Sardinia, while Rome kept the nomination rights for most of the others, including the archbishopric of Naples. Other seats, however, were controlled by the local nobility, who also had considerable rights on benefices. In Tuscany, the grand duke had nomination rights but had to abandon a proposal to restrict ecclesiastical mortmain. It was only for Savoyard and Piedmont seats where the pope had to accept the secular nomination.

36 Gregory XIII successfully appointed apostolic visitors in the dioceses in the Papal States as well as in the Spanish-dominated dioceses of Milan, Brescia, and Cremona; in the Venetian island of Cyprus; in the Venetian territories of Chioggia, Dalmatia, and Istria; in the Gonzaga dioceses of Mantua, Modena, Reggio, and Pescia; in the city of Florence and in most dioceses in Tuscany; in the Farnese territory of Piacenza; in the Spanish dominions in southern Italy at Benevento, Taranto, and other places; in Piedmont at Mondovi; and in the Genoa republic at Noli.

37 The Roman Inquisition operated differently in each state. In the Venetian Republic it was subject to supervision by secular authorities, but it operated uninterruptedly even in moments of great tension between Rome and Venice. In Milan the Roman Inquisition rather than the Spanish operated the tribunal. In Naples the establishment of both an ordinary tribunal and the Spanish Inquisition became impossible and a more complicated system took place based on ad hoc commissions, although the influence from Rome was always strong. Only the tiny Republic of Lucca maintained its own tribunal of faith, unconnected to the Congregation of the Holy Office in Rome. For an overview of the Roman Inquisition, see Romeo 2002 (with bibliography).

CHAPTER 9

1 On the idea of the Christian empire and its connection with Renaissance geography, see Headley 2000.

2 The literature on the Council of Trent is extensive, but for a brief overview and an extensive bibliography, see Jedin 1961; Iserloh–Gladik–Jedin 1980; Hsia 1998; O'Malley 2000; Wright 2000; and Romeo 2002.

3 The first edition of the revised Roman Catechism was *Catechismus ex decreto Concilii Tridentini ad parochos Pii V jussu editus* (Rome, 1566). On the Roman Catechism, see *Catholic Encyclopedia*; and Rodriguez 1982.

4 On the Breviary, see Baumer 1905–67, vol. 2, 160–250.

5 For a general assessment of Gregory XIII's policy see the following studies Pastor 1936–55, vols. 19 and 20; Borromeo 2000; and Borromeo 2002.

6 Vandeville had presented his work to Pius V in 1567, but it was Gregory XIII who used it; he asked the Jesuit Antonio Possevino to revise and summarize it. In fact, Vandeville's proposal is known only through Possevino's revised copy. On Vandeville's proposal, see Donnelly 1988.

7 The German College had been founded already in 1552, but Gregory XIII supported it with the donation of major funds

and a new building; in 1580 he unified it with the Hungarian College founded in 1578. The Greek College was founded in 1577, the English in 1578, and the Maronite in 1584. He also opened numerous seminaries in German lands: at Vienna (1574), Prague (1575), Graz (1578), Olmütz (1579), Klausenburg (1583), Fulda (1584), and Dillingen (1585).

8 The Gregorian Chapel in St. Peter's was also dedicated to the Virgin. Its decoration centered on two interrelated themes: the common roots of the Eastern and Western church and the Eucharist, another central concern of the Council of Trent and of the Boncompagni papacy. The chapel may have functioned as the sacrament chapel of the new basilica. On the Gregorian Chapel, see Rice 1997.

9 On the visit of the Japanese youth in Europe, see Brown 1994.

10 Gregory XIII's restoration of Pius IV's maps is recorded in an inscription located between the map of the British Isles and the map of Spain: "The decolored geographical maps that Pius IV had painted on this wall were reflourished by touching them with their own colors for the order of Gregory XIII Maximus Pontifex, in 1583, the twelfth year of his pontificate" [Geographicae Tabulae Quas Pius IV In Hoc Pariete Pingendas Curavit Decoloratae Gregorii XIII Pont. Max. Iussu Suis Coloribus Iterum Contactae Floruerunt Anno M.DLXXXIII. Pontif. Eius Anno XII.] On Miniato Pitti's remark on Pius IV's map cycle, see Chapter Two. On the Terza Loggia, see: Pastor 1936–55, vol. 20, 557–8. Specifically on the maps of the Terza Loggia, see Banfi 1951 and Almagià 1955. On the vault scenes, see Hess 1967a, vol. 1, 117–25; Redig de Campos 1967, 161–2, 169–74; and Pietrangeli 1992. On the restorations of the Terza Loggia, see Porena 1906, 278–80. Schulz 1990, 99–103, is the only scholar who considered the entire iconography of the Terza Loggia. Headley 2000 offers important insight on the understanding of the relation between Renaissance cartography and the post-Tridentine papacy. On cartography and religion, see also Livingstone 1988.

11 In his seminal study Almagiá 1955, 7–26, identified the cartographic sources of the west wing. Mercator's *Map of Europe* (1554) was the main source of Pius IV's map murals, although it was integrated with other printed maps, such as Lily's *Map of the British Isles* (1546), Thomas Geminus's *Map of Spain* (1555; second edition 1559), Iolivet's *Map of France* (1560), Gastaldi's *Map of Italy* (1561), Gastaldi's *Map of Germany* (1552), and Olaus Magnus's *Map of Northern Europe* (1539). Particularly important are the comments that Chytraeus Nathan wrote in his travel report, *Variorum in Europa itinerum deliciae* (Herborn, 1594), on Pius IV's maps, which he saw during his travels in 1565–7; he mentioned a map of "Islandia insula quam veteres Thulem nominavere," which does not exist anymore (Almagiá 1955, 2). Bragaglia Venuti 2002 analyzed the landscape scenes painted above Pius IV's maps.

12 Like in the Guardaroba Nuova, Danti used Gastaldi's *Map of Africa* (1564), which he integrated with Portuguese sources; Gastaldi's *Maps of Asia* (1559–1561), Jenkins's *Map of Muscovy* (1562), and Ortelius's *Map of the World* (1564). Danti's maps were heavily retouched in the following centuries. Their legends were rewritten in the nineteenth century, although their original version is documented in Agostino Taja's eighteenth-century guide to the Vatican Palace (Taja 1750, 229–343). Taja also reported the following titles for the maps, although it is uncertain whether they were the original ones: *Africa occidentalis*, *Africa orientalis*, *Turcicum extra Europam Dominium*, *Persarum Regnum*, *Utriusque Indiae regiones*, *Sinese Imperium*, *Tartaria Septemtrionalis*, *America*, *Hispania nova*, *Insulae iuxta primum meridianum*. On Danti's maps in the northern wing, see Banfi 1951; Almagiá 1955, 28–33; Porena 1888; Fiorini 1889 (on the world map); and Appendix C.

13 The original city views and plans were very ruined already in the eighteenth century, and Taja himself could identify only a few. He saw images of the cities of Senega, Fessa Nova, and Fessa Vetus below the map *Africa occidentalis*; "Cassumum reginae Candacis sedes" below the map *Africa orientalis*; and Alexandria and Damascus below the map of Turkey. He also reported that city views were painted under each map as well as on the opposite wall, between the openings of the loggia, but only traces of their colors were visible. The city views and personifications that we see today were painted in 1875, when Pius IX commissioned a major restoration to the painter Alessandro Mantovani, on which see Almagiá 1955, 2–6.

14 Danti's letter to Ortelius, 24 December 1580 (Ortelius 1887, letter no. 100: "un mappamondo in piano grande per due volte quello di V[ostra] Signoria.") These world maps are made with a conic projection, corrected in such a way so that the parallels are rectilinear rather than circular. Meridians and parallels are marked for each 15 degrees together with the equator, tropics, the polar circles, and the Zodiac. On these maps, see Fiorini 1889.

15 The inscriptions are quoted in Schulz 1990, 101.

16 Danti's role in the frieze illustrating the translation of Gregory of Nazianzus's body is suggested by a note from the Boncompagni archive: "the paintings in the fourth [loggia] where is depicted the translation of St Gregory of Nazianzus, and some cosmographical maps were ordered by the Most Reverend Bishop of Alatri" [le pitture] della quarta [loggia] dov'é depinta la translazione di San Gregorio Nazianseno, e d'alcune tavole di cosmografia [le ordinó] il R[everendissi]mo Vescovo d'Alatri." (BAV, MS Boncompagni, D 5, fol. 240 (Pastor 1936–55, vol. 20, 918; Pinelli 1994a, 71). On this frieze, see Gori 1868; Kroenig 1972; Pietrangeli 1979; and Valone 1993.

17 This painting has been detached from its original location in the Terza Loggia and is currently hung in the antecham-

ber of the Secretary of State (illustration in Pietrangeli 1992, 160).

18 The reform of the calendar is prominently remembered in a large inscription: "Gregory xiii Maximus Pontifex Pascha suo in perpetuum tempore celebratur rationem anni diuperturbatam modumque adhibuit quo futuris saeculis in preaescriptae definitionis statu sine confusione perseveret anno MDLXXXII Pont. Sui XI."

19 The saints in Paradise could have also been inspired by the Martyrology, the catalogue of saints which was republished in a revised edition during Gregory xiii's papacy (the new edition appeared only in 1584, but a papal commission was at work on it since at least 1580). On the Martyrology, see *Catholic Encyclopedia*; and Paschini 1923.

20 On the definition of the day of Easter and its early history, see Pederson 1983. On the Gregorian reform of the calendar, see: Coyne–Hoskin–Pederson 1983, especially the essay by Moyer, 171–88; Maiello 1989; and Heilbron 1999, 47–80.

21 Maiello 1994, 102–105.

22 Heilbron 1999, 13.

23 Pius v moved the golden numbers, the old system to relate the solar and the moon year, ahead five days, thus eliminating the delay of the lunar cycle on the calculation of Easter. On the relations between the Breviary, the ecclesiastical calendar, and secular methods of reckoning time, see Maiello 1994, 102–05.

24 Initially, Tommaso Gigli, Bishop of Sora and one of Gregory xiii's friend from Bologna, chaired the committee for the reform of the calendar, but his nomination as Bishop of Piacenza in 1576 forced him to leave the papal court, and Sirleto replaced him. Sirleto must have had more than a say in the selection of the other committee members, which included: his friend and fellow citizen Vincenzo Lauro, Bishop of Mondovi; the Spanish Pedro Chacon and the French Seraphim Olivier, representative of the Catholic powers of Europe that had to accept the reform; the Patriarch of Anthioch, Ignatius Na'mattalah, also an expert in mathematics, included in the hope that he would persuade the eastern church to embrace the reform; Giovan Battista Gabio, who acted as a translator from Latin and Italian into Greek for the patriarch, and later translated the papal bull into Greek; Antonio Giglio, who replaced his brother Luigi after his death; Egnazio Danti, and Christoph Clavius. On Luigi Giglio, see Moyer 1983; and Preti in *DBI*. Giglio's original proposal is lost and it is impossible to establish how exactly he proposed to reform the calendar. Thomas Settle has perceptively suggested that Giglio's proposal was finally accepted not only because it celebrated the Constantinian age, but also avoided a more complete and complex reform of astronomy (Settle 1992), "Egnazio Danti as a Builder of Gnomons," paper delivered at Villa I Tatti, Florence 1990.

25 Initially Clavius opted for the calculation of the vernal equinox at the time of Julius Caesar, a fact documented in the papers of the papal committee. However, once the decision was made to calculate the equinox at the time of Constantine, he became the strenuous defender of this choice, which was highly criticized especially in Protestant countries. On Christoph Clavius, see Baldini 1983; Clavius 1992; and Lattis 1994.

26 Danti 1576: "de eodem corrigendo, restituendoq. anno vehementer est solicitus." On this short treatise on the meridian line written by Danti in 1576, see Chapter Six. The report on the motion of trepidation among the papers of the papal commission for the reform of the calendar, which might be attributed to Danti, is in Clavius 1992, vol. 7, 39–46.

27 On Clavius's views on mathematics, see Wallace 1981, 136–41; Lattis 1994; Feldhay 1995, 162–70; Feldhay 1998, 90–100; and Wallace 1998, 32–6.

28 Tolesano's book has often been regarded as emblematic of how the Roman church reacted to Copernicus's theory, especially considering that it was read by the Dominican Tommaso Caccini in preparation for his accusation against Galileo in 1614. But it is equally significant that the book was not printed and that in fact it might have represented an extreme reaction to Copernicus's book, a book that after all was dedicated to the pope, and had been sponsored by another influential Dominican, Cardinal Nicholas Shönberg, who was also a high authority within the Dominican order. On Tolesani, see Garin 1973; Westman 1986; Rosen 1975; and Feldhay 1995, 205–08. On the education in Dominican and Jesuit seminars in the late sixteenth century, see Feldhay 1995, 73–200.

29 On Clavius's wish to introduce a more general reform of astronomy, see Clavius 1992, vol. 2, part 1, 24–5, and vol. 2, part 2, 19–23. On Danti's view on astronomical reform, see Righini Bonelli–Settle 1979.

30 Ciappi 1596, 99: "Inoltre [Gregorio xiii] usó ogni sera dopo che fu riformato il Cale[n]dario di far da due suoi Cappelani su le 23. hore recitare le litanie nella Cappella di Papa Nicola vicino la sala de gli Svizzeri, facendo chiamarvi tutte le genti di Palazzo, non impedite, al suono di campanella; & ció per render gratie á Dio . . . d'haver potuto condurre á perfettione un'opera cosí difficile, & importante."

31 No traces remain of this meridian line in the Gregorian Chapel, although Danti recalled it in his writings: "Quando gli instrumenti saranno grandi, & ben fatti ci mostreranno queste altezze, & li moti delli Pianeti, & delle Stelle fisse più precisamente, che non fanno i calcoli, il che ci ha fatto chiaro conoscere l'Ugnomone della Capella Gregoriana, & quello di S[an] Petronio fatti da noi, ove nel solstizio di bruma si conosce ogni giorno sensibilissimamente la varietá dell'altezza del Sole Meridiana, che da i minutissimi calcoli non se ne há varietá nessuna." (Danti 1583, 56).

32 Luther as quoted in Delano Smith–Morley Ingram 1991, XVI. On maps as bible illustrations, see: Delano Smith 1989;

Delano Smith 1991; and Delano Smith–Morley Ingram 1991. Additionally, Protestants made wall maps showing the Holy Land and the Mediterranean; these maps were autonomous representations that included the necessary elements (verbal and visual, geographical and historical), allowing them to be independent of any particular text. Many bible wall-maps provide abundant internal evidence for their relevance to the Protestant cause. See, for example: W. Wissenburg, *Descriptio Palaestinae Nova*, 1538; T. Stella, *Palestinae sive Totius Terrae Promissionis Nova Descriptio*, 1552; T. Stella, *Itinera Israelitarum ex Aegipto*, 1557 (all reproduced in Nebenzahl 1986).

33 For a list of polyglot and Latin Bibles with maps, see: Delano Smith–Morley Ingram 1991, 132–5. Like Protestants, Catholics also used maps to illustrate books on sacred history and the peregrinations of Jesus and his apostles, and made loose maps of the Holy Land intended for the understanding of religious writings. Typical Catholic maps of the Holy Land and the Mediterranean are the maps *Nova Descriptio Amplissimae Santae Terrae* (1570) and *Peregrinatio Filiorum Dei* (1572) by Christian Sgrooten (both reproduced in Nebenzahl 1986). On Montanus's bible maps, see Shalev 2003. Montanus's maps were also used for a mural map cycle in the library of the convent of San Giovanni Evangelista in Parma, on which see Madonna 1979, *Biblioteca monumentale* 1999; Fiorani 2005 (forthcoming).

34 Loyola 1923, 77 and 85.

35 On Matteo Ricci, see Spence 1984 and Mangani 2003.

36 On Ortelius's sacred cartography, see Mangani 1998a; Mangani 1998b; Melion 1999; Nuti 2003; and Cosgrove 2003b.

Bibliography

Acidini Luchinat 1980a: Cristina Acidini Luchinat, "Trattato di architettura di Gherardo Spini," in Franco Borsi ed., *Il disegno interrotto. Trattati medicei d'architettura*, Florence, 13–201.

Acidini Luchinat 1980b: Cristina Acidini Luchinat, "Niccolò Gaddi collezionista e dilettante del Cinquecento," *Paragone*, cccLIX–cccLXI, 141–75.

Acidini Luchinat 1997: Cristina Acidini Luchinat, ed., *Treasures of Florence. The Medici Collection 1400–1700*, Munich and New York.

Ackerman 1954: James S. Ackerman, *The Cortile of the Belvedere*, Studi e documenti per la storia del Palazzo Vaticano, no. 3, Vatican City.

Adriani 1583: Giovambattista Adriani, *Istoria de' suoi tempi*, Florence.

Affó 1780: Ireneo Affó, *Vita del Cavaliere Bernardino Marliani mantovano*, Parma.

Aiello 1988: V. Aiello, "Aspetti del mito di Costantino in occidente: dalla celebrazione agiografica alla esaltazione epica," *Università di Macerata. Annali della Facoltà di Lettere e Filosofia*, no. 21, 87–116.

Akerman 1995: James R. Akerman, "From Books with Maps to Books as Maps: The Editor in the Creation of the Atlas Idea," in Joan Winearls, ed., *Editing Early and Historical Atlases*, Papers Given at the Twenty-Ninth Annual Conference on Editorial Problems, University of Toronto, 5–6 November 1993, Toronto, Buffalo, and London, 3–48.

Akerman 1998: James R. Akerman, "Atlas, Birth of a Title," in Watelet, 1998, 15–29.

Alberti 1577: Leandro Alberti, *Descrittione di tutta l'Italia & isole pertinenti ad essa*, Bologna. [first edition: Bologna, 1550].

Alberti [1485] 1966: Leon Battista Alberti, *De re aedificatoria*, ed. Paolo Portoghesi, Rome.

Aldrovandi 1907: Ulisse Aldrovandi, "La vita di Ulisse Aldrovandi cominciando dalla sua nativitá sin'á l'etá di 64 anni vivendo ancora," in Ludovico Frati, ed., *Studi intorno alla vita e alle opere di Ulisse Aldrovandi*, Bologna.

Allegri–Cecchi 1980: Ettore Allegri and Alessandro Cecchi, *Palazzo Vecchio e i Medici*, Florence.

Almagià 1922: Roberto Almagià, *"L'Italia" di Giovanni Antonio Magini e la cartografia dell'Italia nei secoli xvi e xvii*, Naples.

Almagià 1929: Roberto Almagià, *Monumenta Italiae cartographicae*, Florence.

Almagià 1942: Roberto Almagià, *L'opera geografica di Luca Holstenio*, Studi e testi, no. 102, Vatican City.

Almagià 1948: Roberto Almagià, *Carte geografiche a stampa di particolare pregio o raritá*, Monumenta cartographica vaticana, vol. 2, Vatican City.

Almagià 1952: Roberto Almagià, *Le pitture murali della Galleria delle Carte Geografiche*, Monumenta cartographica vaticana, vol. 3, Vatican City.

Almagià 1955: Roberto Almagià, *Le pitture geografiche murali della Terza Loggia e di altre sale vaticane*, Monumenta cartographica vaticana, vol. 4, Vatican City.

Almagià 1961: Roberto Almagià, *Scritti geografici*, Rome.

Alpers 1983: Svetlana Alpers, *The Art of Describing: Dutch Art in the Seventeenth Century*, Chicago.

Ambrosini 1981: Federica Ambrosini, "'Descrittioni del Mondo' nelle case venete dei secoli XVI e XVII," *Archivio Veneto*, V, 67–79.

Apian 1533: Peter Apian, *Liber cosmographicus*, Basel.

Aquinas 1922: Thomas Aquinas, *The Summa Theologica of St Thomas Aquinas, Part II (Second Part)*, QQ. LXXX–C, London.

Armenini [1587] 1972: Giovan Battista Armenini, *De' veri precetti della pittura*, Ravenna (facsimile edition: Bologna).

Arrighi 1963: Gino Arrighi, "Note di 'Arithmetica Speculativa' (con una lettera del P. Egnazio Danti)," *Physis*, no. 5, 464–73.

Aujac 1966: Germaine Aujac, *Strabon et la science de son temps*, Paris.

Ayers 1985: John Ayers, "The Early China Trade," in Impey–MacGregor 1985, 259–66.

Babicz 1994: Jozef Babicz, "La resurgence de Ptolemée," Watelet 1994, Brussels, 50–68.

Baccini 1888: G. Baccini, "Un'opera inedita del P. Ignazio Danti da Perugia vescovo di Alatri," *Archivio storico per le Marche e l'Umbria*, IV, nos. 13–14, 82–112.

Baglione 1642: Giovanni Baglione, *Le vite de' pittori, scultori et architetti. Dal pontificato di Gregorio XIII del 1572 in fino a' tempi di papa Urbano Ottavo nel 1642*, Rome.

Baigrie 1992: Brian Baigrie, ed., *Picturing Knowledge: Historical and Philosophical Problems Concerning the Use of Art in Science*, Toronto.

Baldini 1983: Ugo Baldini, "Christoph Clavius and the Scientific Scene in Rome," in Coyne–Hoskin–Pederson 1983, 137–69.

Balsiger 1970: B. J. Balsiger, "The Kunst- und Wunderkammern: A Catalogue raisonné of Collecting in Germany, France and England, 1565–1750," Ph.D. diss., University of Pittsburg.

Banfi 1951: Florio Banfi, "The Cosmographic Loggia of the Vatican Palace," *Imago Mundi*, IX, 23–34.

Barbaro 1568: Daniele Barbaro, *La pratica della perspettiva*, Venice.

Barber 1997: Peter M. Barber, "Maps and Monarchs in Europe 1550–1800," in *Royal and Republican Sovereignty in Early Modern Europe: Essays in Memory of Ragnhild Hatton*, ed. Robert Oresko, G. C. Gibbs and H. M. Scott, London–New York, 75–124.

Barber 1998: Peter M. Barber, "The British Isles," in Watelet 1998, 43–77.

Barocchi 1961: Paola Barocchi, *Trattati d'arte del Cinquecento*, 3 vols., Bari.

Barocchi 1980: Paola Barocchi, "Storiografia e museografia in una mostra medicea," *Bollettino d'arte*, LXV, no. 8, 121–8.

Barocchi–Gaeta Bertelá 1993: Paola Barocchi and Giovanna Gaeta Bertelà, *Collezionismo mediceo: Cosimo I, Francesco I. Documenti, 1540–1587*, Modena.

Barocchi–Gaeta Bertelà 2002: Paola Barocchi and Giovanna Gaeta Bertelà, *Collezionismo Mediceo e storia artistica*, 2 vols., Florence.

Barolsky 1991: Paul Barolsky, *Why Mona Lisa Smiles and Other Stories by Vasari*, University Park, Pa.

Baronio 1588: Cesare Baronio, *Martyrologium Romanum*, Rome.

Baronio 1624: Cesare Baronio, *Annales ecclesiastici*, 12 vols., Cologne.

Baronio e l'arte 1985: *Baronio e l'arte. Atti del convegno internazionale di studi, Sora 10–13 ottobre 1984*, ed. Romeo De Maio, Sora.

Baronio storico 1982: *Baronio storico e la controriforma. Atti del convegno internazionale di studi, Sora, 6–10 ottobre 1979*, ed. Romeo De Maio, Sora.

Bartoli 1564: Cosimo Bartoli, *Del modo di misurare le distanze, le superficie, i corpi, le piante, le provincie, le prospettive, e tutte le altre cose terrene, che possono occorrere à gli huomini, secondo le vere regole d'Euclide e de gli altri più lodati scrittori*, Venice.

Barzman 1989: Karen-edis Barzman, "The Florentine *Accademia del Disegno*: Liberal Education and the Renaissance Artist," in *Academies of Art between Renaissance and Romanticism. Leids Kunsthistorish Jaarboek*, vols. 5–6, ed. Anton Boschloo, Leiden, 14–32.

Barzman 2000: Karen-edis Barzman, *The Florentine Academy and the Early Modern State: The Discipline of Disegno*, Cambridge.

Bassani 1975: Ezio Bassani, "Antichi avori africani nelle collezioni medicee," *Critica d'Arte*, CXLIII, 69–80; and CXLIV, 8–23.

Bassani 1980: Ezio Bassani, "Il collezionismo esotico dei Medici del Cinquecento," in *Le arti del principato mediceo*, Florence, 55–72.

Battisti 1999: Simona Battisti, "Gregorio XIII e la Galleria delle Carte Geografiche," *Ricerche di Storia dell'Arte*, LXVII, 70–80.

Bäumer [1904] 1967: Suitbert Bäumer, *Histoire du bréviaire. Traduction française mise au courant des derniers travaux sur la question par Réginald Biron*, Rome.

Bedini 1995: Silvio Bedini, "The Fate of the Medici-Lorraine Scientific Instruments," *Journal of the History of Collecting*, VII, no. 2, 159–78.

Bellini 1991: Paolo Bellini, *L'opera incisa di Adamo e Diana Scultori*, Vicenza.

Belloni–Drusi 2002: Gino Belloni and Riccardo Drusi, eds., *Vincenzio Borghini. Filologia e invenzione nella Firenze di Cosimo I*, exhibition catalogue, Florence, Biblioteca Nazionale Centrale (Florence).

Bergin 1999: Joseph Bergin, "The Counter-Reformation Church and its Bishops," in *Past and Present*, 165, 30–73.

Berti 1967: Luciano Berti, *Il principe dello Studiolo e la fine del Rinascimento fiorentino*, Florence.

Bertolotti [1886] 1965: Antonio Bertolotti, "Artisti bolognesi, ferraresi ed alcuni altri del già Stato Pontificio in Roma nei secoli XV, XVI e XVII," in *Studi e ricerche tratti dagli archivi romani* (facsimile edition: Bologna).

Biagioli 1989: Mario Biagioli, "The Social Status of Italian Mathematicians, 1450–1600," *History of Science*, XXVII, no. 1, 41–95.

Biagioli 1993: Mario Biagioli, *Galileo Courtier: The Practice of Science in the Culture of Absolutism*, Chicago.

Bianconi–Vassalli 1983: L. Bianconi and A. Vassalli, "Circolazione letteraria e circolazione musicale del madrigale: Il caso di Giovan Battista Strozzi," in Gian Carlo Garfagnini, ed., *Firenze e la Toscana dei Medici nell'Europa del 500*, 3 vols., Florence, vol. 2, 439–55.

Biasutti 1920: R. Biasutti, "La carta dell'Africa di G. Gastaldi (1545–1563) e lo sviluppo della cartografia africana nei secoli XVI e XVII," *Bollettino della Società geografica italiana*, CCCXXVII-CCCXLVI, 387–436.

Bible 1970a: *The New English Bible: The Old Testament*, Oxford and Cambridge.

Bible 1970b: *The New English Bible: The New Testament*, Oxford and Cambridge.

Biblioteca monumentale 1999: *La biblioteca monumentale dell'Abbazia di San Giovanni Evangelista a Parma*,Parma.

Biondo 1453: Flavio Biondo, *Roma restaurata et Italia illustrata*, Rome.

Black 1997: Jeremy Black, *Maps and History: Constructing Images of the Past*, New Haven and London.

Blair 1997: Ann Blair, *The Theater of Nature: Jean Bodin and Renaissance Science*, Princeton.

Boas Hall 1982: Marie Boas Hall, "Renaissance Science and Professionalisation," *Annali dell'Istituto e Museo di storia della scienza di Firenze*, VII, no. 2, 53–64.

Boccazzi 1987: Gaudenzio Boccazzi, "La curiosité du voyageur au XVIe siècle, ou l'art d'apprendre et de se parfaire par les voyages," in *Voyager à la Renaissance. Actes du colloque de Tours 1983*, ed. Jean Céard and Jean Claude Margolin, Paris, 49–64.

Böck 1997: Angela Böck, *Die Sala Regia im Vatikan als Beispiele der Selbdarstellung des Papsttums in der zweiten Häfte des 16. Jahrhunderts*, Hildesheim, Zurich, and New York.

Bolkmann 1777: D. J. J. Bolkmann, *Historisch-kritische Nachrichten von Italien*, Leipzig.

Bolzoni 1985: Lina Bolzoni, "L'invenzione dello Stanzino di Francesco I," in *Le arti del Principato mediceo*, Florence, 255–99.

Bolzoni 1995: Lina Bolzoni, *La stanza della memoria. Modelli letterari e iconografici nell'etá della stampa*, Turin.

Bolzoni 2001: Lina Bolzoni, *The Gallery of Memory: Literary and Iconographic Models in the Age of the Printing Press*, Toronto.

Bonazzi 1959: Luigi Bonazzi, *Storia di Perugia dalle origini al 1860*, Città di Castello.

Borghini 1584: Raffaele Borghini, *Il Riposo*, Florence.

Borromeo 2000: Agostino Borromeo, "Gregorio XIII," *Enciclopedia dei Papi*, Rome.

Borromeo: 2002: Agostino Borromeo, "Gregorio XIII," in *DBI*, vol. 59, 204–19.

Borromeo 2000: Charles Borromeo, *Instructionum fabricae et Supellectilis ecclesiasticae libri II: Caroli Borromei*, ed. and trans., Massimo Marinelli, Vatican City.

Bossy 1983: John Bossy, "The Mass as a Social Institution, 1200–1700," *Past and Present*, C, 29–61.

Bragaglia Venuti 2002: Cristina Bragaglia Venuti, "Etienne Dupérac e i paesaggi della Loggia di Pio IV", *Rivista dell'Istituto Nazionale di Archeologia e Storia dell'Arte*, 57, 279–310.

Bredekamp 1995: Horst Bredekamp, *The Lure of Antiquity and the Cult of the Machine*, Princeton.

Brink 1983: Sonia Brink, "Fra Egnatio Danti. Das Program der Sala Vecchia degli Svizzeri im Vatikan und C. Ripas 'Iconologia'," *Mitteilungen des Kunsthistorischen Institutes in Florenz*, XXVII, no. 2, 223–54.

Brivio 1985: Ernesto Brivio, ed., *Itinerari di san Carlo Borromeo nella cartografia delle visite pastorali*, Milan.

Broc 1986: Numa Broc, *La geografia del Rinascimento*, Modena.

Brosses 1885: Charles De Brosses, *Lettres familiers écrites d'Italie en 1739 et 1740*, 2 vols., Paris.

Brotton 1997: Jerry Brotton, *Trading Territories: Mapping the Early Modern World*, London.

Brotton 1999: Jerry Brotton, "Terrestrial Globalism: Mapping the Globe in Early Modern Europe," in Cosgrove 1999, 71–89.

Brown 1994: Judith C. Brown, "Courtiers and Christians: The First Japanese Emissaries to Europe," *Renaissance Quarterly*, vol. 47, no. 4, 872–906.

Brown 1979: Lloyd A. Brown, *The Story of Maps*, New York (first published: Boston, 1949).

Bryce 1980: Judith Bryce, "Cosimo Bartoli's 'Del modo di misurare le distanze' (1564): A Reappraisal of His Sources," *Annali dell'Istituto e Museo di storia della scienza di Firenze*, v, no. 1, 19–34.

Bryce 1983: Judith Bryce, *Cosimo Bartoli (1503–1572): The Career of a Florentine Polymath*, Geneva.

Bucciantini–Torrini 1997: Massimo Bucciantini and Maurizio Torrini, eds., *La diffusione del copernicanesimo in Italia, 1543–1610*, Florence.

Burckhardt 1992: Jacob Burckhardt, *L'arte italiana del Rinascimento*, vol. 2, *Pittura: I generi*, ed. Maurizio Ghelardi, Venice.

Burden 1996: Philip D. Burden, *The Mapping of North America: A List of Printed Maps, 1511–1670*, Rickmansworth.

Burke 1987: Peter Burke, *The Historical Anthropology of Early Modern Italy*, Cambridge and New York.

Burke 1999: Peter Burke, "Presenting and Re-presenting Charles v," in *Charles v 1500–1558 and His Time*, eds. Hogo Soly and Wim Blockmans, Antwerp, 393–475.

Burmeister 1964: Karl Heinz Burmeister, *Sebastian Münster. Eine Bibliographie mit 22 Abbildungen*, Wiesbaden.

Butters 2002: Suzanne B. Butters, "Ferdinando de' Medici and the Art of the Possible," in *Medici and Michelangelo*, 2002, 66–75.

Buzzi–Zardin 1997: Franco Buzzi and Danilo Zardin, *Carlo Borromeo e l'opera della "grande riforma." Cultura, religione e arti del governo nella Milano del pieno Cinquecento*, Milan.

Camerota 1999: Filippo Camerota, "Il disegno del Territorio e la difesa dello stato. I rilievi di Vincenzo Civitali per una mappa delle fortificazioni lucchesi," *Nuncius: Annali di Storia della Scienza*, 14, 2, 455–70.

Camillo [1550] 1991: Giulio Camillo Delminio, *L'idea del teatro*, Florence (first edition: Florence; facsimile edition: ed. Lina Bolzoni: Palermo 1991).

Campbell 1987: Tony Campbell, *The Earliest Printed Maps*, London.

Carcerieri 1926–9: Luigi Carcerieri, *Cosimo Primo Granduca*, 3 vols., Verona.

Carruthers 1990: Mary Carruthers, *The Book of Memory: A Study of Memory in Medieval Culture*, Cambridge and New York.

Carruthers 1998: Mary Carrruthers, *The Craft of Thought: Meditation, Rhetoric, and the Making of Images, 400–1200*, Cambridge and London.

Casey 1998: Edward Casey, *The Fate of Place: A Philosophical History*, Berkeley, Los Angeles, and London.

Casey 2002: Edward Casey, *Representing Place: Landscape Painting and Maps*, Bloomington, Ind.

Catholic Encyclopedia: *New Catholic Encyclopedia*, 2nd edn., 15 vols., Detroit and Washington D.C., 2003.

Cavanaugh 2001: William T. Cavanaugh, "Eucharistic Sacrifice and the Social Imagination in Early Modern Europe," *Journal of Medieval and Early Modern Studies*, xxxi, no. 3, 585–605.

Cecchi 1977: Alessandro Cecchi, "Pratica, fierezza e terribilità nelle grottesche di Marco da Faenza in Palazzo Vecchio a Firenze," *Paragone*, cccxxvii, 24–54; cccxxix, 6–26.

Cecchini 1978: Giovanni Cecchini, *La Biblioteca Augusta del Comune di Perugia*, Rome.

Cecha Cremades 1999: Fernando Checa Cremades, "The Image of Charles v," in Hugo Soly and W. Blockmans, ed. *Charles v 1500–1558 and His Time*, Antwerp, 477–500.

Cellini 2002: Benvenuto Cellini, *My Life: A New Translation by Julia Conoway Bondanella and Peter Bondanella*, Oxford.

Chandler 1978: Kirwin W. Chandler, "The Life and Drawing Style of Cristoforo Roncalli," *Paragone*, cccxxxv, 18–62.

Cheney 1989: Iris Cheney, "The Galleria delle Carte Geografiche at the Vatican and the Roman Church's View of the History of Christianity," *Renaissance Papers*, 21–37.

Chiusa 2001: Maria Cristina Chiusa, *Parmigianino*, Milan.

Ciappi 1596: Marcantonio Ciappi, *Compendio delle heroiche et gloriose attioni et santa vita di Papa Gregorio xiii*, Rome.

Cieri Via 1988: Claudia Cieri Via, "'Galaria sive loggia': modelli storici e funzionali fra collezionismo e ricerca," in Prinz 1980, vii–xxiii.

Cipriani 1975: Giovanni Cipriani, "Il mito etrusco nella Firenze repubblicana e medicea nei secoli xv e xvi," *Ricerche storiche*, v, 257–309.

Cipriani 1980a: Giovanni Cipriani, "Ideologia politica e 'revival' etrusco," in *Le arti del principato mediceo*, Florence, 9–17.

Cipriani 1980b: Giovanni Cipriani, *Il mito etrusco nel Rinascimento fiorentino*, Florence.

Cirni 1560: Antonio Francesco Cirni, *La reale entrata dell'Ecc.mo Signor Duca e Duchessa di Fiorenza in Siena con la significazione delle latine inscrittioni e con alcuni sonetti*, Rome.

Citolini 1561: Alessandro Citolini, *La Tipocosmia*, Venice.

Clarke 1999: Katherine Clarke, *Between Geography and*

History: Hellenistic Constructions of the Roman World, Oxford.

Clavius 1992: Christoph Clavius, *Corrispondenza*, ed. Ugo Baldini and Pier Daniele Napolitani, 7 vols., Pisa.

Cochrane 1973: Eric Cochrane, *Florence in the Forgotten Centuries, 1527–1800*, Chicago.

Cochrane 1981: Eric Cochrane, *Historians and Historiography in the Italian Renaissance*, Chicago.

Codazzi 1945: Angela Codazzi, *Le edizioni quattrocentesche e cinquecentesche della Geografia di Tolomeo*, Milan and Venice.

Cohen–Drabkin 1958: Morris R. Cohen and I. E. Drabkin, *A Source Book in Greek Science*, Cambridge, Mass.

Coleman 1914: Christopher B. Coleman, *Constantine the Great and Christianity; Three Phases: The Historical, the Legendary, and the Spurious*, New York.

Coleman [1922] 1993: Christopher B. Coleman, *The Treatise of Lorenzo Valla on the Donation of Constantine* (reprint of the first edition: Toronto, Buffalo, and London).

Collaretta 1980: Mario Collaretta, "Review of Allegri–Cecchi", *Bollettino d'Arte*, VIII, 118.

Colonna 1955: Gustavo Brigante Colonna, *Gli Orsini*, Milan.

Comelli 1914: Giovan Battista Comelli, *Piante e vedute della città di Bologna*, Bologna.

Conti 1893: Cosimo Conti, *La prima reggia di Cosimo I de' Medici*, Florence.

Cora–Fanfani 1986: Galeazzo Cora and Angiolo Fanfani, *La porcellana dei Medici*, Milan.

Cormack 1997: Leslie B. Cormack, *Charting an Empire: Geography at the English Universities, 1580–1620*, Chicago and London.

Corte mare mercanti 1980: *La corte il mare i mercanti. La rinascita della scienza, Editoria e società, Astrologia, magia e alchimia*, exhibition catalogue, Orsanmichele, (Florence).

Cortesao–Teixeira da Mota 1960–2: Armando Cortesao and Avelino Teixeira da Mota, *Portugaliae Monumenta cartographica*, Lisbon.

Cortesi 1980: K. W. Garris and J. F. D'Amico, "The Reniassance Cardinal's Ideal Palace: A Chapter from Cortesi *De Cardinalatu*," in *Studies in Italian Art and Architecture: Fifteenth through Eighteenth Centuries*, Rome, 45–119.

Cosgrove 1992: Denis E. Cosgrove, "Mapping New Worlds: Culture and Cartography in Sixteenth-Century Venice," *Imago Mundi*, XLIV, 65–89.

Cosgrove 1993: Denis E. Cosgrove, *The Palladian Landscape: Geographical Change and its Cultural Representation in Sixteenth-Century Italy*, Leicester and London.

Cosgrove 1999: Denis E. Cosgrove, ed., *Mappings*, London.

Cosgrove 2001: Denis E. Cosgrove, *Apollo's Eye: A Cartographic Genealogy of the Earth in the Western Imagination*, Baltimore.

Cosgrove 2003a: Denis E. Cosgrove, "Ptolemy and Vitruvius: Spatial Representation in Sixteenth-Century Texts and Commentaries," in *Architecture and the Sciences: Exchanging Metaphors*, ed. Antoine Picon and Alessandra Ponte, Princeton, 20–51.

Cosgrove 2003b: Denis E. Cosgrove, "Globalism and Tolerance in Early Modern Geography," *Annals of the Association of American Geographers*, XCIII, 852–70.

Costantino il Grande 1992: *Costantino il Grande dall'antichità all'umanesimo*, ed. Giorgio Bonamente and Franca Fusco, Macerata.

Courtright 1990: Nicola Courtright, "The Vatican Tower of the Winds and the Architectural Legacy of the Counter-Reformation," *IL 60: Essays Honoring Irving Lavin on His Sixtieth Birthday*, New York, 117–44.

Courtright 1995: Nicola Courtright, "The Transformation of Ancient Landscape Through the Ideology of Christian Reform in Gregory XIII's Tower of the Winds," *Zeitschrift für Kunstgeschichte*, LVIII, no. 4, 526–41.

Courtright 1997: Nicola Courtright, "Galleria delle carte geografiche in Vaticano," *AB*, LXXIX, no. 1, 156–60.

Courtright 2003: Nicola Courtright, *The Papacy and the Art of Reform in Sixteenth-Century Rome: Gregory XIII's Tower of the Winds in the Vatican*, Cambridge and London.

Couzinet 1996: Marie-Dominique Couzinet, *Histoire et Méthode à la Renaissance. Une Lecture de la "Methodus ad facilem historiarum cognitionem" de Jean Bodin*, Paris.

Cox Rearick 1984: Janet Cox Rearick, *Dynasty and Destiny in Medici Art: Pontormo, Leo X and the Two Cosimos*, Princeton.

Cox Rearick 1993: Janet Cox Rearick, *The Chapel of Eleonora*, Princeton.

Cox Rearick 2002: "Art at the Court of Duke Cosimo I de' Medici (1537–1574)," in *Medici and Michelangelo* 2002, 35–46.

Coyne–Hoskin–Pederson 1983: G. V. Coyne, M. A. Hoskin and O. Pederson, eds., *Gregorian Reform of the Calendar: Proceedings of the Vatican Conference to Commemorate its 400th Anniversary, 1582–1982*, Vatican City.

Crinò 1931: Sebastiano Crinò, *Portolani manoscritti e carte da navigare compilati per la marina medicea. I. I portolani di Bernardo Baroncelli*, Rome.

Cristofani 1979: Mario Cristofani, "Per una storia del collezionismo archeologico nella Toscana Granducale. I: I grandi bronzi," *Prospettiva*, 4–23.

Cristofani 1985: Mario Cristofani, "Per una storia del collezionismo archeologico nella Toscana granducale. Doni e acquisti di statue antiche nella seconda metà del XVI secolo," in *Le arti del principato mediceo*, Florence, 19–30.

Crone 1978: Gerald Roe Crone, *Maps and Their Makers: An Introduction to the History of Cartography*, Hamden.

Crum 1989: Roger J. Crum, "Cosmos: The World of Cosimo," *AB*, LXXI, no. 3, 237–53.

Curatola–Spallanzani 1981: Giovanni Curatola and Marco Spallanzani, *Metalli islamici nelle collezioni granducali*, Florence.

Dacos 1964: Nicole Dacos, *Les peintres belges à Rome au XVIe siècle*, Bruxelles.

Dacos 1986: Nicole Dacos, *Le Logge di Raffaello. Maestro e bottega di fronte all'antico*, Rome (second edition).

Dacos–Giuliano–Pannuti 1972: Nicole Dacos, Antonio Giuliano, and Ulrico Pannuti, eds., *Il tesoro di Lorenzo il Magnifico. Le gemme*, Florence.

Daly Davis 1982: Margaret Daly Davis, "Beyond the 'Primo Libro' of Vincenzo Danti's 'Trattato delle Perfette Proporzioni'," *Mitteilungen des Kunsthistorischen Institutes in Florenz*, XXVI, no. 1, 63–89.

Daly Davis 1985: Margaret Daly Davis, "La Galleria di sculture antiche di Cosimo I a Palazzo Pitti," in *Le arti del principato mediceo*, Florence, 31–54.

D'Amico 1983: John F. D'Amico, *Papal Humanism in Renaissance Rome: Humanists and Churchmen on the Eve of the Reformation*, Baltimore.

Danti 1567: Vincenzo Danti, *Il Primo Libro del Trattato delle perfette proporzioni*, Florence.

Dandelet 2001: Thomas J. Dandelet, *Spanish Rome, 1500–1700*, New Haven and London.

Danti 1569: Egnazio Danti, *Trattato dell'uso et della fabbrica dell'astrolabio con la giunta del planisferio del Rojas*, Florence.

Danti 1571a: Egnazio Danti, *La sfera di messer Giovanni Sacrobosco tradotta, emendata et distinta in capitoli da Piervincentio Dante de' Rinaldi, con molte et utili annotazioni del medesimo, rivista da Frate Egnatio Danti Cosmografo del Granduca di Toscana*, Florence.

Danti 1571b: Egnazio Danti, *Trattato della sfera ridotto in cinque tavole*, Florence.

Danti 1573a: Egnazio Danti, *La sfera di Proclo Liceo tradotta da maestro Egnatio Danti Cosmografo del serenissimo Gran Duca di Toscana, con le annotazioni, et con l'uso della sfera del medesimo*, Florence.

Danti 1573b: Egnazio Danti, *La prospettiva di Euclide nella quale si tratta di quelle cose che per raggi diritti si veggono, et di quelle che con raggi riflessi nelli specchi appariscono. Tradotta dal R.P.M. Egnatio Danti Cosmografo del serenissimo Gran Duca di Toscana con alcune sue annotazioni de' luoghi più importanti. Insieme con la prospettiva di Eliodoro Larisseo cavata dalla Libreria Vaticana, e tradotta dal medesimo novamente data in luce*, Florence.

Danti 1574: Egnazio Danti, *La sfera di messer Giovanni Sacrobosco tradotta, emendata et distinta in capitoli da Piervincentio Dante de' Rinaldi, con molte et utili annotazioni del medesimo, rivista da Frate Egnatio Danti Cosmografo del Granduca di Toscana*, 2nd edn., Florence.

Danti 1576: Egnazio Danti, *Usus et tractatio gnomonis magni quem Bononiae ipse in Divi Petronii templo ex illustrium Senatorum comitis Joannis Pepuli perpetui illius fabricae praesidis, et collegarum autorictate*, Bologna.

Danti 1577: Egnazio Danti, *Le scienze matematiche ridotte in tavole*, Bologna.

Danti 1578a: Egnazio Danti, *Trattato dell'uso et della fabbrica dell'astrolabio con la giunta del planisferio del Rojas*, 2nd edn., Florence.

Danti 1578b: Egnazio Danti, *Anemographia M. Egnatii Dantis Mathematicarum artium in almo bononiensis gymnasio Professoris in Anemoscopium verticale instrumentum ostensorem ventorum. His accessit ipsius instrumenti constructio ut nihl in hac materia amplius desideretur*, Bologna.

Danti 1583: Egnazio Danti, *Trattato del radio latino inventato dall'Ill.mo et Ecc.mo Signor Latino Orsini. con li commentari del R.P.M. Egnatio Danti da esso di nuovo ricorretto et ampliato con molte nuove operationi*, Rome.

Danti 1585: Egnazio Danti, *Breve narrazione della traslazione del Corpo di S. Sisto I Papa e martire alla cattedrale di Alatri nell'anno 1132 e dopo 452 anni ritrovato nell'istessa chiesa nell'anno 1584 scritta per ordine di Mons. Egnazio Danti*, Rome.

Danti 1586: Egnazio Danti, *Trattato del radio latino inventato dall'Ill.mo et Ecc.mo Signor Latino Orsini. con li commentari del R.P.M. Egnatio Danti da esso di nuovo ricorretto et ampliato con molte nuove operationi*, 2nd edn., Rome.

Danti P. 1571: Piervincento Danti, *La Sfera di messer Giovanni Sacrobosco*, Florence.

Danti-Bonsignori 1995: *Le tavole geografiche della Guardaroba Medicea di Palazzo Vecchio in Firenze*, ed. Gemmarosa Levi-Donati, Perugia.

Davidson 1985: Bernice F. Davidson, *Raphael's Bible: A*

Study of the Vatican Logge, University Park, Pa. and London.

Davis 1985a: Charles Davis, "Frescos by Vasari for Sforza Almeni, 'Coppiere' to Duke Cosimo I," *Mitteilungen des Kunsthistorischen Institutes in Florenz*, XXIX, no. 2, 127–202.

Davis 1985b: Charles Davis, "Working for Vasari. Vincenzo Danti in Palazzo Vecchio," in *Giorgio Vasari tra decorazione ambientale e storiografia artistica*, ed. Gian Carlo Garfagnini, Florence, 205–71.

Dee [1570] 1975: John Dee, *The Mathematical Preface to the Elements of Geometrie of Euclid of Megara.* ed. Allen G. Debus, New York, 1975.

De Grazia 1984: Diane De Grazia, *Prints and Related Drawings by the Carracci Family*, Bologna.

Delano Smith 1985: Catherine Delano Smith, "Cartographic Signs on European Maps and their Explanation Before 1700," *Imago Mundi*, XXXVII, 9–29.

Delano Smith 1989: Catherine Delano Smith, "Maps as Art *and* Science: Maps in Sixteenth-Century Bibles", *Imago Mundi*, XLII, 65–83.

Delano Smith 1991: Catherine Delano Smith, "Geography or Christianity? Maps of the Holy Land before AD 1000," *Journal of Theological Studies*, XLII, no. 1, 143–52.

Delano Smith–Morley Ingram 1991: Catherine Delano Smith and Elizabeth Morley Ingram, *Maps in Bibles 1500–1600: An Illustrated Catalogue*, Geneva.

Del Badia 1881: Iodico Del Badia, *Egnazio Danti cosmografo e matematico e le sue opere in Firenze*, Florence.

Delumeau 1957: Jean Delumeau, *Vie économique et sociale de Rome dans la séconde moitié du XVIe siècle*, Paris.

Dempsey 1977: Charles Dempsey, *Annibale Carracci and the Beginnings of Baroque Style*, Gluckstadt.

Denucé 1916: Jean Denucé, *Correspondance de Christophe Plantin vol. 6*, Antwerp.

De Pace 1993: Anna De Pace, *Le matematiche e il mondo. Ricerche su un dibattito nella seconda metà del Cinquecento*, Milan.

Diaz 1976: Furio Diaz, *Il Granducato di Toscana. I Medici*, Turin.

Dictionnaire de Droit Canonique 1937, ed. R. Naz, 6 vols., Paris.

Dilke 1985: Oswald A. W. Dilke, *Greek and Roman Maps*, London.

Donnelly 1998: John Patrick Donnelly S. J., "Antonio Possevino's Plan for World Evangelization," *The Catholic Historical Review*, LXXIV, no. 2, 179–98.

Dubourg Glatigny 2002: Pascal Dubourg Glatigny, "Egnazio Danti O.P. (1536-1586): Itinéraire d'un mathématicien parmi les artistes", in *Melanges de l'Ecole Française de Rome. Italie et Méditerranée*, 114, 543-605.

Dürst 1998: A. Dürst, "The Map of Europe," in Watelet, 1998, 31–42.

Eamon–Paheau 1984: William Eamon and Françoise Paheau, "The Accademia Segreta of Girolamo Ruscelli: a Sixteenth-Century Italian Scientific Society," *Isis*, LXXV, 327–42.

Edson 1997: Evelyn Edson, *Mapping Time and Space: How Medieval Mapmakers Viewed Their World*, London.

Ehrlich 2002: Tracy L. Ehrlich, *Landscape and Identity in Early Modern Rome: Villa Culture at Frascati in the Borghese Era*, Cambridge and London.

Eisenbichler 2001: Konrad Eisenbichler, ed., *The Cultural Politics of Duke Cosimo I de Medici*, Aldershot, Burlington, Singapore and Sydney.

Eisler 1983: W. Eisler, "The Impact of Emperor Charles V upon Italian Visual Culture 1529–1533," *Arte Lombarda*, LXV, no. 2, 93–110.

Eitel-Porter 1995: Rhoda Eitel-Porter, "Cesare Nebbia at the Vatican: The *Sale dei Foconi*," *Apollo*, CXLII, 19–24.

Elliot 1987: James Elliot, ed., *The City in Maps: Urban Planning to 1900*, London.

Encyclopedia of Early Christianity 1997: *Encyclopedia of Early Christianity*, Everett Ferguson, ed., 2nd edn., New York and London.

Eriksen–Malmanger 2001: Roy Eriksen and Magne Malmanger, eds., *Basilike Eikon, Renaissance Representations of the Prince*, Rome.

Eusebius 1999: Eusebius, *Life of Constantine*, introduction, translation, and commentary by Averil Cameron and Stuart C. Hale, Oxford.

Fagiolo 1980: Marcello Fagiolo, ed., *La città effimera e l'universo artificiale del giardino*, Rome.

Fahy 1993: Conor Fahy, "The Venetian Ptolemy of 1548," in *The Italian Book 1475–1800: Studies Presented to Dennis E. Rhodes on His 70th Birthday*, ed. Denis V. Reidy, London, 89–116.

Falguières 1992: Patricia Falguières, "Fondation du Théâtre ou Méthode de l'exposition universelle. Le 'Inscriptiones' de Samuel Quicchelberg (1565)," in *Cahiers du Musée national d'art moderne*, XL, summer, 91–115.

Fanti 1967: Mario Fanti, *Ville, castelli e chiese bolognesi da un libro di disegni del Cinquecento*, Bologna.

Fantuzzi [1781–94] 1965: Giovanni Fantuzzi, *Notizie degli scrittori bolognesi*, 9 vols., (facsimile edition: Bologna 1965).

Farago 1995: Claire Farago, ed., *Reframing the Renaissance: Visual Culture in Europe and Latin America 1450–1650*, New Haven and London.

Federzoni 2002: Laura Federzoni , "La carta degli Stati estensi di M. Pasi," in *Alla scoperta del mondo. L'arte della cartografia da Tolomeo a Mercatore*, introduction by Francesco Sicilia, texts by Mauro Bini, Ernesto Milano, Annalisa Battini, and Laura Federzoni, Modena, 241–85.

Feest 1990: Christian F. Feest, "Vienna's Mexican Treasures: Aztec, Mixtec, and Tarascan Works from Sixteen-Century Austrian Collections," *Archiv für Völkerkunde*, XLIV, 1–64.

Fehl 1974: Philip Fehl, "Vasari's Extirpation of the Huguenots and the Challenge of Pity and Fear," *Gazette des Beaux-Arts*, LXXXIV, nos. 1270–1, 257–84.

Feinberg 2002: Larry J. Feinberg, "The Studiolo of Francesco I Reconsidered," in *Medici and Michelangelo* 2002, 47–66.

Feldhay 1995: Rivka Feldhay, *Galileo and the Church: Political Inquisition or Critical Dialogue?*, London and New York.

Feldhay 1998: Rivka Feldhay, "The Use and Abuse of Mathematical Entities: Galileo and the Jesuits Revisited," in *The Cambridge Companion to Galileo*, ed. Peter Machamer, Cambridge and New York, 80–145.

Ferri 1994: Rolando Ferri, "Una '*passeggiata per l'Italia*.' L'anonima 'Ambulatio gregoriana'," in Gambi–Pinelli 1994, vol. 1, 73–82.

Field 1988: Judith Field, "What Is Scientific about a Scientific Instrument?," *Nuncius: Annali dell'Istituto e Museo di storia della scienza di Firenze*, II, 3–26.

Findlen 1994: Paula Findlen, *Possessing Nature: Museums, Collecting, and Scientific Culture in Early Modern Italy*, Berkeley.

Findlen 1999: Paula Findlen, "The Formation of a Scientific Community: Natural History in Sixteenth-Century Italy," in *Natural Particulars: Nature and the Disciplines in Renaissance Europe*, ed. Anthony Grafton and Nancy Siraisi, Cambridge and London.

Fiorani 1996: Francesca Fiorani, "Post-Tridentine 'Geographia Sacra': The Galleria delle Carte Geografiche in the Vatican Palace," *Imago Mundi*, XLVIII, 124–48.

Fiorani 2001: Francesca Fiorani, "Maps, Politcs, and the Grand Duke of Florence: The Sala della Guardaroba Nuova of Cosimo I de Medici," in Eriksen–Malmanger 2001, 73–102.

Fiorani 2003: Francesca Fiorani, "Danti Edits Vignola: The Formation of a Modern Classic on Perspective," in *The Treatise on Perspective: Published and Unpublished*, ed. Lyle Massey, Studies in the History of Art, 59, Center for Advanced Study in the Visual Arts. Symposium Papers XXXVI, National Gallery of Art, Washington, D.C., 127–59.

Fiorani 2005: Francesca Fiorani, "Painted Map Cycles in the Renaissance," in *History of Cartography*, Vol. 3. *The Renaissance*, ed. David Woodward, Chicago, forthcoming.

Fiorini 1889: Mario Fiorini, *Sfere terrestri e celesti di autore italiano oppure fatte o conservate in Italia*, Rome.

Fischer 1932: Joseph Fischer, *Claudii Ptolemaei Geographiae codex Urbinas graecus 82*, Leipzig.

Flowering of Florence 2002: *The Flowering of Florence: Botanical Art for the Medici*, ed., Lucia Tongiorgi Tomasi and Gretchen A. Hirschauer, exhibition catalogue National Gallery of Art, Washington D.C.

Forster 1971: Kurt W. Forster, "Metaphors of Rule: Political Ideology and History in the Portraits of Cosimo I de' Medici," *Mitteilungen des Kunsthistorischen Institutes in Florenz*, XV, no. 1, 65–104.

Foucault 1970: Michel Foucault, *The Order of Things: An Archaeology of the Human Sciences*, New York.

Foucault 1994: Michel Foucault, "Questions à Michel Foucault sur la géographie," in *Dits et écrits, 1954–1988*. Edition établie sous la direction de Daniel Defert et François Ewald avec la collaboration de Jacques Lagrange, 4 vols., Paris, vol. 3, 28–40.

Frangenberg 1988: Thomas Frangenberg, "Egnatio Danti's Optics: Cinquecento Aristotelianism and the Medieval Tradition," *Nuncius: Annali dell'Istituto e del Museo di storia della scienza di Firenze*, III, no. 1, 3–38.

Frangenberg 1994: Thomas Frangenberg, "Chorographies of Florence: The Use of City Views and City Plans in the Sixteenth Century," *Imago Mundi*, XLVI, 41–64.

Franzoni 1994: Claudio Franzoni, "I restauri delle carte geografiche," in Gambi–Pinelli 1994, vol. 1, 169–75.

Frasca Spada–Jardine 2000: Marina Frasca-Spada and Nick Jardine, eds., *Books and the Sciences in History*, Cambridge and New York.

Freedberg 1982a: David Freedberg, "Johannes Molanus on Provocative Paintings: 'De Historia sanctarum imaginum et picturarum,' Book II, Chapter 42", *JWCI*, XXXIV, 229–45.

Freedberg 1982b: David Freedberg, "The Hidden God: Image and Interdiction in the Netherlands in the Sixteenth Century," *Art History*, V, no. 2, 133–53.

Freedberg 2002: David Freedberg, *The Eye of the Lynx:*

Galileo, His Friends, and the Beginnings of Modern Natural History, Chicago.

Freiberg 1991: Jack Freiberg, "The Lateran Patronage of Gregory XIII and the Holy Year 1575," *Zeitschrift für Kunstgeschichte*, LIV, 66–87.

Freiberg 1995: Jack Freiberg, "In the Sign of the Cross: The Image of Constantine in the Art of Counter-Reformation Rome," *Studies in the History of Art*, XLVIII, 66–87.

Frey 1940: H. W. Frey, *Neue Briefe von Giorgio Vasari*, Hopfer.

Frey–Frey 1923–30: K. Frey and H. W. Frey, *Der literarische Nachlass Giorgio Vasaris*, 2 vols., Munich.

Fubini 1999: Riccardo Fubini, "Baronio e la tradizione umanistica. Note su di un libro recente," *Cristianesimo nella Storia*, XX, 147–59.

Furttenbach [1627] 1971: Joseph Furttenbach, *Newes Itinerarium Italiae*, Ulm (facsimile edition: Hildesheim and New York).

Gaeta Bertelà 1997: Giovanna Gaeta Bertelà, *La Tribuna di Ferdinando I de' Medici. Inventari 1589–1631*, Modena.

Gaeta Bertelà–Petroli Tofani 1969: Giovanna Gaeta Bertelà and Annamaria Petrioli Tofani, eds., *Feste e apparati medicei da Cosimo I a Cosimo II. Mostra di disegni e incisioni*, exhibition catalogue, Gabinetto Disegni e Stampe degli Uffizi, Florence.

Galluzzi 1980: Paolo Galluzzi, "Il mecenatismo mediceo e le scienze," in *Idee, istituzioni, scienza ed arti nella Firenze dei Medici*, ed. Cesare Vasoli, Florence, 189–217.

Gambi 1977: Lucio Gambi, "Per una rilettura di Biondo e Alberti geografi," *Il Rinascimento nelle corti padane. Società e cultura*, Bari.

Gambi 1994: Lucio Gambi, "Egnazio Danti e la Galleria delle Carte Geografiche," in Gambi–Pinelli 1994, vol. 1, 83–96.

Gambi–Pinelli 1994: Lucio Gambi and Antonio Pinelli, *La Galleria delle Carte Geografiche*, 3 vols., Modena.

Garin 1973: Eugenio Garin, "Alle origini della polemica anticopernicana," *Colloquia Copernicana*, II Studia Copernicana, vol. 6, Wrokclaw, 31–42.

Gautier-Dalché 1991: Patrik Gautier-Dalché, "L'espace de l'histoire: le rôle de la geographie dans les chroniques universelles," *L'Historiographie médiévale en Europe. Paris, 29 mars-1 avril 1989*, Paris, 287–300.

Ghim 1969: Walter Ghim, ed., " 'Vita Mercatoris'," in A. S. Osley, *Mercator*, London.

Ghirardi 1990: Anna Ghirardi, *Bartolomeo Passerotti pittore (1529–1592)*, Rimini.

Ghizzoni 2003: Manuela Ghizzoni, "L'immagine di Bologna nella veduta vaticana del 1575," in *Imago Urbis. L'immagine della città nella storia d'Italia*, ed. Francesca Bocchi and Rosa Smurra, Rome.

Giardino dei Semplici 1991: *Giardino dei Semplici. L'Orto Botanico di Pisa dal XVI al XX secolo*, ed. Fabio Garbari, Lucia Tongiorgi Tomasi, and Alessandro Tosi, Pisa.

Gilio [1564] 1986: Giovanni Andrea Gilio, *Due dialoghi*, Camerino (facsimile edition: Florence).

Gillies 1994: John Gillies, *Shakespeare and the Geography of Difference*, Cambridge, New York and Melbourne.

Ginori Conti 1936: P. Ginori Conti, *L'apparato per le nozze di Francesco de' Medici e Giovanna d'Austria nelle narrazioni del tempo e da lettere inedite di Vincenzo Borghini e di Giorgio Vasari*, Florence.

Giovio 1556: Paolo Giovio, R*agionamentio sopra i motti e disegni d'arme e d'amore che comunemente chiamano imprese*, Venice.

Goethe [1740] 1932: Johann Caspar Goethe, *Viaggio in Italia 1740*, Rome.

Goffart 1995: Walter Goffart, "Breaking the Ortelian Pattern: Historical Atlases with a New Program," in *Editing Early and Historical Atlases*, Papers Given at the Twenty-Ninth Annual Conference on Editorial Problems, University of Toronto, 5–6 November 1993, Toronto, Buffalo, and London, 49–81.

Goffart 1998: Walter Goffart, "Christian Pessimism on the Walls of the Vatican Galleria delle carte geografiche," *Renaissance Quarterly*, LI, 788–828.

Goffart 2003: Walter Goffart, *Historical Atlases: The First Three Hundred Years, 1570–1870*, Chicago.

Goldstein 1988: Carl Goldstein, *Visual Fact over Verbal Fiction: A Study of the Carracci and the Criticism, Theory, and Practice of Art in Renaissance and Baroque Italy*, Cambridge and New York.

Goldstein 1996: Carl Goldstein, *Teaching Art: Academies and Schools from Vasari to Alpers*, Cambridge.

Gonzaga [1529–30] 1892: Luigi Gonzaga, *Cronaca del soggiorno di Carlo V in Italia, 1529–30*, ed. Giacinto Romano, Milan.

Gori 1868: F. Gori, "Una delle più singolari processioni del XVI secolo in Roma," *Il Buonarroti*, s. 2, 3, 41–50.

Grafton 1991: Anthony Grafton, *Defenders of the Text: The Tradition of Scholarship in an Age of Science, 1450–1800*, Cambridge, Mass.

Grafton 1992: Anthony Grafton, *New Worlds, Ancient Texts: The Power of Tradition and the Shock of Discovery*, Cambridge, Mass.

Grafton 1997: Anthony Grafton, *Commerce with the Clas-*

sics: Ancient Books and Renaissance Readers, Ann Arbor, Mich.

Grafton 1999: Anthony Grafton, *Cardano's Cosmos: The World and Works of a Renaissance Astrologer*, Cambridge, Mass.

Grafton 2000: Anthony Grafton, *Leon Battista Alberti: Master Builder of the Italian Renaissance*, New York.

Greenblatt 1980: Stephen J. Greenblatt, *Renaissance Self-Fashioning from More to Shakespeare*, Chicago.

Grendler 1989: Paul F. Grendler, *Schooling in Renaissance Italy: Literacy and Learning, 1300–1600*, Baltimore.

Grendler 2002: Paul F. Grendler, *The Universities of the Italian Renaissance*, Baltimore and London.

Haardt 1952: Robert Haardt, "The Globe of Gemma Frisius," *Imago Mundi*, XI, 109–110.

Hajos 1963: E. M. Hajos, "References to Giulio Camillo in Samuel Quicchelberg: 'Inscriptiones vel tituli theatri amplissimi'," *Bibliothèque de l'Humanisme et de la Renaissance*, XXV.

Hale 1977: John R. Hale, *Florence and the Medici: The Pattern of Control*, London.

Hall 1992: Bert S. Hall, "The Didactic and the Elegant: Some Thoughts on Scientific and Technological Illustrations in the Middle Ages and Renaissance," in Baigrie 1992, 3–39.

Hall 1979: Marcia B. Hall, *Renovation and Counter-Reformation: Vasari and Duke Cosimo in Sta Maria Novella and Sta Croce, 1565–1577*, London.

Hantzsch 1898: V. Hantzsch, *Sebastian Münster, Leben, Werk, wissenschaftliche Bedeutung*, Leipzig.

Harley 2001: Brian J. Harley, *The New Nature of Maps: Essays in the History of Cartography*, Baltimore.

Harley–Woodward 1987: Brian J. Harley and David Woodward, eds., *History of Cartography*, vol. 1, Chicago.

Haskell–Penny 1981: Francis Haskell and Nicholas Penny, *Taste and the Antique: The Lure of Classical Sculpture, 1500–1900*, New Haven and London.

Hauger 1991: Harriet Hauger, "Samuel Quiccheberg: 'Inscriptiones vel tituli theatri amplissimi.' "Uber die Entstehung der Museen und das Sammeln," in *Universität und Bildung. Festschrift Laetitia Boehm zum 60. Geburstag*, ed. Winfried Mueller, Wolfang J. Smolka and Helmut Zedelmaier, Munich, 129–39.

Headley 1963: John M. Headly, *Luther's View of Church History*, New Haven and London.

Headley 2000: John M. Headley, "Geography and Empire in the Late Renaissance: Botero's Assignment, Western Universalism, and the Civilizing Process," in *Renaissance Quarterly*, LIII, vol. 4, 1119–55.

Headley–Tomaro 1988: John M. Headley and John B. Tomaro, eds., *San Carlo Borromeo. Catholic Reform and Ecclesiastical Politics in the Second Half of the Sixteenth Century*, Washington D.C.

Hecht 1997: Christian Hecht, *Katholische Bildentheorie im Zeitalter von Gegenreformation und Barock: Studien zu Traktaten von Johannes Molanus, Gabriele Paleotti und anderern Autoren*, Berlin.

Heikamp 1963: Detlef Heikamp, "Zur Geschichte der Uffizien Tribuna und der Kunstschränke in Florenz und Deutschland," *Zeitschrift für Kunstgeschichte*, XXVI, 193–268.

Heikamp 1970: Detlef Heikamp, "L'antica sistemazione degli strumenti scientifici nelle collezioni fiorentine," *Antichità viva*, IX, 3–25.

Heikamp 1972: Detlef Heikamp, *Mexico and the Medici*, Florence.

Heikamp 1974: Detlef Heikamp, ed., *Il Tesoro di Lorenzo il Magnifico. I vasi*, Florence.

Heikamp 1998: Detlef Heikamp, "Il Nuovo Mondo," in *Magnificenza alla corte dei Medici*, exhibition catalogue, Florence, Palazzo Vecchio, 399–402.

Heilbron 1999: J. L. Heilbron, *The Sun in the Church: Cathedrals as Solar Observatories*, London and Cambridge.

Herz 1986: Alexandra Herz, "Vasari's Massacre Series in the Sala Regia: The Political, Juristic and Religious Background," *Zeitschrift für Kunstgeschichte*, XLIX, 41–54.

Hess 1967a: Jacob Hess, *Kunstgeschichte Studien zur Renaissance und Barok*, 2 vols., Rome.

Hess 1967b: Jacob Hess, "On Some Celestial Maps and Globes of the Sixteenth Century," *JWCI*, 1967, 406–09.

Heyer 1889: Alfons Heyer, "Drei Mercatorkarten in der Breslauer Stadtbibliothek," *Zeitschrift für wissenschaftliche Geographie*, VII, 379–89, 474–87, 507–28.

Hogden 1954: Margaret Hogden, "Sebastian Münster, a Sixteenth-Century Ethnographer," *Osiris*, 504–29.

Hogden 1964: Margaret Hogden, *Early Anthropology in the Sixteenth and Seventeenth Centuries*, Philadelphia.

Hope 1985: Charles Hope, "Historical Portraits in the 'Lives' and in the Frescoes of Giorgio Vasari," in Gian Carlo Garfagnini, ed., *Giorgio Vasari tra decorazione ambientale e storiografia artistica*, Florence, 321–38.

Hope 1995: Charles Hope, "Can We Trust Vasari?" *New York Review of Books*, 5 October, 10–13.

Howell 2002: Kenneth J. Howell, *God's Two Books: Copernican Cosmology and Biblical Interpretation in Early Modern Science*, Notre Dame.

Hsia 1998: R. Po-Chia Hsia, *The World of Catholic Renewal 1540–1770*, Cambridge and London.

Impey–MacGregor 1985: Oliver Impey and Arthur Mac-

Gregor, eds., *The Origins of Museums. The Cabinet of Curiosities in Sixteenth- and Seventeenth-Century Europe*, Oxford.

Iserloh–Gladik–Jedin 1980: Erwin Iserloh, Joseph Grazik and Hubert Jedin, eds., *History of the Church: Reformation and Counter-Reformation*, vol. 5, New York.

Italian Drawings 1983: *Italian Drawings in the Department of Prints and Drawings in the British Museum: Artists Working in Rome c.1550 to c.1640*, ed. John A. Gere, Philip Pouncey and Rosalind Wood, London.

Jacks 1989: Philip Jacks, "A Sacred Meta for Pilgrims in the Holy Year 1575," *Architettura*, XIX, 137–65.

Jacks 1998: Philip Jacks, ed., *Vasari's Florence: Artists and Literati at the Medicean Court*, Cambridge and New York.

Jacob 1992: Christian Jacob, *L'Empire des cartes. Approche théorique de la cartographie à travers l'histoire*, Paris.

Jansen 1982: Dirk Jacob Jansen, "Jacopo Strada (1515–1588). Antiquario della Sacra Cesarea Maesta," *Leids Kunsthistorish Jaarboek*, I, 57–69.

Jansen 1987: Dirk Jacob Jansen, "Jacopo Strada et le commerce d'art," *La Revue de l'art*, LXXVII, 11–21.

Jedin 1961: Hubert Jedin, *A History of the Council of Trent*, English translation, vols. 1 and 2, London and New York (original edition in German: Freiburg 1957).

Jedin 1982: Hubert Jedin, *Il cardinale Cesare Baronio*, Rome.

Jedin–Alberigo 1985: Hubert Jedin and Giuseppe Alberigo, *Il tipo ideale di vescovo secondo la riforma cattolica*, Brescia.

Johns 1998: Adrian Johns, *The Nature of the Book: Print and Knowledge in the Making*, Chicago.

Jones 2000: Alexander Jones, "Introduction," in Ptolemy 2000.

Jones 1995: Pamela M. Jones, "Art Theory as Ideology: Gabriele Paleotti's Hierarchical Notion of Painting's Universality and Reception," in Farago 1995, 127–39.

Jones–Galison 1998: Caroline A. Jones and Peter Galison, eds., *Picturing Science, Producing Art*, New York.

Jong 1998: Jan L. de Jong, "Papal History and Historical 'invenzione': Vasari's Frescos in the Sala Regia," in Jacks 1998, 220–37.

Julien 1948: Charles A. Julien, "La papauté et le partage du monde," in *Les Voyages de découverte et les premiers établissements* (XVe–XVI siècles), in *Histoire de l'expansion et de la colonisation françaises*, Paris.

Kahn-Rossi–Franciolli 1999: Manuela Kahn-Rossi and Marco Franciolli, eds., *Il giovane Borromini. Dagli esordi a San Carlo alle Quattro Fontane*, Lugano.

Karl V 2000: *Kaiser Karl V. 1500–1558. Macht und Ohnmacht Europas*, exhibition catalogue, Kunsthistorisches Museum, Vienna (Geneva and Milan).

Karrow 1993: Robert W. Karrow, Jr., *Mapmakers of the Sixteenth Century and Their Maps: Bio-Bibliography of the Cartographers of Abraham Ortelius, 1570*, Chicago.

Kaufmann 1993: Thomas DaCosta Kaufmann, *The Mastery of Nature: Aspects of Art, Science and Humanism in the Renaissance*, Princeton.

Kemp 1990: Martin Kemp, *The Science of Art: Optical Themes in Western Art from Brunelleschi to Seurat*, New Haven and London.

Kemp 1991: Martin Kemp, "Intellectual Ornaments: Style, Function and Society in Some Instruments of Art," in *Interpretation and Cultural History*, eds. Joan H. Pittock and Andrew Wear, New York, 135–52.

Kemp 1992: Martin Kemp, "Temples of the Body and Temples of the Cosmos: Vision and Visualization in the Vesalian and Copernican Revolutions," in Baigire 1992, 40–85.

Kemp 1995: Martin Kemp, " 'Wrought by No Artist's Hand': The Natural, the Articifical, the Exotic, and the Scientific in Some Artifacts from the Renaissance," in Farago 1995, 177–96.

Kemp 2000: Martin Kemp, *Visualizations: The Nature Book of Art and Science*, Oxford.

Kenney 1974: Edward J. Kenney, *The Classical Text: Aspects of Editing in the Age of the Printed Book*, Berkeley, Los Angeles, and London.

Kent 2000: Dale Kent, *Cosimo de' Medici and the Florentine Renaissance: The Patron's Oeuvre*, New Haven and London.

Kish 1951: George Kish, "The Japan on the 'Mural Atlas' of the Palazzo Vecchio, Florence," *Imago Mundi*, VIII, 1951, 52–4.

Kliemann 1993: Julian-Matthias Kliemann, *Gesta dipinte. La grande decorazione nelle dimore italiane dal Quattrocento al Seicento*, Milan.

Koeman 1964: Cornelius Koeman, *The History of Abraham Ortelius and his 'Theatrum Orbis Terrarum,'* New York.

Krönig 1972: Wolfang Krönig, "Storia di una veduta di Roma," *Bollettino d'arte*, LVII, 165–98.

Kuntz 1998: Margaret A. Kuntz, "Federico Zuccari, Gregory XIII, and the Vault Frescoes of the Cappella Paolina," in *Römisches Jahrbuch der Bibliotheca Hertziana*, XXXII, 221–9.

Kusukawa 1997: Sachiko Kusukawa: "Leonhart Fuchs on the Importance of Pictures," *Journal of the History of Ideas*, LVIII, no. 3, 403–27.

Lalande 1786: Joseph Jérôme de Lalande, *Voyage d'un Fran-*

cois en Italie, fait dans les années 1765–1766, Paris.
Langendijk 1981–7: Karla Langendijk, *The Portraits of the Medici, Fifteenth to Eighteenth Centuries*, 3 vols., Florence.
Lattis 1994: James M. Lattis, *Between Copernicus and Galileo: Christoph Clavius and the Collapse of Ptolemaic Cosmology*, Chicago and London.
Le Mollé 1995: Ronald Le Mollé, *Giorgio Vasari. L'homme des Médicis*, Paris.
Lensi 1929: Alfredo Lensi, *Palazzo Vecchio*, Milan.
Lestrignant 1991: Frank Lestrignant, *L'atelier du cosmographe ou l'image du monde à la Renaissance*, Paris.
Levesque 1994: Catherine Levesque, *Journey Through Landscape in Seventeenth-Century Holland: The Haarlem Print Series and Dutch Identity*, University Park, Pa.
Levi-Donati 2003: G. Levi-Donati, "Four Hundred and Twenty-Five Years Later: Egnazio Danti's Anemoscope," in *Nuncius: Annali di storia della scienza*, 18, 2, 467–80.
Lewis–Wigen 1997: Martin W. Lewis and Karen E. Wigen, *The Myth of Continents: A Critique of Metageography*, Berkeley, Los Angeles, and London.
Licini 1988: Patrizia Licini, *La Moscovia Rappresentata. L'immagine "Capovolta" della Russia nella cartografia rinascimentale europea*, Milan.
Liebenwein 1988: Wolfgang Liebenwein, *Studiolo. Storia e tipologia di uno spazio culturale*, ed. Claudia Cieri Via, Modena (original edition in German: Berlin 1977).
Lightbown 1969: Ronald Lightbown, "Oriental Art and the Orient in Late Renaissance and Baroque Italy," *JWCI*, XXXII, 228–79.
Lippincott 1990: Kristen Lippincott, "Two Astrological Ceilings Reconsidered: The Sala di Galatea in the Villa Farnesina and the Sala del Mappamondo at Caprarola," *JWCI*, LIII, 185–207.
Lippincott 2002: Kristen Lippincott, "Power and Politics: The Use of the Globe in Renaissance Portraiture," *Globe Studies*, XLIX–L, 3–20.
Litteri 2000: Tiziana Litteri, "Paris Nogari, pittore manierista romano," in *Storia dell'arte*, XCIX, 23–54.
Livingstone 1988: David N. Livingstone, "Science, Magic and Religion: A Contextual Reassessment of Geography in the Sixteenth and Seventeenth Centuries," *History of Science*, XXVI, no. 3, 269–94.
Lorenzoni 1912: Antonio Lorenzoni, *Carteggio artistico inedito di Don Vincenzo Borghini*, Florence.
Loyola 1923: *The Spiritual Exercises of St. Ignatius Loyola*, Spanish and English with a Continuous Commentary by Joseph Rickaby S. J., 2nd edn., London.
Lugli 1990: Adalgisa Lugli, *Naturalia et mirabilia: il collezionismo enciclopedico nelle Wunderkammern d'Europa*, Milan.
Mack 1974: Rosamond E. Mack, "Girolamo Muziano e Cesare Nebbia at Orvieto," *AB*, LVI, no. 3, 319–35.
Madonna 1979: Maria Luisa Madonna, "La biblioteca: 'Theatrum mundi e Theatrum sapientiae' ", in *L'abbazia benedettina di S. Giovanni Evangelista a Parma*, Parma.
Maffei 1506: Raffaele Maffei, *Commentariorum urbanorum*, Rome.
Magnificenza 1997: *Magnificenza alla corte dei Medici. Arte a Firenze alla fine del Cinquecento*, exhibition catalogue, Florence, Palazzo Pitti.
Maiello 1989: Francesco Maiello, "Tempo, potere e cosmologia: la riforma del Calendario," in *Dimensioni e problemi della ricerca storica*, I, 102–37.
Maiello 1994: *Storia del calendario. La misurazione del tempo, 1450–1800*, Turin.
Mâle 1958: Emile Mâle, *Religious Art from the Twelfth to the Eighteenth Century*, New York.
Maltese 1980: Corrado Maltese, "La prospettiva curva di Leonardo da Vinci e uno strumento di Baldassarre Lanci," in *Prospettiva rinascimentale. Codificazioni e trasgressioni*, ed. Marisa Dalai Emiliani, Florence, 407–19.
Mancinelli–Casanovas 1980: Fabrizio Mancinelli and Juan S. Casanovas, *La Torre dei Venti in Vaticano*, Vatican City.
Mancini 1956: Giulio Mancini, *Considerazioni sulla pittura*, ed. Adriana Marucchi, 2 vols., Rome.
Mangani 1998a: Giorgio Mangani, *Il "mondo" di Abramo Ortelius. Misticismo, geografia e collezionismo nel Rinascimento dei Paesi Bassi*, Modena.
Mangani 1998b: Giorgio Mangani, "Abraham Ortelius and the Hermetic Meaning of the Cordiform Projection," *Imago Mundi*, L, 59–83.
Mangani 2003: Giorgio Mangani, "Misurare, calcolare, pregare. Il mappamondo ricciano come strumento meditativo," in *Padre Matteo Ricci. L'Europa alla corte dei Ming*, exhibition catalogue, Macerata, Auditorium San Paolo (Milan).
Marchese 1879: Vincenzo Marchese, *Memorie dei più insigni pittori, scultori e architetti domenicani*, 2 vols., Bologna.
Marchis–Dolza 2002: Vittorio Marchis and Luisa Dolza, eds., *L'album fiorentino dei disegni artificiali raccolti da Jacopo e Ottaviano Strada*, Florence.
Marconi–Cipriani–Valeriani 1974: Paolo Marconi, Angela Cipriani, and Enrico Valeriani, eds., *I disegni di architettura dell'Archivio storico dell'Accademia di San Luca*, Rome.
Margolin 1991: Jean Claude Margolin, "Le Théâtre de

mémoire de G. Camillo," in *L'Encyclopedisme*. Actes du Colloque de Caen, 12–16 janvier 1987, ed. Annie Becq, Paris, 459–82.

Marías 2000: Fernando Marías, "La Casa real en la Alhambra: Letras, armas y architectura entre Roma y Granada," in *Carlos v. Las armas y las letras*, exhibition catalogue, Hospital Real, Granada.

Markus 1994: R. A. Markus, "How on Earth Could Places Become Holy? Origins of the Christian Idea of Holy Places," *Journal of Early Christian Studies*, II, 257–72.

Masini 1666: Antonio di Paolo Masini, *Bologna perlustrata*, Bologna.

Mason Rinaldi 1984: Simona Mason Rinaldi, *Palma il Giovane. L'opera completa*, Milan.

Massinelli 1991: Anna Maria Massinelli, *Bronzetti e anticaglie dalla Guardaroba di Cosimo I*, Florence.

Massinelli–Tuena 1992: Anna Maria Massinelli and Filippo Tuena, eds., *Treasures of the Medici*, New York.

Mazzolini 1993: Renato G. Mazzolini, *Non-Verbal Communication in Science Prior to 1900*, Biblioteca di Nuncius. Studi e Testi XI, Florence.

Mazzone 1997: Umberto Mazzone, "Dal primo Cinquecento alla dominazione napoleonica", in *Storia della Chiesa di Bologna*, ed. Paolo Prodi and Lorenzo Paolini, Bologna.

McCuaig 1989: William McCuaig, *Carlo Sigonio: The Changing World of the Late Renaissance*, Princeton.

McCuaig 1994: William McCuaig, "The Tridentine Ruling on the Vulgate and Ecclesiastical Censorship," *Renaissance and Reformation*, XVIII, 43–56.

McGrath 1985: Elizabeth McGrath, "'Il senso nostro': The Medici Allegory applied to Vasari's Mythological Frescoes in the Palazzo Vecchio," in *Giorgio Vasari tra decorazione ambientale e storiografia artistica*, ed. Gian Carlo Garfagnini, Florence, 117–34.

McHugh 1991: J. F. McHugh, "The Sacrifice of the Mass at the Council of Trent," in *Sacrifice and Redemption: Durham Essays in Theology*, ed. S. W. Sykes, Cambridge 1991, 157–85.

Mcllwaine 1996: John Mcllwaine, *Maps and the Mapping of Africa: A Resource Guide*, London.

McTavish 1980: David McTavish, "Pellegrino Tibaldi's 'Fall of Phaethon' in the Palazzo Poggi", Bologna, *BM*, CXXII, 186–8.

Medici and Michelangelo 2002: *The Medici, Michelangelo, and the Art of Late Renaissance Florence*, exhibition catalogue, Chicago, Art Institue (New and Haven London).

Melion 1999: Walter S. Melion, "*Ad ductum itineris et dispositionem mansionum ostendendam*: Meditation, Vocation, and Sacred History in Abraham Ortelius's Parergon," *The Journal of the Walters Art Gallery*, LVII, 49–72.

Mercator 1891: *Drei karten von Gerhard Mercator: Europa, Britische Inseln, Weltkarte. Facsimile-lichtdruck nach den Originalen der Stadtbibliothek zu Breslau*, Berlin.

Mercator 1961: *Gerard Mercator's Map of the World (1569)*, in the form of an atlas in the Maritiem Museum "Prins Hendrik" at Rotterdam, introduction by B. Van T' Hoff, *Imago Mundi*, supplement No. 2, Rotterdam.

Merisalo 1999: Outi Merisalo, *Le collezioni medicee nel 1495. Deliberazioni degli ufficiali dei ribelli*, Florence.

Meucci 1876: Ferdinando Meucci, *La sfera armillare di Tolomeo costruita da Antonio Santucci*, Florence.

Meurer 1998: Peter H. Meurer, "Ortelius as the Father of Historical Geography," in Van der Broecke–Van der Krogt–Meurer 1998, 133–60.

Mignolo 1995: Walter D. Mignolo, *The Darker Side of the Renaissance: Literacy, Territoriality and Colonization*, Ann Arbor, Mich.

Milanesi 1994: Marica Milanesi, "Le ragioni del ciclo delle carte goegrafiche," in Gambi–Pinelli 1994, vol. 1, 97–124.

Minella 1993: Massimo Minella, *Il mondo ritrovato: le tavole sudamericane di Giacomo Gastaldi. Con la riproduzione anastatica del libro, "La universale descrittione del mondo" (1561)*, Genoa.

Miniati 1991: Mara Miniati, ed., *Museo di Storia della Scienza. Catalogo*, Florence.

Minor–Mitchell 1968: A. C. Minor and Bonner Mitchell, *A Renaissance Entertainment: Festivities for the Marriage of Cosimo I, Duke of Florence in 1539*, Columbia, Mich.

Mitchell 1979: Bonner Mitchell, *Italian Civic Pageantry in the High Renaissance*, Florence.

Montaigne 1955: Michel de Montaigne, *Journal du Voyage en Italie*, ed. Marcel Rat, Paris.

Montesquieu 1971: Charles de Secondat, Baron de Montesquieu, *Viaggio in Italia*, ed. Giovanni Macchia and Massimo Colesanti, Bari.

Morel 1982: Philip Morel, "Le Studiolo de Francesco I de' Medici et l'économie symbolique du pouvoir au Palazzo Vecchio," in *Symboles de la Renaissance*, 2, *Arts et langage*, Paris, 185–205.

Morel 1993: Philip Morel, "L'Etat médicéen au XVIe siècle: de l'allégorie à la cartographie," *Melanges de l'Ecole Française de Rome. Italie et Méditerranée*, CV, 93–131.

Mosco 1980: Marilena Mosco, "Una 'Descrittione dell'Apparato delle Stanze del Palazzo de' Pitti in Fiorenza' edita a Venezia nel 1577," *Antichità Viva*, XIX, 18–21.

Mostra Vespucciana 1954: *Mostra Vespucciana*, Palazzo Vecchio, Florence (Milan).

Moyer 1983: G. Moyer, "Aloisius Lilius and the 'Compendium Novae Rationis Restituendi Kalendarium'," in Coyne–Hoskin–Pederson 1983, 171–88.

Muccini 1991: Ugo Muccini, *The Apartments of Cosimo in Palazzo Vecchio*, Florence.

Muccini 1992: Ugo Muccini, *The Apartments of the Priori in Palazzo Vecchio*, Florence.

Münster 1550: Sebastian Münster, *Cosmographia universalis*, Basel.

Müntz 1888: Eugène Müntz, *Les collections des Médicis au* XV*e siècle. Le Musée, la bibliothèque, le mobilier*, Paris.

Müntz 1895: Müntz, "Les collections d'antiques formées par les Médicis aux XVIe siècle", *Memoires de l'academie des inscriptions et belles lettres*, XXXV, 1895.

Nagler 1964: Alois Maria Nagler, *Theatre Festivals of the Medici, 1539–1637*, New Haven and London.

Nebenzahl 1986: Kenneth Nebenzahl, *Maps of the Holy Land: Images of Terra Sancta through Two Millennia*, New York.

Nicolet 1988: Claude Nicolet, *L'inventaire du monde: géographie et politique aux origines de l'Empire romain*, Paris.

North 1983: J. D. North, "The Western Calendar: 'Intolerabilis, Horribilis, et Desiderabilis'," in Coyne–Hoskin–Pederson 1983, 75–115.

Nuti 1988: Lucia Nuti, "The Mapped Views by Georg Hoefnagel: The Merchant's Eye, the Humanist's Eye," *Word and Image*, IV, no. 2, 545–70.

Nuti 1994: Lucia Nuti, "The Perspective Plan in the Sixteenth Century: The Invention of a Representational Language," *AB*, LXXVI, no. 1, 105–28.

Nuti 1999: Lucia Nuti, "Mapping Places: Chorography and Vision in the Renaissance," in Cosgrove 1999, 90–108.

Nuti 2003: Lucia Nuti, "The World Map as an Emblem: Abraham Ortelius and the Stoic Contemplation," *Imago Mundi*, LV, 38–55.

Olmi 1992: Giuseppe Olmi, *L'inventario del mondo. Catalogazione della natura e luoghi del sapere nella prima età moderna*, Bologna.

O'Malley 2000: John W. O'Malley, *Trent and All That: Renaming Catholicism in the Early Modern Era*, Cambridge, Mass.

Ong 1958: Walter J. Ong, S. J., *Ramus and the Decay of Dialogue: From the Art of Discourse to the Art of Reason*, Cambridge, Mass.

Orbaan 1920: Johannes Albertus Franciscus Orbaan, *Documenti sul Barocco in Roma*, Miscellanea della Società romana di storia patria, Rome.

Orella y Unze 1976: José L. de Orella y Unze, *Respuestas católicas a las Centurias de Magdeburgo (1559–1588)*, Madrid.

Ortelius 1887: *Abrahami Ortelii epistolae et virorum eruditorum ad eundem et ad Jacobum Colium*, ed. H. Hessel, Cambridge.

Ostrow 1996: Steven F. Ostrow, *Art and Spirituality in Counter-Reformation Rome: The Sistine and Pauline Chapels in S. Maria Maggiore*, Cambridge and New York.

Pagliara 1986: Pier Nicola Pagliara, "Vitruvio da testo a canone," in *Memoria dell'antico nell'arte italiana*, ed. Salvatore Settis, 3 vols., Turin vol. 3, 7–87.

Palazzo Vecchio 1980: *Palazzo Vecchio. Committenza e Collezionismo Medicei*, exhibition catalogue, Palazzo Vecchio, Florence.

Paleotti [1582] 1990: Gabriele Paleotti, *Discorso intono alle imagini sacre et profane*, Bologna (facsimile edition with preface by Paolo Prodi: Bologna).

Palmesi 1899: Vincenzo Palmesi, "Ignazio Danti," *Bollettino della reale deputazione di storia patria per l'Umbria*, V, 81–125.

Paratore 1976: Ettore Paratore, *La Moscovia nelle carte italiane della rinascenza*, Pubblicazioni dell'Istituto di Geografia, serie B, Geostorica, Rome.

Partner 1980: Peter Partner, "Papal Financial Policy in the Renaissance and Counter-Reformation," *Past and Present*, LXXXVIII, 17–63.

Partridge 1983: Loren Partridge, "Review of Quednau 1979," *AB*, 65, no. 3, 515–20.

Partridge 1995: Loren Partridge, "The Room of Maps at Caprarola, 1573–1575," *AB*, LXXVII, 413–44.

Partridge–Starn 1990: Loren Partridge and Randolph Starn, "Triumphalism and the Sala Regia in the Vatican," in *All the World's a Stage: Art and Pageantry in the Renaissance and Baroque.* Part 1 *Triumphal Celebrations and the Rituals of Statecraft*, ed. Barbara Wisch and Susan Scott Munshower, Papers in Art History from the Pennsylvania State University, 23–82.

Partridge–Starn 1993: Loren Partridge and Randolph Starn, *Arts of Power: Three Halls of State in Italy, 1300–1600*, Berkeley and Los Angeles.

Paschini 1923: Pio Paschini, "La riforma gregoriana del Martirologio," *Scuola cattolica*, LI, 198–210, 274–84.

Passerini 1869: Luigi Passerini, *Gli Alberti di Firenze. Genealogia storia e documenti*, 2 vols., Florence.

Pastor [1936–55]: Ludwig von Pastor, *The History of the Popes from the Close of the Middle Ages* (English translation, 40 vols.), London.

Pastoureau 1998: Mireille Pastoureau, "The 1569 World Map," in Watelet, 1998, 79–88.

Pedersen 1983: Olaf Pedersen, "The Ecclesiastical Calendar and the Life of the Church," in Coyne–Hoskin–Pederson 1983, 17–73.

Pedley 1993: Mary Pedley, " 'I Due Valentuomini Indefessi': Christopher Maire and Roger Boscovich and the Mapping of the Papal States (1750–1755)," *Imago Mundi*, XLV, 59–76.

Pedretti 1977: Carlo Pedretti, *The Literary Works of Leonardo da Vinci*, compiled and edited from the original manuscript by Jean Paul Richter and commentary by Carlo Pedretti, Berkeley and Los Angeles.

Pepe 1996: Luigi Pepe, ed., *Copernico e la questione copernicana in Italia*, Florence.

Pietrangeli 1979: Carlo Pietrangeli, "Roma 1580," *Strenna dei Romanisti*, 457–68.

Pietrangeli 1992: Carlo Pietrangeli, *Il Palazzo Apostolico Vaticano*, Florence.

Pillsbury 1974: Edmund Pillsbury, "Jacopo Zucchi in S. Spirito in Sassia," *BM*, CXVI, no. 857, 434–44.

Pinelli 1994a: Antonio Pinelli, "Il '*bellissimo spasseggio*' di papa Gregorio XIII Boncompagni," in Gambi–Pinelli 1994, vol. 1, 9–71.

Pinelli 1994b: Antonio Pinelli, "Sopra la terra, il cielo. Geografia, storia, e teologia: il programma iconografico della volta," in Gambi–Pinelli 1994, vol. 1, 125–53.

Pinelli 1996: Antonio Pinelli, "Geografia della fede: L'Italia della Controriforma unificata sulla carta," in *Cartographiques. Actes du colloque de l'Academie de France à Rome, 19–20 mai 1995*, ed. Marie-Ange Brayer, Paris, 63–94.

Pittura in Emilia 1996: *La pittura in Emilia e in Romagna. Il Cinquecento*, 2 vols., Milan.

Pittura in Italia. Il Cinquecento 1988: *La pittura in Italia. Il Cinquecento*, 2 vols., Milan.

Plaisance 1973: Michel Plaisance, "Une prèmiere affirmation de la politique culturelle de Côme Ier: la transformation de l'Académie des *Humidi* en Académie Florentine (1540–1542)," in *Les écrivains et le pouvoir en Italie à l'époque de la Renaissance*, ed. A. Rochon, Première Série, 361–438.

Plaisance 1974: Michel Plaisance, "Culture et politique à Florence de 1542 à 1551," in *Les écrivains et le pouvoir en Italie à l'époque de la Renaissance*, ed. A. Rochon, Deuxième Série, 148–242.

Plaisance 1992: Michel Plaisance, "Les dédicaces a Côme 1er: 1546–1550," in *L'écrivan face à son public en France et en Italie à la Renaissance*, eds. Charles Adelin Fiorato and Jean-Claude Margoline, Paris, 173–187.

Pliny 1949: Pliny, *Natural history*, trans. H. Rackham, Cambridge, Mass.

Pomian 1990: Krzysztof Pomian, *Collectors and Curiosities: Paris and Venice 1580–1800*, Cambridge.

Porena 1888: Filippo Porena, "La geografia in Roma e il mappamondo vaticano," *Bollettino della società geografica italiana*, XXV, 221–38, 311–39, 427–53.

Porena 1906: Filippo Porena, "Una confessione sulle carte geografiche dipinte nella Loggia occidentale del Cortile di San Damaso in Vaticano," *Rivista geografica italiana*, XIII, 278–80.

Potere e spazio 1980: *Il potere e lo spazio. La scena del principe*, exhibition catalogue, Forte di Belvedere, Florence.

Prinz 1988a: Wolfram Prinz, *Galleria*, ed. Claudia Cieri Via, Modena (original edition in German: Berlin, 1977).

Prinz 1988b: Wolfram Prinz, "La serie gioviana o la collezione dei ritratti degli uomini illustri," in *Gli Uffizi. Catalogo generale*, Florence, 603.

Prinz–Beyer 1987: Wolfram Prinz and Andreas Beyer, eds., *Die Kunst und das Studium der Natur vom 14. Zum 16. Jahrhundert*, Weinheim.

Prodi 1957: Paolo Prodi, "San Carlo Borromeo e le trattative tra Gregorio XIII e Filippo II sulla giurisdizione ecclesiastica," *Rivista di storia della Chiesa in Italia*, XI, 195–240.

Prodi 1959–69: Paolo Prodi, *Il cardinale Gabriele Paleotti (1522–1596)*, 2 vols., Rome.

Prodi 1962: Paolo Prodi, "Ricerca sulla teorica delle arti figurative nella riforma cattolica," *Archivio italiano per la storia della Pietà*, IV (2nd edition: Bologna, 1984).

Prodi 1977: Paolo Prodi, "Storia sacra e controriforma. Nota sulle censure al commento di Carlo Sigonio a Sulpicio Severo," *Annali dell'Istituto storico italo-germanico in Trento*, III, 75–104.

Prodi 1982: Paolo Prodi, *Il sovrano pontefice*, Bologna.

Ptolemy 1475: Claudius Ptolemy, *Cosmographia*, trans. Jacopo d'Angelo, Vicenza.

Ptolemy 1482: Claudius Ptolemy, *Geographia*, trans. Francesco Berlinghieri, Florence (facsimile edition, ed. Raleigh A. Skelton: Amsterdam, 1966).

Ptolemy 1540: Claudius Ptolemy, *Geographia*, trans. Sebastian Münster, Basel (facsimile edition, ed. Raleigh A. Skelton: Amsterdam, 1966).

Ptolemy 1548: Claudius Ptolemy, *La geografia*, trans. Giovanni Battista Pedrezzano, maps by Giacomo Gastaldi, Venice.

Ptolemy 1561: Claudius Ptolemy, *Geographia*, trans. Girolamo Ruscelli, maps by Giacomo Gastaldi, Venice.

Ptolemy 1932: Claudius Ptolemy, *The Geography*, trans. and ed. Edward Luther Stevenson, with an introduction by Joseph Fischer, London (reprint: Toronto, 1991).

Ptolemy 2000: *Ptolemy's Geography: An Annotated Translation of the Theoretical Chapters*, ed. J. Lennart Berggren and Alexander Jones, Princeton.

Pullapilly 1975: Cyriac K. Pullapilly, *Cesare Baronio, Counter-Reformation Historian*, London.

Quednau 1979: Rolf Quednau, "Die Sala di Costantino im Vatikanischen Palast," Ph.D. diss., University of Munich, Hildesheim and New York.

Quicchelber 1565: Samuel Quiccheberg, *Inscriptiones vel tituli theatri amplissimi*, Munich.

Quinlan-McGrath 1997: Mary Quinlan-McGrath, "Caprarola's Sala della Cosmografia," *Renaissance Quarterly*, L, 1045–1100.

Raffaello 1993: *Raphael in the Apartments of Julius II and Leo X*, ed. Guido Cornini, Milan.

Ramusio 1978–83: Giovan Battista Ramusio, *Navigazioni e viaggi*, ed. Marica Milanesi, 4 vols., Turin.

Razzi 1596: Serafino Razzi, *Historia degli huomini illustri, così nelle prelature come nelle dottrine del sacro ordine de Predicatori*, Lucca.

Redig de Campos 1967: Decio Redig de Campos, *I Palazzi Vaticani*, Bologna.

Relano 1995: Francesco Relano, "Against Ptolemy: The Significance of the Lopes-Pigafetta Map of Africa," *Imago Mundi*, XLIX, 49–65.

Relano 2001: Francesco Relano, *The Shaping of Africa: Cosmographic Discourse and Cartographic Science in Late Medieval and Early Modern Europe*, Burlington.

Ricci 2001: Antonio Ricci, "Lorenzo Torrentino and the Cultural Programme of Cosimo I de' Medici," in Eisenbichler 2001, 103–20.

Rice 1997: Louise Rice, *The Altars and Altarpieces of New St. Peter's: Outfitting the Basilica, 1621–1666*, Cambridge and New York.

Richelson 1974: Paul W. Richelson, "Studies in the Personal Imagery of Cosimo I de' Medici, Duke of Florence," Ph.D. diss., Princeton University.

Righini 1977: Guglielmo Righini, "Il grande astrolabio del Museo della scienza di Firenze," *Annali dell'Istituto e del Museo di storia della scienza di Firenze*, II, no. 2, 45–66.

Righini 1951: Maria Luisa Righini, "The Armillary Sphere of Antonio Santucci delle Pomarance," *Imago Mundi*, VIII, 99–100.

Righini Bonelli 1978: Maria Luisa Righini Bonelli, "Di alcuni manoscritti inediti di Antonio Santucci delle Ripomarance," *Annali dell'Istituto e Museo di storia della scienza di Firenze*, III, 59–67.

Righini Bonelli–Settle 1979: Maria Luisa Righini Bonelli and Thomas Settle, "Egnatio Danti's Great Astronomical Quadrant," *Annali dell'Istituto e museo di storia della scienza di Firenze*, IV, nos. 2, 3–13.

Robertson 1992: Claire Robertson, *Il "gran cardinale": Alessandro Farnese Patron of the Arts*, New Haven and London.

Rodríguez 1982: Pedro Rodríguez, *El Catecismo romano. Fuentes e historia del texto y de la redacción. Bases criticas para el estudio teológico del Catecismo del Concilio de Trento (1566)*, Navarra.

Roma 1300–1875 1985: *Roma 1300–1875. La città degli Anni Santi. Atlante*, ed. Marcello Fagiolo and Maria Luisa Madonna, exhibition catalogue, Palazzo Venezia, Rome (Milan).

Roma Sancta 1985: *Roma Sancta. La città delle basiliche*, ed. Maurizio Fagiolo and Maria Luisa Madonna, Rome.

Rombai 1993: Leonardo Rombai, "La nascita e lo sviluppo della cartografia a Firenze e nella Toscana granducale," in *Imago et descriptio Tusciae. La Toscana nella geocartografia dal XV al XIX secolo*, ed. Leonardo Rombai, Venice, 83–193.

Romeo 2002: Giovanni Romeo, *L'Inquisizione nell'Italia moderna*, Bari.

Ronchi 1958: Vasco Ronchi, *Il cannochiale di Galileo e la scienza del seicento*, Turin.

Ronchini 1873: Amadio Ronchini, ed., *Lettere di Bernardino Baldi cavate dagli autografi che sono a Parma nell'Archivio di Stato*, Parma.

Rose 1973: Paul L. Rose, "Humanist Culture and Renaissance Mathematics," *Studies in the Renaissance*, XX, 46–105.

Rose 1975: Paul L. Rose, *The Italian Renaissance of Mathematics from Petrarch to Galileo*, Geneva.

Rosen 1975: Edward Rosen, "Was Copernicus's Revolution Approved by the Pope?", *Journal of the History of Ideas*, XXXVI, no. 3, 531–42.

Rosen 2003: Mark Rosen, "Don Miniato Pitti and the Second Life of a Scientist's Tools in Cinquecento Florence," in *Nuncius: Annali di storia della scienza*, 18, 1, 3–24.

Rosenthal 1971: E. S. Rosenthal, "Plus Ultra, Non Plus Ultra and the Columnar Device of Emperor Charles V," *JWCI*, XXXIV, 204–28.

Rosenthal 1973: E. S. Rosenthal, "The Invention of the Columnar Device of Emperor Charles V at the Court of Burgundy in Flanders in 1516," *JWCI*, XXXVI, 199–211.

Rosenthal 1974: E. S. Rosenthal, "Plus Oultre: The Idea Imperial of Charles V in His Columnar Device on the

Alhambra," in *Hortus Imaginum: Essays in Western Art*, ed. Robert Engass and Marilyn Stokstad, Lawrence, Kans.

Rossi 1980: Sergio Rossi, *Dalle botteghe alle accademie. Realtà sociale e teorie artistiche a Firenze dal XIV al XVI secolo*, Milan.

Rossi 1983: Paolo Rossi, *Clavis universalis. Arti della memoria e logica combinatoria da Lullo a Leibniz*, Bologna.

Rossi 1999: Federico Alessandro Rossi, *Carlo Borromeo. I tre volti della riforma cattolica*, Milan.

Rostagno 1923: Enrico Rostagno, ed., *Mostra storica di geografia*, Florence.

Röttgen 1975: Herwarth Röttgen, "Zeitgeschichtliche Bildprogramme der katholischen Restauration unter Gregor XIII, 1572–1585," *Münchner Jahrbuch der bildende Kunst*, XXVI, 89–117.

Roversi 1974: Giancarlo Roversi, "Il patrimonio fondiario dei Tanari a Gaggio Montano e nel Belvedere," *Strenna storica bolognese*, XXIV, 235–78.

Rowland 1998: Ingrid D. Rowland, *The Culture of the High Renaissance: Ancients and Romans in Sixteenth-Century Rome*, Cambridge and New York.

Rubin 1995: Patricia Lee Rubin, *Giorgio Vasari: Art and History*, New Haven and London.

Rubinstein 1967: Nicolai Rubinstein, "Vasari's Painting of the Foundation of Florence in the Palazzo Vecchio," in *Essays in the History of Architecture Presented to Rudolf Wittkower*, ed. Douglas Fraser, Howard Hibbard, and Milton Lewine, London, 64–73.

Rubinstein 1995: Nicolai Rubinstein, *Palazzo Vecchio, 1296–1532: Government, Architecture and Imagery in the Civic Palace of the Florentine Republic*, Cambridge and New York.

Ruscelli 1562: Girolamo Ruscelli, *Lettere di Principi, le quali ò si scrivono da Principi, ò a Principi, ò ragionan di Principi*, Venice.

San Carlo e il suo tempo 1986: *San Carlo e il suo tempo.* Atti del convegno internazionale nel iv centenario della morte (Milano 21–26 maggio 1984), 2 vols., Rome.

Sapori 1981: Giovanna Sapori, "Perugia 1565–1575: Girolamo Danti," *Bollettino d'arte*, LXVI, nos. 1–12.

Scafi 1995: Alessandro Scafi, Review of Gambi–Pinelli 1994, *Imago Mundi*, XLVII, 193–4.

Scafi 1999: Alessandro Scafi, "Mapping Eden: Cartographies of the Earthly Paradise," in Cosgrove 1999, 24–49.

Scaglia 1992: Gustina Scaglia, *Francesco di Giorgio: Checklist and History of Manuscripts and Drawings in Autographs and Copies from ca.1470 to 1687 and Renewed Copies (1764–1839)*, London.

Scalini 1996: Mario Scalini, "Oggetti rari e curiosi nelle collezioni medicee: esotica e naturalia," *Antichità viva*, XXXV, nos. 2–3, 59–67.

Scalini 1997: Mario Scalini, "Curios and Exotica in the Medici Collections," in Acidini Luchinat 1997, 145–56.

Scalini 1998: Mario Scalini, "L'armeria europea e orientale," in *Magnificenza* 1998, 396–8.

Scavizzi 1989: Giuseppe Scavizzi, "Storia ecclesiastica e arte nel secondo Cinquecento," *Storia dell'arte*, XXIX, 29–46.

Scavizzi 1992: Giuseppe Scavizzi, *The Controversy on Images from Calvin to Baronius*, New York.

Schaefer 1976: S. Schaefer, "The Studiolo of Francesco I de' Medici in the Palazzo Vecchio in Florence," Ph.D. diss., Bryn Mawr College.

Schilder 1981: Günther Schilder, "The Cartographical Relationships Between Italy and the Low Countries in the Sixteenth Century," *Map Collector*, XVII, 3–7.

Schilder 1986: Günther Schilder, *Monumenta cartographica Neerlandica*, II, Alphen aan den Rijn.

Schmitt 1981: Charles B. Schmitt, *Studies in Renaissance Philosophy and Science*, London.

Schmitt 1983: Charles B. Schmitt, *Aristotle and the Renaissance*, Cambridge and London.

Schramm 1958: Percy E. Schramm, *Sphaira, Globus, Reichsapfel. Wanderung und Wandlung eines Herrschaftszeichens von Caesar bis zu Elizabeth II*, Stuttgart.

Schuddeboom 1996: Cornelius Schuddeboom, *Philips van Winghe (1560–1592). En het ontstaan van de Christelijke Archeologie*, Leiden.

Schulz 1990: Jürgen Schulz, *La cartografia tra scienza e arte. Carte e cartografi nel Rinascimento italiano*, Modena.

Schütte 1993: Margret Schütte, *Die Galleria delle Carte Geografiche im Vatikan. Eine ikonographische Betrachtung des Gewölbeprogramms*, Bonn.

Scorza 1981: Richard Scorza, "Vincenzo Borghini and 'Invenzione': The Florentine 'Apparato' of 1565," *JWCI*, XLIV, 57–75.

Scorza 1988: Richard Scorza, "Imprese and Medals. 'Invenzioni all'antica' by Vincenzo Borghini," *Medal*, XIII, no. 2, 18–32.

Scorza 1989: Richard Scorza, "Vincenzo Borghini and the Impresa," *JWCI*, LII, 85–110.

Scorza 1998: Richard Scorza, "Vasari's Painting of the Terzo Cerchio in the Palazzo Vecchio: A Reconstruction of Medieval Florence," in Jacks 1998, 182–205.

Sebastian Münster 1988: *Sebastian Münster*, exhibition catalogue, Altes Rathaus, Ingelheim am Rhein.

Sereno 1994: Paola Sereno, "La fortuna del ciclo delle carte geografiche," in Gambi–Pinelli 1994, vol. 1, 155–68.

Settis 1983: Salvatore Settis, "Origine e significato delle gallerie in Italia," in *Gli Uffizi. Quattro secoli di una galleria.* Atti del convegno internazionale di studi, Firenze 20–24 settembre 1982, Florence, 309–17.

Settle 1984: Thomas Settle, "Antonio Santucci, His 'New Tractatus on Comets,' and Galileo," in *Novità celesti e crisi del sapere*, ed. Paolo Galluzzi, Florence, 229–38.

Settle 1990: Thomas Settle, "Egnazio Danti and Mathematical Education in Late Sixteenth-Century Florence," in *New Perspectives on Renaissance Thoughts: Essays in the History of Science, Education and Philosophy in Memory of Charles B. Schmitt*, ed. J. Henry and S. Hutton, London, 24–37.

Sframeli 1997: Maria Sframeli, "The Symbols of Power: Dukes and Grand Dukes," in Acidini Luchinat 1997, 103–06.

Shalev 2003: Zur Shalev, "Sacred Geography, Antiquarianism, and Visual Erudition: Benito Arias Montano (1527–1598) and the Maps of the Antwerp Polyglot Bible," *Imago Mundi*, LV, 56–80.

Shelton 1994: Anthony Alan Shelton, "Cabinets of Transgression: Renaissance Collecting and the Incorporation of the New World," in *Cultures of Collecting*, ed. John Elsner and Roger Cardinal, London, 177–203.

Shirley 1983: Rodney W. Shirley, *The Mapping of the World: Early Printed World Maps, 1472–1700*, London.

Shirley 1991: Rodney W. Shirley, *Early Printed Maps of the British Isles, 1477–1650: A Bibliography*, 3rd edition, Somerset.

Shirley 1998: Rodney Shirley, "The Title Pages to the 'Theatrum and Parergon'," in Van der Broecke–Van der Krogt–Meurer 1998, 161–70.

Shirley and Hoeniger 1985: *Science and the Arts in the Renaissance*, ed. John W. Shirley and F. David Hoeniger, Washington, D. C.

Simeoni 1947: Luigi Simeoni, *Storia dell'università di Bologna*, Vol. 2, *L'età moderna*, Bologna.

Smith 1977: Graham Smith, *The Casino of Pius IV*, Princeton.

Spallanzani 1978: Marco Spallanzani, *Ceramiche orientali a Firenze nel Rinascimento*, Florence.

Spallanzani 1980a: Marco Spallanzani, "Ceramiche nelle raccolte medicee da Cosimo I a Ferdinando I," in *Le Arti del principato mediceo*, Florence, 73–94.

Spallanzani 1980b: Marco Spallanzani, "Metalli islamici nelle collezioni medicee dei secoli XV–XVI," in *Le arti del principato Mediceo*, Florence, 95–116.

Spallanzani 1994: Marco Spallanzani, *Ceramiche alla Corte dei Medici nel Cinquecento*, Modena.

Spallanzani 1996: Marco Spallanzani, *Inventari Medicei, 1417–1465. Giovanni di Bicci, Cosimo e Lorenzo di Giovanni, Piero di Cosimo*, Florence.

Spence 1984: Jonathan D. Spence, *The Memory Palace of Matteo Ricci*, New York.

Spini 1570: Gherardo Spini, *Annotazioni intorno al Trattato dell'Astrolabio et del Planisfero universale del R. P. Ignazio Danti*, Florence.

Spini 1980: Giorgio Spini, *Cosimo I e l'indipendenza del principato mediceo*, Florence.

Stafford 1991: Barbara Maria Stafford, *Body Criticism: Imagining the Unseen in Enlightenment Art and Medicine*, Cambridge, Mass.

Stafford 1994: Barbara Maria Stafford, *Artful Science: Enlightenment, Entertainment, and the Eclipse of Visual Education*, Cambridge, Mass.

Stafford 1999: Barbara Maria Stafford, *Visual Analogy: Consciousness as the Art of Connecting*, Cambridge, Mass.

Strabo 1472: *Straboni Geographia, latine ex interpretatione Guarini Veroniensis et Gregorii Typhernatis*, Venice.

Strauss 1959: Gerald Strauss, "A Sixteenth-Century Encyclopedia: Sebastian Münster's 'Cosmography' and its Editions," in *From the Renaissance to the Counter-Reformation: Essays in Honor of Garret Mattingly*, ed. Charles H. Carter, New York, 145–63.

Strauss 1965: Gerald Strauss, *Sixteenth-Century Germany: Its Topography and Topographers*, Madison, Wisc.

Strinati 1980: Claudio Strinati, "Roma intorno al 1600. Studio di pittura," *Ricerche di Storia dell'arte*, 15–48.

Strong 1973: Roy Strong, *Art and Power: Renaissance Festivals, 1450–1650*, Berkeley and Los Angeles.

Strozzi 1975: Giovan Battista Strozzi, *Madrigali inediti*, ed. Marco Ariani, Urbino.

Suarez 1999: Thomas Suarez, *Early Mapping of South East Asia*, Singapore.

Summers 1969: David Summers, "The Sculpture of Vincenzo Danti: A Study in the Influence of Michelangelo and the Ideals of the Maniera," Ph.D. diss., Yale University (2nd edition: New York, 1979).

Summers 1987: David Summers, *The Judgment of Sense. Renaissance Naturalism and the Rise of the Aesthetics*, Cambridge and New York.

Summers 2003: David Summers, *Real Spaces: World Art*

and the Rise of Western Modernism, London.

Taja 1750: Agostino Taja, *Descrizione del Palazzo apostolico vaticano*, Rome.

Tarde [1593] 1984: Jean Tarde, *A la rencontre de Galilée. Deux voyages en Italie, préface et notes de Francois Moureau. Texte établi par Francois Moureau et Marcel Tetel*, Geneva.

Testaverde Matteini 1983: Anna Maria Testaverde Matteini, "La biblioteca erudita di Vincenzo Borghini," in *Firenze e la Toscana dei Medici nell'Europa del '500*, 3 vols., Florence, vol. 3, 611–43.

Thomassy 1852–3: R. Thomassy, "Les papes géographes et la cartographie du Vatican," in *Nouvelles Annales des voyages et de sciences géographiques*, XXXII, 57–129; XXXIII, 151–72; XXXIV, 7–47; XXXV, 266–9.

Thornton 1997: Dora Thornton, *The Scholar in His Study: Ownership and Experience in Renaissance Italy*, New Haven and London.

Tinagli 2000: Paola Tinagli, "The Identity of the Prince: Cosimo de' Medici, Giorgio Vasari and the 'Ragionamenti'," in *Fashioning Identities in Renaissance Art*, ed. Mary Rogers, Aldershot, 189–96.

Tinagli 2001: Paola Tinagli, "Claiming a Place in History: Giorgio Vasari's 'Ragionamenti' and the Primacy of the Medici," in Eisenbichler 2001, 63–76.

Tinagli Baxter 1985: Paola Tinagli Baxter, "Rileggendo i 'Ragionamenti'," in *Giorgio Vasari tra decorazione ambientale e storiografia artistica*, Florence, 83–93.

Tiraboschi 1833: Girolamo Tiraboschi, *Storia della letteratura italiana*, vol. 3, Florence.

Tolias 2002: George Tolias, "'Erudite Pleasures': Two Portable Painted Maps in Greek Collections," in *Moyseio Mpenake* (Athens, Greece), II, 89–109.

Tongiorgi Tomasi–Garbari 1995: Lucia Tongiorgi Tomasi and Fabio Garbari, *Il Giardiniere del Granduca. Storia e immagini del Codice Casabona*, Pisa.

Tooley 1939: Roland V. Tooley, "Maps in Italian Atlases of the Sixteenth Century, Being a Comparative List of the Italian Maps by Lafreri, Forlani, Duchetti, Bertelli and Others, Found in Atlases," *Imago Mundi*, III, 12–47.

Topper 1992: David Topper, "Towards an Epistemology of Scientific Illustrations," in Baigrie 1992, 215–49.

Trento 1981: Dario Trento, "Cartografi e periti nella campagna bolognese," in *Paesaggio: immagine e realtà*, exhibition catalogue, Rome (Milan), 23–4.

Trionfi Onorati 1985: Maria Trionfi Onorati, "Note sui Maestri Legnaioli,' in *Le arti del principato mediceo*, Florence, 371–82.

Turcic 1981: L. Turcic, "Cristofano Roncalli in the Sala della Meridiana," *BM*, CXXIII, no. 943, 614–7.

Turner 1995: Gerald L'E Turner, "The Florentine Workshop of Giovanni Battista Giusti," in *Nuncius: Annali di Storia della scienza*, 10, 1, 131–72.

Utz 1971: Hildegard Utz, "The Labors of Hercules and Other Works by Vincenzo de' Rossi," *AB*, LIII, 346–66.

Valone 1977: Carolyn Valone, "A Note on the Collection of Niccolò Gaddi," *Critica d'Arte*, CLI, 205–07.

Valone 1993: Carolyn Valone, "The Pentecost: Image and Experience in Late Sixteenth-Century Rome," *Sixteenth-Century Journal*, XXIV, no. 4, 801–28.

Van den Broecke–Van der Krogt–Meurer 1998: M. Van den Broecke, P. Van der Krogt, and Peter Meurer, eds., *Abraham Ortelius and the First Atlas. Essays Commemorating the quadricentennial of his death, 1598–1998*, Utrecht.

Van der Broecke–Günzburger 1998: Marcel Van der Brocke and Deborah Günzburger, "The Wandering of Patriarch Abraham," in Van der Broecke–Van der Krogt–Meurer 1998, 319–30.

Van der Gucht 1994: Alfred Van der Gucht, "Gerard Mercator: une biographie," in *Le cartographe Gerard Mercator, 1512–1594*, exhibition catalogue, Bibliothèque Royale Albert Ier, Brussels, 139–52.

Van Mander 1604: Carol Van Mander, *Het Schilder Boeck*, Haarlem.

Varchi 1963: Benedetto Varchi, *Storia fiorentina*, Florence.

Vasari 1568: Giorgio Vasari, *Le vite de' più eccellenti pittori scultori et architettori*, Florence.

Vasari–Barocchi 1987: Giorgio Vasari, *Le vite de' più eccellenti pittori scultori et architetti*, ed. Paola Barocchi and R. Bottarini, 6 vols., Florence.

Vasari–Milanesi 1906: Giorgio Vasari, *Le vite de' più eccellenti pittori scultori et architettori*, ed. Gaetano Milanesi, 9 vols., Florence.

Vasari–de Vere, 1996: *The Lives of the Painters, Sculptors and Architects*, trans. Gaston du C. de Vere, 2 vols., New York.

Vasari–Del Vita 1938: Giorgio Vasari, *Zibaldone*, ed. Alessandro Del Vita, Rome.

Veen 1998: Henk Th. Van Veen, "Circles of Sovereignty: The Tondi of the Sala Grande in Palazzo Vecchio and the Medici Crown," in Jacks 1998, 206–19.

Vignola–Danti 1583: *Le due regole dello prospettiva pratica di M. Iacopo Barozzi da Vignola con i comentarj del R. P. M. Egnatio Danti dell'ordine dei Predicatori, Matematico dello Studio di Bologna*, Rome.

Vitruvius 1999: Vitruvius, *Ten Books On Architecture*, trans.

Ingrid D. Rowland, commentary and illustrations by Thomas Noble Howe, with additional commentary by Ingrid D. Rowland and Michael J. Dewar, Cambridge and New York.

Voragine 1993: Jacobus de Voragine, *The Golden Legend: Readings of the Saints*, trans. William Granger Ryan, 2 vols., Princeton.

Wallace 1981: William A. Wallace, *Prelude to Galileo: Essays on Medieval and Sixteenth-Century Sources of Galileo's Thought*, Boston.

Wallace 1998: William A. Wallace, "Galileo's Pisan Studies in Science and Philosophy," in *The Cambridge Companion to Galileo*, ed. Peter Machamer, Cambridge and New York, 27–52.

Wallis 1997: Helen Wallis, "Sixteenth-Century Maritime Manuscript Atlases for Special Presentation," in *Images of the World: The Atlas through History*, ed. John A. Wolter and Ronald E. Grim, Washington, 3–30.

Warner 1971: Deborah Jean Warner, "The Celestial Cartography of Giovanni Antonio Vanosino," *JWCI*, 1971, 336–7.

Warnke 1993: Martin Warnke, *The Court Artist: On the Ancestry of the Modern Artist*, New York and Cambridge.

Wasserman 1966: Jack Wasserman, *Ottaviano Mascherino and His Drawings in the Accademia di San Luca*, Berlin.

Watelet 1994: Marcel Watelet, ed., *Gerard Mercator Cosmographe. Le temps et l'espace*, Brussels.

Watelet 1998: Marcel Watelet, ed., *The Mercator Atlas of Europe*, Pleasant Hill, Oreg.

Waterschoot 1979: W. Waterschoot, "The Title-Page of Ortelius' 'Theatrum orbis terrarum'," *Quaerendo*, IX, 43–68.

Wazbinski 1987: Zygmunt Wazbinski, *L'accademia medicea del disegno a Firenze nel Cinquecento. Idea e Istituzione*, 2 vols., Florence.

Westman 1980: Robert S. Westman, "The Astronomer's Role in the Sixteenth Century: A Preliminary Study," *History of Science*, XVIII, no. 2, 105–47.

Westman 1986: Robert S. Westman, "The Copernicans and the Churches," in *God and Nature: Historical Essays on the Encounter between Christianity and Science*, ed. David C. Lindberg and Ronald L. Numbers, Berkeley, Los Angeles and London, 76–114.

Williams 1998: Robert Williams, "The Sala Grande in the Palazzo Vecchio and the Precedence Controversy Between Florence and Ferrara," in Jacks 1998, 163–81.

Winkler–Van Helden 1992: Winkler and Albert Van Helden, "Representing the Heavens: Galileo and Visual Astronomy," *Isis*, LXXXIII, 197–217.

Wisch 1990: Barbara Wisch, "The Roman Church Triumphant: Pilgrimage, Penance, and Processions Celebrating the Holy Year of 1575," in *All the World's a Stage: Art and Pageantry in the Renaissance and Baroque*, Part 1, *Triumphal Celebrations and the Rituals of Statecraft*, ed. Barbara Wisch and Susan Scott Munshower, Papers in Art History from the Pennsylvania State University, 83–118.

Wohl 1992: Hellmut Wohl, "New Light on the Artistic Patronage of Sixtus V," *Arte cristiana*, LXXX, no. 749, 123–34.

Wood Ruby 1999: Louisa Wood Ruby, *Paul Bril: The Drawings*, Brepols.

Woodward 1987: David Woodward, "Medieval Mappaemundi," in Harley–Woodward 1987, 286–370.

Woodward 1990: David Woodward, *The Maps and Prints of Paolo Forlani: A Preliminary Catalogue*, Chicago.

Woodward 1996: David Woodward, *Maps as Prints in the Italian Renaissance: Makers, Distributors and Consumers*, London.

Woodward 1997: David Woodward, "Italian Composite Atlases of the Sixteenth Century," in *Images of the World: The Atlas through History*, ed. John A. Wolter and Ronald E. Grim, Washington, 51–70.

Woodward 2005: David Woodward, ed., *History of Cartography*. Volume Three: *The Renaissance*, Chicago (forthcoming).

Wright 2000: Anthony D. Wright, *The Early Modern Papacy: From the Council of Trent to the French Revolution, 1564–1789*, London.

Yates 1966: Francis Yates, *The Art of Memory*, London.

Zaccagnini 1930: Giuseppe Zaccagnini, "Storia dello Studio di Bologna durante il Rinascimento," *Biblioteca dell'Archivium romanicum*, I, no. 14.

Zen 1994: Stefano Zen, *Baronio storico. Controriforma e crisi del metodo umanistico*, Naples.

Zeri [1957] 1997: Federico Zeri, *Pittura e Controriforma. L'arte senza tempo di Scipione Pulzone*, Rome.

Index

Note Page reference in **bold** indicate illustrations

Aaron, 220, 225
Abel, 216, 217, 218, 222, 223,
Abraham, 202, 219, 220, 224, 225, 226, 250
Adam, 187, 219–20, 218, 224, 249
Adriani, Giovambattista, 17, 20
Adriatic Sea, 2, 172, 184, 188, 195
Aelian, 12, 151
Aenaeas, 202
Africa, 83, 85–87, 107, 108, 113–15, 116, 117, 118, 119, 125, 201, 233, 234, 237, 238, 239, 255, 260–61; *see also* Gastaldi, Ortelius, and Vehlo
Agrippa, map of Rome, 12, 42
Ailly, Pierre d', 52
Alatri, 11, 174, 226
Alberti, Leandro, 187–91, 194–98, 205, 210
Alberti, Leon Battista, 12, 46
Albicante, Alberto, 39
Alcibiades, 151
Aldrovandi, Ulisse, 27, 29–30, 88, 150, 153, 154
Alexander VI, Pope, 125
Alexander the Great, 42, 202
Alexander the Molossian, 196
Alfonso, King of Spain, 50, 54, 246
Almeni, Sforza, 29
Alpers, Svetlana, 278 n. 10
Alps, 172, 184
Altissimo, Cristofano dell', 26, 67, 132
Americas, 85–87, 107, 126, 127, 129, 128, 233, 237, 238, 244, 263–64; *see also* Ortelius
Anacletus, Pope, 246
Ancona, 171, 186
anamorphosis, 224–25
anemoscope, 49–50, 134, 153, 154
Angelo, Jacopo d', 78, 99
antiquarianism *see* mapping and antiquarianism
Apian, Peter, 38, 50, 99, 100, 102, 159, 186
Apostolic Chamber, *see* papal administration
Appenines, 2, 168, 172, 184, 231
Apulia, 102, 174, 187, 188, 193, 198, 201, 212
Aquinas, Thomas, 11, 148, 219, 247
Arabia, 70
Archimedes, 47
Aristippus, 111
armillary sphere, 52, 53, 54, 154
Ariosto, Ludovico, 88
Aristotle, 41–42, 51
Armenia, 6, 75, 79
Armenini, Giovanni Battista, 13
Asia, 68, 69, 70, 76, 85–87, 107, 108, 109–10, 115–25, 120–25, 151, 234, 238, 243, 244 261–263; *see also* Gastaldi and Ortelius
astrolabe, 48, 49, 51
astrology, 28, 153
astronomy, 41–54, 142, 147, 153, 191–192, 205, 233, 244–49, 253; universal church and, 233, 244–49
atlas, 93, 108, 135, 201–02; *see also* Ortelius and Lafrery
Attila, 188, 194, 198
Augustine, 45, 148, 151, 155
Augustus, 36, 40, 42, 200
Avignon, 171, 180
Avvisi di Roma, 171, 227

Baglione, Giovanni, 175, 178, 183
Bandinelli, Baccio, 19
barbarian invasions, 188, 194–95, 198–200
Barbaro, Daniele, 50
Barga, Pietro degli Angeli, 184
Baroncelli, Bernardo, 108, 238
Baronio, Cesare, 182, 198, 199, 217–18, 228–29, 237
Barros, Joan, do, 119–120
Bartoli, Cosimo, 27, 34, 36–37, 46, 50, 65, 159

Battista del Tasso, 19
Beatus of Seville, 98
Bellarmato, Girolamo, 186, 190
Berlinghieri, Francesco, 85, 99
Bertelli, Fernando, 110, 113
Bethlehem, 151, 250
Bible, 88, 202, 217–21, 220, 224, 225, 226, 228, 249–52
  Deuteronomy, Book of, 228, 229
  Exodus, Book of, 88, 202, 220, 221, 225
  Genesis, Book of, 181–82, 219, 220, 223, 225, 226
  Hebrews, Book of, 219
  John, Gospel of, 182
  Letter to the Hebrews, 219, 224, 225
  Leviticus, Book of, 181, 220–23
    map, 88, 248, 249–50
  Psalms, 217
Biblical sacrifices, 172, 179, 180–85, 187, 206, 209, 218–30, 221–23, 232, 234,
  Eucharist and, 224–25
    ecclesiastical supremacy and, 225–29; *see also* mass and offering
Biondo, Flavio, 187, 188, 200, 205, 210
Boccaccio, Giovanni, 11, 34
Bodin, Jean, 89
Bologna, 11, 138, 142–48, 144–45, 148–50, 149, 153–58, 161, 182, 194, 196, 197, 212, 233
  Danti and, 153–157
  maps of, 138, 144–145, 149, 157–158, 160, 194, 196
  Paleotti and, 148–150
  university of, 54, 143, 153, 154, 163, 199, 246
Bologna, Archivio Isolani, 154
Bologna, churches, 54, 153–154
  San Domenico, 54, 153
  San Petronio, 54, 153
  Santo Stefano, 154
Bologna, palaces: 138, 143, 149, 153–154
  Boncompagni palace, 138, 143
  Episcopal palace, 149, 153–54
Bolognetti, Francesco, 49, 190
Boncompagni, family of Bologna, 143–48, 153, 174
Boncompagni, Cristoforo, 143
Boncompagni, Giacomo, 156–57, 162, 174, 191
Boncompagni, Ugo, 136, 143; *see also* Gregory XIII, Pope
Boniface VIII, Pope, 143
Bonsignori, Stefano, 25, 32, 89
  Guardaroba Nuova and, 105–08, 105–36
  Maps: Northern Asia, 125, 262–63
    The Strait of Magellan, 25, 265
    Parte dell'Agisamba (Côte d'Avoire), 75, 83, 259
    Libia interiore, 114–15, 119, 259
    North Pole, 130
    The Old Dominion of Florence, 132
    The Sienese State, 132, 133
Borghini, Vincenzo, 20–21, 36, 37, 65, 72–73, 135, 156–57, 164
Borromeo, Charles, Cardinal, 148, 149, 231
Borromeo, Federico, Cardinal, 88
botanical drawings, 7, 27–28, 29, 134, 150–152
Brahe, Thyco, 42, 43, 53
Bramante, 141
Braun, Georg, 186, 190
Bredekamp, Horst, 5
Bril, Mathijs, 141, 170, 178, 179, 239, 242
British Isles, 70, 109–10, 112, 113, 202, 257
Bronzino, Agnolo, 19, 65
Brosses, Charles de, 204
Buonarroti, Michelangelo, 139, 180
Burckhardt, Jacob, 11, 251

Cain, 216, 217, 218, 219, 220, 223, 224
Calabria, 188, 245
calendar, reform of, 51–54, 153, 184, 212, 244, 243–249
California, 126–27, 128
Callippus, 54
Camillo, Giulio, 91–92
Campania, 174, 176, 177, 178, 191
Cannae, 102, 187, 193, 195
Cape of Good Hope, 25, 70, 113, 127, 260
Cappellino, Rocco, 190
Capponi, Piero di Gherardo, 135
Caprarola, Palazzo Farnese, 11, 20, 143–44, 147, 238–39
Carracci, Agostino, 149
Cartaro, Mario, 157, 159, 163
cartography, 6–8, 105–08, 155–68, 233, 237–44, 248–54
  politics and, 6–8, 38–42, 131–34
  religion and, 249–254
  universal church and, 233, 234, 237–44, 249–52 *see also* mapping and maps
Casey, Edward, 5, 8–10
Cataldi, Pietro Antonio, 154
Cavalieri, Giovan Battista, 139, 140
Cavalli, Serafino, 136
Cecchi, G.M., 135
Cellini, Benvenuto, 63, 65
Charlemagne, 188, 194
Charles V, Emperor, 18, 21, 33, 38–41, 54, 131
  cartography and, 38–39, 40–42, 131
Cosimo I and, 18, 21
Charles VIII, King of France, 194
Charles of Austria, 135
Chersonese, 103–04, 118–19
China, 70, 75, 76, 115, 119–21, 123, 233, 237, 250; 263
chorography, 13, 96–102, 99, 155–68, 169, 186–94, 191, 254 *see also* cosmography, geographical description, geography, and topography
Chrysoloras, Manuel, 78, 98
church history, 169, 179, 180–185, 188, 197–200, 205, 214–216, 251
Cicero, 251
Cini, Giovan Battista, 135
Cirni, Antonio Francesco, 40
Citolini, Alessandro, 90–91
Civitavecchia, 171, 186
Clarici, Giovan Battista, 191
Clavius, Christoph, 46, 245–249
Clement VII, Pope, 19, 39, 40, 204
collecting, 1–2, 5, 63–73, 91–92, 131–34
  art of memory and, 89–92
  cabinets of art and curiosities, 1–2, 22, 63–67, 91–92, 131–134
  geographical order and, 67–73

Kunstkammer, 5, 90–92; *see also* Medici collections, Medici guardaroba, and exotica
Commandino, Federico, 47, 50
Compagnoni, Antonio, 29
congregations: see papal congregations
Constantine, 139–40, 171, 173, 180, 200, 207, 209–16, 231, 246, 247
donation of, 210–14, 212, 215
Italy and, 211–14
Constantinople, 171, 216, 233, 235
Copernicus, 42, 45, 51, 52, 247
Corfu, 171, 180, 186, 232
Corsica, 180, 212, 232
Cortesi, Paolo, 12
Cortez, Hernando, 74, 126–27
Cosgrove, Denis, 5
cosmography, 41–42, 43–59, 96–102, 99, 135
Augustus and, 42
Charles V and, 38, 42
Renaissance courts and, 41–42
Guardaroba Nuova and, 55–59
instruments of, 48–51
Medici court and, 43–54, 135
Ptolemy and, 96–102
teaching of, 43–48 *see also* chorography, geographical description, geography, and topography
cosmos, 41–42, 55–59
Cosimo I and, 55–59
Côte d'Avoire, 75, 83
Council of Nicaea, 245, 246
Council of Trent, 136, 142, 145, 148, 150, 207, 220, 224, 225, 226, 230, 231, 232, 234, 243, 245–46
Croce, Baldassarre, 178
Cuba, 126, 129, 263
Cuzco, 127

Danti, Egnazio, 10–11, 13, 23–26, 28–29, 31, 42–59, 62, 68, 100–03, 135–36, 153–54, 182–84, 245, 246
as court cosmographer, 10, 42–59, 103, 135–36, 153, 182–84, 246
chorography and, 100–03, 186–87, 190–204, 191
Cosimo I and, 43, 131–132
education of, 29, 315 n. 29
Gallery of Maps and, 170–205, 266–72
geography and, 98, 100–03
globes and, 25–26, 49, 120, 130–32, 239–40
Gregory XIII and, 156–57, 167–68
Guardaroba Nuova and, 23–26, 28, 68, 89; 105–36, 256–72
mixed sciences and, 44–48, 155
Paleotti and, 150, 153–155, 251
reform of the calendar and, 51–54, 184, 245–49
scientific instruments and, 48, 49–52, 53–54, 184, 248
surveying and, 153, 155–168
Terza Loggia and, 238–44, 273–77
Maps in the Guardaroba Nuova: *Arabia*, 24, 262;
Armenia, 75, 79, 261
China, 70, 76, 119–21, 261
China and Japan, 119–21, 123, 263
Gronlandia, 130, 131, 265
Indochina, 70, 76, 263
Indostan fuori in Gange, 68, 69, 70, 116, 125, 262
Isola di San Lorenzo, 113–14, 116, 261
Isole Britanniche, 70, 109–10, 112, 257
L'Isole Moluche, 124, 263
Livonia et Lituania, 109–11, 114, 258
Mexican Gulf, 126–27, 129, 264
Moschovia, 95, 110, 258
Norvegia Gotiae, 109, 110, 258
Nuova Spagna, 75, 81, 126, 264
Parte del India dentro il Gange, 121, 262
Sogdiana, 3, 261
Strait of Magellan, 126–27, 265
Le ultime parti dell'Indie occidentali, 126–27, 128, 265
Maps in the Gallery of Maps: *Apulia*, 102, 174, 187, 188, 193, 198, 201, 210, 269
Avenionensis ditio, 171, 180
Bononiensis ditio, 157–58, 160, 197, 70
Campania, 174, 176, 177, 178, 191, 268
Corsica, 180, 210, 232
Flaminia, 157, 161, 194, 198, 199, 225, 226, 270
Genoa, 171, 173, 181, 186, 266
Ilva, 171, 186, 204
Malta, 171, 180, 181, 186, 204, 232
Mediolanensis ducatus, 191–92, 194–95, 196, 271
Perusinus ac Tifernas, 157, 158, 159, 267
Picenum, 195, 270
Sallentina Hydrunti Terra, 193, 269
Sardinia, 180, 181, 190, 228, 231, 232
Sicilia, 180, 190, 210, 225, 231, 232
Tremitae, 171, 186, 191, 204
Venice, 106–07, 171, 173, 180, 186, 188, 190–91, 210, 213, 223
Maps in the Terzia Loggia: South America, 239
Eastern Hemisphere 241
Western Hemisphere 240
Treatises: *Anemografia*, 154
Le due regole della prospettiva pratica di Giacomo Vignola, 10, 143, 225
Le scienze matematiche ridotte in tavole, 44, 101–02, 165–65
La sfera di messer Giovanni Sacrobosco, 45, 46
Trattato del Radio Latino, 159–65, 164–65
Trattato dell'uso et della fabbrica dell'astrolabio, 46, 50–51, 55, 54, 153, 159
Trattato della sfera ridotto in cinque tavole, 44
Usus et tractatio gnomonis magni, 246
MS Gozzadini 171, 155, 163–165, 166–167
Danti, Girolamo, 156, 178
Danti, Giulio, 102, 163, 165, 187
Danti, Piervincenzo, 47
Danti, Vincenzo, 36, 64, 65
David, King, 218
Dee, John, 13, 42
Delano Smith, Catherine, 7
Della Bella, Stefano, 38
Della Penna, Cesare, 45, 47
Deuteronomy, see Bible, Deuteronomy
Doetichum, editor, 194
Dominicans, Order, 10, 188, 191, 233, 246–47

Donatello, 65
Du Perac, 29, 238

ecclesiastical revenues, *see* papal administration 209, 218, 226–29, 232
Egypt, 74, 201, 220
Egyptian kings, 42, 51, 54
Elba, 171, 186, 204
Eleanor of Toledo, 17, 19, 65
Elizabeth I, 42
empire, Roman, 202
Christian, 205, 207, 233, 244, 248, 252
Holy Roman, 231
England, 228, 235, 239
Esau, 218
Este d', family of Ferrara, 101
Este, Alfonso II d', 191
Ethiopia, 237
Eucharist, 181, 185, 205, 219, 224–26, *see also* biblical sacrifices, mass, and offering
Euclid, 44–45, 47, 153
Europe, 11, 85–87, 108, 109–10, 111, 112–15, 180, 235, 257–58
Eusebius, 216–218
Eve, 218, 249
Exodus, see Bible, Exodus, Book of
exotica, 60, 64, 65, 72–76, 77, 78, 80, 82, 131–134 *see also* Medici, Cosimo I, exotica
explorations, 70, 73–74, 113, 115, 119, 121, 125, 126, 233
Portuguese, 70, 73, 113, 115, 119, 121, 125, 126, 233
Spanish, 74, 126–127 233

Fantoni, Filippo, 44
Farnese, Alessandro, Cardinal, 143–44, 183, 237
Ferrara, 157, 186, 204
Fillastre, Guillaume, Cardinal, 85
Finé, Oronce, 50, 51, 159
Flaminia, 157, 161, 194, 198, 199, 227, 226, 270
Florence, 11, 20–21, 32, 58–59, 62, 156, 189
Florence, academies, 46, 135;
Accademia del Disegno, 46;
Accademia della Crusca, 46;
Accademia fiorentina, 46, 135
Florence, churches, 31, 43, 47, 51–52, 53, 54, 154, 247
Monastery of San Marco, 31, 247
Sta Maria del Fiore, 43
Sta Maria Novella, 31, 47, 51–52, 53, 54, 154
Florence, Ducal Palace, 16, 17–22, 26, 34–35, 36–37, 56–57, 58–59, 61–62, 63–65, 66–67, 70–71, 72–73, 75, 92
Apartment of the Elements, 19, 34–35, 36–37, 64
Apartment of Leo X, 19, 62
Guardaroba Nuova: *see* Florence, Ducal Palace
Guardaroba Nuova; Sala dei Duecento, 19
Sala dei Gigli, 26, 57
Sala dell'Udienza, 19
Sala Grande, 19–22, 56, 58–59
Studiolo di Calliope, 63–65, 75
Studiolo of Francesco I, 70–71, 72–73, 92
Tesoretto, 65–66, 67
Florence, Ducal Palace, Guardaroba Nuova, 1–2, 14, 22, 23–24, 28–32, 55–57, 58–59, 61–83, 89–92, 93–104, 105–136, 131–134, 141, 252, 256–265
cartographic sources, 93, 105–36
collecting and, 61–62, 67–83, 131–34
Cosimo I and, 55–59, 89, 131–34
geographical order and, 89–92, 108–09
iconography, 28–32, 68–72
model of, 31
orientation, 1–2
Ptolemy's *Geography* and, 93–104; use of, 23, 131–134, 282 ns. 20 and 21
Vasari's description of, 24–25, 281 n. 15
Bonsignori's Map: Northern Asia, 125, 262–63
The Strait of Magellan, 25, 258
Parte dell'Agisamba (Côte d'Avoire), 75, 83, 259
Libia interiore, 114–115, 119, 259
North Pole, 130
Danti's Maps : *Arabia*, 24, 262
Armenia, 75, 79, 261
China, 70, 76, 119–21, 263
China and Japan, 119–21, 123, 263
Gronlandia, 130, 131, 265
Indochina, 70, 76, 263
Indostan fuori in Gange, 68, 69, 70, 116, 125, 262
Isola di San Lorenzo, 113–14, 116, 261
Isole Britanniche, 70, 109–10, 112, 257
L'Isole Moluche, 124, 263
Livonia et Lituania, 109–11, 114, 258
Mexican Gulf, 126–27, 129, 264
Moschovia, 95, 110, 258
Norvegia Gotiae, 109, 110, 258
Nuova Spagna, 75, 81, 126, 264
Parte del India dentro il Gange, 121, 262
Sogdiana, 3, 261
Strait of Magellan, 126–27, 265
Le ultime parti dell'Indie occidentali, 126–27, 128, 265
Florence, Fortezza da Basso, 18
Florence, palaces: 17–18, 26, 43, 65, 67, 92, 132–33, 134
Medici Palace, 17–18
Pitti Palace: 43, 65, 67, 92
Palazzo Vecchio: *see* Florence
Ducal Palace; Uffizi Palace: Loggia della Cosmografia, 26, 132–33
Stanza delle matematiche, 48, 132
Tribuna, 28, 50, 134
Fontana, Giovanni, 204
Fontana, Prospero, 150, 154
Forma urbis Romae, 12
Fornovo, 194
Foucault, Michel, 279 n. 11
Fra Mauro, 98
France, 73, 78, 181, 215, 231, 239, 248
Franciscans, Order, 235
Frisius, Gemma, 38, 42, 159
Frizzolio, Lorenzo, 184, 185

Gaddi, Nicoló, 135
Galilei, Galileo, 45, 135, 247
Gastaldi, Giacomo, 85, 86–87, 89, 100, 102, 103, 106–07, 108, 114, 115, 118,

119, 120, 121, 122, 126, 186, 190
edition of Ptolemy's *Geography*: 85, 86–87, 89, 100, 102, 106, 108, 293 n. 38
Cosmographia universalis: 106–107, 114, 115, 126
Africa: 107, 114–115, 118;
Asia: 107, 115–116, 119, 120, 121, 122
Gelli, Giambattista, 65
Genesis, see Bible, Genesis, Book of,
Genoa, 171, 173, 181, 186
Geography, 13, 41–59, 92, 96–102, 99 *see also* chorography, cosmography, geographical description, and topography
geographical description, 3, 13, 92, 97–98, 96–104, 111, 186, 254 *see also* chorography, cosmography, geography, and topography
geographical order, 67–92 *see also* Ptolemy and collecting
Germany, 73, 78, 86, 215, 235, 239
Ghebellino, Stefano, 190
Ghirlandaio, Ridolfo, 19
Ghislieri, Giovanni Pietro, 156–57
Giambullari, Pier Francesco, 65
Gigante, Antonio, 153
Giglio, Luigi, 245, 247
Giovanni Antonio da Varese (known as Vanosino), 29, 143, 147, 238, 243
Giovio, Paolo, 26, 110
Giulio Romano, 213, 215
globe, 35–41
Charles V and, 38, 39–40, 41
Cosimo I and, 35–36, 37, 38–41, 49, 59
Goethe, Johannes Caspar, 204
Goethe, Wolfang, 204
Gonzaga, family of Mantua, 11, 20, 101
Gonzaga, Guglielmo, 189
Gonzaga, Luigi, 39
Grafton, Anthony, 4
Granada, New Palace of Charles V, 38–39
Greece, 202, 233, 235, 239, 258
Greenland, 127–129, 130–131, 265
Gregory XIII, Pope, 1–2, 9, 10, 108, 136, 142–50, 153, 156–57, 167–68, 170–205, 209–16, 225–38, 246–54
Bologna and, 142–48
cartography and, 146–47, 246–54
Danti and, 156–57, 167–168, 246–52
ecclesiastical revenues and, 226–29
Gallery of Maps and, 168, 170–205
Holy Year and, 139–40
Italy and, 205, 209–16
map cycles and, 1–2, 108, 254–55
mixed sciences and, 146–48, 246–54
Paleotti and, 148–150
reform of the calendar and, 52, 153, 246–54
universal church and, 229–37, 254
Gregory Nazianzius, 155, 236, 239, 242, 244
grid, 7, 8, 78, 82, 94–97, 193, 205
grotesque, 154–55, 156, 252
Guarino Veronese, 99, 101
Guastavillani, Filippo, Cardinal 153
Gueroult, Guillaume, 189

Hadrian, 217, 218
Hannibal, 102, 188, 192–93, 194–95
Harley, Brian, 7
Heliodorus, 44, 47
Herodotus, 88
Hindustan, 68–69, 70, 116, 125, 262
Hipparcos, 54
Hogenberg, Franz, 186, 190
Holste, Lukas, 159, 174–75, 204, 210, 272–274
Maps: *Italia antiqua*, 174
Italia nova, 175
Latium, 159, 161, 186, 205, 267
Holy Land, 201, 202, 235, 249, 250–51, 252
Holy Year, 139, 140–41, 146, 152
Hungary, 233
Husserl, Edmund, 8, 9
Hydrography, 102

Ignatius of Loyola, 250, 252
Ilarius, Pope, 246
In Coena Domini, 230–32
India, 118–19, 121, 151
Indochina, 70, 118–19, 122
Innocent VIII, Pope, 246
Isola di San Lorenzo, see Madagascar
isolari, 188, 232
Italy, 2–3, 168, 180–81, 185, 186–97, 209–32, 233, 248, 252, 254, 259
maps of, 169, 172, 174–75, 180–81
primacy of, 209–32
universal church and, 230–33

Jacob, 218, 219–20, 223, 224
Japan, 115, 119–21, 123, 233, 236
Jenkinson, Anthony, 111, 115
Jerome, St., 155
Jerusalem, 229, 235, 249, 252
Jesuits, Order, 237, 247, 250
Johanna of Austria, 20, 32, 36, 135
Joshua, 220
Julius II, Pope, 246
Julius Caesar, 42, 43, 51, 52, 54, 194, 195, 199, 202, 212, 244, 246

Kant, Immanuel, 5
Kemp, Martin, 5, 48
Kepler, Johannes, 42
Kunstkammer *see* collecting

Lafrery, Antonio, 88, 91
Lanci, Baldassarre, 48, 135
Landi, Antonio, 58
Latium, 159, 161, 162–63, 186, 205, 267
Laureti, Tommaso, 153, 212–13, 214
Laureto, Marco, 234
Leo I, Pope, 198, 246
Leo X, Pope, 19, 40, 75, 246
Leonardo da Vinci, 11, 64
Lepanto, Battle of, 180

Ligozzi, Jacopo, 28
Liguria, 181, 186, 188, 191, 212
Lily, George, 110, 113
Livonia and Lithuania, see Muscovy
Livy, Titus, 12, 102, 187
Lombardelli, 178
Lombardy, 215
London, 11
Loreto, House of, 199
Ludius, 12
Luther, Martin, 214, 249
Lyceus, Proclus, 44, 47

Macao, 237
Madagascar, 113–14, 115
Madrid, 11
Maffei, Raffaele, 187, 188, 200, 205, 210
Magdeburg, *Centuriae*, 216–17
Magini, Giovanni Antonio, 205
Malta, 171, 180–81, 186, 204, 232
Manila, 237
Mantua, 11, 198
map cycles, 1–3, 9, 11–13, 152, 253–55
mapping, Renaissance, 3–10, 48–54, 103–08, 155–68, 235, 239–46, 251–53
  philology and, 4, 7, 103–04
  antiquarianism and, 3–4, 103–104, 169, 187, 189–190, 249
  walking and, 8–10
  techniques of, 105–08, 155–18
  instruments of, 48–54, 164–65, 166–68
  drawing and, 163–164, 166–67
  semiotics and, 7–8
  politcs and, 3, 11–13
  religion and, 235, 239–46, 251–54
Maps: cadastral maps, 7, 153, 157
  display and use, 3–4, 7, 9, 12–13, 105, 108, 155–57
  historical vignettes and, 172, 177, 192–193, 194–198, 199–200, 203
  landscape features and, 172, 176, 179, 193
  language of, 7, 164–65, 172, 192–97, 253–55
  legends and cartouches, 68–72, 116, 118–20, 125, 185
  mnemonics and, 67–92
  printed, 7, 105–108
  regional , 24–28, 155–167, 185
  religion and, 150–155, 172, 239–46, 251–53
  sky maps, 28, 142–145, 146, 147
  wall, 105–108; *see also* mapping, chorography, geography, and topography
  world, 103–104, 106–07, 111–12, 126, 240–41, 250–51
  Marcellus, 194
Marches, 186, 188, 191
Marco da Faenza, 178
Marco Polo, 103, 115–25
Margarita of Austria, 18, 62,
Marinus of Tyra, 94, 101
Martini, Bernardino, 54–55, 59
Martini, Francesco di Giorgio, 11
Mascherino, Ottaviano, 143, 147, 170, 174
mass, 225–232, *see also* biblical sacrifices and offering
mathematics, 42, 44–45, 47–48, 147, 244–47, 253–54
Matilda of Canossa, 216, 230
Matteo da Siena, 178
Matteo da Verona, 38
Mattioli, Pietro Andrea, 29
Maximilian, Emperor, 42
Medici, family of Florence, 1–22, 34, 40–41, 55–57, 59, 61–78, 131–34
  coat of arms, 35, 41, 55–57, 59
  collections, 2, 18, 22, 61–78, 131–34, 290–91, ns. 2–6
  guardaroba, 23, 62–63, 290–91, ns. 2–5
  palace, 17–18
Medici, Alessandro de', 18
Medici, Cosimo the Elder, 17–18, 19, 135
Medici, Cosimo I de', 1–2, 9, 10, 17–23, 26–37, 40–59, 65, 73–78, 89–92, 105, 107–08, 119–21, 131–35, 141, 169, 186, 235, 237, 238, 239, 254
  Atlas and, 36–37
  Augustus and, 36, 40
  Capricorn and, 33–36, 37, 40, 65, 285 n 2
  Charles V and, 18, 21, 33, 38–41
  cosmography and, 33, 42–54, 135
  Cosmos cosmoi cosmos, 33, 55–59, 131
  ducal palace and, 17–23
  exotica and, 60, 61, 64, 73–74, 75–76, 77–78, 92
  globes and, 35–36, 37–41, 49, 59
  Guardaroba Nuova and, 28–32, 55–59, 89–92, 93, 94, 105, 107–08, 119–21, 131–34, 141
  Hercules and, 33–37, 59
  map cycles and, 1–2, 135
  reform of the calendar and, 51–54
Medici, Cosimo II, 135
Medici, Cosimo III, 37
Medici, Ferdinando, 26, 40, 45, 48, 50, 74, 132, 136, 185, 237
Medici, Francesco I, 20, 26, 28, 32, 33–34, 36, 45, 70–73, 74, 92, 132, 136, 157
Medici, Giovanni dalla Bande Nere, 18, 19
Medici, Isabella Orsini, 45
Medici, Lorenzo the Magnificent, 18, 19, 26, 36, 63, 65, 74
Medici, Lorenzo (known as Lorenzaccio), 18
Mediterranean Sea, 201, 203, 233, 247, 7.26
Mela, Pomponius, 7, 101, 187
Melanchton, Philipp, 224
Melchizedeck, 225–26, 229
Mercator, Gerard, 38, 103104, 107, 109–11, 114, 118, 121, 125, 127, 130
  geographical description and, 103–04
  globes and, 38, 103–104
  Maps: British Isles, 109–10
    Europe, 107, 109–11
    World, 103–04, 114, 118, 121, 125, 127, 130
meridian line, 54, 153, 184, 244, 246
metaphysics, 148–152, 246
Mexico, 73–74, 126–127, 129, 264
Milan, 148, 180, 191, 192, 194, 195, 196
Miltiades, 216, 217, 218
Missal, see Roman liturgy
mixed sciences (*scientiae mediae*), 6, 23, 44–48, 135–36, 146–52, 155, 244, 246–49
religion and, 244–52, 253
Moluccas Islands, 85, 113, 124, 125
Montaigne, Michel de, 86, 230
Montanus, Benedictus Arias, 249–51, 251
Montefeltro, family of Urbino, 101
Monti della Luna, 114

Moses, 218, 220 221, 222, 223, 250; law of, 224
Mountain of Hyperborea, 111
Münster, Sebastian, 85–87, 88–90, 100, 102, 188, 211
Muret, Antoine, 184
Muscovy, 95, 109–111, 114–15, 235, 258, 297 n. 36
Muziano, Girolamo, 174, 175, 179

Na'mattalah, Ignatius, 235
Naples, 11, 196
navigational charts, 7, 107, 113–15, 121, 125, 126–27, 296 n. 27
Nebbia, Cesare, 175, 178, 182, 183–84
Negro, Domenico Mario, 101
Neri, Bernardo, 121
Neri, Filippo, 183
Neri, Giovanni, 27
Nicholas of Cusa, 52
Nicola Pisano, 64
Nicolaus Germanus, 94–95
Nigetti, Dionigi di Matteo, 26, 31
Noah, 218, 219, 220, 224, 247
Nobili, Pietro de', 140, 141
Nogari, Paris, 178
North Pole, 127–129, 130–31, 265
Nozzolino, Giuseppe, 44

offering, 225, 227–28, *see also* mass and tithe
optics, 44–45, 47, 286 n. 33, 287 n. 45
Order of the Golden Fleece, 21, 55, 57
Orléans, Marguerite-Louise d', 37
Oronce, 98
Orsini, Latino, 162, 182–83, 184, 185, 250,
Ortelius, Abraham, 68, 87–89, 103, 105, 107–08, 111–12, 114, 116, 115, 118–19, 121, 125–26, 189, 190, 201–02, 205, 238, 250–252;
  Maps: Asia, 116, 118, 119, 121, 125
    Theatrum orbis terrarum, 87–89, 90, 103, 105, 108, 111–12, 114, 115, 189, 90, 201–02, 205, 238, 250–252
    World, 68, 107, 114, 126, 248–50
Ostia, 204

Paleotti, Camilla, 49
Paleotti, Gabriele, Cardinal, 13, 142, 148–54, 199, 217, 230, 246, 251–53
  Bologna and, 148–50
  Gregory XIII and, 142
  scientific images and, 148–52
  Danti and, 153–55
  Discorso intorno alle immagini sacre e profane, 150–54, 251–52
Panigarola, Francesco, 236
Papal administration, 157, 207, 219, 226–30, 233, 237–39; Apostolic Chamber, 157, 226, 227, 233
  Congregation of Propaganda Fide, 235
  Congregazione ai confini, 227
  Dataria, 226
  ecclesiastical revenues, 209, 218, 226–30, 233
  Greek Congregation, 236, 237
  Ufficio degli spolii, 204
Papal States, 62, 155–68, 180, 191, 216, 218, 226–27, 229, 230, 243, 244, 252;
  surveying of 155–57, 158–63, 164–68
Parma and Piacenza, Duchy, 11, 194
Parmigianino, 40
Pasi, Marc'Antonio, 186, 191
Passerotti, Bartolomeo, 11, 153, 154
Patrimony of St. Peter, 159, 215 *see also* Latium
Paul, St., 181, 184, 203, 250
Paul III, Pope, 52, 234, 246
Paul V, Pope, 224, 227
Paul of Middleburg, 52
Pedro of Toledo, 18
Pepin, 188, 217
Pepoli, Giovanni, 153
Pereira, Benedict, 247
Persia, 70, 75
Perugia, 6, 11, 102, 156–157, 158–59, 161, 163, 187, 192, 195
Peter, St., 182, 184, 200
Peter II of Aragona, 217
Philip II, 39, 144, 202, 228, 235
Philippines, 237
Piceno, 195, 199, 212
Piedmont, 174, 181, 194
Piero di Cosimo, 27
Pisa, conquest of, 20–21;
  university of, 43–44, 153
Pietro del Massaio, 85
Pigafetta, Filippo, 191, 205
Pitti, Miniato, 28–29, 31, 238
Pius I, Pope, 246
Pius IV, Pope, 20, 29, 93, 141, 145, 180, 233, 237, 238, 239–42, 243
Pius V, Pope, 215, 226, 230, 236, 244
place, 5, 8–10, 98, 186, 254
Plato, 186
Plantin, Editor, 189
Pliny, 12, 64, 78, 89, 101, 187, 210
Podiani, Prospero, 135
Poland, 235, 239, 246
Polybius, 102, 187
Pontormo, 19
Possevino, 235
Ptolemy, 3, 6–7, 13, 23–25, 29, 44, 47, 50, 52, 54, 78–92, 94–102, 107, 113–14, 115, 118–19, 127, 130, 153, 161, 180, 186, 189, 200, 238, 253–54
  geographical order of, 25, 29, 78–92, 238
  geographical theory, 13, 78, 82, 94–102, 186, 253–54
  geography, 3, 6, 25, 44, 47, 78–89, 92, 93–102, 127, 130, 186 Renaissance editions, 25, 47, 84, 85, 86–87, 89, 94–95, 96–102, 189
  Maps: Map of the known world, 82, 84
    tables of, 82, 84, 86, 92–93, 94, 95–102, 295 n. 34
    tabulae modernae, 82, 84, 87, 95

quadrant, 52, 53, 54
Quicchelberg, Samuel, 91–92

radius latinus, 161–63, 164, 165–67
Ramusio, Giovan Battista, 103, 106, 115
Raphael, 65, 141–142, 218
Razzi, Serafino, 29,

Regiomontanus, 52
Ricci, Giuliano, de, 135
Ricci, Matteo, 250
Rojas de Sarmiento, Juan, 50
Roman liturgy, 221, 222, 232, 241, 234, 236, 242
  Breviary, 233, 243, 244
  Catechism, 224, 225, 233, 234, 243
  Missal, 233
Romanelli, Giovan Battista, 181–182
Rome, 11, 139–140, 148, 151, 157, 180, 183, 185, 187, 189, 190, 192, 202, 212, 213, 216–18, 230, 239, 242, 243, 252
  Acqua Vergine, 139
  Rome, churches: San Giovanni in Laterano, 140, 141, 212, 216, 217
    San Paolo, 218
    Sta Maria in Campo Marzio, 235
    Sta Maria Sopra Minerva, 11;
    Greek College of Sant' Anastasio, 236
  Temple of Peace, 12
  university of, 244
Rossi, Vincenzo de, 37
Rowland, Ingrid, 4
Rudolph II, 42
Ruscelli, Girolamo, 85, 89, 100–02, 130, 186, 189–90
  edition of Ptolemy's *Geography*, 85, 89, 100–02, 130
Russia, *see* Muscovy

Sabatini, Lorenzo, 29, 141–143
Sacrobosco, Giovanni, 44, 46, 51, 153, 245, 247
Salomon, King, 220
Salviati, Francesco, 19
Santucci, Antonio, 26, 130
Saracens, 194, 196
Sardinia, 180, 181, 190, 231, 232; map in the shape of, 12
Savelli, Orazio, 225
Scandinavian Peninsula, 109–10, 233
scientiae mediae, *see* mixed sciences
Schulz, Jurgen, 280 n. 10
Scipio, Cornelius, 194
Semenza, Jacopo, 178
Seneca, 109, 251
Sicily, 180, 190, 212, 228, 230, 232,
Siena, conquest of 19–21
Sigonio, Carlo, 150, 154, 199, 216–18
Sirleto, Guglielmo, Cardinal, 47, 182–83, 217, 245
site, 5, 8–10, 98, 252
Sixtus IV, Pope, 52
Sixtus V, Pope, 237
Socrates, 12, 151
Soderini, Pietro, 17
Soliman the Magnificent, 99
Sorte, Cristoforo, 186
Sosigenes, 42, 54
Spain, 33, 215, 218, 228, 229, 232, 236, 246
Spini, Gherardo, 46, 51, 54
Stabius, 42
Stafford, Barbara, 5
Stella, Giacomo, 178
Stella, Giovanni Battista, 171
Strabo, 7, 98, 100–102, 107, 180, 187, 200, 212–13
Strada, Jacopo, 189–90
Strait of Anian, 115, 121, 123
Strait of Magellan, 25, 126–27
Strozzi, Filippo, 58
Strozzi, Giovambattista, 58–59, 135
Sylvester I, Pope, 140, 212–215
Symmachus, Pope, 181
Sweden, 235

Taja, Agostino, 239
Tarde, Jean, 224
Tempesta, Antonio, 141, 178, 179, 239, 242
Toscanelli, Paolo, 54
tithe, 209, 219–20, 224–28 *see also* biblical sacrifices, offering, and papal administration
Tolesani, Giovanni Maria, 245
topography, 98, 187 *see also* chorography, cosmography, geographical description, and geography
Tremiti, 171, 186, 191, 204
trepidation, motion of, 51, 245, 317 n. 26
Turks, 21, 46, 180, 228, 230, 233, 234, 236, 237
Tuscany, 20–21, 35, 58–59, 186, 188, 190, 204
Tyrrhenian Sea, 2, 172, 188, 194

Urban VIII, Pope, 182, 204, 227
Umbria, 161, 194
Urbino, 204
universities, 44–45, 54, 143, 153, 154, 163, 199, 248;
  of Bologna, 54, 143, 153, 154, 163, 199, 248;
  of Pisa, 43–44, 153;
  of Rome, 245

Valier, Agostino, 217
Valla, Lorenzo, 214
Vandeville, Jean de, 234–236
Varchi, Benedetto, 46
Varolio, Carlo, 150
Vasari, Giorgio, 19–21, 23–26, 27, 28–29, 31–36, 40, 46, 63, 65, 67, 72, 89;
  Works: *Saturn Castrating Heaven*, 34–35
    Saturn Receiving the Tribute of the Earth, 34–35
    Studiolo di Calliope, 63–65
    Guardaroba Nuova, 14, 19–21, 22, 23–26, 28–29, 31–36, 57, 89, 93, 94–97, 130–31
    Sala Grande, 19–22, 56, 59–59
    Sala Regia, 140–41
  Writings: *Lives*, 23–24, 27, 40, 46, 63, 67, 135
    Ragionamenti, 33–36, 58, 285 n. 3
Vatican, Gallery of Maps, 2–3, 155, 157, 159, 164, 168, 169–205, 234, 235, 238, 243, 252, 254, 266–72
  orientation, 2–3, 172, 190
  cartographic sources, 190
    use of, 170, 204–205;
  construction and decoration, 169–79
  iconography, 172, 180–185
  geographical order, 180–185, 190
  Biblical Sacrifices: *Adam's Sacrifice of Praise*, 187
    The Priest Leading the Bullock to the Temple for the Communal

Sacrifice, 218–30, 221
The Priest Cutting the Throat of the Bullock and Collecting its Blood in a Basin, 218–30, 221
The Priest Sprinkling the Blood on the Curtain of the Sanctuary, 218–30, 221
The Priest Smearing the Blood on the Corners of the Altar, 218–30, 222
The Priest Pouring the Remaining Blood at the Foot of the Altar, 216–28, 220
The Priest Burning the Fat, Kidneys, and Lobes of the Liver on the Altar, 218–30, 222
The Priest Burning the Carcass and the Meat Outside the Camp, 218–30, 223
Ceiling Paintings: *The Exposition of the Holy Shroud*, 174, 181
St. Ambrose Refusing Emperor Theodosius Admittance to Church, 174
St. Francis Appearing to St. Anthony of Padua, 181
The Translation of St. John the Baptist's Ashes to Turin, 181
The Manna Coming from St. Andrew's Body, 181
Pope Symmachus Sending Aid to the Exiled Bishops, 181
The Miracle of the Blood of St. Januarius in Naples, 174, 179
Stopping the Lava on Aetna By Holding St. Agata's Veil, 182 183
St. Michael the Archangel Appearing on Monte Gargano, 174, 198, 200
The Hermit St. Peter of Morrone Elected Pope, 198
Countess Matilda Donating her Possessions to the Roman Church, 198
Feed My Sheep, 181–82;
St. Sylvester Baptizing Emperor Constantine, 211–212
Emperor Constantine Building the Basilica of St. Peter's, 211–212
Emperor Constantine Building the Basilica of St. Paul, 211–212
Emperor Constantine Holding the Reins of St. Sylvester's Horse, 211–212, 213
Constantine's Vision of the Cross Before the Battle at the Milvian Bridge, 211–213
Maps: *Apulia*, 102, 174, 187, 188, 193, 198, 201, 212, 270;
Avenionensis ditio, 171, 180
Bononiensis ditio, 157–58, 160, 197, 270
Campania, 174, 176, 177, 178, 191, 268
Corsica, 180, 212, 232
Flaminia, 157, 161, 194, 198, 199, 227, 228, 270
Genoa, 171, 173, 181, 186, 267
Ilva, 171, 186, 204
Italia antiqua, 174, 210, 272
Italia nova, 174, 210, 267
Malta, 171, 180, 181, 186, 204, 232
Mediolanensis ducatus, 191–92, 194–195, 196, 271
Perusinus ac Tifernas, 157, 158, 159, 267
Picenum, 195, 270
Sallentina Hydrunti Terra, 193, 269
Sardinia, 180, 181, 190, 231, 232
Sicilia, 180, 190, 212, 228, 230, 232
Tremitae, 171, 186, 191, 204
Venice, 106–07, 171, 173, 180, 186, 188, 190–91, 212, 215, 225
Vatican Palace, 2, 29–30, 31, 65, 108, 138, 139–40, 141–43, 145–49, 151–52, 155, 165, 168–70, 180, 182, 184, 212–14, 239, 248, 250
Belvedere Villa and Courtyard, 65, 152, 168, 169–70
Chapel of Nicholas V, 248
Gallery of Maps, *see* Vatican, Gallery of Maps;
Palace of San Damaso, 141–42
Pauline Chapel, 140
Raphael's Loggias, 180, 220
Raphael's Stanze, 170
Sala Bologna, 138, 142, 143, 144–46, 148, 149, 151, 152, 182, 184, 233, 238, 250
Sala degli Svizzeri, 184
Sala dei Palafrenieri, 184
Sala di Costantino, 170, 180, 212–214, 214
Sala Regia, 140–41, 180, 236, 237
Sistine Chapel, 180
Terza Loggia, 29–30, 31, 108, 141–42, 145, 180, 234, 235, 236–41, 250–52
Tower of the Winds, 54, 170, 184, 248
Vatican, St. Peter's, 54, 139, 140, 213, 230, 236, 239
Gregorian Chapel, 175, 236–37, 239, 242, 248
Velho, Bartholomeo, 107, 114, 117, 121
Venice, 106–07, 171, 173, 180, 186, 188, 190, 191, 212, 215, 226
ducal palace, 11, 106
city view, 171
Vignola, Giacomo, 10, 11, 44, 223
Vitelli, Alessandro, 18
Vitruvius, 42, 49, 78, 89
Volpaia, Girolamo della, 48
Volpaia, Lorenzo della, 24, 26, 48, 132

Winghe, Philip van, 159, 162, 205
Woodward, David, 7
Wright, A.D., 309 n. 1

Zeno, brothers and map, 130
Zimmermann, Matthias, 38
zoological drawings, 27–28, 29, 134, 150–52, 284 n. 23

Photograph Credits

Fototeca dei Musei Comunali di Firenze: 1.1, 1.2, 2.3, 2.5, 2.6, 3.16, 4.3, 4.6, 4.9, 4.11, 4.13, 4.15, 5.2, 5.7, 5.8, 5.10, 5.12, 5.17, 5.18, 5.19, 5.20, 5.24, 5.25, 5.27; Archivio Fotografico, Musei Vaticani: 1.3,.1.4, 2.9, 3.16, 6.3, 6.5, 6.6, 6.7, 6.12, 6.13, 6.15, 6.16, 7.1, 7.8, 7.9, 7.11, 7.13, 7.14, 7.15, 7.16, 7.17, 7.20, 7.21, 7.22, 7.23, 8.3, 8.4, 8.5, 8.6, 8.7, 8.8, 8.9, 8.10, 8.11, 8.12, 8.13, 8.14, 8.15, 8.16, 8.17, 8.18, 8.19, 9.2, 9.3, 9.4, 9.5, 9.6, 9.7, 9.8; Alinari/Art Resource, N.Y.: 3.4, 3.5 7.2, 7.3, 7.4, 7.5, 7.6, 7.7, 7.10, 7.18, 7.19, 7.24 8.1, 8.2; Scala/Art Resource, N.Y.: 2.1, 2.2, 3.14, 3.1, 3.2, 3.3, 3.15, 4.1, 4.4 4.7, 5.15, 5.16, 5.28, 5.29, 5.30; Courtauld Institute: 7.12; Leiden, Universiteitsbibliotheek: 6.17; Biblioteca Apostolica Vaticana: 6.18; The Folger Shakespeare Library: 6.19, 6.20, 7.26; Bologna, Biblioteca dell'Archiginnasio: , 6.21, 6.22, 6.23, 6.24; John Carter Brown Library. Brown Library: 7.25; Fabbrica di San Pietro: 9.1; Yale Center for British Art, Paul Mellon Collection: 2.7; Cooper-Hewitt, National Design Museum Smithsonian Institution/Art Resource, N.Y.: 3.45; Yale University Library: 5.6; Oronoz: 3.6; Bibliothèque nationale de France: 3.7, 5.4, 6.14; The British Library: 5.5; Bologna, Biblioteca Universitaria: 2.8; Stiebel, ltd: 3.8; Florence, Museo di Storia della Scienza: 3.9, 3.10, 3.11, 3.12, 5.13; Houghton Library, Harvard University: 3.13, 4.17, 4.18, 4.19; The University of Virginia Library: 4.20, 5.3, 5.11, 5.26; Antonio Quattrone: 4.2, 4.8, 4.10, 4.14; Ministero dei Beni e le attivitá culturali: 4.12, 6.8, 6.9; Bibliotheca Hertziana: 6.2; Öffentliche Bibliothek, Universität Basel: 4.5, 5.14, 5.16, 5.21, 5.22; The Newberry Library: 4.21, 5.9; Princeton University Library: 4.16, 5.1, 9.9, 9.10; Brest: 6.10; Florence, Biblioteca nazionale: 6.11; Adriano Scognamillo: 3.16, 5.23; British Museum: 6.1.